Statistical Thermodynamics and Kinetic Theory

Charles E. Hecht

Hunter College
City University of New York

DOVER PUBLICATIONS, INC.
Mineola, New York

Bibliographical Note

This Dover edition, first published in 1998, is an unabridged republication of the work originally published by W. H. Freeman and Company, New York, in 1990.

Library of Congress Cataloging-in-Publication Data

Hecht, Charles E.
　　Statistical thermodynamics and kinetic theory / Charles E. Hecht. — Dover ed.
　　　　p.　　cm.
　　Originally published : New York : W. H. Freeman, c1990.
　　Includes bibliographical references and index.
　　ISBN 0-486-40457-9 (pbk.)
　　1. Statistical thermodynamics.　2. Kinetic theory of gases.　I. Title.
[QC311.5.H43　　1998]
530.13—dc21　　　　　　　　　　　　　　　　　　　　　　98-24687
　　　　　　　　　　　　　　　　　　　　　　　　　　　　　　CIP

Manufactured in the United States of America
Dover Publications, Inc., 31 East 2nd Street, Mineola, N.Y. 11501

Dedicated to the memory of
Joseph E. Mayer
5 February 1904–15 October 1983
Eminent scientist and teacher of teachers

Contents

Preface

This text may serve for all or part of a one-semester graduate course in chemistry or for part of a semester in a two- or three-semester, rigorous undergraduate physical chemistry sequence. It should also be useful for self-study by physical chemists, particularly the sections on negative absolute temperatures, supersonic molecular beams, critical phenomena, and all of Chap. 7, which introduces three new fields of currently exciting development.

It is assumed that the reader will have a general acquaintance with quantum mechanics at the level presented in an undergraduate physical chemistry course. Although the mathematics in the text is in some instances detailed (so that the reader should be able to reproduce the results), it is not esoteric. Many examples are fully worked out in the body of the text. They are accompanied by proposed exercises that serve to further involve the reader in the development of the material. Cumulatively there are some 230 end-of-chapter problems, with answers and comments on the solutions for most of them provided in Appendix III.

The core of this text is statistical mechanics, a complicated subject with a long history and one that is often presented with such a large amount of complex mathematics, classical physics, and even philosophical musings that students cannot discern its fundamental principles. I have tried to present these fundamentals in a direct, even brisk, manner with some asides to the reader regarding the strategy of my approach, especially in Chaps. 1 and 3. I hope that most readers will find my presentation both congenial and accessible.

Charles E. Hecht
New York
August 1989

1

Statistical Mechanics

1-1
Introduction

Thermodynamics provides us with correct relationships between various macroscopic properties of bulk matter, but in no case can it provide us with numerical values for a given property. For example, the Clapeyron equation relates the vapor pressure of a substance to the temperature and the enthalpy of vaporization. However, thermodynamics cannot give us the means of calculating a priori the value of this enthalpy for any substance nor give us a reason why, for example, water molecules should have a larger enthalpy of vaporization at 25°C than carbon tetrachloride molecules. To obtain such information the properties of microscopic mechanical models based on quantum or classical mechanics must be averaged. The methods of this averaging, which will predict average values for macroscopic properties, constitutes the science of statistical mechanics. Statistical mechanics is the bridge connecting microscopically defined quantities (usually particle energies) and thermodynamic variables.

In broad outline what we will do is interpret the laws of thermodynamics to be consequences of the principles of quantum mechanics applied to systems of N particles where N is very large (Sec. 1-2). Then we will derive an equation by means of which a single thermodynamic function of a system may be calculated in principle from the microscopic properties of the constituent molecules of the system. We do this for the entropy function in Sec. 1-2 and for other functions in Sect. 1-3. Then we rely on the standard procedures of thermodynamics to manipulate this single function so as to derive equations for any other thermodynamic functions of interest.

To understand the methods of statistical mechanics we need to introduce some ideas of probability theory, which deals with the properties of any collection of elements each of which has certain characteristics. We consider first characteristics that can be represented by discrete values, for example age in years at last birthdate for a collection of people. If N is the total number of elements and $N(l_i)$ is the number that have the discrete value l_i ot characteristic i, the probability, $P_i(l_i)$, that characteristic i has the value l_i is defined by

$$P_i(l_i) = \frac{N(l_i)}{N} \tag{1-1}$$

It is usual to normalize the probabilities such that the sum of $P_i(l_i)$ over all possible values of the parameter is unity:

$$\sum_{l_i} P_i(l_i) = \sum_{l_i} \frac{N(l_i)}{N} = \frac{N}{N} = 1 \tag{1-2}$$

Each member of a collection is characterized by some value l_i of parameter i. To obtain the average value of any function of l_i, say $b(l_i)$, in the collection (which average is written $\bar{b}$) we have

$$\bar{b} = \frac{1}{N} \sum_{l_i} N(l_i) b(l_i) = \sum_{l_i} P_i(l_i) b(l_i) \tag{1-3}$$

If for example we consider a collection of four people of ages 18, 29, 37, and 52 years, the average age $\bar{a}$ and the average of the squares of the ages $\overline{a^2}$ would be

$$\bar{a} = \frac{1}{4}(136) = 34 \text{ yr}$$
$$\overline{a^2} = \frac{1}{4}(5{,}238) = 1{,}309.5 \text{ (year)}^2$$

Once the average is known, the deviation from this average (given the symbol $\delta b(l_i)$) may be computed for any value of l_i

$$\delta b(l_i) = [b(l_i) - \bar{b}] \tag{1-4}$$

The average of the deviations is of course zero

$$\overline{\delta b} = \sum_{l_i} P_i(l_i)[b(l_i) - \overline{b}]$$

$$= \sum_{l_i} P_i(l_i)b(l_i) - \overline{b} \sum_{l_i} P_i(l_i)$$

$$= \overline{b} - \overline{b} = 0 \tag{1-5}$$

A useful measure of the average deviation of individual values from their mean is given by the average of the square of $\delta b(l_i)$, which is called the variance and is usually denoted by σ^2 (σ is the root mean square deviation from the mean):

$$\sigma_b^2 \equiv \overline{(\delta b)^2} = \sum_{l_i} P_i(l_i)[b(l_i) - \overline{b}]^2$$

$$= \sum_{l_i} P_i(l_i)[b^2(l_i) - 2b(l_i)\overline{b} + (\overline{b})^2]$$

$$= \overline{b^2} - 2(\overline{b})^2 + (\overline{b})^2 = \overline{b^2} - (\overline{b})^2 \tag{1-6}$$

The left-hand side of Eq. (1-6) must be positive unless $\delta b(l_i)$ is zero for all l_i, which can only occur if $b(l_i)$ always has the same value. Thus

$$\overline{b^2} \geq (\overline{b})^2 \tag{1-7}$$

with the equality holding only for the case of b being a constant. In our example

$$\overline{(\delta a)^2} = 1,309.5 - (34)^2 = 153.5 \text{ (year)}^2$$

$$= \frac{(-16)^2 + (-5)^2 + (3)^2 + (18)^2}{4}$$

If a characteristic of a collection of elements can take on a continuum of values and thus have any one of an infinite number of values, the probability that a parameter i has some particular discrete value is zero and we must work with the probability that its value lies in a range X_i to $X_i + \Delta X_i$. This probability may be defined as the fraction of members of the collection that have this parameter between X_i and $X_i + \Delta X_i$. If we shrink the interval ΔX_i to an infinitesimal value dX_i, this probability may be used to define a probability density f_i:

$$\frac{dN_i}{N} = f_i(X_i)\, dX_i \tag{1-8}$$

where dN_i is the number of elements with parameter X_i between X_i and $X_i + dX_i$. Note that $f_i(X_i)$ must have dimensions of X_i^{-1}. Then in terms of f_i, the probability that parameter i has a value between finitely differing values X_1 and X_2 is given by the integral

$$P(X_1 \rightarrow X_2) = \int_{X_1}^{X_2} f_i(X_i)\, dX_i \qquad (1\text{-}9)$$

Our preceding equations with discrete parameters may be taken over for the continuous case by replacing summations by integrations. In particular if we normalize the probability density we have

$$\int f_i(X_i)\, dX_i = 1 \qquad (1\text{-}10)$$

where the limits of integration must be the complete range of values possible for X_i. Also, the average value of any function of X_i becomes

$$\bar{b} = \int b(X_i) f_i(X_i)\, dX_i \qquad (1\text{-}11)$$

An often encountered continuous probability density is the Gaussian distribution function:

$$f_G(X) = \frac{1}{(2\pi\sigma^2)^{1/2}} \exp\left[-\frac{(X - \bar{X})^2}{2\sigma^2} \right] \qquad -\infty \leq X \leq \infty \qquad (1\text{-}12)$$

This function has a maximum at $X = \bar{X}$ that becomes higher and more narrow the smaller σ becomes. In the limit $\sigma \rightarrow 0, f_G(X)$ becomes a δ function (see Appendix II).

■ ───

Exercise 1-1

Show that $f_G(X)$ is properly normalized

$$\int_{-\infty}^{+\infty} f_G(X)\, dX = 1$$

and that

$$\int_{-\infty}^{+\infty} (X - \bar{X})^n f_G(X)\, dX = \frac{\sigma^n (n)!}{2^{n/2}(n/2)!} \qquad n \text{ even}$$

$$= 0 \qquad n \text{ odd}$$

The case of $n = 2$ gives the variance defined in Eq. (1-6).

─── ■

Now let us turn to the simplest microscopic model of a gas. This is the kinetic theory of N monatomic molecules each of mass m. The fundamental assumptions of kinetic theory are that molecules exist and are in continuous random motion. We assume the gas is at rest (no net mass flow in any direction) and at macroscopic equilibrium, meaning the properties of the collection of molecules do not change with time. We further assume that the molecules exert no forces on each other, except during binary (two-body) collisions. In an *elastic* collision there will be conservation of the translational kinetic energy and momentum of the *pair*, but generally a change from one to the other of individual kinetic energy and momentum. Monatomic molecules lack internal energy states of vibration and rotation that can easily take up or emit energy in a collision and so their collisions can be regarded as elastic. These collisions are the means by which a gas reaches equilibrium. However, once the gas has attained equilibrium these collisions can give rise to no net change in the molecular speed probability density, which as we shall see determines the pressure and temperature of a gas. Any single binary collision can only instantaneously make a negligible fluctuation in this probability density since N is so large in any case of practical interest. By a fluctuation we mean here a change in the relative numbers of molecules in different speed ranges. Hence these collisions may be neglected for the purpose of calculating equilibrium gas properties. By the same reasoning we could also omit the restriction to monatomic molecules and thus to elastic collisions in the equilibrium gas.

We now proceed to calculate the pressure at equilibrium for our model gas considered as contained in a cube of side l with cube faces perpendicular in pairs to the x, y, and z Cartesian coordinate axes. For the equilibrium situation we may treat particle collisions with the walls as elastic; that is, its one velocity component perpendicular to the wall is reversed while its two velocity components parallel to the wall are unchanged, as shown in Fig. 1-1. Consider one particular molecule,

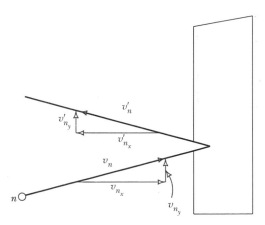

FIGURE 1-1 Elastic collision of the nth particle in a gas at one face of its containing cube perpendicular to the x axis. The initial velocity is $\boldsymbol{v}_n$ with components v_{n_x}, v_{n_y}, and v_{n_z} (not shown). After collision the velocity is $\boldsymbol{v}'_n$, with components $v'_{n_x} = -v_{n_x}$, $v'_{n_y} = v_{n_y}$, and $v'_{n_z} = v_{n_z}$ (not shown).

the nth, moving in the cube with x component of velocity v_{n_x} before collision with the face perpendicular to the positive x direction. Its change of momentum on collision is $2mv_{n_x}$ (from $+mv_{n_x}$ to $-mv_{n_x}$). It will undergo a similar change in magnitude of momentum on the face perpendicular to the negative x direction, and the time interval Δt between collisions on one or the other of these two faces will be l/v_{n_x}. Thus the force exerted by the nth particle on the two faces perpendicular to the x axis will be the rate of momentum change on these faces, or

$$\frac{2mv_{n_x}}{\Delta t} = \frac{2mv_{n_x}^2}{l}$$

Because of its y and z velocity components, similar terms will arise for the force exerted by the nth particle on the two faces perpendicular to the y axis and the two faces perpendicular to the z axis. This total force is thus exerted on a total area of $6l^2$, and the pressure due to the nth particle is

$$P_n = \frac{\text{force}}{\text{area}} = \frac{2m}{l}\,(v_{n_x}^2 + v_{n_y}^2 + v_{n_z}^2)/6l^2 \tag{1-13.1}$$

$$= \frac{mc_n^2}{3V} \tag{1-13.2}$$

The quantity c_n is the magnitude of the velocity vector, that is, the speed of the nth particle, which by the usual vector relation is given by

$$c_n^2 = v_{n_x}^2 + v_{n_y}^2 + v_{n_z}^2 \tag{1-14}$$

We have introduced the volume $V\;(=l^3)$ for the cube. (In Sec. 5-1 we will give a better kinetic theory derivation for particle pressure in an ideal gas that does not depend on the gas being in a cube.) For a total of $\mathcal{N}$ molecules the total pressure will be

$$P = \frac{m}{3V}\,(c_1^2 + c_2^2 + \ldots + c_n^2 + \ldots + c_N^2)$$

$$= \frac{m\mathcal{N}}{3V}\,\overline{c^2} \tag{1-15}$$

in which we introduce the average of the square of the speed:

$$\overline{c^2} = \frac{\displaystyle\sum_{i=1}^{N} c_i^2}{\mathcal{N}} \tag{1-16}$$

Since speeds may take on continuous values from zero to infinity, use of Eq. (1-11) gives

$$\overline{c^2} = \int_0^\infty c^2 f(c)\,dc \tag{1-17}$$

where $f(c)$ is the probability density for molecular speeds in an ideal gas. We will obtain an explicit form for $f(c)$ in Sec. 1-9, but here we may note that the average kinetic energy of translation, $\bar{U}_{trans}$, is by definition given by

$$\bar{U}_{trans} = \mathcal{N} \int_0^\infty \tfrac{1}{2}mc^2 f(c)\, dc \tag{1-18}$$

which from Eqs. (1-15) and (1-16) becomes

$$\bar{U}_{trans} = \frac{m\mathcal{N}}{2} \bar{c^2} = \tfrac{3}{2}PV \tag{1-19}$$

Next we realize that the kinetic theory model of a gas of noninteracting molecules at equilibrium is the classical mechanical microscopic model for the thermodynamic ideal gas, for which

$$PV = nRT = \mathcal{N}kT \tag{1-20}$$

where $k = R/\mathcal{N}_A$, and thus we have

$$\bar{U}_{trans}/\mathcal{N} = \tfrac{3}{2}kT \tag{1-21}$$

This is an extremely important result, for it provides us with a mechanical interpretation of temperature, namely, that temperature is proportional to the average translational kinetic energy of random motion per particle. It is a statistical concept, an average property of a huge number of particles. In addition, Eq. (1-21) is our first example of a relation between the average of a microscopic property (molecular kinetic energy) and a thermodynamic variable (the temperature). As derived here, Eq. (1-21) applies only to monatomic ideal gases. As we will show in later sections, it is also generally true for interacting polyatomic gases and liquids, and even for solids as long as classical physics is valid.

1-2
Ensembles: The Microcanonical Case

We now proceed to present in a formal way the postulates and methodology of averaging that comprise equilibrium statistical mechanics. There are three types of equilibrium systems most often used as a basis for developing links between thermodynamic variables and molecular properties. These are:

1. Isolated systems of fixed energy U containing $\mathcal{N}$ particles of known type or types (if several species a, b, . . . of particles are present, $\mathcal{N}$ will represent the number set $\mathcal{N}_a$ of type a, $\mathcal{N}_b$ of type b, etc.) in a fixed volume V.

2. Systems in volume V containing number set $\mathcal{N}$ of particles in thermal contact with a heat reservoir characterized by a temperature T.

3. Systems in volume V in thermal contact with a heat reservoir characterized by a temperature T and open to exchange of matter with the reservoir, which is also considered as a source of particles characterized by a set of chemical potentials, μ (μ_a for species a, μ_b for species b, etc.).

In all these cases the thermodynamic state is specified by only a few macroscopic properties. Any observed property, such as the pressure, is actually a time average over a finite time interval sufficiently long that the recorded average is not dependent on the magnitude of the time interval. Recall our method of calculating pressure in Sec. 1-1. On a very short time scale the momentum reversals of particles at a wall will vary appreciably, whereas for longer time intervals τ the total momentum reversal at a wall during τ will become proportional to τ, such that dividing by τ the time average rate of momentum reversal becomes independent of τ. To attempt an a priori mechanical (classical or quantum) calculation of pressure by following molecular trajectories in a gas sample over a long enough time period to finally obtain results independent of time interval is an impossible task, owing to the huge number N of particles involved. There would not be paper enough to record all the necessary data. This statement has not been made obsolete by the fact that computers can be programmed to follow systems of up to 10^3 particles in simulations of molecular motion (as will be discussed in Chap. 3). The results of such simulations are numerical data (not analytical expressions) that must be systematized using just those theoretical concepts that we are developing here.

We will bypass the impossible calculation by following the methods of J. Willard Gibbs, the brilliant American polymath* who introduced in 1902 the concept of an ensemble of η replicas of the thermodynamic system of interest. An ensemble is an example of a collection of elements to which probability theory as discussed in Sec. 1-1 can be applied. Each element in this collection (called a replica or copy) is an entire thermodynamic system *specified by the same few macroscopic conditions*. When these conditions are that each replica has fixed volume V, fixed total energy U, and fixed number set N, the collection is called the *microcanonical ensemble*. We must realize that specifying just these few conditions does not exactly fix the value of any other thermodynamic variable ϕ that will be subject to fluctuations. In the microcanonical ensemble different replicas will have different temperatures. In the canonical (fixed V, N, T) ensemble a replica can be imagined as prepared by bringing a system of fixed V and N into thermal contact with a huge reservoir at temperature T. When equilibrium is established after some energy exchange, the replica may in principle have any possible energy. Thus the replicas of the canonical ensemble will have different energies. However as we will see, the possible fluctuations are small, indeed usually negligible. The crucial idea of Gibbs was based on the principle that at equilibrium the properties of a system become time independent and so it should be appropriate to replace a time average on one

---

* In 1863 Gibbs earned the first Ph.D. in engineering (the second in science) ever given in the United States and set an impossibly high standard for those who were to follow.

actual system with an instantaneous time independent ensemble average over η replicas of such a system, each thermodynamically representative of the actual one of interest. This will be our first postulate:

Postulate I   The time average of any thermodynamic variable ϕ in an actual system is equal to the ensemble average of ϕ in the limit as the number of replicas η in the ensemble goes to infinity, provided that the members of the ensemble copy precisely the thermodynamic state and environment of the actual system, that is, adhere to the specified macroscopic conditions.

This postulate is called the *ergodic postulate* or hypothesis. There is a vast and mathematically quite complex literature, most of which is irrelevant to physics, that treats of the necessary and sufficient conditions for this postulate to be valid. We take its validity as having been established from the fact that calculations of thermodynamic properties based on its use are in accord with experiment, as shown in Chaps. 2–4.

There is a fantastically large number of micromolecular ways W by which a system of $\mathcal{N}$ moving and interacting molecules can be arranged and still satisfy the restriction of fixed total U in a fixed volume V. We will present a quantum mechanical approach from the outset because it is fundamentally correct and heuristically easier to understand than a classical physics treatment. We will refer to W as the number of microstates available to the system. These are quantum states of a macroscopic system with $\mathcal{N}$ particles. Every quantized energy eigenstate of a single molecule will be split into $\mathcal{N}$ eigenstates of the system as a whole. Then the average energy difference between successive states in the system will be of order $1/\mathcal{N}$ the already small energy spacing in a single molecule. Although we specify fixed U in the microcanonical ensemble, there must in practice always be some experimental uncertainty in U and the energy of all members of the ensemble will fall in the range U to $U + \Delta U$. This ΔU will always be far larger than the energy spacing in the macroscopic system, perhaps roughly by a factor of $\mathcal{N}^2$. One factor $\mathcal{N}$ will bring us to single molecule size energy spacing, and another factor $\mathcal{N}$ (when $\mathcal{N}$ is of order 10^{20} or more) will bring us up to ordinary laboratory-scale size uncertainties in the energy input or output of a system we attempt to isolate for experimental observation. Then the number of microstates available subject to our experimental uncertainty might better be taken as $\mathcal{N}^2$ times the previous W, which was the degeneracy of the exactly specified $\mathcal{N}$ molecule energy level of value U. However as we shall see, the thermodynamic entropy function is proportional to the logarithm of the number of microstates available, and since entropy is an extensive property, $\ln W$ is of order $\mathcal{N}$ and W is of order $e^{\mathcal{N}}$. Then

$$\ln \mathcal{N}^2 W = \ln W + 2 \ln \mathcal{N}$$

and the second term is only 111 even with $\mathcal{N} = 10^{24}$ and is negligible compared to $\ln W$. Hence any increase in the number of microstates available because of the size of ΔU cannot affect thermodynamic properties and need not be more precisely treated. However the fact that ΔU cannot be made exactly zero is very important, for it means that no thermodynamic system can be truly isolated. Any random force (electromagnetic or any other) or random emission or absorption of energy by atoms in the walls surrounding a member of the microcanonical ensemble will be sufficient to cause transitions between the W microstates available to the system. For macroscopically large systems these transitions will be practically continuous, such that any ordinary measurement will record an average of some property over a huge number of microstates. From the random origin of the transitions it is reasonable to suppose that every possible microstate occurs with equal probability. This is the motivation for the second fundamental postulate of equilibrium statistical mechanics:

> Postulate II   The many body systems of the ensemble are distributed with equal a priori probabilities over all the possible quantum states of the many-body system consistent with the few macroscopic conditions.

In the language of quantum mechanics, there is a set of W-fold degenerate N-body wave functions Ψ_l (where l ranges from 1 to W) all with the same energy U. By postulate II the probability P_l of finding a member of the microcanonical ensemble in the state l is the same for all l and is, by noting Eq. (1-1),

$$P_l = \frac{1}{W}$$

since then we will have the correct normalization of Eq. (1-2):

$$\sum_{l=1}^{W} P_l = \sum_{l=1}^{W} \left(\frac{1}{W} \right) = W \left(\frac{1}{W} \right) = 1$$

Every state l will be found the same (large) number of times among the replicas of the microcanonical ensemble.*

What kinds of changes are possible for an isolated N-body system at constant V and U? Such a system may be very large, composed of many parts, and need not be homogeneous at the start. It may contain sources and sinks of energy and/or particles that may be permitted to exchange energy or particles between themselves. There will generally be internal parameters or conditions of constraint, Y_α, labeled by a number α that counts subsets of microstates W_α available to one replica at constant N, V, and U. Consider an initial condition of a replica with $W = W_i$. Upon removal or relaxation of a constraint Y_α, the set of W_α new microstates becomes available to the replica such that by postulate II all $W_i + W_\alpha = W_f$ final microstates must eventually be occupied with equal probability. At the moment the constraint was first released—for example by removing a partition between a

---

* The presentation in this and the next section now follows that given by J. E. Mayer [1]; see also Berry et al. [2].

gas-filled part and a vacuum part of the replica—the replica was no longer in an equilibrium condition. Transitions into the newly available microstates will occur until a new equilibrium condition is reached. In this example more microstates are available with increased gas volume because the number of permitted quantum states of translation increases with volume increase, as will be shown in Example 1-3. By definition, a constraint precludes certain sets of microstates. Removal or relaxation of a constraint thus means more states become available to the many body system. Then by Postulate II all these microstates will share uniform probability of occupation. Thus W can only increase when a constraint is removed or relaxed. Such a process is an irreversible change. Notice that if we imagine reinserting the partition with no expenditure of work (in the idealized *reversible* limit) the number of microstates W available will not change. A reversible change is one for which W remains the same at all stages of the process, and so detailed equilibrium (meaning at no stage is there net flow into previously unoccupied sets of microstates) is always maintained. Notice also that reinserting the partition does not restore the original constraint of gas occupation of only a part of the replica. This constraint could be restored by the input of work from outside (e.g., by moving a section of wall to sweep the molecules back into a fraction of the replica's total fixed volume), but this would violate the condition of constant U.

This behavior of W for possible changes for a system of fixed N, V, and U motivates the definition of the quantity S as

$$S = k \ln W \tag{1-22}$$

In what follows S will be shown to be identical with the thermodynamic entropy function. Subject to whatever set of internal parameters of constraint exists, W and thus S will rise to a maximum for systems of fixed U, V, N. For a reversible change W is a function only of U, V, N; S is constant. For an irreversible change W is a completely undefined function of not merely U, V, N but of any of the infinite number of contingent ways that conditions of restraint might be removed or relaxed. Obviously changes of S can only be calculated along reversible paths. All of this is in accord with our understanding of the entropy function of thermodynamics.

The logarithmic function is necessary in Eq. (1-22) to assure that if a homogeneous system be divided into any number of parts with common number density and common energy per molecule, the sum of the S values of the parts will equal the S value of the total system. If a system is divided into two parts of respective values $S_1 = k \ln W_1$ and $S_2 = k \ln W_2$, then the total $S = S_{12}$ is

$$S_{12} = k \ln W_{12} = S_1 + S_2 \tag{1-23}$$

since each of the microstates W_1 could be associated with each of the microstates W_2 and obviously

$$W_{12} = (W_1)(W_2) \tag{1-24}$$

Thus S is an extensive variable and must be proportional to the number of particles N. Then from Eq. (1-22) W is stupendously large, being of order e^N.

The correlation length is an average maximum distance in a system of inter-acting particles over which changes in particle number at one point in space can affect changes in particle number at another point. The fundamental reason for the extensive nature of S is that the correlation length is almost always very short (no more than a few average molecule-molecule separations). The correlation length becomes of macroscopic size in the region of a critical point of a phase transition (see Chap. 3) but then the number density fluctuations will be so large as to preclude treating a phase as homogeneous.

It is convenient to introduce a classification of the microstates of a macroscopic system that is intermediate between the inordinately detailed classification of speci-fic quantum number enumerations and the very broad one of the thermodynamic state. Such a classification is called a distribution, α. Many microstates will corre-spond to one distribution, but the total number of distributions consistent with fixed N, V, U will be a number M of order of magnitude N. For example, α could label the number of molecules in one part of a set of replicas composed of two equal volume parts. Then we have

$$W = \sum_{\alpha=1}^{M} W_\alpha \tag{1-25}$$

and

$$S = k \ln W = k \ln \sum_{\alpha=1}^{M} W_\alpha \tag{1-26}$$

where the sum is over distributions α that satisfy whatever constraints apply.

So far we have considered an ensemble of interacting molecules in general terms. Now for ease of concrete illustration we treat independent particles with no forces acting between them. Then the total energy can be written in terms of the single-molecule energy states ε_i:

$$U = \sum_i N_i \varepsilon_i \tag{1-27}$$

and

$$N = \sum_i N_i \tag{1-28}$$

Here by a distribution we mean that N_1 molecules have energy ε_1, N_2 have energy $\varepsilon_2, \ldots, N_i$ have energy ε_i, and so on. By combinatorial mathematics the number of ways W_α to arrive at a distribution of N particles that can be distinguished from one another (given labels $a, b, \ldots,$ etc.) and put into states is

$$W_\alpha = \frac{N!}{N_1! N_2! \ldots N_i!} = \frac{N!}{\prod_i (N_i)!} \tag{1-29}$$

W_α is also called the weight of the distribution. The notation $\prod_i$ means the re-peated product of all terms i. Note that $0! = 1$ for consistency. In Fig. 1-2 we illus-

$$W = \frac{5!}{3!2!} = 10$$

State (1)	State (2)	State (1)	State (2)
a b c	d e	a b d	e c
a b e	d c	a c d	e b
b c e	d a	b c d	e a
a c e	d b	a d e	c b
		b d e	c a
		c d e	b a

FIGURE 1-2 The number of ways of distributing five distinguishable particles, a–e, into two states of three and two, respectively. The order of filling left to right does not give rise to a new way of arriving at the distribution; for example, $bca | de$ is the same as the first arrangement on the left, $abc | de$.

trate the use of Eq. (1-29) for the example of distributing five particles labeled a, b, c, d, e into two states, one with three particles, the other with two.

We assert, based on our previous discussion, that the quantity S can only increase (or remain essentially unchanged) for every change that is possible for an isolated $\mathcal{N}$-body system at constant V and U. As an example consider a trivial system of two molecules of type A, two of type B, and three of type C at fixed total $U = 86\varepsilon$ where ε is the zero point energy of species C. Two allowed distributions for this system are shown in Fig. 1-3a based on the arbitrarily chosen

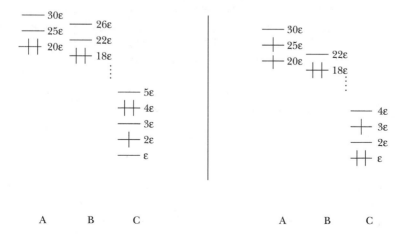

FIGURE 1-3a Two allowed distributions for two molecules of type A, two of type B, and three of type C at constant $U = 86\varepsilon$. The molecular energy spacings shown are arbitrarily assigned.

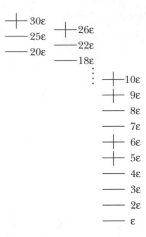

A   B   C

FIGURE 1-3b An example of a new distribution allowed to the system in Figure 1-3a if a catalyst is uncovered that enables the chemical reaction A + B $\rightleftarrows$ C to proceed.

energy state spacings for the three species (which have ground states for both A and B much higher than that of C). If now a catalyst is inserted in the system that enables the chemical reaction A + B $\rightleftarrows$ C to occur (at fixed $U = 86\varepsilon$) then distributions with changed numbers of molecules such as the one shown in Fig. 1-3b with one molecule of A, one molecule of B, and four of C are newly allowed. Thus S will increase. Incidentally, it is interesting to note in this example that when the catalyst is present no C can react to give A and B since $3(20\varepsilon) + 3(18\varepsilon)$ already exceeds 86ε. In other words, molecule C is the favored species when chemical equilibrium is reached for the low constant U chosen.

By Eq. (1-1) the probability of finding a member of our microcanonical ensemble with the distribution α after equilibrium is

$$P_\alpha = \frac{W_\alpha}{W} \tag{1-30}$$

There will always be one value of α, say $\alpha = m$, for which W_m is a maximum

$$P_m \geq P_\alpha, \qquad W_m \geq W_\alpha \qquad \text{for all } \alpha \tag{1-31}$$

Since the terms in the sum in Eq. (1-26) are all positive, the sum is larger than any single term and of course less than the value of the largest term multiplied by the total number of terms, M. Hence we can bracket $S = k \ln W$ in value between

$$k \ln W_m \leq S \leq k \ln (MW_m)$$

that is,

$$S_m \leqq S \leqq S_m \left(1 + \frac{\ln M}{\ln W_m} \right) \tag{1-32}$$

Since M is of order N while W_m is of order e^N, the additive term on the right-hand side of Eq. (1-32) is of order $N^{-1} \ln N$ and is numerically negligible for macroscopic systems of thermodynamic interest, for which $N \geq 10^{18}$ usually. Hence we have found the important result that the quantity S at equilibrium is numerically equal, with minor error, to that for the system restricted to be in its most probable distribution. Thus for practical purposes we may calculate S using W_m only, that is:

$$S \cong S_m = k \ln W_m \tag{1-33}$$

Furthermore, if we were to reimpose a constraint, such as that of no chemical reaction possible (remove catalyst), in our above example the probability of trapping a member of our ensemble in a distribution α measurably different from the most probable one corresponding to complete reaction equilibrium would be entirely negligible in any real system of thermodynamic size. If the system has distribution α, the value of $S_\alpha = k \ln W_\alpha$ would be less than $S_m = k \ln W_m$ by the amount

$$\Delta S_\alpha = S_m - S_\alpha = k \ln (W_m/W_\alpha)$$

which by Eq. (1-30) would occur with probability

$$P_\alpha = P_m \exp - (\Delta S_\alpha/k) \tag{1-34}$$

To be measurable ΔS_α must be some tiny (but not zero) fraction of S_m, say $10^{-6} S_m \cong 10^{-6} kN$, which for $N \geq 10^{18}$ gives

$$P_\alpha \leq P_m \exp (-10^{12})$$

which is certainly a negligible probability.

Example 1-1

Consider an ensemble of $N = 18$ distinguishable molecule systems of fixed energy $U = 6\varepsilon$ and of fixed volume such that the only individual molecule energy states are fixed as

$$\underline{\hspace{3cm}} \ 2\varepsilon$$

$$\underline{\hspace{3cm}} \ \varepsilon$$

$$\underline{\hspace{3cm}} \ 0$$

If N_i is the number of molecules in state i ($= 0, 1, 2$), show by explicit calculation that the full ensemble average and the average in the most probable distribution for (N_i/N) are practically identical.

Solution:

It is easily seen that there are only four distributions that satisfy the two constraints

$$N_0 + N_1 + N_2 = 18$$

$$N_1 + 2N_2 = 6$$

These distributions are

2ε	___ 3	___ 2	___ 1	___ 0
ε	___ 0	___ 2	___ 4	___ 6
0	___ 15	___ 14	___ 13	___ 12
	α	β	γ	δ

with

$$W_\alpha = \frac{18!}{15!3!0!} = 816 \qquad W_\beta = \frac{18!}{14!2!2!} = 18,360$$

$$W_\gamma = \frac{18!}{13!4!1!} = 42,840 \qquad W_\delta = \frac{18!}{12!6!0!} = 18,564$$

$$W = W_\alpha + W_\beta + W_\gamma + W_\delta = 80,580$$

Distribution γ is the most probable, and even in this few-molecule case, $\ln W = 11.3$ and $\ln W_\gamma = 10.7$ do not differ greatly and both are of order $N = 18$. The number of distributions ($M = 4$) is also of order N.

In the most probable distribution

$$\frac{N_0}{N} = \frac{13}{18} = 0.722, \qquad \frac{N_1}{N} = \frac{4}{18} = 0.222, \qquad \frac{N_2}{N} = \frac{1}{18} = 0.056$$

In this trivial case the full ensemble average may be calculated. The total number of molecules in all the replicas of our system is $18W = 1,450,440$, of which the number in the zero state are

$$(816)(15)(1) + (18,360)(14)(1) + (42,840)(13)(1) + (18,564)(12)(1) = 1,048,968$$

The inclusion of the factors (1) in the above is to remind the reader of Postulate II: all the ways of occupying the quantum states that satisfy the constraints are taken with equal a priori probability (factors of unity), but the weights W_i differ because of the great difference in the number of micromolecular ways in which these modes of occupation can be realized.

Thus the ensemble average for $\mathcal{N}_0/\mathcal{N}$ is

$$\frac{\overline{\mathcal{N}_0}}{\mathcal{N}} = \frac{1,048,968}{1,450,440} = 0.723$$

Similarly,

$$\frac{\overline{\mathcal{N}_1}}{\mathcal{N}} = 0.220 \qquad \text{and} \qquad \frac{\overline{\mathcal{N}_2}}{\mathcal{N}} = 0.057$$

These ensemble results for even this few-molecule system are practically the same as the results obtained by using only the one most probable distribution, W_γ. This illustrates that the principle embodied in Eq. (1-33) is very accurate.

We now complete our development showing that the quantity S may be identified with the thermodynamic entropy function. Consider a microcanonical ensemble of two-part systems of fixed total energy $U = U_1 + U_2$. The parts are not in thermal contact and they each have different fixed volumes and numbers of particles. The ensemble will include replicas with every possible amount of energy in part 1 from 0 to U. To keep α (number of distributions) as an integer we define $\delta\varepsilon$ as some fixed minimum amount of energy that can be detected by hypothetical experiment and set $\alpha = U_1/\delta\varepsilon$. In the limit that $U_1 \to U$, and if $\delta\varepsilon$ could be made as small as the average energy per molecule, the maximum value of α will be $\mathcal{N}$. Then the probability of having a particular energy subdivision or distribution is, by Eq. (1-24),

$$P_\alpha = \frac{W_1(U_1)W_2(U - U_1)}{\sum\limits_\alpha W_1(U_1)W_2(U - U_1)}$$

To obtain the most probable distribution we maximize P_α:

$$\frac{\partial}{\partial \alpha} P_\alpha = \delta\varepsilon \frac{\partial}{\partial U_1} P_\alpha = \delta\varepsilon \frac{(\partial W_1/\partial U_1)W_2 + W_1(\partial W_2/\partial U_1)}{\sum\limits_\alpha W_1 W_2} = 0$$

Since $U_2 = U - U_1$ and $\partial/\partial U_1 = (\partial U_2/\partial U_1)(\partial/\partial U_2) = -\partial/\partial U_2$, this condition for the most probable distribution becomes

$$W_2 \frac{\partial W_1}{\partial U_1} = W_1 \frac{\partial W_2}{\partial U_2}$$

and dividing both sides by $W_1 W_2$, this becomes

$$\frac{\partial \ln W_1}{\partial U_1} = \frac{\partial \ln W_2}{\partial U_2}$$

which, from Eq. (1-26) and noting explicitly the constant V, $\mathcal{N}$ condition stated originally, reads

$$\left(\frac{\partial S_1}{\partial U_1}\right)_{V_1, N_1} = \left(\frac{\partial S_2}{\partial U_2}\right)_{V_2, N_2} \tag{1-35}$$

We may identify this condition for the most probable distribution as the condition of equilibrium for the ensemble of two-part systems. This is because S and functions of S at equilibrium are only negligibly different from the values attained by the many-body system in its most probable distribution (see Problem 1). Other distributions of energy between the two parts are possible for which the entropy derivatives of the two parts are not equal, but by Eq. (1-34) these distributions will be extremely improbable. Note that in this ensemble there is no energy exchange between the two parts of each replica, yet we find that in the most probable distribution the two parts have equal temperatures [see Eq. (1-37) below]. This kind of result is the reason (see Sec. 1-3) that fundamentally all ensembles are thermodynamically equivalent. It is in fact crucial for our above argument that U_1 and U_2 be time independent (and not subject to ongoing energy exchange between themselves) in order to talk about different energy distributions between the two parts. This is related to the physical fact that it is impossible to measure the properties of a system in a certain distribution unless there be some inhibition against transition into other distributions during the time of the experiment. This analysis can be generalized. If each replica were to consist of any number of isolated parts among which a fixed amount of energy is divided, the most probable distribution is the one characterized by equality of temperature in all the parts.

Next we shift our point of view and consider a single isolated two-part system of respective fixed volumes V_1 and V_2 and respective fixed numbers of particles $\mathcal{N}_1$ and $\mathcal{N}_2$, and with fixed total $U = U_1 + U_2$. If we permit an energy exchange between the parts such that $dU_1 = -dU_2$, we have from Eq. (1-23) and our general result, $dS \geq 0$,

$$dS_{12} = dS_1 + dS_2 = \left(\frac{\partial S_1}{\partial U_1}\right)_{V_1, N_1} dU_1 + \left(\frac{\partial S_2}{\partial U_2}\right)_{V_2, N_2} dU_2 \geq 0 \tag{1-36}$$

Now we define T by the equation

$$T^{-1} = \left(\frac{\partial S}{\partial U}\right)_{V,N} \tag{1-37}$$

such that Eq. (1-36) becomes

$$(T_1^{-1} - T_2^{-1})\, dU_1 \geq 0 \tag{1-38}$$

This means that energy flows from 2 to 1 with $dU_1 > 0$ if $T_2 > T_1$ and in the reverse direction if $T_1 > T_2$, and that there is equilibrium with $dS_{12} = 0$ if $T_1 = T_2$. Hence T must be a scale of temperature and is consistent with the result in Eq. (1-35). We will find later that T is the usual absolute temperature scale if k in the defining Eq. (1-22) is taken to be the Boltzmann constant (R/N_A). Similar arguments (see Problem 2) can show that

$$P = T\left(\frac{\partial S}{\partial V}\right)_{U,N} \tag{1-39}$$

and

$$\mu_a = -T\left(\frac{\partial S}{\partial N_a}\right)_{V,U,N_b} \tag{1-40}$$

have the meaning of pressure and chemical potential per particle of species a, respectively. Thus at constant number set N, by the relation for a total differential we have

$$dS_N(U, V) = \left(\frac{\partial S}{\partial U}\right)_{V,N} dU + \left(\frac{\partial S}{\partial V}\right)_{U,N} dV$$

$$= T^{-1}\, dU + P/T\, dV \tag{1-41}$$

or

$$dU = T\, dS_N - P\, dV \tag{1-42}$$

From Eq. (1-42), and recalling from thermodynamics (Appendix I) that in a reversible change the work done on a system (if only expansion-compression work is possible) is $-P\, dV$, we see that $T\, dS$ must have the meaning of a heat input in a *reversible* process. Thus the quantity S as defined for the microcanonical ensemble in Eq. (1-22) can be identified as the thermodynamic entropy, and Eq. (1-41), along with $dS \geq 0$ at constant U, V, N, is the expression of the Second Law of thermodynamics in this ensemble. The fact that S is proportional to the logarithm of an integer (W), which at least conceivably can be as low as unity if a many-body system is constrained by macroscopic conditions to a single quantum state, leads to the Third Law of thermodynamics.

We may return to the consideration of a microcanonical ensemble of two-part systems (which is generalizable to systems of any number of parts) in which not only is a fixed amount of energy distributed between the two parts, but also a fixed wall occurs in every replica that thus divides the fixed volume V into two parts, V_1 and V_2. Maximizing the W_{12} of Eq. (1-24) separately with respect to variations in U_1 and then with respect to variations in V_1 (see Problem 3) leads to the results that in the most probable distribution

$$T_1 = T_2 \tag{1-43}$$

and

$$\left(\frac{\partial S_1}{\partial V_1}\right)_{U_1, N_1} = \left(\frac{\partial S_2}{\partial V_2}\right)_{U_2, N_2} \tag{1-44}$$

This last, by the definition given in Eq. (1-39), means

$$P_1 = P_2 \tag{1-45}$$

This procedure can be further extended to the distribution of a fixed number $\mathcal{N}_a$ of molecules of species a between the two parts. Then if we investigate the condition of the maximization of W_{12} with respect to $\mathcal{N}_{a_1}$, we find

$$\left(\frac{\partial S_1}{\partial \mathcal{N}_{a_1}}\right)_{V_1, U_1} = \left(\frac{\partial S_2}{\partial \mathcal{N}_{a_2}}\right)_{V_2, U_2} \tag{1-46}$$

This means, by definition of Eq. (1-40) and the result in Eq. (1-43),

$$\mu_{a_1} = \mu_{a_2} \tag{1-47}$$

The generalization of these results is that for a microcanonical ensemble of systems of any number of parts:

1. If the total energy is constant and distributed among the parts, the most probable distribution will have equal temperatures in all parts

2. If the total volume is constant and distributed among the parts by insertion of partitions, the most probable distribution will have equal pressures in all parts

3. If the total number set is constant and distributed among the parts the most probable distribution will have equal chemical potentials for all species in all of the parts

1-3
Other Ensembles and Thermodynamic Identifications

Having gained confidence in treating an ensemble in Sec. 1-2, we are ready to use a more general method that will rederive the major result of that section and lead to new relationships between microscopic properties and thermodynamic functions other than the entropy.

Consider a supersystem of fixed volume V_t, fixed energy U_t, and number set N_t (which really stands for $\mathcal{N}_{a_t}, \mathcal{N}_{b_t}, \ldots$ the total number of particles of species $a, b, \ldots$) where subscript t denotes total. One supersystem is made up of $(\mathcal{J} + 1)$ connected many-body parts—one unique part called the reservoir and $\mathcal{J}$ other parts in contact with the reservoir and through the reservoir with each other. Imagine an ensemble of supersystems. Since V_t, U_t, N_t are fixed, this is a microcanonical ensemble of supersystems. We know from Sec. 1-2 that at equilibrium the entropy of the supersystem S_t will be a maximum as a function of a variable specifying the distribution of one supersystem. A convenient variable for this is the number $\mathcal{J}_K$, which counts the number of many-body parts of the supersystem in energy quantum state K of fixed volume V, energy U_K, and number set N. For fixed $\mathcal{J}_K$ the total number of quantum states $W_t(\mathcal{J}_K)$ of the supersystem will be the product of the number of ways to pick out $\mathcal{J}_K$ from $\mathcal{J}$, which is

$$\frac{\mathcal{J}!}{(\mathcal{J} - \mathcal{J}_K)!\mathcal{J}_K!}$$

times a number $W_r(\mathcal{J}_K)$, which represents the number of states allowed to the reservoir and the $(\mathcal{J} - \mathcal{J}_K)$ systems in energy states different from K. Thus

$$W_t(\mathcal{J}_K) = \frac{\mathcal{J}!W_r(\mathcal{J}_K)}{(\mathcal{J} - \mathcal{J}_K)!\mathcal{J}_K!} \tag{1-48}$$

Notice that subscript K is a concise notation for a huge number of quantum numbers for a particular detailed energy microstate of the interacting number set N of molecules. If we had wanted to write the number of ways to pick out the energy *level* K of degeneracy W_K, then the expression for W_t in Eq. (1-48) would have had to be multiplied by $(W_K)^{\mathcal{J}_K}$ and $\mathcal{J}_K$ would count the number of many-body parts of the supersystem in the degenerate energy level K. W_K itself would be the total W of a fixed energy case as given by Eq. (1-25). In the method used here there is only *one* choice of detailed quantum number set for each of the $\mathcal{J}_K$ particular energy states.

The subscript r denotes the reservoir *and* the other $\mathcal{J} - \mathcal{J}_K$ systems. Clearly this region of the supersystem has

$$S_r = k \ln W_r(\mathcal{J}_K)$$
$$V_r = V_t - \mathcal{J}_K V$$
$$U_r = U_t - \mathcal{J}_K U_K \tag{1-49}$$
$$N_r = N_t - \mathcal{J}_K N$$

In addition, by virtue of contact via the reservoir this region (as well as the set of systems in energy state K) by the principles of thermodynamics will have common T, P, and set $\boldsymbol{\mu}$ (which stands for $\mu_a, \mu_b, \ldots$) that are given by Eq. (1-37), (1-39), and (1-40) as

$$\left(\frac{\partial S_r}{\partial U_r}\right)_{V,N} = T^{-1}$$

$$\left(\frac{\partial S_r}{\partial V_r}\right)_{U,N} = \frac{P}{T} \tag{1-50}$$

$$\left(\frac{\partial S_r}{\partial \mathcal{N}_a}\right)_{V,U,N_b} = -\frac{\mu_a}{T}$$

We now determine the equilibrium value of $\mathcal{J}_K = \bar{\mathcal{J}}_K$, which is such that

$$S_t(\mathcal{J}_K) = k \ln W_t(\mathcal{J}_K) = k \ln \frac{\mathcal{J}!}{(\mathcal{J} - \mathcal{J}_K)! \mathcal{J}_K!} + S_r(\mathcal{J}_K) \tag{1-51}$$

is a maximum:

$$\left(\frac{dS_t}{d\mathcal{J}_k}\right)_{\mathcal{J}_K = \bar{\mathcal{J}}_K} = 0 = k \ln \left(\frac{\mathcal{J} - \bar{\mathcal{J}}_k}{\bar{\mathcal{J}}_k}\right) + \left(\frac{dS_r}{d\mathcal{J}_k}\right)_{\mathcal{J}_K = \bar{\mathcal{J}}_K} \tag{1-52}$$

In obtaining Eq. (1-52), Stirling's approximate formula:

$$\ln \mathcal{N}! = \ln 1 + \ln 2 + \cdots + \ln \mathcal{N} \cong \int_1^N \ln x \, dx \cong \mathcal{N} \ln \mathcal{N} - \mathcal{N} \tag{1-53}$$

which is accurate for large $\mathcal{N}$ as shown in Appendix II, has been used for $\ln (\mathcal{J}_K)!$ and $\ln (\mathcal{J} - \mathcal{J}_K)!$

Exercise 1-2

Use a pocket calculator with $\mathcal{N}!$ program to calculate the percent error in using Eq. (1-53) for $\mathcal{N} = 20, 40, 60, 80$.

Exercise 1-3

Derive Eq. (1-52) from Eq. (1-51) using the Stirling approximation.

In Eq. (1-52) we can neglect $\bar{\mathfrak{J}}_K$ with respect to $\mathfrak{J}$, for we may assume that our supersystem is practically infinite in size and the particular energy state K will be realized by only a tiny fraction of the total systems present. This fraction $\bar{\mathfrak{J}}_K/\mathfrak{J}$ is in fact the probability P_K of occurrence of energy state K among what we may now term the ensemble of many-body systems in one average supersystem. In other words, P_K is the probability of observing quantum energy state K if the energy is determined for a many-body system chosen at random from among the $\mathfrak{J}$ in the supersystem.

Rewriting Eq. (1-52) as

$$k \ln (P_K^{-1}) + \left(\frac{dS_r}{d\mathfrak{J}_K}\right) = 0$$

we have

$$P_K = \exp \frac{1}{k}\left(\frac{dS_r}{d\mathfrak{J}_K}\right) \tag{1-54}$$

with

$$dS_r = \left(\frac{\partial S_r}{\partial U_r}\right)dU_r + \left(\frac{\partial S_r}{\partial V_r}\right)dV_r + \sum_a \left(\frac{\partial S_r}{\partial N_{ar}}\right)dN_a$$

in which the sum on a is over types of species present. Use of Eqs. (1-50) and (1-49) in the form $dU_r/d\mathfrak{J}_K = -U_K$ and so on gives

$$P_K = \exp \frac{1}{kT}(-PV + \mathbf{N} \cdot \boldsymbol{\mu} - U_k) \tag{1-55}$$

in which

$$\mathbf{N} \cdot \boldsymbol{\mu} = \sum_a N_a \mu_a \tag{1-56}$$

is a sum over types of species present and μ_a is the chemical potential per particle of a.

From Eq. (1-55) we may obtain all the fundamental formal expressions for calculating thermodynamic properties from U_K values by specifying what actual conditions of interaction (or lack of it) exist via the reservoir among the $\mathfrak{J}$ many-body systems of one supersystem. We have used the concept of equality of T, P, and μ (in Eq. 1-50) via interaction with the reservoir for the $\mathfrak{J}$ systems in one supersystem. However, as discussed at the end of Sec. 1-2 the most probable overall distribution of a microcanonical ensemble of many-part systems will always have equality of T, P, and μ throughout its parts if a fixed amount of total volume, total energy, and total number of particles is distributed among the parts even though no actual volume, energy, or mass transfer is permitted between the parts! To the extent necessary, in what follows we assume we are dealing with the most probable distribution with respect to volume, energy, and particle distribution. We also use from thermodynamics [see Eq. (I-13)] the integrated form for the energy of a system:

$$U = TS - PV + \mathbf{N} \cdot \boldsymbol{\mu} = TS - PV + G \qquad (1\text{-}57)$$

in which G is the Gibbs free energy.

Our first specification of ensemble will be that of the microcanonical case for the $\mathcal{J}$ systems, that is, all contact cut off, all with fixed volume V, fixed number set $\mathbf{N}$, and all with energy $U_K \equiv U$. Hence, using Eq. (1-57) the P_K from Eq. (1-55) is the same for all states:

$$P_K = e^{-S/k} \qquad (1\text{-}58)$$

Thus P_K must be the reciprocal of the total number of states W_K allowed and we obtain our previous result:

$$S = k \ln W_K \qquad (1\text{-}59)$$

Our second specification, called the canonical ensemble, is that the systems $\mathcal{J}$ have fixed volume V and number set $\mathbf{N}$, but the walls of the the systems $\mathcal{J}$ may conduct heat and so all energies U_K are permitted. Since $-PV + G = A$, the Helmholtz free energy, we find

$$P_K = \exp \left(A - U_K\right)/kT \qquad (1\text{-}60)$$

By normalization of probabilities we have

$$\sum_K P_K = 1 = \exp \left(A/kT\right) \sum_K \exp \left(-U_K/kt\right)$$

which leads to

$$A = -kT \ln Q_N(V, T) \qquad (1\text{-}61)$$

in which

$$Q_N(V, T) = \sum_K \exp \left(\frac{-U_K}{kT}\right) \qquad (1\text{-}62)$$

is termed the canonical partition function.

Our final specification, called the grand canonical ensemble, is that the systems $\mathcal{J}$ have fixed volumes V but with walls that permit molecules of all species to enter or leave as well as permit heat flow. States for all possible values of $\mathbf{N}$ as well as of

U_K occur. To normalize we must sum over all $\mathbf{N}$ as well as over all K. Thus

$$\sum_N \sum_K P_K = 1 = \exp\left(\frac{-PV}{kT}\right) \sum_N \sum_K \exp\left(\frac{\mathbf{N} \cdot \boldsymbol{\mu}}{kT}\right) \exp\left(\frac{-U_K}{kT}\right)$$

which leads to

$$PV = kT \ln \mathcal{Q}(\mu, V, T) \tag{1-63}$$

in which

$$\mathcal{Q}(\mu, V, T) = \sum_N \sum_K \exp\left(\frac{\mathbf{N} \cdot \boldsymbol{\mu}}{kT}\right) \exp\left(\frac{-U_K}{kT}\right) = \sum_N \left(\exp\frac{\mathbf{N} \cdot \boldsymbol{\mu}}{kT}\right) Q_N(V, T) \tag{1-64}$$

is termed the grand canonical partition function.

Only for independent particles is it easy to compute W_K and so to use Eq. (1-59) to obtain thermodynamic results. For interacting particles we must use Eq. (1-61) or Eq. (1-63). In each choice of ensemble, having made the fundamental indentification as one of the above three relations, all other quantities are computed using the equations of thermodynamics. In the canonical ensemble for example

$$P = -\left(\frac{\partial A}{\partial V}\right)_{N,T} = \frac{kT\left(\frac{\partial Q_N}{\partial V}\right)_{N,T}}{Q_N} \tag{1-65}$$

and

$$S = -\left(\frac{\partial A}{\partial T}\right)_{N,V} = k \ln Q_N + \frac{kT}{Q_N}\left(\frac{\partial Q_N}{\partial T}\right)_{N,V} \tag{1-66}$$

Since by definition

$$U = \sum_K P_K U_K \tag{1-67}$$

and using Eq. (1-55)

$$-k \ln P_K = \frac{1}{T}(PV - \mathbf{N} \cdot \boldsymbol{\mu} + U_K)$$

$$\sum_K -kP_K \ln P_K = \frac{1}{T}(PV - \mathbf{N} \cdot \boldsymbol{\mu})\sum_K P_K + \frac{1}{T}\sum_K P_K U_K$$

$$= S \tag{1-68}$$

noting that

$$\sum_K P_K = 1$$

and Eq. (1-57). This means that the ensembles all interpret entropy as the average of $-k \ln P_K$, which becomes larger and larger the more states K there are available to a system. This is the basis of the general connection between entropy increase and the increase of randomness or disorder in a system.

If we differentiate Eq. (1-68) we have

$$dS = -k \sum_K (dP_K + \ln P_K \, dP_K) = -k \sum_K \ln P_K \, dP_K \qquad (1-69)$$

since

$$\sum_K P_K = 1 \quad \text{and} \quad \sum_K dP_K = 0 \qquad (1-70)$$

Substituting from Eq. (1-55) and again using Eq. (1-70) we have

$$T \, dS = dq_{\text{rev}} = \sum_K (PV - \mathbf{N} \cdot \boldsymbol{\mu} + U_K) \, dP_K = \sum_K U_K \, dP_K \qquad (1-71)$$

Then differentiate Eq. (1-67) to find

$$dU = \sum_K U_K \, dP_K + \sum_K P_K \, dU_K = T \, dS + \sum_K P_K \, dU_K \qquad (1-72)$$

From Eq. (1-72) and the Second Law of thermodynamics we must interpret reversible work done on a system as

$$dW_{\text{rev}} = \sum_K P_K \, dU_K \qquad (1-73)$$

Reversible work is a change of the U_K values themselves without change of the probabilities of their occurrence, as for example by reversibly changing the volume of a system of independent particles that *will* change their U_K values, which go as $V^{-2/3}$. This follows from the formula for the quantized states of translational energy of center of mass (m) motion (quantum number set K becomes merely the integers n_x, n_y, n_z) in a three-dimensional box of sides with lengths a, b, c, parallel respectively to the x, y, z axes (see any discussion of quantum mechanics):

$$U_K \equiv \varepsilon_{\text{trans}} = \frac{h^2}{8m} \left(\frac{n_x^2}{a^2} + \frac{n_y^2}{b^2} + \frac{n_z^2}{c^2} \right) \qquad (1-74)$$

since each squared length is some numerical factor times the common value of $V^{2/3}$. In Eq. (1-74) h is Planck's constant. From Eq. (1-71) on the other hand, reversible heat input involves a change in the probability of occurrence of fixed value energy states.

Table 1-1 provides a summary of our ensemble results. The same fundamental relation can be shown to hold (see reference [1] and Appendix I) in every ensemble: $-kT$ times the logarithm of the partition function is in each case a function which

TABLE 1-1 Ensemble Summary

Ensemble Name	Independent Variables	Type of Contact to Surroundings	Partition Function	Fundamental Thermodynamic Identification
Grand canonical	$V, \boldsymbol{\mu}, T$	Thermal, material	$\mathcal{Q} = \sum_{NK} \exp\left[\dfrac{\boldsymbol{N} \cdot \boldsymbol{\mu} - U_K}{kT}\right]$	$-kT \ln \mathcal{Q} = -PV$
Canonical	$V, \boldsymbol{N}, T$	Thermal	$Q_N = \sum_K e^{-U_K/kT}$	$-kT \ln Q_N = A$
Microcanonical	$V, \boldsymbol{N}, U$	None	$W = \sum_K (1)$	$-kT \ln W = -TS$

at equilibrium is a minimum for all possible variations that hold constant the independent variables of the ensemble.

All the ensembles are thermodynamically equivalent because as long as we are dealing with systems containing large numbers of molecules, the most probable distribution is so overwhelmingly more probable than any other that it is virtually the average distribution. From Eq. (1-64) the grand canonical partition function is a sum over number set $\boldsymbol{N}$

$$\mathscr{Q} = \sum_{N} t_{N} \tag{1-75}$$

with

$$t_{N} = Q_{N}(V, T) \exp{(\boldsymbol{N} \cdot \boldsymbol{\mu}/kT)} \tag{1-76}$$

Its logarithm (which is equal to PV/kT), which is of order $\mathcal{N}$, is completely dominated by the single term using the number set $\boldsymbol{N^*} = \bar{\boldsymbol{N}}$ that maximizes $\ln t_{N}$. Thus the grand canonical ensemble is essentially identical to a canonical ensemble that has *fixed* number set $\boldsymbol{N}$ equal to the set that is most probable for the grand canonical ensemble. Similarly, although the canonical ensemble partition function involves a sum over energies, there is a single most probable energy $U_K^* = \bar{U}_K$ that is the only term significantly contributing to the logarithm of Q_N from which thermodynamic properties are calculated [see Eqs. (1-65) and (1-66)]. Thus, in turn the canonical ensemble is essentially identical to a microcanonical ensemble with fixed energy $= \bar{U}_K$ and *fixed* number set equal to that of the canonical ensemble.

We close this section with a technical illustration of the above remarks by taking a Taylor series expansion of $\ln t_N$ about its maximum value. From Eq. (1-76)

$$\ln t_{N} = \ln Q_{N}(V, T) + \frac{\mathcal{N}\mu}{kT}$$

$$\left(\frac{\partial \ln t_{N}}{\partial \mathcal{N}} \right)_{V,T,\mu} = \left[\frac{\partial \ln Q_{N}(V, T)}{\partial \mathcal{N}} \right]_{V,T} + \frac{\mu}{kT}$$

$$\left(\frac{\partial^2 \ln t_{N}}{\partial \mathcal{N}^2} \right)_{V,T,\mu} = \left[\frac{\partial^2 \ln Q_{N}(V, T)}{\partial \mathcal{N}^2} \right]_{V,T}$$

The first derivative vanishes for the $\mathcal{N} \equiv \mathcal{N^*} = \bar{\mathcal{N}}$ that maximizes $\ln t_N$. Thus

$$\left(\frac{\partial \ln t_{N}}{\partial \mathcal{N}} \right)_{V,T,\mu,N=N^*} = 0 = \left[\frac{\partial \ln Q_{N}(V, T)}{\partial \mathcal{N}} \right]_{V,T,N=N^*} + \frac{\mu}{kT}$$

or

$$\left[\frac{\partial \ln Q_{N^*}(V, T)}{\partial \mathcal{N^*}} \right]_{V,T} = -\frac{\mu}{kT}$$

and

$$\left(\frac{\partial^2 \ln t_N}{\partial \mathcal{N}^2}\right)_{V,T,\mu,N=N^*} = \frac{\partial}{\partial \mathcal{N}^*}\left[\frac{\partial \ln Q_{N^*}(V, T)}{\partial \mathcal{N}^*}\right]_{V,T} = \frac{\partial}{\partial \mathcal{N}^*}\left(-\frac{\mu}{kT}\right)_{V,T}$$

Retaining terms through the quadratic only

$$\ln t_N = \ln t_{N^*} + \frac{1}{2}\left(\frac{\partial^2 \ln t_N}{\partial \mathcal{N}^2}\right)_{V,T,\mu,N=N^*}(\mathcal{N} - \mathcal{N}^*)^2 + \cdots$$

such that t_N assumes a Gaussian form [see Eq. (1-12)]

$$t_N = t_{N^*}\exp\left[\frac{-(\mathcal{N} - \mathcal{N}^*)^2}{2kT(\partial \mathcal{N}^*/\partial \mu)_{V,T}}\right] \tag{1-77}$$

with $\mathcal{N}^* = \bar{\mathcal{N}}$ and variance

$$\sigma_N^2 = kT\left(\frac{\partial \bar{\mathcal{N}}}{\partial \mu}\right)_{V,T} \tag{1-78}$$

The grand partition function sum in Eq. (1-75) can then be accurately replaced by an integral:

$$\mathcal{Q} = t_{N^*}\int_{-\infty}^{+\infty}\exp-\left(\frac{X^2}{2\sigma_N^2}\right)dX = t_{N^*}\sqrt{2\pi}\,\sigma_N$$

using $X = \mathcal{N} - \bar{\mathcal{N}}$. The logarithm then becomes:

$$\ln \mathcal{Q} = \ln t_{N^*} + \ln\left(\sqrt{2\pi}\sigma_N\right) \tag{1-79}$$

It can be shown [see Eqs. (3-134) and (3-138)] that σ_N is of order $\mathcal{N}^{1/2}$ such that only the first term in Eq. (1-79) is numerically significant.

1-4
Ideal Gases in the Microcanonical Ensemble; Bosons and Fermions

For gases of identical species (in detail identical isotopically), say all H_2 molecules, no two molecules are distinguishable since the entire volume is available to them. They are termed nonlocalized identical particles. For this case. Eq. (1-29) cannot be used for the weight of a distribution. Before we describe how to compute for this situation it is useful to note that identical particles can still be distinguished by their "addresses"—their equilibrium coordinates—if they exist in a crystal. These are termed localized identical particles. In an ideal crystal, even though the identical particles can and do (slowly) change their coordinates we can describe the crystal by giving the particle quantum state on a well-defined (localized) position in space. This question is discussed further in reference [3], pages 36–39.

For an ideal gas, unless the temperature is almost at absolute zero the separation of successive quantized energy states of translation from Eq. (1-74) is so small compared to kT that these states are practically continuous. For H_2 molecules, for example, in a cube of side 0.01 m the separation of the first two levels is $\sim 10^{-37}$ J whereas at 300 K kT is 10^{-21} J, a factor 10^{16} larger. For purposes of calculating the W of Eq. (1-25) for the microcanonical ensemble we can collect the states into groups (called levels) of g_j states all with practically the same energy E_j. Thus g_j

---

Exercise 1-4

Verify the above energy separation of the first two allowed levels of translation for an H_2 molecule in a cube of side 0.01 m.

---

plays the role of the degeneracy of level E_j with $\mathcal{N}_j$ particles assigned to the level. Then it will be consistent with the nature of Eq. (1-74) for $g_j \gg \mathcal{N}_j$, while $\mathcal{N}_j$ itself remains very large. This will validate the use of the Stirling approximation in what follows. In Fig. 1-4 we schematically represent this assignment of particles to levels and to states within the levels, with n_i representing the number of particles in state i of a given level. We must satisfy the overall constraints (summing over levels):

$$U = \sum_j \mathcal{N}_j E_j \tag{1-80}$$

$$\mathcal{N} = \sum_j \mathcal{N}_j \tag{1-81}$$

and the particular constraints for a given distribution in every level (summing over states)

$$\mathcal{N}_j = \sum_i^{g_j} n_i \tag{1-82}$$

State labels	1	2	3	4	5	6	7	8	9	10	11	12	13	14	15	16
n_i	0	1	0	2	0	0	1	1	0	0	1	0	0	0	2	1

Level labels $\qquad\qquad g_1 = 10, \quad \mathcal{N}_1 = 5 \qquad\qquad\qquad g_2 = 6, \quad \mathcal{N}_2 = 4$

FIGURE 1-4 A representation of the assignment of particles to quantized energy levels of translation, including the internal assignment of particles to states that collectively make up the level. In actual practice both g_j and $\mathcal{N}_j$ are very large, with $g_j \gg \mathcal{N}_j$.

For a given allowed distribution α (specify all $\mathcal{N}_j$, E_j) the weight $W(\mathcal{N}_j)$ will be the product over j (symbol π_j) of the ways W_j of arranging each level j:

$$W_\alpha(\mathcal{N}_j) = \pi_j W_j \tag{1-83}$$

Then the W of Eq. (1-25) will be the sum of W_α over distributions α satisfying Eqs. (1-80) and (1-81), but we will replace this sum with the one distribution for which W_α is a maximum.

If our gas were made up of classical distinguishable particles, each W_j would be

$$(W_j)_{c\ell} = \sum_n \frac{\mathcal{N}_j!}{(n_1)!(n_2)! \ldots (n_{g_j})!} \tag{1-84}$$

where the sum over n denotes a sum over all arrangements that satisfy Eq. (1-82) and subscript $c\ell$ denotes classical.[*] However for $\mathcal{N}_j$ indistinguishable particles there will be $\mathcal{N}_j!$ permutations that are all identical as far as physics is concerned. We must divide through by $(\mathcal{N}_j)!$ to obtain what is termed the Boltzmann weights (subscript Bo)

$$(W_j)_{\text{Bo}} = \frac{1}{(\mathcal{N}_j)!} \sum_n \frac{(\mathcal{N}_j)!}{(n_1)!(n_2)! \ldots (n_{g_j})!} \equiv \frac{1}{(\mathcal{N}_j)!} \sum_n \frac{(\mathcal{N}_j)!(1)^{n_1} \ldots (1)^{n_{g_j}}}{(n_1)!(n_2)! \ldots (n_{g_j})!}$$

$$= \frac{g_j^{\mathcal{N}_j}}{(\mathcal{N}_j)!} \tag{1-85}$$

To prove Eq. (1-85) we use the mathematical identity called the multinomial expansion (see Appendix II)

$$(a_1 + a_2 + \cdots + a_{g_j})^{\mathcal{N}_j} = \sum_n \frac{(\mathcal{N}_j)!(a_1)^{n_1}(a_2)^{n_2} \ldots (a_{g_j})^{n_{g_j}}}{(n_1)!(n_2)! \ldots (n_{g_j})!} \tag{1-86}$$

where the sum extends over all sets of nonnegative integers n_i that satisfy $n_1 + n_2 + \ldots + n_{g_j} = \mathcal{N}_j$. In the application of this to Eq. (1-85) all the a's are unities and there are a total of g_j of them such that our result follows. The binomial expansion

$$(x + y)^{\mathcal{N}} = \sum \frac{\mathcal{N}!}{(\mathcal{N} - M)!M!} x^M y^{\mathcal{N} - M}$$

which is familiar from elementary mathematics, is a special case of Eq. (1-86). Equation (1-85) is an example of classical or Boltzmann statistics, where statistics means here the counting of possible arrangements.

It is a surprising fact that Boltzmann statistics are never exactly correct because all particles follow quantum mechanical laws and Eq. (1-85) turns out to be merely an approximate interpolation formula between the two types of quantum statistics. The physics of a set of $\mathcal{N}$ indistinguishable particles must remain unchanged by any of the total of $\mathcal{N}!$ possible permutations of the particles. This means

*See note on page 74.

every physically allowed N-body wave function must transform in a mathematical sense (see a reference on group theory, e.g., [4a] or [4b]) as one of the irreducible representations of the permutation group of order N! All elementary particles so far discovered actually belong to one or the other of the two simplest (both one-dimensional) irreducible representations of the permutation group. Elementary particles with spin values of half an odd integer (such as electrons, protons, and neutrons) have many-body wave functions (called antisymmetric wave functions) that are merely multiplied by -1 (i.e., change sign) if the permutation is odd, and are multiplied by $+1$ if the permutation is even. Every permutation can be classified as odd (if an odd number of *pairs* of identical particles exchange position coordinates, that is, switch places) or even (if an even number of pairs is so permuted). Such elementary particles are called fermions and are said to follow Fermi-Dirac (FD) statistics. Elementary particles with integer spin values (such as photons of spin 1 and pions of spin 0) have many-body wave functions (called symmetric wave functions) that are always only multiplied by $+1$, that is, are unchanged for every permutation. These particles are called bosons and are said to follow Bose-Einstein (BE) statistics. Any compound particle of spin $\frac{1}{2}$ elementary particles (such as nuclei, atoms, or molecules) will be a fermion if it has an odd number of them (e.g., $^7_3\text{Li}^{+3}$ nuclei, ^3_2He atoms, HD molecules), and a boson if it has an even number (e.g., $^4_2\text{He}^{+2}$ nuclei, ^4_2He atoms, H_2 molecules). Thus all particles, whether elementary or compound, are either fermions or bosons. That this is so and that many-body wave functions transforming in more complicated ways (as the higher dimensional irreducible representations of the permutation group) are not needed to explain nature is both a profound mystery and a simplification. But let us be thankful for small favors! (See reference [5] pages 14–21, for further discussion.)

The detailed way the requirement of antisymmetric or symmetric wave functions changes the counting of W_j for Eq. (1-83) is in the rules that result for allowed occupation numbers n_i of individual states in a level where i enumerates all possible sets of quantum numbers including spin projection—"up" or "down" for spin $\frac{1}{2}$ particles. We mean that state i_1 and state i_2 with identical n_x, n_y, n_z values for Eq. (1-74) count as different states if one has spin "up," the other spin "down." The rules are that for fermions n_i can only be 0 or 1 and for bosons n_i can take any value $n_i \geq 0$. The former rule is just the Pauli exclusion principle for fermions. To illustrate this, recall that for independent particles simple products of one-particle functions are solutions of the N-body Schroedinger equation, and as linear combinations to satisfy the symmetry requirements one can take a single determinant for fermions and a single permanant (the determinant with all signs $+$) for bosons. Then in a three-particle case expand

$$\boldsymbol{\Psi} = \begin{matrix} \psi_{i_1} \\ \psi_{i_2} \\ \psi_{i_3} \end{matrix} \begin{array}{c} \quad (1) \qquad (2) \qquad (3) \\ \left| \begin{array}{ccc} \psi_{i_1}(1) & \psi_{i_1}(2) & \psi_{i_1}(3) \\ \psi_{i_2}(1) & \psi_{i_2}(2) & \psi_{i_2}(3) \\ \psi_{i_3}(1) & \psi_{i_3}(2) & \psi_{i_3}(3) \end{array} \right| \end{array} \qquad (1\text{-}87)$$

to obtain with $-$ signs for FD:

$$\boldsymbol{\Psi}_{\substack{\text{FD} \\ \text{BE}}} = \psi_{i_1}(1)\psi_{i_2}(2)\psi_{i_3}(3) + \psi_{i_1}(2)\psi_{i_2}(3)\psi_{i_3}(1) + \psi_{i_1}(3)\psi_{i_2}(1)\psi_{i_3}(2)$$
$$\mp \psi_{i_1}(2)\psi_{i_2}(1)\psi_{i_3}(3) \mp \psi_{i_1}(1)\psi_{i_2}(3)\psi_{i_3}(2) \mp \psi_{i_1}(3)\psi_{i_2}(2)\psi_{i_3}(1) \quad (1\text{-}88)$$

We leave it as an exercise for the reader to show that Eq. (1-88) has all the correct symmetry properties for bosons and for fermions and that if state $i_1 =$ state i_2, $\boldsymbol{\Psi}_{\text{FD}}$ vanishes.

Exercise 1-5

Derive Eq. (1-88) from Eq. (1-87) and demonstrate the properties required above.

The number of ways W_j of arranging the jth level for $\mathcal{N}_j$ bosons among g_j states is easily computed by imagining the bosons as points on a line that are separated into all possible arrangements by $(g_j - 1)$ partitions, thus arriving at g_j sets (the states); see Fig. 1-5. Since permutations of the points among themselves or the partitions among themselves does not give new arrangements, we have

$$(W_j)_{\text{BE}} = \frac{(\mathcal{N}_j + g_j - 1)!}{(\mathcal{N}_j)!(g_j - 1)!} \quad (1\text{-}89)$$

For fermions the states are either empty or occupied in one way only (singly), so W_j is just the number of ways to pick out $\mathcal{N}_j$ occupied states from a total of g_j:

$$(W_j)_{\text{FD}} = \frac{g_j!}{(g_j - \mathcal{N}_j)!(\mathcal{N}_j)!} \quad (1\text{-}90)$$

FIGURE 1-5 Two ways to use eight partitions to arrive at nine sets of points (the bosons) containing six points in total.

Example 1-2

Distribute three particles fictitiously labeled a, b, c among four degenerate states 1, 2, 3, 4 in one level. Write down explicitly the 64 arrangements possible if the particles were truly distinguishable and from these pick out (by grouping together indistinguish-

TABLE 1-2   Possible Arrangements of Three Particles (a, b, c) among Four Degenerate States (1, 2, 3, 4) of One Energy Level

1	2	3	4		1	2	3	4	
abc				1BE	c			ab	1BE
	abc			1BE	b			ac	1BE
		abc		1BE	a			bc	
			abc	1BE		c		ab	
ab	c					b		ac	1BE
ac	b			1BE		a		bc	
bc	a						c	ab	
ab		c					b	ac	1BE
ac		b		1BE			a	bc	
bc		a			a	b	c		
ab			c		a	c	b		
ac			b	1BE	b	a	c		1BE or 1FD
bc			a		b	c	a		
c	ab				c	b	a		
b	ac			1BE	c	a	b		
a	bc				a	b		c	
	ab	c			a	c		b	
	ac	b		1BE	b	a		c	1BE or 1FD
	bc	a			b	c		a	
	ab		c		c	b		a	
	ac		b	1BE	c	a		b	
	bc		a		a		b	c	
c		ab			a		c	b	
b		ac		1BE	b		a	c	1BE or 1FD
a		bc			b		c	a	
	c	ab			c		b	a	
	b	ac		1BE	c		a	b	
	a	bc				a	b	c	
		ab	c			a	c	b	
		ac	b	1BE		b	a	c	1BE or 1FD
		bc	a			b	c	a	
						c	b	a	
						c	a	b	

able arrangements) the actual number of different arrangements allowed for bosons and the actual number allowed for fermions.

Solution:

The required listing is given in Table 1-2. The different degenerate states are distinguishable since they would correspond to different sets of quantum numbers (which all give the same energy). This is a one-level case with $g_j = 4$, $N_j = 3$, and our above formulas check the results found in Table 1-2:

$$(W_j)_{cl} = \frac{3!4^3}{3!} = 64 \qquad (W_j)_{BE} = \frac{6!}{3!3!} = 20$$

$$(W_j)_{Bo} = \frac{4^3}{3!} = 10\tfrac{2}{3} \qquad (W_j)_{FD} = \frac{4!}{1!3!} = 4$$

Note that the number of arrangements given by the Boltzmann counting $(10\tfrac{2}{3})$ is a mathematical interpolation between the two possible physical answers. It is not physically correct, for what is two-thirds of an arrangement? However, whenever $g_j \gg N_j$ it will be accurate.

Taking the natural logarithm of Eq. (1-83) for our three cases, the repeated product becomes a sum over j, and using the Stirling approximation for the factorials of all large numbers and neglecting -1 in the expressions $(N_j + g_j - 1)$ and $(g_j - 1)$ in Eq. (1-89), we find

$$\ln W_{Bo} = \sum_j N_j \left[\ln \left(\frac{g_j}{N_j} \right) + 1 \right] \tag{1-91.1}$$

$$\ln W_{BE} = \sum_j \left\{ (N_j + g_j) \ln \left[\left(\frac{g_j}{N_j} \right) + 1 \right] - g_j \ln \left(\frac{g_j}{N_j} \right) \right\} \tag{1-91.2}$$

$$\ln W_{FD} = \sum_j \left\{ (N_j - g_j) \ln \left[\left(\frac{g_j}{N_j} \right) - 1 \right] + g_j \ln \left(\frac{g_j}{N_j} \right) \right\} \tag{1-91.3}$$

Exercise 1-6

Derive Eq. (1-91.1), Eq. (1-91.2), Eq. (1-91.3) from Eq. (1-83) and Eqs. (1-85), (1-89), and (1-90).

The above equations give the logarithm of W for one distribution only [we have omitted the subscript α in Eqs. (1-91)], but we are going to find the set $\mathcal{N}_j$ that maximizes this W, and as we know from Sec. 1-2 the entropy is given with negligible error by $k \ln W_m$. To find the maximum for Eqs. (1-91) we set the derivative equal to zero to find

$$d \ln W = \sum_j \left[\ln \left(\frac{g_j}{\mathcal{N}_j} + c \right) \right] d\mathcal{N}_j = 0 \qquad (1\text{-}92)$$

$$c_{\text{Bo}} = 0, \qquad c_{\text{BE}} = +1, \qquad c_{\text{FD}} = -1$$

Exercise 1-7

Derive Eq. (1-92) from Eqs. (1-91).

The level occupation numbers $\mathcal{N}_j$ are not all independent since we have the two constraints of Eq. (1-80) and Eq. (1-81), which in differential form read as

$$dN = \sum_j d\mathcal{N}_j = 0 \qquad (1\text{-}93)$$

$$dU = \sum_j E_j \, d\mathcal{N}_j = 0 \qquad (1\text{-}94)$$

Introducing two parameters, α multiplying dN, and $-\beta$ multiplying dU, we have

$$d \ln W + \alpha \, dN - \beta \, dU = 0$$

such that

$$\sum_j \left[\ln \left(\frac{g_j}{\mathcal{N}_j} + c \right) + \alpha - \beta E_j \right] d\mathcal{N}_j = 0 \qquad (1\text{-}95)$$

This is Lagrange's method of undetermined multipliers, which is useful since in Eq. (1-95) we can take all the $\mathcal{N}_j$ as independent because the two parameters can be used later to enforce the satisfaction of our two constraints. This will also give us the physical meaning of α and β. Now the only way for Eq. (1-95) to be satisfied is for each bracket term to be zero. Hence for the $\bar{\mathcal{N}}_j$ distribution that makes $W(\mathcal{N}_j) = W(\bar{\mathcal{N}}_j)$ a maximum, we have for all j:

$$(\bar{\mathcal{N}}_j)_{\text{Bo}} = g_j \exp (\alpha) \exp (-\beta E_j) \qquad (1\text{-}96.1)$$

$$(\bar{\mathcal{N}}_j)_{\text{BE}} = \frac{g_j}{\exp (-\alpha) \exp (+\beta E_j) - 1} \qquad (1\text{-}96.2)$$

$$(\mathcal{N}_j)_{\text{FD}} = \frac{g_j}{\exp{(-\alpha)}\exp{(+\beta E_j)} + 1} \tag{1-96.3}$$

We see that both the BE and FD distributions become equal to the Boltzmann distribution when $e^{-\alpha}e^{+\beta E_j} \gg 1$, which from Eq. (1-96.1) means when $\bar{\mathcal{N}}_j/g_j \ll 1$. This makes good physical sense, because if the number of particles is very small compared to the number of quantum states available to receive them, almost all the states will be empty and only a small fraction will be occupied singly. Thus the question of multiple occupation of a state (possible for bosons only) will have no importance and so the difference between bosons and fermions will be insignificant.

With reference to Eq. (1-6), the mean square fluctuation in level occupation

$$\overline{(\delta \mathcal{N}_j)^2} = \overline{\mathcal{N}_j^2} - (\bar{\mathcal{N}}_j)^2 \tag{1-97}$$

is a measure of the deviation of level occupations from the mean values calculated using the one most probable distribution. We will now show that this fluctuation is indeed negligible. By the definition of average in Eq. (1-3)

$$\bar{\mathcal{N}}_j = \frac{\sum\limits_{N_j} \mathcal{N}_j W(\mathcal{N}_j)}{\sum\limits_{N_j} W(\mathcal{N}_j)} = \frac{\sum\limits_{N_j} \mathcal{N}_j W(\mathcal{N}_j)}{W}$$

or

$$\bar{\mathcal{N}}_j W = \sum_{N_j} \mathcal{N}_j W(\mathcal{N}_j) \tag{1-98}$$

Now we use a mathematical trick: we differentiate Eq. (1-98) with respect to one particular g_j value and use the Boltzmann relations, Eqs. (1-83) and (1-85):

$$W(\mathcal{N}_j) = \prod_j \frac{g_j^{N_j}}{\mathcal{N}_j!}$$

$$\frac{\partial W(\mathcal{N}_j)}{\partial g_j} = \frac{\mathcal{N}_j}{g_j} W(\mathcal{N}_j)$$

and so differentiation gives [$W = \sum_{N_j} W(\mathcal{N}_j)$ also depends on g_j]:

$$\frac{\partial \bar{\mathcal{N}}_j}{\partial g_j} W + \bar{\mathcal{N}}_j \sum_{N_j} \frac{\mathcal{N}_j}{g_j} W(\mathcal{N}_j) = \sum_{N_j} \frac{\mathcal{N}_j^2}{g_j} W(\mathcal{N}_j)$$

Multiply through by g_j/W to obtain

$$g_j \frac{\partial \bar{\mathcal{N}}_j}{\partial g_j} = \frac{\sum\limits_{N_j} \mathcal{N}_j^2 W(\mathcal{N}_j)}{W} - \frac{\bar{\mathcal{N}}_j \sum\limits_{N_j} \mathcal{N}_j W(\mathcal{N}_j)}{W}$$

$$g_j \frac{\partial \bar{\mathcal{N}}_j}{\partial g_j} = \overline{\mathcal{N}_j^2} - (\bar{\mathcal{N}}_j)^2 \tag{1-99}$$

From Eq. (1-96.1) the left-hand side is just $\bar{N}_j$. The quantity of interest is the fractional fluctuation

$$\frac{\overline{N_j^2} - (\bar{N}_j)^2}{(\bar{N}_j)^2} = \frac{1}{\bar{N}_j} \tag{1-100}$$

which is seen to be negligible if $\bar{N}_j$ is large, which it is. This is a further ex post facto justification of our methods. For more on fluctuations see Problem 16 and Secs. 2-1, 3-7, and 3-8.

1-5

Details for the Ideal Classical (Boltzmann) Gas in the Microcanonical Ensemble

We focus our attention now on the Boltzmann distribution, Eq. (1-96.1), for a classical ideal gas, treating only the translational energy given by Eq. (1-74). Internal energy levels, such as those of rotation and vibration, will be treated in Chap. 2.

First we eliminate e^α from Eq. (1-96.1) using one constraint

$$N = \sum_j \bar{N}_j = e^\alpha \sum_j g_j e^{-\beta E_j} \equiv e^\alpha q \tag{1-101}$$

defining q as the single-molecule partition function:

$$q = \sum_j g_j e^{-\beta E_j} \tag{1-102}$$

Hence

$$\bar{N}_j = \frac{N g_j e^{-\beta E_j}}{q} \tag{1-103}$$

and from our second constraint

$$U = \sum_j \bar{N}_j E_j = \sum_j \frac{N g_j E_j e^{-\beta E_j}}{q} \tag{1-104}$$

Our fundamental thermodynamic identification is for the entropy:

$$S/k = \ln W_{\mathbf{Bo}} = \sum_j \bar{N}_j [\ln (g_j/\bar{N}_j) + 1] \qquad \text{using Eq. (1-91.1)}$$

$$= \sum_j \bar{N}_j \left[\ln \left(\frac{q e^{+\beta E_j}}{N} \right) + 1 \right] \qquad \text{using Eq. (1-103)}$$

$$= N \ln q + N - N \ln N + \beta U$$

$$= N \ln \sum_j g_j e^{-\beta E_j} + N - N \ln N + \beta U \tag{1-105}$$

We identify α using the thermodynamic relation

$$\left(\frac{\partial S}{\partial \mathcal{N}}\right)_{U,V} \equiv -\frac{\mu}{T} = -k \ln\left(\frac{\mathcal{N}}{\sum_j g_j e^{-\beta E_j}}\right) \quad \text{from Eq. (1-105)}$$

$$= -k \ln\left(e^{\alpha}\right) \quad \text{from Eq. (1-101)}$$

Thus

$$\alpha = \mu/kT \tag{1-106}$$

where μ is the chemical potential per molecule.

From Eq. (1-105) at constant $\mathcal{N}$ and V (which means constant E_j), S is a function of U and possibly β, since we have not yet determined what β is. Hence, using at constant $\mathcal{N}$, V

$$dS = \left(\frac{\partial S}{\partial U}\right)_{\beta} dU + \left(\frac{\partial S}{\partial \beta}\right)_{U} d\beta$$

$$\left(\frac{\partial S}{\partial U}\right)_{V,N} = \left(\frac{\partial S}{\partial U}\right)_{\beta,V,N} + \left(\frac{\partial S}{\partial \beta}\right)_{U,V,N}\left(\frac{\partial \beta}{\partial U}\right)_{V,N} \tag{1-107}$$

From Eq. (1-105)

$$\left(\frac{\partial S}{\partial \beta}\right)_{U,V,N} = k\left(-\mathcal{N}\sum_j \frac{E_j g_j e^{-\beta E_j}}{q} + U\right) = 0 \quad \text{from Eq. (1-104)}$$

Thus, using the thermodynamic relation,

$$\left(\frac{\partial S}{\partial U}\right)_{V,N} = \frac{1}{T} = \left(\frac{\partial S}{\partial U}\right)_{\beta,V,N} = k\beta$$

$$\beta = \frac{1}{kT} \tag{1-108}$$

Having identified α and β, we go through some further derivations to obtain all the thermodynamic functions explicitly in terms of the partition function q, which is our link to molecular properties. We now know

$$q = \sum_j g_j e^{-E_j/kT} \tag{1-109}$$

From the definition of the Helmholtz free energy and Eq. (1-105)

$$A = U - TS = -\mathcal{N}kT \ln q + kT(\mathcal{N}\ln \mathcal{N} - \mathcal{N}) \tag{1-110}$$

which on using the Stirling approximation in an inverse sense gives

$$A = -kT \ln (q^N/N!)$$ (1-111)

which will be useful in comparing with later results [see Eq. (1-127)].

We evaluate q for the translational states by replacing the sum over levels j with the triple sum over the states themselves labeled by the quantum number triplet from Eq. (1-74)

$$q_{trans} = \left[\sum_{n_x=1}^{\infty} \exp\left(\frac{-h^2 n_x^2}{8ma^2 kT}\right) \right]\left[\sum_{n_y=1}^{\infty} \exp\left(\frac{-h^2 n_y^2}{8mb^2 kT}\right) \right]\left[\sum_{n_z=1}^{\infty} \exp\left(\frac{-h^2 n_z^2}{8mc^2 kT}\right) \right]$$ (1-112)

Since the states of translation are so closely spaced it will be essentially exact to replace each summation by an integral:

$$\sum_{n_x=1}^{\infty} \exp\left(\frac{-h^2 n_x^2}{8ma^2 kT}\right) \cong \int_0^{\infty} \exp\left(\frac{-h^2 n_x^2}{8ma^2 kT}\right) dn_x = \frac{a}{h} \sqrt{2\pi mkT}$$ (1-113)

The integral is a standard form. This replacement will be precise even at temperatures below which it is no longer possible to use Boltzmann statistics; that is, as long as a gas behaves classically this integration is precise. Using $abc = V$, the volume of the gas, Eq. (1-112) becomes

$$q_{trans} = V/\lambda^3$$ (1-114)

where λ has the dimensions of a length and is called the thermal wavelength:

$$\lambda = \frac{h}{\sqrt{2\pi mkT}}$$ (1-115)

For numerical calculations

$$\lambda^3 = \frac{5.321 \times 10^{-27}}{(MT)^{3/2}} \text{ m}^3$$ (1-116)

with the molecular weight M in g/mol.

Exercise 1-8

Except for a minor numerical factor, λ in Eq. (1-114)–(1-116) is the de Broglie wavelength of a free particle with kinetic energy equal to kT. Show that this is true and determine the factor.

Qualitatively, all partition functions are a measure of the number of states available to a system or (as here) to a molecule at the temperature in question. To good accuracy the partition function counts the number of states between the lowest allowed energy up to a state of energy kT. We can derive this by taking $\exp(-E_j/kT) = 1$ for $E_j < kT$ and $\exp(-E_j/kT) \approx 0$ for $E_j > kT$ in Eq. (1-109). This is crude but not unreasonable.

Example 1-3

a. Derive an expression for $\Phi(\varepsilon')$, the number of translational energy states available to a particle of mass m in a cube of side a from the lowest permitted energy to $\varepsilon_{trans} = \varepsilon'$. Use Eq. (1-74) and consider the appropriate fraction of a spherical volume in quantum number space of "radius" n given by

$$n = \sqrt{n_x^2 + n_y^2 + n_j^2}$$

b. For $\varepsilon' = kT$ evaluate the ratio of $\Phi(kT)$ to q_{trans}. For a ^{4}He atom at 100 K in a cube of side 1.00×10^{-2} m evaluate both quantities numerically.

a. *Solution:*

Writing Eq. (1-74) for a cube of side a, and $\varepsilon_{trans} = \varepsilon'$,

$$n^2 \equiv n_x^2 + n_y^2 + n_j^2 = \frac{8m\varepsilon' a^2}{h^2}$$

such that

$$n = \sqrt{n_x^2 + n_y^2 + n_z^2} = \sqrt{\frac{8m\varepsilon' a^2}{h^2}}$$

is mathematically a radius in quantum number space of a sphere that contains all states with energies equal to or less than ε'. Every point in this space with *positive* integers for the n_x, n_y, n_z coordinates is an allowed state. So, if ε' is large compared to the spacing of the quantized states, the volume of the sphere in the positive octant will be an accurate count of the allowed states:

$$\Phi(\varepsilon') = \frac{1}{8}\left(\frac{4}{3}\pi n^3\right) = \frac{\pi}{6}\left(\frac{8ma^2\varepsilon'}{h^2}\right)^{3/2} = \frac{\pi}{6}V\left(\frac{8m\varepsilon'}{h^2}\right)^{3/2}$$

with $V = a^3$, the volume.

b. *Solution:* If $\varepsilon' = kT$, for the number of allowed states of translation out to this value

$$\Phi(kT) = \frac{\pi}{6}\left(\frac{V}{h^3}\right)(8mkT)^{3/2}$$

whereas from Eq. (1-114)

$$q_{trans} = \frac{V}{h^3} (2\pi mkT)^{3/2}$$

$$\frac{\Phi(kT)}{q_{trans}} = \frac{4}{3\sqrt{\pi}} = 0.7523$$

that is, $\Phi(kT)$ and q_{trans} are practically equal.

For a ^{4}He atom at 100 K in $V = 1.00 \times 10^{-6}$ m^3, Eq. (1-114) gives

$$q_{trans} = \frac{(1.00 \times 10^{-6})(4.00 \times 100)^{3/2}}{5.321 \times 10^{-27}} = 1.50 \times 10^{24}$$

$$\Phi(kT) = 1.13 \times 10^{24}$$

■

With q known we may evaluate any thermodynamic property of our ideal classical gas. From Eqs. (1-110) and (1-114)

$$A_{trans} = -NkT \ln \left[\frac{V}{N} \left(\frac{2\pi mkT}{h^2} \right)^{3/2} \right] - NkT \qquad (1\text{-}117)$$

$$P = -\left(\frac{\partial A}{\partial V} \right)_{T,N} = \frac{NkT}{V} \qquad (1\text{-}118)$$

Equation (1-118) is the ideal gas law and demonstrates that the constant k first used in Eq. (1-22) must be the Boltzmann constant R/N_A. Furthermore, since (as we shall see in Chap. 2) for an ideal classical gas only translation of the center of mass contributes to the pressure, we do not include a subscript *trans* on P.

$$U_{trans} = -T^2 \frac{\partial}{\partial T} \left(\frac{A_{trans}}{T} \right)_{V,N} = \frac{3}{2} NkT \qquad (1\text{-}119)$$

$$(C_V)_{trans} = \left(\frac{\partial U_{trans}}{\partial T} \right)_V = \frac{3}{2} Nk \qquad (1\text{-}120)$$

$$S_{trans} = -\left(\frac{\partial A_{trans}}{\partial T} \right)_{V,N} = \frac{U_{trans} - A_{trans}}{T}$$

$$= Nk \left[\ln \left(\frac{V}{N} \right) + \frac{3}{2} \ln T + \frac{3}{2} \ln \left(\frac{2\pi mk}{h^2} \right) + \frac{5}{2} \right] \qquad (1\text{-}121.1)$$

$$= Nk \left\{ \frac{5}{2} \ln T - \ln P + \ln \left[\left(\frac{2\pi m}{h^2} \right)^{3/2} k^{5/2} \right] + \frac{5}{2} \right\} \qquad (1\text{-}121.2)$$

Equation (1-121.2) follows from Eq. (1-121.1) using the ideal gas equation to eliminate (V/N). Equation (1-121) is the famous Sackur-Tetrode equation for the
entropy of translation of a classical gas. It was first derived in a rather ad hoc way
in 1913 before the development of quantum mechanics, but after Planck's work on
quantum theory, since we note that h occurs in the equation!

Exercise 1-9

Derive Eq. (118)–(121) from Eq. (1-117) using the indicated thermodynamic
relations.

Exercise 1-10

Show that in terms of M, molecular weight in g/mol, and P', pressure in atm,
$(S_m)_{trans}/R$, which of course is dimensionless, is given by

$$(S_m)_{trans}/R = \tfrac{5}{2} \ln T + \tfrac{3}{2} \ln M - \ln P' - 1.1649$$

1-6
Ideal Gases in the Canonical Ensemble

To evaluate the canonical partition function of Eq. (1-62):

$$Q_N(V, T) = \sum_K \exp\left(-U_K/kT\right) \qquad (1\text{-}122)$$

we need not think about grouping large numbers of states into various levels but
treat directly the actual individual stationary states of energy ε_i themselves each
with occupation number n_i (which on average will be very small) such that

$$\sum_i n_i = N \qquad (1\text{-}123)$$

and

$$U_K \equiv \sum_i \varepsilon_i n_i \qquad (1\text{-}124)$$

The sum over K becomes a sum over distributions that satisfy Eqs. (1-123) and (1-124), with the exponential for each distribution multiplied by the weight or the effective number of ways of arranging that distribution, which for the Boltzmann case from Eq. (1-85) has the form:

$$\frac{1}{\mathcal{N}!} \frac{\mathcal{N}!}{\prod_i (n_i)!}$$

Thus

$$Q_N(V, T) = \frac{1}{\mathcal{N}!} \sum_n \left[\frac{\mathcal{N}! \exp\left(-\sum_i \varepsilon_i n_i / kT\right)}{\prod_i (n_i)!} \right]$$

$$= \frac{1}{\mathcal{N}!} \sum_n (\mathcal{N})! \left\{ \left[\frac{\exp\left(-\varepsilon_0 n_0 / kT\right)}{n_0!} \right] \left[\frac{\exp(-\varepsilon_1 n_1 / kT)}{n_1!} \right] \cdots \right\}$$

$$= \frac{1}{\mathcal{N}!} \sum_n (\mathcal{N})! \prod_i \left[\frac{\exp\left(-\varepsilon_i / kT\right)^{n_i}}{(n_i)!} \right] \tag{1-125}$$

where we have written the exponential of a sum as a product of individual terms and we recognize, because of the constraint of Eq. (1-123), that the multinomial theorem of Eq. (1-86) gives us the exact result:

$$Q_N(V, T) = \frac{1}{\mathcal{N}!} \left[\exp\left(-\varepsilon_0 / kT\right) + \exp\left(-\varepsilon_1 / kT\right) + \cdots \right]^N$$

$$= \frac{1}{\mathcal{N}!} \left[\sum_i \exp\left(-\varepsilon_i / kT\right) \right]^N = \frac{1}{\mathcal{N}!} q_{trans}^N \tag{1-126}$$

where the q_{trans} is just the individual molecule partition function for the translational states already evaluated in Sec. 1-5. For classical noninteracting particles the $\mathcal{N}$-body partition function is the one-particle partition function raised to the power $\mathcal{N}$ and divided by $\mathcal{N}!$ to take account approximately of the quantum indistinguishability of the particles. Naturally, for interacting particles everything will be much more complex (see Sec. 3-4).

In the canonical ensemble the fundamental thermodynamic identification is to A via Eq. (1-61):

$$A_{trans} = -kT \ln Q_N(V, T) = -kT \ln \frac{q_{trans}^N}{\mathcal{N}!} \tag{1-127}$$

which is identical to our previous results in Eqs. (1-111) and (1-117).

Why have we spent so much time with the more cumbersome microcanonical ensemble when in the canonical ensemble the same results may be obtained so much more directly? The answer is, partly for historical reasons but more pertinently because the microcanonical approach is less abstract and because some explicit results for ideal bosons and fermions (Eqs. 1-96.2, 1-96.3) can also be derived in the microcanonical ensemble. Because we cannot use a multinomial expansion trick to exactly evaluate Q_N even for ideal bosons or fermions, we will not treat them in the canonical ensemble but will discuss them in Sec. 1-7 using the grand canonical partition function.

One result, which will be important in later applications, is the simple generalization of Eq. (1-126) for a mixture of ideal classical gases of species $a, b, \ldots$. Since in such a mixture

$$U = \sum_i n_{ia}\varepsilon_{ia} + \sum_i n_{ib}\varepsilon_{ib} + \cdots$$

$$N = N_a + N_b + \cdots$$

and all the species will have a common temperature, we have

$$Q_N(V, T) = \frac{1}{N_a!}\sum_{n_a} N_a! \frac{\prod_i \exp(-n_{ia}\varepsilon_{ia}/kT)}{\prod_i (n_{ia})!} \cdot \frac{1}{N_b!}\sum_{n_b} N_b! \frac{\prod_i \exp(-n_{ib}\varepsilon_{ib}/kT)}{\prod_i (n_{ib})!} \cdots$$

$$= \left(\frac{q_a^{N_a}}{N_a!}\right)\left(\frac{q_b^{N_b}}{N_b!}\right)(\cdots) \tag{1-128}$$

because of the conditions

$$\sum_i n_{ia} = N_a, \qquad \sum_i n_{ib} = N_b, \qquad \text{etc.}$$

Then for such a mixture (using Stirling's approximation)

$$A = -N_a kT \ln q_a + N_a kT(\ln N_a - 1) \\ - N_b kT \ln q_b + N_b kT(\ln N_b - 1) - \cdots \tag{1-129}$$

From which we obtain the chemical potential per molecule of species a in the form

$$\mu_a = \left(\frac{\partial A}{\partial N_a}\right)_{T,V,N_b} = -kT \ln\left(\frac{q_a}{N_a}\right) \tag{1-130}$$

1-7
Ideal Gases in the Grand Canonical Ensemble; Virial Expansion for Bosons and Fermions

We have treated the classical ideal gas sufficiently and in this section will describe only Fermi-Dirac (FD) and Bose-Einstein (BE) ideal gases. Although there are no actual forces acting between particles of such gases, the particles are not really independent of one another because of the occupation number rules, which do not permit doubling up of fermions in the same spin state but do permit this for bosons.

Hence their properties cannot be given merely in terms of a single-molecule partition function because of these subtle correlations between particles. Even when the quantum effects are small these gases will have nonzero second, third, and succeeding virial coefficients.

The grand canonical partition function from Eq. (1-64) and the discussion and notation of Sec. 1-6 is

$$\mathscr{Q}(\mu, V, T) = \sum_{N \geq 0} \exp(\mathscr{N} \cdot \mu/kT) Q_N(V, T)$$

$$= \sum_{N \geq 0} \sum_{\boldsymbol{n}} \exp\left(\mu/kT \sum_i n_i\right) \exp\left(-1/kT \sum_i \varepsilon_i n_i\right)$$

$$= \sum_{N \geq 0} \sum_{\boldsymbol{n}} \prod_i [\exp(\mu/kT) \exp(-\varepsilon_i/kT)]^{n_i} \qquad (1\text{-}131)$$

in which the $\boldsymbol{n}$ notation under the second summation symbol is to remind the reader that the sum is over *all* sets n_i that satisfy $\sum_i n_i = \mathscr{N}$. There is no added factor $1/(n_i)!$ in Eq. (1-131) because this equation does not apply to the Boltzmann gas but may be used for either BE or FD according to how we restrict the individual n_i. Since $\mathscr{N}$ can assume any value—that is, since we are using the grand canonical method—the above expression is mathematically equivalent to summing every individual n_i over all its permitted values (0 to ∞ for bosons, only 0 to 1 for fermions):

$$\mathscr{Q}(\mu, V, T) = \sum_{n_0=0}^{n_0 \text{ max}} \sum_{n_1=0}^{n_1 \text{ max}} \sum_{n_2=0}^{n_2 \text{ max}} \cdots \prod_i [\exp(\mu/kT) \exp(-\varepsilon_i/kT)]^{n_i}$$

$$= \prod_i \sum_{n_i=0}^{n_i \text{ max}} [\exp(\mu/kT) \exp(-\varepsilon_i/kT)]^{n_i} \qquad (1\text{-}132)$$

Hence for BE, where all n_i values are possible, using the well-known series

$$\sum_{n=0}^{\infty} X^n = \frac{1}{1-X} \qquad \text{for } X < 1$$

we obtain

$$\mathscr{Q}(\mu, V, T) = \prod_i \frac{1}{1 - e^{\mu/kT} e^{-\varepsilon_i/kT}} \qquad (\text{BE}) \qquad (1\text{-}133)$$

and for FD, where n_i is 0 and 1 only,

$$\mathscr{Q}(\mu, V, T) = \prod_i (1 + e^{\mu/kT} e^{-\varepsilon_i/kT}) \qquad (\text{FD}) \qquad (1\text{-}134)$$

In the grand canonical ensemble the fundamental thermodynamic identification is Eq. (1-63)

$$\frac{PV}{kT} = \ln \mathscr{Q}(\mu, V, T) \qquad (1\text{-}135)$$

such that

$$\frac{PV}{kT} = -\sum_i \ln\left(1 - e^{\mu/kT}e^{-\varepsilon_i/kT}\right) \qquad \text{(BE)} \qquad (1\text{-}136)$$

$$\frac{PV}{kT} = \sum_i \ln\left(1 + e^{\mu/kT}e^{-\varepsilon_i/kT}\right) \qquad \text{(FD)} \qquad (1\text{-}137)$$

These are two exact expressions for the equation of state, but they are inconvenient because they contain a function of μ. By deriving further general relations we shall discover how to eliminate the terms in μ between pairs of equations.

From Eqs. (1-55) and (1-63) the probability of being in state K in the grand canonical ensemble is

$$P_K = e^{-PV/kT}e^{N\mu/kT}e^{-U_K/kT}$$
$$= (1/\mathcal{Q})e^{N\mu/kT}e^{-U_K/kT} \qquad (1\text{-}138)$$

such that the average value of any quantity f in this ensemble is by definition

$$\bar{f} = \frac{\sum\limits_N f e^{N\mu/kT}\sum\limits_K e^{-U_K/kT}}{\mathcal{Q}} = \frac{\sum\limits_N f e^{N\mu/kT}Q_N(V,\,T)}{\mathcal{Q}} \qquad (1\text{-}139)$$

Define

$$z \equiv e^{\mu/kT} \qquad (1\text{-}140)$$

and note that

$$\left(z\frac{\partial}{\partial z}\right)_{V,T}\ln\mathcal{Q}(\mu,\,V,\,T) = \frac{\left(z\dfrac{\partial}{\partial z}\right)_{V,T}\sum\limits_N z^N Q_N(V,\,T)}{\mathcal{Q}}$$

$$= \frac{\sum\limits_N N z^N Q_N(V,\,T)}{\mathcal{Q}} = \bar{N} \qquad (1\text{-}141)$$

using Eq. (1-139). Then by explicitly differentiating Eq. (1-136) and eq. (1-137) with respect to z we find

$$\bar{N} = \sum_i \frac{z e^{-\varepsilon_i/kT}}{1 - z e^{-\varepsilon_i/kT}} \qquad \text{(BE)} \qquad (1\text{-}142)$$

$$\bar{N} = \sum_i \frac{z e^{-\varepsilon_i/kT}}{1 + z e^{-\varepsilon_i/kT}} \qquad \text{(FD)} \qquad (1\text{-}143)$$

In this ensemble, where all values of N are permitted, we can only speak of an average value of the total number of molecules (equivalent to the most probable value) and this is given by these equations. We should suspect that the summands in Eqs. (1-142) and (1-143) represent average values of individual state occupation numbers n_i, and this is true as we now prove.

By definition

$$\bar{n}_i = \frac{1}{\mathcal{Q}} \sum_N z^N \sum_n n_i \exp\left(-\frac{1}{kT}\sum_i \varepsilon_i n_i\right) \tag{1-144}$$

and by differentiation of the defining Eq. (1-131)

$$\bar{n}_i = -kT\frac{\partial}{\partial \varepsilon_i}\ln \mathcal{Q} \tag{1-145}$$

Thus by differentiating Eqs. (1-136) and (1-137) with respect to ε_i we find

$$\bar{n}_i = \frac{ze^{-\varepsilon_i/kT}}{1 - ze^{-\varepsilon_i/kT}} = \frac{1}{e^{-\mu/kT}e^{+\varepsilon_i/kT} - 1} \quad \text{(BE)} \tag{1-146}$$

$$\bar{n}_i = \frac{ze^{-\varepsilon_i/kT}}{1 + ze^{-\varepsilon_i/kT}} = \frac{1}{e^{-\mu/kT}e^{+\varepsilon_i/kT} + 1} \quad \text{(FD)} \tag{1-147}$$

such that for both bosons and fermions

$$\bar{N} = \sum_i \bar{n}_i \tag{1-148}$$

Exercise 1-11

Check the derivations of Eqs. (1-142)–(1-147).

Equations (1-146) and (1-147) are respectively equivalent to Eqs. (1-96.2) and (1-96.3) when the latter—which give average numbers in a level—are divided by the level degeneracy. Our derivation here is superior because it is in terms of the individual state occupation numbers directly, without need of using Stirling's approximation and because we have additional equations at hand to work out all details of their equation of state.

Some interesting qualitative results can be obtained at once from Eqs. (1-146) and (1-147). Both give the Boltzmann result of extremely tiny $\bar{n}_i$ value if $e^{-\mu/kT}e^{+\varepsilon_i/kT} \gg 1$. A way to satisfy this is for μ to be extremely negative (i.e., z be very small). Any value of μ will be possible for fermions but at the absolute zero, $\mu = \varepsilon_{max}$ where ε_{max} is the maximum allowed kinetic energy—that is, the kinetic energy of the uppermost filled state such that for states below ε_{max} Eq. (1-147) will give $\bar{n}_i = 1$ and for states above ε_{max} Eq. (1-147) will give $\bar{n}_i = 0$. We speak of a filled Fermi sphere of states at the absolute zero. This will be discussed further in Chap. 4.

For bosons, since negative $\bar{n}_i$ values are impossible μ is restricted to such values that if ε_0 is the lowest allowed kinetic energy $\mu \to \varepsilon_0$ in such a way that

$$e^{-\mu/kT + \varepsilon_0/kT} \to 1 + \frac{c}{\bar{N}} \qquad (1\text{-}149)$$

[because the denominator of Eq. (1-146) must remain positive] where c is a constant such that taking logarithms and expanding $\ln(1 + c/\bar{N}) \cong c/\bar{N}$,

$$\mu \to \varepsilon_0 - \frac{kT}{\bar{N}} c$$

which at absolute zero gives $\mu \to \varepsilon_0$ from negative values. For convenience we may set $\varepsilon_0 = 0$. Actually c is somewhat temperature dependent and must be equal to 1 at absolute zero so that $\bar{n}_0 = \bar{N}$ and $\bar{n}_i = 0$ for $i > 0$ at absolute zero. This is the very interesting situation of Bose-Einstein condensation of all ideal bosons into the lowest allowed energy state at absolute zero. This phenomenon, which is important for understanding superfluidity (of liquid ^4_2He) and other many-body quantum effects, will not be discussed further in this text. References [6, 7, 8] provide a good introduction to these matters.

Both of the ordered situations at absolute zero described above lead to an entropy of zero at absolute zero in accordance with the Third Law of thermodynamics. The entropy formula for a Boltzmann gas, Eq. (1-121), does not behave properly at absolute zero since the assumptions on which it is based do not hold there.

In the rest of this section we will work out the equation of state in the weakly degenerate case of small z, first for fermions and then for bosons.

Fermions

Equations (1-137) and (1-143) involve sums over energy states i for (PV/kT) and $\bar{N}$, respectively. In addition we have the obvious equation for U,

$$U = \sum_i \varepsilon_i n_i = \sum_i \frac{z\varepsilon_i e^{-\beta\varepsilon_i}}{1 + ze^{-\beta\varepsilon_i}} \qquad (1\text{-}150)$$

in which we will use

$$\beta = 1/kT \qquad (1\text{-}151)$$

We will replace these sums over states by integrations over ε multiplying by $\omega(\varepsilon)$, the number of translational energy states between ε and $\varepsilon + d\varepsilon$ or an effective degeneracy at ε. From Example 1-3 we found $\Phi(\varepsilon)$, the number of translational states out to ε, to be given by

$$\Phi(\varepsilon) = \frac{\pi}{6} V \left(\frac{8m\varepsilon}{h^2}\right)^{3/2} \qquad (1\text{-}152)$$

Then

$$\omega(\varepsilon) = \frac{d\Phi(\varepsilon)}{d\varepsilon} = 2\pi \left(\frac{2m}{h^2}\right)^{3/2} V\varepsilon^{1/2} \tag{1-153}$$

This replacement of sum by an integral will remain accurate irrespective of the temperature in the sense of what is called the thermodynamic limit $\bar{N} \to \infty, V \to \infty$; but $\bar{N}/V = \rho$ the fixed number density. Then the density of states is ∞ and they can be taken accurately as continuous. To check this technique we evaluate q_{trans} for a Boltzmann gas again from Eq. (1-109):

$$\begin{aligned}
q_{trans} &= \sum_i g_i e^{-\beta \varepsilon_i} = \int_0^\infty \omega(\varepsilon) e^{-\beta \varepsilon} \, d\varepsilon \\
&= 2\pi \left(\frac{2m}{h^2}\right)^{3/2} V \int_0^\infty \varepsilon^{1/2} e^{\beta \varepsilon} \, d\varepsilon \\
&= 2\pi \left(\frac{2m}{h^2}\right)^{3/2} V \int_0^\infty 2y^2 \, dy \, e^{-\beta y^2} \qquad \text{using } y = \varepsilon^{1/2} \\
&= V \left(\frac{2\pi mkT}{h^2}\right)^{3/2} \tag{1-154}
\end{aligned}$$

which agrees with our result in Eq. (1-114).

Since the number of spin $\frac{1}{2}$ fermion states is twice the number of translational states (spin up and spin down states) and generally $(2s + 1)$ times the number of translational states for spin s, we use for spin $\frac{1}{2}$:

$$\underset{\text{FD}}{\omega(\varepsilon)} = 4\pi \left(\frac{2m}{h^2}\right)^{3/2} V\varepsilon^{1/2} \tag{1-155}$$

However for the calculations here we will assume a spinless fermion (perhaps a composite particle) and obtain from Eqs. (1-137) and (1-143)

$$\frac{PV}{kT} = 2\pi \left(\frac{2m}{h^2}\right)^{3/2} V \int_0^\infty \varepsilon^{1/2} \ln \left(1 + ze^{-\beta \varepsilon}\right) d\varepsilon \tag{1-156}$$

$$\bar{N} = 2\pi \left(\frac{2m}{h^2}\right)^{3/2} V \int_0^\infty \frac{z\varepsilon^{1/2} e^{-\beta \varepsilon} \, d\varepsilon}{1 + ze^{-\beta \varepsilon}} \tag{1-157}$$

These integrals cannot be evaluated in closed form, but for small z we can expand as follows:

$$\ln \left(1 + ze^{-\beta \varepsilon}\right) = \sum_{j=1}^\infty \frac{(-1)^{j-1}}{j} z^j e^{-\beta j\varepsilon} \tag{1-158}$$

$$\frac{1}{1 + ze^{-\beta \varepsilon}} = \sum_{j=0}^\infty (-1)^j z^j e^{-\beta j\varepsilon} \tag{1-159}$$

and integrate term by term to find

$$\frac{P}{kT} = \frac{1}{\lambda^3} \sum_{j=1}^{\infty} \frac{(-1)^{j-1} z^j}{j^{5/2}} \tag{1-160}$$

$$\frac{\bar{N}}{V} = \rho = \frac{1}{\lambda^3} \sum_{j=1}^{\infty} \frac{(-1)^{j-1} z^j}{j^{3/2}} \tag{1-161}$$

Explicitly these series begin

$$P/kT = \frac{1}{\lambda^3} \left(z - \frac{z^2}{2^{5/2}} + \frac{z^3}{3^{5/2}} - \cdots \right) \tag{1-162}$$

$$\rho = \frac{1}{\lambda^3} \left(z - \frac{z^2}{2^{3/2}} + \frac{z^3}{3^{3/2}} - \cdots \right) \tag{1-163}$$

We may assume z is itself a power series in ρ, which from Eq. (1-163) begins as $\lambda^3 \rho + \ldots$. So we write

$$z = \lambda^3 \rho + a_1 \rho^2 + a_2 \rho^3 + \cdots$$

take powers of z, and insert into Eq. (1-163) and equate the coefficients of ρ^2, ρ^3, etc., to zero (since the left-hand side is exactly ρ), thus determining $a_1, a_2, \ldots$ The result is:

$$z = \lambda^3 \rho + \frac{(\lambda^3 \rho)^2}{2^{3/2}} + \left(\frac{1}{4} - \frac{1}{3^{3/2}} \right) (\lambda^3 \rho)^3 + \cdots \tag{1-164}$$

This and its powers are inserted into Eq. (1-162) to obtain P/kT as the desired power series in ρ:

$$P/kT = \rho + \frac{\lambda^3 \rho^2}{2^{5/2}} + \lambda^6 \rho^3 \left(\frac{1}{8} - \frac{2}{3^{5/2}} \right) + \cdots \tag{1-165}$$

This equation is in the form of a virial expansion for the pressure of the ideal fermion gas

$$P/kT = \rho + B_2(T) \rho^2 + B_3(T) \rho^3 + \cdots \tag{1-166}$$

where $B_i(T)$ is a function of temperature only and is called the ith virial coefficient. We find

$$B_2(T) = \frac{\lambda^3}{2^{5/2}} = 0.1768 \lambda^3 \tag{1-167}$$

$$B_3(T) = \left(\frac{1}{8} - \frac{2}{3^{5/2}} \right) \lambda^6 = -0.0033 \lambda^6 \tag{1-168}$$

The effective interaction between ideal fermions due to the requirement of an anti-symmetric $\mathcal{N}$-body wave function is repulsive, as is shown by a positive leading correction term, B_2, to the classical ideal gas result.

The methods of this section will be useful in treating imperfect classical gases (see Chap. 3).

Bosons

The sums in this case that must be changed to integrals are

$$\frac{PV}{kT} = -\sum_i \ln\left(1 - ze^{-\beta\varepsilon_i}\right) \tag{1-169}$$

$$\bar{N} = \sum_i \frac{ze^{-\beta\varepsilon_i}}{\left(1 - ze^{-\beta\varepsilon_i}\right)} \tag{1-170}$$

Clearly both of the summands will diverge when $ze^{-\beta\varepsilon_0} \to 1$, where ε_0 is the energy of the lowest translational state. Although from Eq. (1-74) $\varepsilon_0 = 3h^2/(8m)V^{2/3}$, we may for convenience measure all our energies relative to this one and effectively put ε_0 at zero. No physical results can depend on where one sets the zero of a set of energy states. Hence in each sum we must separate out the $i = 0$ term and replace the rest of the series by an integral. This problem did not arise with fermions since their summands never could diverge for any ε_i state. We have

$$P/kT = -\frac{1}{V}\ln\left(1 - z\right) - 2\pi\left(\frac{2m}{h^2}\right)^{3/2}\int_0^\infty \varepsilon^{1/2}\ln\left(1 - ze^{-\beta\varepsilon}\right)d\varepsilon \quad (1\text{-}171)$$

$$\frac{\bar{N}}{V} = \rho = \frac{z}{V(1-z)} + 2\pi\left(\frac{2m}{h^2}\right)^{3/2}\int_0^\infty \frac{z\varepsilon^{1/2}e^{-\beta\varepsilon}\,d\varepsilon}{\left(1 - ze^{-\beta\varepsilon}\right)} \tag{1-172}$$

These equations with the terms in $(1/V)$ needed for the very degenerate situation in which $z \to 1$ can be simplified for our discussion here for small z. For V of macroscopic size (indeed infinite in the thermodynamic limit), unless $z \to 1$ the terms in $1/V$ may be discarded. Exactly as with the FD case, the integrands can be expanded in powers of z and integrated term by term to give

$$P/kT = \frac{1}{\lambda^3}\sum_{j=1}^\infty \frac{z^j}{j^{5/2}} \tag{1-173}$$

$$\frac{\bar{N}}{V} = \rho = \frac{1}{\lambda^3}\sum_{j=1}^\infty \frac{z^j}{j^{3/2}} \tag{1-174}$$

The power series in ρ for z from Eq. (1-174) is

$$z = \lambda^3\rho - \frac{(\lambda^3\rho)^2}{2^{3/2}} + \left(\frac{1}{4} - \frac{1}{3^{3/2}}\right)(\lambda^3\rho)^3 + \cdots \tag{1-175}$$

which, when inserted in Eq. (1-173), gives as virial expansion

$$P/kT = \rho - \frac{\lambda^3 \rho^2}{2^{5/2}} + \lambda^6 \rho^3 \left(\frac{1}{8} - \frac{2}{3^{5/2}} \right) + \cdots \tag{1-176}$$

with

$$B_2(T) = -\frac{\lambda^3}{2^{5/2}} \tag{1-177}$$

$$B_3(T) = \left(\frac{1}{8} - \frac{2}{3^{5/2}} \right) \lambda^6 \tag{1-178}$$

The effective interaction between ideal bosons is attractive, as is indicated by the negative second virial coefficient. This B_2 is the negative of that for spinless fermions.

In Fig. 1-6 we plot PV/NkT from the respective virial series through B_3 as a function of $\lambda^3 \rho$ for ideal bosons and ideal fermions. Because of the dominating B_2

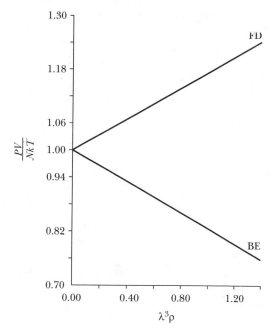

FIGURE 1-6 PV/NkT from the respective virial series through B_3 plotted against $\lambda^3 \rho$ for ideal bosons and ideal fermions.

values these equations are practically straight lines of opposite slope for $\lambda^3 \rho$ values, for which an expansion in powers of z (as done here) is reasonable.

Finally, we note that both equations of state will become identical to that of the classical ideal gas if

$$\lambda^3 \rho = \left(\frac{h^2}{2\pi m k T} \right)^{3/2} \rho \to 0$$

This can be arrived at in three independent ways:

$$\rho \to 0 \qquad \text{at fixed } T \text{ (so } P \to 0 \text{ also)}$$

$$T \to \infty \qquad \text{at fixed } \rho$$

$$h \to 0 \qquad \text{i.e., no quantum mechanics!}$$

1-8
The Semiclassical Partition Function; Equipartition of Energy Theorem

We have so far presented statistical mechanics as based on quantum mechanics. Historically it developed first from a classical mechanics view of the $\mathcal{N}$-body problem expressed in terms of the center of mass position coordinates $x_1, y_1, z_1, \ldots,$ x_N, y_N, z_N and their conjugate momenta $p_{x_1} \ldots p_{z_N}$, that is, $6\mathcal{N}$ coordinates in all, constituting what was termed the *phase space* of the system. The classical energy of the $\mathcal{N}$-body system is written noting $p_{x_i} = m v_{x_i}$ where v_{x_i} is the x component of the velocity of the ith particle as

$$U = \frac{1}{2m} \sum_{j=1}^{N} (p_{x_i}^2 + p_{y_i}^2 + p_{z_i}^2) + V_N(x_1 \ldots z_N) \qquad (1\text{-}179)$$

in which V_N is a term depending on the position coordinates alone [usually a sum of $\mathcal{N}(\mathcal{N} - 1)/2$ pair interaction terms]. The original analog to the canonical partition function Q_N of Eq. (1-62) was an integration of $e^{-U/kT}$ over the full continuous range of values allowed to all the position and momentum coordinates. We shall not present the proof of the validity of this identification here; it may be found in many readily available references [6, 7, 9]. The integral must be divided by $\mathcal{N}!$ because in the real world (which fundamentally follows quantum physics) non-localized identical particles cannot be distinguished. Finally, to accord with the uncertainty principle that the phase space for a conjugate momentum-coordinate pair cannot be subdivided more finely than a size of order h, we must divide each $dp_{x_i} dx_i$ factor of the phase volume element by h. This leads, as we shall see, to correct results. Notice also that it makes the phase integral form $Q_N{}^{cl}$ dimensionless, as it must be if it is to represent a partition function (i.e., h has dimensions of momentum times length). Thus the semiclassical prescription is

$$Q_N{}^{cl}(V, T) = \frac{1}{N!h^{3N}} \int \ldots \int \exp \frac{-1}{kT} \left[\frac{1}{2m} \sum_{i=1}^{N} (p_{x_i}^2 + p_{y_i}^2 + p_{z_i}^2) + V_N \right]$$
$$\times \, dx_1 \ldots dp_{z_N} \tag{1-180}$$

The range of integration of the position variables x_i is defined by the volume of the system. For mathematical convenience in doing the momentum integrals—which can be seen to separate out as the product of $3N$ identical ones over a common variable, say p)—we can let p range from $-\infty$ to $+\infty$. Since $\exp(-p^2/2mkT)$ falls off to zero so rapidly for large p^2, details of the limits for finite energy have no effect on the value of the integral.

Thus we find

$$Q_N{}^{cl}(V, T) = \frac{1}{N!h^{3N}} \left[\int_{-\infty}^{+\infty} \exp \left(\frac{-p^2}{2mkT} \right) dp \right]^{3N} \int \ldots \int \exp \left(\frac{-V_N}{kT} \right) dx_1 \ldots dz_N$$

$$= \left(\frac{2\pi mkT}{h^2} \right)^{3N/2} \frac{Z_N}{N!} = \frac{Z_N}{\lambda^{3N} N!} \tag{1-181}$$

in which we have defined Z_N, the classical configuration integral or partition function for N particles as

$$Z_N = \int \ldots \int \exp \left(\frac{-V_N}{kT} \right) dx_1 \ldots dz_N \tag{1-182}$$

and λ is the thermal wavelength of Eq. (1-115). Equation (1-181) is the starting point for all current work on interacting classical systems such as imperfect gases and liquids (see Chap. 3). It incorporates only the gross quantum effects and so is termed a semiclassical form. It cannot be used when the difference between boson and fermion behavior is significant.

Exercise 1-12

Check the integration required to obtain Eq. (1-181).

In Eq. (1-181), if we have a noninteracting case then $V_N = 0$ and each triple integration $\iiint dx_i \, dy_i \, dz_i$ gives a factor V, which is the volume of the system such that $Z_N = V^N$, and for the ideal classical gas

$$Q_N{}^{cl}(V, T) = \left(\frac{V}{\lambda^3} \right)^N \frac{1}{(N)!} = \frac{q_{trans}^N}{N!} \tag{1-183}$$

which checks Eq. (1-126) and shows that our prescription presented in Eq. (1-180) is indeed correct in this case. It is of interest to go further with the classical ideal gas and include internal degrees of freedom and focus on the partition function of one molecule made up of s atoms. Such a molecule is describable classically (in three dimensions) in terms of $3s$ momenta (symbols $p_1 \ldots p_{3s}$) and $3s$ coordinates (symbols $q_1 \ldots q_{3s}$). It is always more convenient to change from ordinary Cartesian coordinates of the atoms to three each of momenta and position of the center of the mass:

$$p_1 = p_x \qquad q_1 = x$$
$$p_2 = p_y \qquad q_2 = y$$
$$p_3 = p_z \qquad q_3 = z$$

and $(3s - 3)$ each of coordinates and momenta for rotation and vibration relative to the center of mass, which in fact leave it fixed. Linear molecules will have two degrees of rotational motion (since rotation about the line of atomic masses will not change any atomic positions) leaving $(3s - 5)$ vibrational coordinates and momenta, whereas nonlinear molecules will have three degrees of rotational motion leaving $(3s - 6)$ vibrational coordinates and momenta. Then the energy of one molecule can be written as

$$\varepsilon(p, q) = \frac{p_x^2}{2m} + \frac{p_y^2}{2m} + \frac{p_z^2}{2m} + \varepsilon_f(x, y, z) + \varepsilon_r(q_4 \cdots p_{3s}) \qquad (1\text{-}184)$$

ε_f is a term depending on the center of mass position coordinates only and will exist if the molecule is in some external field (electrical, gravitational, or centrifugal) and has a potential energy by virtue of this. ε_r represents the energy due to the rotations and vibrations. For every vibrational degree of freedom treated approximately as a harmonic oscillation there will be terms in ε_r of the form $(i \geq 4)$

$$\frac{p_i^2}{2\mu_i} + \frac{f_i}{2} q_i^2$$

in which μ_i is a generalized reduced mass, that is, a function of two or more of the atomic masses that contribute to the full molecular mass m. The f_i is the force constant for the oscillation and q_i is the normal mode displacement coordinate. Also, in ε_r for linear and nonlinear molecules there will occur two or three terms of the type $p_j^2/2\mu_j$ in which the μ_j besides having a mass dependence will also depend on some of the q_j, which will be angles describing the rotations about the mass center. Consult references [6, 7, 9] for further details.

By analogy to our result in Eq. (1-180), the one-molecule classical partition function is

$$q^{cl} = \frac{1}{h^{3s}} \int \cdots \int \exp\left[-\beta\varepsilon(p, q)\right] dp\, dq \qquad (1\text{-}185)$$

where $dp\, dq = dp_1\, dp_2 \ldots dp_{3s}\, dq_1\, dq_2 \ldots dq_{3s}$ and the fractional number of molecules with phase space coordinates in the range p to $p + dp$, q to $q + dq$ is

$$\frac{dN}{N} = \frac{\exp\left[-\beta\varepsilon(p, q)\right](dp\, dq/h^{3s})}{q^{cl}} = \frac{\exp\left[-\beta\varepsilon(p, q)\right] dp\, dq}{\int \ldots \int \exp\left[-\beta\varepsilon(p, q)\right] dp\, dq} \tag{1-186}$$

such that the average energy per molecule is

$$\bar{\varepsilon} = \int \varepsilon\, \frac{dN}{N} = \frac{\int \ldots \int \varepsilon \exp\left(-\beta\varepsilon\right) dp\, dq}{\int \ldots \int \exp\left(-\beta\varepsilon\right) dp\, dq} \tag{1-187}$$

From Eq. (1-187) we may prove the classical equipartition of energy theorem, which states that for every squared term appearing in the $\varepsilon(p, q)$ expression there is a contribution to $\bar{\varepsilon}$ of $kT/2$ and thus a contribution to the heat capacity at constant volume C_V of $k/2$ per molecule. We may take the most difficult case of a rotational squared term, say $(p_4^2/2\mu_4)$ in which μ_4 may also depend on q_4. Let

$$\varepsilon'(p, q) = \varepsilon(p, q) - \frac{p_4^2}{2\mu_4}; \qquad dp'\, dq' = \frac{dp\, dq}{dp_4\, dq_4}$$

Then $\bar{\varepsilon}$ is a sum of terms, the one coming from $(p_4^2/2\mu_4)$ being

$$\overline{\varepsilon p_4} = \frac{\int \ldots \int \exp\left[-\beta\varepsilon'(p, q)\right] dp'\, dq' \int dq_4 \int_{-\infty}^{+\infty} dp_4\left[\exp\left(-\beta p_4^2/2\mu_4\right)\right](p_4^2/2\mu_4)}{\int \ldots \int e^{-\beta\varepsilon'(p,q)}\, dp'\, dq' \int dq_4 \int_{-\infty}^{+\infty} dp_4[\exp\left(-\beta p_4^2/2\mu_4\right)]} \tag{1-188}$$

by the factorization property of the exponential of $-\beta\varepsilon$ with ε given by Eq. (1-184). Since

$$\int_{-\infty}^{+\infty} \exp\left(-\beta p_4^2/2\mu_4\right) dp_4 = \left(\frac{2\pi\mu_4}{\beta}\right)^{1/2}$$

$$\int_{-\infty}^{+\infty} \frac{p_4^2}{2\mu_4} \exp\left(-\beta p_4^2/2\mu_4\right) dp_4 = \frac{1}{2\beta}\left(\frac{2\pi\mu_4}{\beta}\right)^{1/2}$$

after the integrations over p_4 the remaining manyfold integrals cancel exactly and the factor $1/2\beta$ in the numerator remains such that

$$\overline{\varepsilon p_4} = \left(\frac{1}{2\beta}\right) = \frac{kT}{2} \tag{1-189}$$

Since normal mode displacement coordinates as well as their momenta have integration ranges from $-\infty$ to $+\infty$, the proof carries through for their contributions as well.

The predictions of this theorem for the molar constant volume heat capacity of gaseous molecules are the following *temperature independent* values:

Monatomic $(C_V)_m = (\frac{3}{2})R$

Linear $(C_V)_m = (\frac{3}{2})R + R + (3s - 5)R = (3s - \frac{5}{2})R$

Nonlinear $(C_V)_m = (\frac{3}{2})R + (\frac{3}{2})R + (3s - 6)R = (3s - 3)R$

since each vibrating mode with a squared momentum and a squared coordinate term contributes R per mole. These predictions are always wrong except for monatomic gases at temperatures for which only their electronic ground state is appreciably occupied. The observed strong temperature dependence of polyatomic $(C_V)_m$ values was one of the earliest experimental indications of the failure of classical physics.

Equation (1-189), applied to the three squared terms associated with translation of the mass center alone, gives per molecule:

$$\overline{\varepsilon_{trans}} = \frac{\overline{p_x^2}}{2m} + \frac{\overline{p_y^2}}{2m} + \frac{\overline{p_z^2}}{2m} = \frac{3}{2}kT \qquad (1\text{-}190)$$

This is identical to the result in Eq. (1-21). We see it holds for any type of molecule, monatomic or polyatomic. Moreover, as long as the interaction energy V_N of Eq. (1-181) does not depend on particle momenta (i.e., on velocities), we can always separate out the momenta of translation for separate averaging and arrive at this same result as long as classical physics holds and momenta and coordinates are commuting variables.

Thus for real interacting liquids and gases Eq. (1-190) holds accurately at atmospheric pressure until the temperature is well below 10 K (see Problem 1-14). As will be explained in Chap. 4, for some solids this relation will be inaccurate at or not much below 298 K.

1-9

The Distribution of Molecular Speeds and Velocities in a Classical Fluid

In Sec. 1 we introduced a function representing the probability density of speeds in a classical gas. In this section we will derive its explicit form and also that of related probability densities or *distribution functions* as they are also called.

From Eq. (1-186) for a classical ideal gas we may derive probability density functions with respect to any subset of the variables in $dp\,dq$ by integrating over all those not in the subset, for we will obtain in every case a properly normalized expression for dN/N in terms of just those variables we are interested in. Furthermore, the integrations over those variables not of interest will automatically be canceled by the appropriate factors of the complete integration in the denominator, just as occurred in Eq. (1-188). This is very nice, because these integrals (for internal coordinates) are complex. If, for example, we are only interested in the distribution of center of mass momenta, we integrate as follows using Eq. (1-184) in Eq. (1-186):

$$\frac{dN}{N} = \frac{\exp\left[-(p_x^2 + p_y^2 + p_z^2)/2mkT\right] dp_x\, dp_y\, dp_z \int \cdots \int \exp\left(-\varepsilon_f/kT\right) dx\, dy\, dz \int \cdots}{\int\!\!\int\!\!\int_{-\infty}^{+\infty} \exp\left[-(p_x^2 + p_y^2 + p_z^2)/2mkT\, dp_x\, dp_y\, dp_z \int \cdots \int \exp\left(-\varepsilon_f/kT\right) dx\, dy\, dz \int \cdots \right.}$$

$$\frac{\int \exp\left[-\varepsilon_r(q_4 \cdots p_{3s})/kT\right] dq_4 \cdots dp_{3s}}{\int \exp\left[-\varepsilon_r(q_4 \cdots p_{3s})/kT\right] dq_4 \cdots dp_{3s}} \tag{1-191.1}$$

to obtain

$$\frac{dN}{N} = \frac{\exp\left[-(p_x^2 + p_y^2 + p_z^2)/2mkT\right] dp_x\, dp_y\, dp_z}{\int\!\!\int\!\!\int_{-\infty}^{+\infty} \exp\left[-(p_x^2 + p_y^2 + p_z^2)/2mkT\right] dp_x\, dp_y\, dp_z} \tag{1.191.2}$$

To convert this into a velocity distribution, we change variables $p_x = mv_x$, and so on:

$$\frac{dN}{N} = \frac{\exp\left[-m(v_x^2 + v_y^2 + v_z^2)/2kT\right] dv_x\, dv_y\, dv_z}{\int_{-\infty}^{+\infty} \exp\left(-mv_x^2/2kT\right) dv_x \int_{-\infty}^{+\infty} \exp\left(-mv_y^2/2kT\right) dv_y \int_{-\infty}^{+\infty} \exp\left(-mv_z^2/2kT\right) dv_z}$$

Doing the standard integrations in the denominator, we find:

$$\frac{dN}{N} = \left(\frac{m}{2\pi kT}\right)^{3/2} \exp\left[-m(v_x^2 + v_y^2 + v_z^2)/2kT\right] dv_x\, dv_y\, dv_z \tag{1-192}$$

Equation (1-192) gives the fraction of the molecules at temperature T in a classical fluid that simultaneously have velocity components in the ranges:

$$v_x \text{ to } v_x + dv_x$$

$$v_y \text{ to } v_y + dv_y$$

$$v_z \text{ to } v_z + dv_z$$

We use the term fluid because by the arguments given following Eq. (1-190) the same expression would hold if the molecules had nonzero interactions between them, that is, if the molecules constituted an imperfect gas or liquid.

We may go on and consider a single-direction velocity component distribution, say of v_x, by integrating Eq. (1-192) over dv_y and dv_z to obtain

$$f(v_x) \equiv \frac{dN}{N\,dv_x} = \left(\frac{m}{2\pi kT}\right)^{1/2} \exp-(mv_x^2/2kT) \qquad (1\text{-}193)$$

Exercise 1-13

Review the properties of a probability density f as given in Eqs. (1-9)–(1-11) and show explicitly that $\int_{-\infty}^{+\infty} f(v_x)\,dv_x = 1$.

The function $f(v_x)$ is shown in Fig. 1-7 for three temperatures for which we have defined T_0 as that temperature at which $(2kT_0/m)^{1/2} = 1.00 \times 10^3$ m·s^{-1} (1920 K for O_2 molecules). All the curves have a maximum at zero velocity (which is the most probable velocity at any temperature) that rises and narrows as temperature decreases. At any temperature

$$v_x = \int_{-\infty}^{+\infty} v_x f(v_x)\,dv_x = 0 \qquad (1\text{-}194)$$

This is physically understandable since velocity has a direction, and for a gas at rest there must be as many molecules going in the positive direction as there are going in the negative direction for every $|v_x|$. Mathematically, the integral vanishes because the positive area under the curve from 0 to ∞ cancels by symmetry the negative area under the curve from $-\infty$ to 0.

Exercise 1-14

$\overline{(v_x^2)} \neq 0$. Calculate its value by integration and show that it fits what the principle of equipartition of energy would predict.

To obtain a distribution with respect to speeds c we return to Eq. (1-192) and change variables from Cartesian velocity coordinates to spherical velocity space coordinates. Since

$$c^2 = v_x^2 + v_y^2 + v_z^2$$

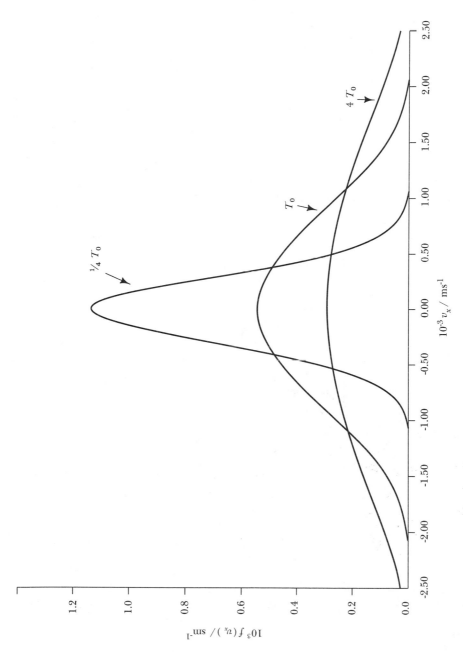

FIGURE 1-7 The one-component velocity function $f(v_x) = (1/N)(dN/dv_x)$ as a function of v_x for temperatures $\frac{1}{4}T_0$, T_0, and $4T_0$ where T_0 is that temperature for which $(2kT_0/m)^{1/2} = 1.00 \times 10^3$ m·s^{-1}. The area under each curve is unity

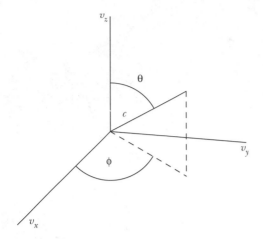

FIGURE 1-8 Spherical coordinates in velocity space.

speed is a radius vector in velocity space, as shown in Fig. 1-8. The volume elements are related by

$$dv_x \, dv_y \, dv_z = c^2 \, dc \, \sin \theta \, d\theta \, d\phi$$

If we are not interested in the direction of molecular motion we integrate over all θ and ϕ using

$$\int_0^\pi \sin \theta \, d\theta = 2, \qquad \int_0^{2\pi} d\phi = 2\pi$$

to obtain from Eq. (1-180)

$$f(c) = \frac{1}{N} \frac{dN}{dc} = 4\pi c^2 \left(\frac{m}{2\pi k T} \right)^{3/2} e^{-mc^2/2kT} \qquad (1\text{-}195)$$

This is the famous Maxwell-Boltzmann speed distribution for a three-dimensional classical fluid. This distribution is a primary determining factor in controlling the rates of chemical reactions and is shown in Fig. 1-9 for three different temperatures. The area under each curve is unity, corresponding to the fact that *all* molecules have speeds between zero and infinity (omitting relativity effects). As the temperature increases, the speed c_p at the maximum [see Eq. (1-199)] increases but the value of $f(c_p)$ drops until we may imagine at infinite temperature that the curve disappears into the c-axis. This means that all speeds become equally likely at infinite temperature, that is, we have a state of maximum disorder. Conversely, as the temperature decreases the height of the maximum increases as c_p decreases, until at 0 K the maximum is infinity (a delta function like that discussed in Appendix II)

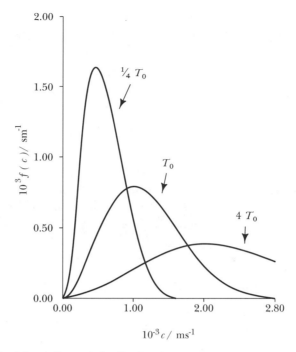

FIGURE 1-9 The Maxwell speed distribution for several temperatures. Conventions as in Figure 1-7.

at $c_p = 0$. This means only the speed zero is permitted and perfect order exists at absolute zero as the curve disappears into the $f(c)$ axis. Of course this is a classical result that does not take into account the existence of zero point motion.

Since by Eq. (1-11) the average value $\overline{F(c)}$ of any function of molecular speed is given by

$$\overline{F(c)} = \int_0^\infty F(c)f(c)\,dc \qquad (1\text{-}196)$$

on doing the standard integrals we find for the average speed

$$\bar{c} = \left(\frac{8kT}{\pi m}\right)^{1/2} \qquad (1\text{-}197)$$

and for the square root of the average of the square of the speed, the root mean square speed

$$c_{rms} = (\overline{c^2})^{1/2} = \left(\frac{3kT}{m}\right)^{1/2} \qquad (1\text{-}198)$$

Another significant speed is the value c_p for which $f(c)$ in Eq. (1-195) is a maximum. This is termed the most probable speed and is found by setting $df(c)/dc$ equal to zero:

$$c_p = \left(\frac{2kT}{m}\right)^{1/2} \tag{1-199}$$

Exercise 1-15

Do the required integrations to obtain Eq. (1-196) and Eq. (1-197) and the differentiation to obtain Eq. (1-199). Show that these speeds are in the ratios $c_p : \bar{c} : c_{rms}$ as $1.000 : 1.128 : 1.225$.

Example 1-4

Rewrite Eq. (1-195) in terms of the dimensionless variable $y = c/c_p$ and obtain an expression for the fraction of the molecules that have a speed greater than sc_p (where s is any numerical value) in terms of the error function, erf (s), where

$$\text{erf}(s) = \frac{2}{\sqrt{\pi}} \int_0^s e^{-y^2}\,dy$$

and is listed in numerical tables. By referring to such tables, give this fraction for $s = 0.2, 1.0, 2.0, 3.0, 5.0$. Note that this fraction is independent of T and of m (molecular mass); c_p depends on both. Evaluate c_p at $25°C$ for H_2 and O_2.

Solution:

Using $c = yc_p$, $dc = c_p\,dy$, the fraction of molecules with speeds greater than sc_p (F_s) is given by the integral

$$F_s = \int_{sc_p}^{\infty} \left(\frac{1}{N}\frac{dN}{dc}\right) dc = \int_s^{\infty} \left(\frac{m}{2\pi kT}\right)^{3/2} 4\pi y^2 c_p^2 \exp\left[-(mc_p^2 y^2)/2kT\right]c_p\,dy$$

which by use of Eq. (1-199) becomes

$$F_s = \frac{4}{\sqrt{\pi}} \int_s^{\infty} y^2\,dy\,e^{-y^2}$$

independent of T and of m.

The integral above may be integrated by parts

$$\int_s^\infty y\,dv = vy\Big]_{y=s}^{y=\infty} - \int_s^\infty v\,dy \qquad \text{where } v = \int e^{-y^2}y\,dy = -\tfrac{1}{2}e^{-y^2}$$

Since $\lim_{y=\infty} (-\tfrac{1}{2}e^{-y^2}y) = 0$, we have

$$\int_s^\infty y\,dv = \tfrac{1}{2}se^{-s^2} + \tfrac{1}{2}\int_s^\infty e^{-y^2}\,dy$$

and

$$F_s = \frac{2}{\sqrt{\pi}} se^{-s^2} + \frac{2}{\sqrt{\pi}}\int_s^\infty e^{-y^2}\,dy$$

$$= \frac{2}{\sqrt{\pi}} se^{-s^2} + \frac{2}{\sqrt{\pi}}\int_0^\infty e^{-y^2}\,dy - \frac{2}{\sqrt{\pi}}\int_0^s e^{-y^2}\,dy$$

$$= \frac{2}{\sqrt{\pi}} se^{-s^2} + 1 - \text{erf}\,(s)$$

in which we used the known standard integral for the second term. A brief listing of the erf (s) function is:

$s = 0$	0.20	0.40	0.60	0.80	—
erf $(s) = 0$	0.2227	0.4284	0.6039	0.7421	—
$s = 1.00$	1.20	1.40	1.60	1.80	2.00
erf $(s) = 0.8427$	0.9103	0.9523	0.9763	0.9891	0.9953

SOURCE: M. Abramowitz and I. A. Stegun. *Handbook of Mathematical Functions*. Dover.

For large s one can show

$$1 - \text{erf}\,(s) = \frac{s}{\sqrt{\pi}} e^{-s^2}\left(\frac{1}{s^2} - \frac{1}{2s^4} + \cdots\right)$$

Thus

$$F_{0.2} = 0.2168 + 1 - 0.2227 = 0.9941$$

$$F_{1.0} = 0.4151 + 1 - 0.8427 = 0.5724$$

$$F_{2.0} = 0.0413 + 1 - 0.9953 = 0.0460$$

For $s > 2$ use large s form:

$$F_{3.0} = 4.18 \times 10^{-4} + 0.219 \times 10^{-4} = 4.40 \times 10^{-4}$$

$$F_{5.0} = 78.35 \times 10^{-12} + 1.54 \times 10^{-12} = 7.99 \times 10^{-11}$$

$$c_p = \left(\frac{2kT}{m}\right)^{1/2} = \left(\frac{2RT}{M}\right)^{1/2} = \left(\frac{2 \times 8.3144 \times T}{M/1000}\right)^{1/2} = 129.0(T/M)^{1/2}$$

yields c_p in m·s^{-1} with T in K and M in g/mol, the molecular weight at 298.2 K.

$$\text{For } H_2 \quad c_p = (129.0)\left(\frac{298.2}{2.016}\right)^{1/2} = 1569 \text{ m·s}^{-1}$$

$$\text{For } O_2 \quad c_p = (129.0)\left(\frac{298.2}{32.00}\right)^{1/2} = 398.8 \text{ m·s}^{-1}$$

With the development of molecular beam techniques (see Sec. 5-1 and Appendix 5.B) it has been possible to experimentally confirm the Maxwell speed distribution in some detail. One such experiment is well described by Morse [10].

References and Recommended Reading

[1] J. E. Mayer. *Equilibrium Statistical Mechanics*. Pergamon Press, Oxford, 1968. Chapters 1 and 8.

[2] R. S. Berry, S. A. Rice, and J. Ross. *Physical Chemistry*. Wiley, New York, 1980. Chapters 15 and 19.

[3] G. S. Rushbrooke. *Introduction to Statistical Mechanics*. Oxford University Press, Oxford, 1949.

[4a] D. S. Schonland. *Molecular Symmetry*. D. Van Nostrand, Princeton, 1965.

[4b] A. J. Coleman in *Advances in Quantum Chemistry*, vol. 4, P. O. Lowdin, ed. Academic Press, New York, 1968, pp. 83–108.

[5] H. A. Bethe and R. Jackiw. *Intermediate Quantum Mechanics*. Addison Wesley, Reading, Mass., 2d ed., 1968.

[6] D. A. McQuarrie. *Statistical Mechanics*. Harper & Row, New York, 1976.

[7] A. Munster. *Statistical Thermodynamics*, vol. 1. Springer Verlag, Berlin, 1969.

[8] J. Wilks. *The Properties of Liquid and Solid Helium*. Oxford University Press, Oxford, 1967.

[9] N. Davidson. *Statistical Mechanics*. McGraw-Hill, New York, 1962.

[10] P. M. Morse. *Thermal Physics*. Benjamin & Co., New York, 2d ed., 1969, pp. 172–174; experimental verification of Maxwell speed distribution.

Problems

1. Consider a binomial distribution weight

$$W(\mathcal{N}_1) = \frac{\mathcal{N}!}{\mathcal{N}_1!(\mathcal{N} - \mathcal{N}_1)!}$$

which could apply for example to the distribution of distinguishable non-interacting molecules between two equal volume parts of a container. $\mathcal{N}_1$, which can vary from 0 to $\mathcal{N}$, labels the possible distributions. Show that the most probable distribution has $\mathcal{N}_1 \equiv \mathcal{N}_1^* = \mathcal{N}/2$. By Taylor series expansion through quadratic terms show that $W(\mathcal{N}_1)$ assumes the form of a Gaussian distribution [Eq. (1-12)] centered at $\mathcal{N}_1 = \mathcal{N}_1^*$ with root mean square deviation, $\sigma = \sqrt{\mathcal{N}}/2$ such that $\sigma/\mathcal{N}_1^*$ is of order $\mathcal{N}^{-1/2}$ and is negligible for thermodynamic systems. Use the binomial expansion formula to show that the exact average $\bar{\mathcal{N}}_1$ is also $\mathcal{N}/2$ and that $\sum_{\mathcal{N}_1=0}^{\mathcal{N}} W(\mathcal{N}_1) = 2^{\mathcal{N}}$. Compare this latter value with the single term $W(\mathcal{N}_1^*)$ by use of the Stirling approximation given in Appendix II.

2. Consider a system of two parts at equilibrium with respect to heat flow such that $T_1 = T_2 = T$ (temperature) with fixed total volume $V = V_1 + V_2$ but with a wall separating them that can move. Defining $P = T(\partial S/\partial V)_{U,N}$, provide the argument analogous to that for Eq. (1-38) that shows that P must have the meaning of pressure.

3. Consider a microcanonical ensemble of two-part systems divided by a fixed wall that separates volume V into V_1 and V_2. Each volume has fixed number set, respectively $\mathcal{N}_1$ and $\mathcal{N}_2$, and fixed energy U_1 and U_2 (you may assume equilibrium with respect to heat flow and thus $T_1 = T_2$), $\mathcal{N}_1 \neq \mathcal{N}_2$, and $U_1 \neq U_2$ in general). Each replica has a different V_1 value, but $V_1 + V_2 = V$ constant for all replicas
a. Show that in the most probable distribution $P_1 = P_2$.
b. Your result in part **a** means that the ensemble average

$$\overline{(P_1 - P_2)} = 0$$

Will $[\overline{(P_1 - P_2)^2}]^{1/2}$ also be zero? Explain.
c. Suppose the ensemble is of two-part systems but with a frictionless sliding wall separating V_1 and V_2 in each replica.
What will be $\overline{(P_1 - P_2)}$ and $[\overline{(P_1 - P_2)^2}]^{1/2}$ for this ensemble?

4. By the methods of this chapter express for a two-part otherwise isolated system (W'/W) for the case in which an amount of energy in the form of heat $|Q|$ flows from part 1 to part 2, which are sufficiently large to maintain their temperatures fixed at T_1 and T_2, respectively. W' is the number of micromolecular states available to the two-part system after the energy exchange. Express your answer numerically if $|Q| = 1.00$ J, $T_2 = 296$ K, $T_1 = 300$ K.

5. In how many ways may ten distinguishable particles be distributed among an energy state set in such a manner that four are in the first state, three in the second, two in the third, and one in the fourth?

6. Consider an infinite set of nondegenerate energy states starting at zero and spaced ε apart (i.e., $\varepsilon_0 = 0$; $\varepsilon_1 = \varepsilon$; $\varepsilon_2 = 2\varepsilon$; etc.). By trial and error work out all the distributions permissible and their W values if we distribute four distinguishable particles among these states such that the total energy $U = 4\varepsilon$.

7. Consider the general case of the energy state set of Problem 6. The Boltzmann formula for the set $\bar{N}_j$ that maximizes W for distinguishable particles remains that of Eq. (1-103) because division by $N!$ (a constant) is immaterial to the maximization process and its results. In this case we can perform the full sum exactly by use of the geometric series:

$$q = \sum_{i=0}^{\infty} e^{-i\beta\varepsilon} = \frac{1}{1 - e^{-\beta\varepsilon}}$$

Show that U from Eq. (1-104) in this case becomes

$$U = -\frac{N}{q}\left(\frac{\partial q}{\partial \beta}\right) = \frac{N\varepsilon e^{-\beta\varepsilon}}{1 - e^{-\beta\varepsilon}}$$

If we introduce a parameter p such that $U = Np\varepsilon$ show that in this Boltzmann counting

$$\beta\varepsilon = \ln\left(\frac{1+p}{p}\right)$$

8. Take the set of states of Problem 6 but now with nine distinguishable particles and $U = 9\varepsilon$ (such that p of Problem 7 is 1). Start to work out the allowed distributions, in this case beginning with the one with $N_0 = 8$, $N_9 = 1$. You will find it tedious to determine all 30 possible distributions but it is not hard to convince yourself that the actual physical distribution of maximum W has $N_0 = 4$, $N_1 = 2$, $N_2 = 2$, $N_3 = 1$. Use the equations derived in Problem 7 to work out the nonphysical Boltzmann results for the N_i's (now no longer integers) that mathematically maximize W. Compare (N_i/N) for the first ten states in the physical distribution and in the Boltzmann distribution that maximize W.

9. Assume the following *limited* set of energy levels with degeneracies as given are available to a set of N localized (i.e., distinguishable) particles.

$$\underline{\quad}\ \underline{\quad}\ \underline{\quad}\quad \varepsilon_3 = 2a,\, g_3 = 3$$

$$\underline{\quad}\ \underline{\quad}\quad \varepsilon_2 = a,\, g_2 = 2$$

$$\underline{\quad}\quad\quad \varepsilon_1 = 0,\, g_1 = 1$$

a. Given that the average energy per particle is $\frac{2}{3}a$, assuming a Boltzmann distribution write down four independent equations for the four unknowns

$$w = e^{a/kT}, \quad x = \frac{\overline{N_1}}{N}, \quad y = \frac{\overline{N_2}}{N}, \quad z = \frac{\overline{N_3}}{N}$$

b. Solve these equations by expressing all variables in terms of one (say z) and obtaining its numerical value.

c. If $N_A a = 5.4 \text{ kJ·mol}^{-1}$, find the temperature T. (Usually we would be given T and asked to solve for the average energy per particle. In this case we have the reverse problem.)

d. Determine x, y, z if T is infinite. The average energy per particle will now be greater than $\frac{2}{3}a$. What will it be? Interpret your result in terms of the principle that at infinite temperature all states become equally occupied.

10. By making use of the definition of the factorial, show explicitly that if $g_j \gg N_j$ both Eq. (1-89) and Eq. (1-90) become identical to Eq. (1-85).

11. Calculate the molar translational entropy at 100°C and 1 atm pressure for the gases He, Ar, UF_6.

12. Explain in physical terms why at constant temperature and pressure the molar translational entropy is larger the larger is the mass per molecule.

13. For He gas demonstrate that the Sackur-Tetrode equation for translational entropy will clearly lead to impossible results at sufficiently low temperature and sufficiently high number density. Calculate the temperature below which at 1 atm pressure the equation gives a negative entropy. Also calculate the number density above which at 300 K the equation gives a negative entropy.

14. In more general terms for classical statistics to hold, N = number of particles must be much less than $\Phi(kT) = (\pi/6)(V/h^3)(8mkT)^{3/2}$; that is,

$$(N/V) \frac{6}{\pi} \frac{h^3}{(8mkT)^{3/2}} \ll 1$$

or the ratio

$$R' \equiv \frac{\rho'}{M^{5/2} T^{3/2}} \left(\frac{6}{\pi}\right) h^3 \frac{N_A^{5/2}}{(8k)^{3/2}} \ll 1$$

For the following systems with given ρ' (g/cm³) calculate R' (assume $P = 1$ atm). Gaseous densities may be calculated from the ideal gas law; liquids are at temperatures corresponding to their boiling point at 1 atm.

System	ρ' g/cm^3
Liquid He^4 at 4.20 K	0.1249
Gaseous He^4 at 4.2 K	
Gaseous He^4 at 300 K	
Liquid He^3 at 3.22 K	0.0568
Gaseous He^3 at 3.22 K	
Liquid H$_2$ at 20 K	0.0708
Gaseous H$_2$ at 20 K	
Liquid Ne at 27.1 K	1.207
Electron gas in Na at 300 K	(0.971)*

* ρ' is for solid Na; you need to convert to e^-
density assuming 1 "free" e^- per Na atom.

15. Repeated independent trials are called Bernoulli trials if there are only two
outcomes for each trial with fixed probabilities p for one result (A) and probability $q = (1 - p)$ for the other result (B). The most familiar example is (honest)
coin tossing with $p = q = \frac{1}{2}$. If in N trials the results are N_1 A's and $(N - N_1)$
B's, the probability for this is

$$P(N_1) = \frac{N!}{(N - N_1)!(N_1)!} \, (p)^{N_1}(1 - p)^{N - N_1}$$

since the number of ways that this outcome could be realized is the number
of ways of picking N_1 A's out of N results, which is $(N!)/(N - N_1)!(N_1)!$.
N_1 can assume any value from 0 to N. This distribution is called the binomial distribution. Refer to the binomial expansion formula to prove

$$\sum_{N_1 = 0}^{N} P(N_1) = 1$$

By differentiating this normalization condition with respect to p once and then
a second time, work out and prove the following

$$\overline{N_1} = \sum_{N_1 = 0}^{N} N_1 P(N_1) = Np$$

$$\overline{N_1^2} = \sum_{N_1 = 0}^{N} N_1^2 P(N_1) = (Np)^2 + Np(1 - p)$$

16. Apply the formulas of the preceding problem to the case of inserting $\mathcal{N}_1$ ideal gas molecules into a region of volume ω in a total volume V with $p = \omega/V$ the probability of the molecule being in ω. Obtain an expression for the relative deviation in particle number $(\overline{\delta\mathcal{N}_1^2})^{1/2}/\mathcal{N}_1$ [see Eq. (1-6)] in volume ω. What is its value for $\omega = 1.0$ mm³ in an ideal gas at STP? For what volume ω will the relative deviation be unity for a gas at STP? Repeat these calculations for a gas at $0°C$ and $P = 10^{-20}$ atm.

17. For some purposes a fourth ensemble called the isothermal-isobaric ensemble is used. This is defined according to the methods of Sec. 1-3 as being made up of many-body systems each with fixed number set $\mathbf{N}$ but which can exchange energy and volume (at fixed pressure) among themselves; that is, the replicas can have different volumes as well as different energies. Write down Δ, the partition function for this ensemble in terms of certain sums. Show that

$$-kT \ln \Delta = \mathbf{N} \cdot \boldsymbol{\mu} = G$$

18. Use Eq. (1-128) for the canonical partition function of a classical ideal gas mixture of species $a, b, \ldots$ to derive Dalton's law of partial pressures, $P = P_a + P_b + \ldots$.

19. In the grand canonical ensemble for the Boltzmann ideal gas we may write

$$\mathcal{Q}(\mu, v, T) = \sum_{N>0} z^N Q_N(V, T)$$

with $z = e^{\mu/kT}$ and $Q_N(V, T) = q_{trans}^N/N!$ from Eq. (1-126). Use the thermodynamic identifications of the grand canonical ensemble to obtain the classical ideal gas equation of state and an expression for μ. Note the exact summation formula

$$e^x = \sum_{i=0}^{\infty} \frac{(x)^i}{(i)!}$$

20. Use the definition, Eq. (1-139), of average in the grand canonical ensemble to prove

$$\bar{U} = -\left(\frac{\partial}{\partial\beta}\right)_z \ln \mathcal{Q}$$

21. Since $\ln \mathcal{Q} = PV/kT$, make use of the result of Problem 20 and Eqs. (1-160) and (1-173) to show that for both ideal bosons and ideal fermions $\bar{U} = \frac{3}{2} PV$. This relation also holds for the classical ideal gas.

22. Using the result of Problem 21 and other equations from the text, obtain expressions for the entropy of translation for ideal bosons and ideal fermions in the (small z) weakly degenerate case. Show that the equations read

$$\frac{S}{Nk} = \left(\frac{S}{Nk}\right)_{c\ell} \pm \frac{\lambda^3 \rho}{2^{7/2}} + \cdots$$

with fermions having the higher entropy and the leading term being the Sackur-Tetrode entropy.

23. From the equations given in the text, calculate $z = e^{\mu/kT}$ as a function of $\lambda^3 \rho$ for ideal bosons and ideal fermions in the range $\lambda^3 \rho = 0$ to 1.4. Sketch your results on a graph. For what value of $\lambda^3 \rho$ does μ pass through zero (z become greater than 1) for fermions?

24. At 100°C $\bar{c}_p$ for phosphorous vapor is $0.582 \text{ J} \cdot \text{K}^{-1} \cdot \text{g}^{-1}$. Show that this result means the vapor cannot be monatomic. If the actual molecules are P_4 (tetrahedral), calculate the actual vibrational contribution to the $(C_P)_m$ of this vapor at 100°C.

25. From experiment on NH_3 gas at 25°C, $(C_V)_m/R = 3.29$. What should be its value on the basis of the classical equipartition of energy principle?

26. If an ideal gas is flowing in a tube parallel to the x axis with a constant average flow velocity v_{0_x}, how will the distribution of molecular velocities v_x differ from that of a stationary ideal gas at the same temperature? Sketch both these distributions on one diagram (see Fig. 1-7).

27. Change Eq. (1-193) for velocity component distribution into a one-dimensional speed distribution. Use $c_1 = $ 1D speed, $c_1 = |v_x|$, $dv_x = 2dc_1$, since c_1 can only vary from 0 to ∞ whereas v_x can vary 0 to ∞ and 0 to $-\infty$ Check the normalization of the distribution and calculate $\bar{c}_1$ and $\bar{c}_1^2$. If we define $c_{1\rightarrow}$ as speed in one direction only along an axis, what is the value of $\bar{c}_{1\rightarrow}$?

28. Calculate the number of molecules in 1.0 mole of He gas at 1000 K that have speeds between 816 m/s and 2.85×10^3 m/s. Hint: Use the methods of Example 1-4.

29. Derive an equation for the average time it takes for a molecule in a classical fluid to move a distance L. Hint: Use time $= L/c$)

30. Derive an expression for the ratio of the number of molecules in an infinitesimal range of speeds at sc_p to those in such a range of speeds at c_p. This is a simpler problem than that treated in Example 1-4. Evaluate this ratio for $s = 2, 3, 4$.

31. Calculate $[\overline{(c - \bar{c})^2}]^{1/2}$ for molecules. Comment on its dependence on mass and temperature.

32. Reaction rates for many reactions at room temperature increase roughly by a factor of two for only a 10°C rise in temperature. This is related to doubling the fraction of the molecules that have high enough kinetic energy (speeds) to react. If the molecules at room temperature must have speeds of at least sc_p

(where c_p is the most probable speed) in order to react, use the methods of Example 1-4 to estimate s, which you may assume is larger than 2. Assume that the room temperature is 25°C, the higher temperature is 35°C, and note that $sc_p = s'c_p'$, where primes refer to the higher temperature. What is the additional quantity of kinetic energy per mole carried by molecules with speeds sc_p over that carried by molecules with speeds c_p at 25°C?

33. Show that in a classical fluid the distribution function for kinetic energy (in three dimensions) for translation of the center of mass is

$$f(\varepsilon) = \frac{1}{N}\frac{dN}{d\varepsilon} = \frac{2}{\sqrt{\pi}}\frac{\varepsilon^{1/2}}{(kT)^{3/2}}\exp\left(-\varepsilon/kT\right)$$

and use it to calculate $\bar{\varepsilon}$ and the most probable kinetic energy, ε_p.

34. Suppose that at some initial time all the molecules in a container have the *same* kinetic energy of translation, ε_0 per molecule. After a long enough time the energies will be distributed in a Maxwellian way.
a. Express the equilibrium temperature in terms of ε_0.
b. What fraction of the molecules will at equilibrium have their kinetic energies in the range 0.999 to $1.001\varepsilon_0$? You need not integrate to answer this—use a differential approximation.

35. Neglecting the curvature of a large planetary body's surface, the potential energy of a molecule in the gravitational field of the body is mgz where z is taken perpendicular to the surface and g is the gravitational acceleration. From Eq. (1-186), using the field term of Eq. (1-184) as $\varepsilon_f = mgz$, obtain the distribution function for the z components of molecular position assuming constant T and g. Use this function to obtain an expression for the average z of molecules in a cube of side l resting on the surface of the body. By expanding

$$e^x = 1 - x + \frac{x^2}{2} - \frac{x^3}{6} + \cdots$$

show

$$\bar{z} = \frac{l}{2}\left(1 - \frac{1}{6}al\right) \qquad \text{where } a = \frac{mg}{kT}$$

What is the value of the term multiplying $l/2$ for N_2 molecules at 298 K in a cube of side 1 m on the earth ($g = 9.81 \text{ m·s}^{-2}$)?

36. A tube 80 cm long, supported on a vertical axis midway between its ends, is rotated in a horizontal plane at a rate of 3000 revolutions per minute (rpm) in surroundings at 25°C. The tube is charged with an equimolar mixture of H_2 and O_2 gases. Obtain the distribution law for the molecules in the centrifugal force field (force $= m\omega^2 r$, potential energy $= -\frac{1}{2}m\omega^2 r^2$, with ω the angular velocity in units of radians per second, each revolution being 2π radians, and r the distance of molecules from the origin on the axis of rotation). Use the law to calculate the ratio of the concentration at the ends of the tube to that on the axis for each of the gases, assuming equilibrium is established. Assuming that at the center of the tube the gaseous mixture is kept equimolar and at a total pressure of 1.0 atm (for example by passing a current of the mixture vertically through the axis of rotation that has openings into the tube arms), compute the mole fractions of H_2 and O_2 at the ends of the tube (40 cm distant from the axis of rotation). Repeat all calculations for a rotation rate of 30,000 rpm.

*Note added for 1998 Dover edition.

Equation (1-84) only applies to a *one* level classical distinguishable particle case. In the classical counting, since interchanging particles in different levels leads to a new state of the system as a whole, we must consider all $\mathcal{N}$ particles together. Hence we do not use a product of $(W_j)_{cl}$ given by Eq. (1-84), but we have at once for the total number of ways classically to arrange the distribution α

$$[W_\alpha(\mathcal{N}_j)]_{cl} = \left[\frac{\mathcal{N}!}{\pi_j(\mathcal{N}_j)!} \right] [\pi_j(g_j)^{N_j}]$$

where the first bracket is the number of ways to put a total of $\mathcal{N}$ particles into levels j with $\mathcal{N}_j$ in the jth and the second bracket arises since in the jth level the first and every succeeding particle can be put into g_j states there are $(g_j)^{N_j}$ ways for $\mathcal{N}_j$ particles to occupy the states of level j. Then only one overall division by $\mathcal{N}!$ is needed to obtain the correct Boltzmann counting formula that is the repeated product of the result in Eq. (1-85).

2

Elementary Statistical Thermodynamics

In this chapter we will calculate, from atomic and molecular spectroscopic data, the thermodynamic properties of substances in the classical ideal gas state, including equilibrium constants for reactions of ideal gases. In addition we will obtain the statistical thermodynamics of electromagnetic radiation in an evacuated volume, the so-called blackbody radiation, by treating this radiation as an ideal boson gas of photons.

2-1
General Principles: Internal Energy and Its Separability; Energy Fluctuations in the Canonical Ensemble

We shall use for our calculations only the canonical partition function $Q_N(V, T)$ for the classical ideal gas, which can be simply expressed in terms of the single-molecule partition function q

$$Q_N(V, T) = \frac{[q(V, T)]^N}{N!} \qquad (2\text{-}1)$$

in which q from Chap. 1 is

$$q = \sum_j g_j e^{-(\beta E_j)} \tag{2-2}$$

with $\beta = 1/kT$. Now we recognize explicitly that as long as Boltzmann statistics holds we may *exactly* separate the E_j into a sum over translational and internal energy contributions. Then the partition function will be a product of separate sums over translational and internal energy states:

$$q(V, T) = q_{trans}(V, T) q_{int}(T) \tag{2-3}$$

with

$$q_{trans}(V, T) = \frac{V}{\lambda^3} \tag{2-4}$$

and q_{int}, as we shall see, is a function only of temperature. This leads to the usual ideal gas equation of state irrespective of the details of the internal energy levels.

Exercise 2-1

Show from Eq. (2-3) and $P = -(\partial A/\partial V)_{T,N}$ that the ideal gas law results no matter what the form for $q_{int}(T)$ is.

The factor $1/N!$ in Eq. (2-1) is taken with q_{trans}, and the translational results have been worked out in Chap. 1. From our thermodynamic identifications the additive contributions of the internal energy levels are as follows, beginning with the Helmholtz free energy, A:

$$A = -kT \ln Q_N(V, T)$$

$$= -kT \ln \frac{[q_{trans}(V, T)]^N}{N!} + A_{int} \tag{2-5.1}$$

with additive term

$$A_{int} = -NkT \ln q_{int}(T) \tag{2-5.2}$$

Then for the energy, entropy, constant volume heat capacity, and chemical potential the additive terms are

$$U_{int} = -T^2 \left(\frac{\partial}{\partial T}\right)_{V,N} \left(\frac{A_{int}}{T}\right) = +\frac{NkT^2}{q_{int}}\frac{\partial}{\partial T}(q_{int}) \tag{2-6}$$

$$S_{int} = \frac{U_{int} - A_{int}}{T} = \frac{U_{int}}{T} + Nk \ln q_{int}(T) \tag{2-7}$$

$$(C_V)_{int} = \left(\frac{\partial U_{int}}{\partial T}\right)_V = \frac{2NkT(\partial/\partial T)q_{int}}{q_{int}} + \frac{NkT^2(\partial^2/\partial T^2)q_{int}}{q_{int}} - \frac{NkT^2[(\partial/\partial T)q_{int}]^2}{q_{int}^2} \tag{2-8}$$

$$\mu_{int} = \left(\frac{\partial A_{int}}{\partial N}\right)_{V,T} = -kT \ln q_{int}(T) \tag{2-9}$$

For cases in which we do not have a simple closed form for q_{int}, as will occur when we have only a table of experimental energy level values, it is more convenient to rewrite some of the above equations in terms of the sums

$$q'_{int} = \sum_j g_j(\beta E_j)e^{-\beta E_j} = T\left(\frac{\partial q_{int}}{\partial T}\right) \tag{2-10}$$

and

$$q''_{int} = \sum_j g_j(\beta E_j)^2 e^{-\beta E_j} \tag{2-11}$$

to obtain

$$U_{int} = NkT\frac{q'_{int}}{q_{int}} \tag{2-12}$$

$$(C_V)_{int} = Nk\left[\frac{q''_{int}}{q_{int}} - \left(\frac{q'_{int}}{q_{int}}\right)^2\right] \tag{2-13}$$

Exercise 2-2

Derive Eqs. (2-12) and (2-13) from Eqs. (2-6) and (2-8).

We can to good approximation write the internal energy of a molecule as a sum of rotational, vibrational, electronic, and nuclear contributions.

$$E_{int} \simeq E_{rot} + E_{vib} + E_{el} + E_{nu} \tag{2-14}$$

such that

$$q_{int} \cong q_{rot} q_{vib} q_{el} q_{nu} \tag{2-15}$$

The separation of the vibrational and rotational energies is accurate. It only neglects a coupling of these modes of motion in which the moment of inertia of a molecule changes slightly as the molecule vibrates. Certain quantum symmetry restrictions under some circumstances prevent the separation of the rotational and nuclear energy states. This is discussed in Appendix 2.A. If, as only rarely occurs in practice, one or two electronic excited states are sufficiently close to the electronic ground state to have a significant effect on the thermodynamic properties, then the rotational and vibrational constants of these states will differ from that of the ground state and we must write

$$\begin{aligned} q_{int} = q_{nu}[g_{e_0}(q_{rot})_0(q_{vib})_0 \exp{(-\beta E_{e_0})} \\ + g_{e_1}(q_{rot})_1(q_{vib})_1 \exp{(-\beta E_{e_1})} + \ldots] \end{aligned} \tag{2-16}$$

We will illustrate how the distribution of molecules for truly separable or uncoupled energy sets works by considering an energy level dependent on two quantum numbers i, j associated with energy sets a and b, respectively. By the Boltzmann distribution, the number of particles $\mathcal{N}_{ij}$ in this level is given by

$$\frac{\mathcal{N}_{ij}}{\mathcal{N}} = \frac{g_i g_j e^{-\beta E_{ij}}}{q_a q_b} \tag{2-17}$$

with

$$E_{ij} = E_{ia} + E_{jb}$$

$$q_a = \sum_i g_i e^{-\beta E_{ia}}$$

$$q_b = \sum_j g_j e^{-\beta E_{jb}}$$

By summing over all j, we find for the fraction of molecules in the ith level for the a set, irrespective of the b set (or any other separable set):

$$\sum_j \frac{\mathcal{N}_{ij}}{\mathcal{N}} = \frac{g_i e^{-\beta E_{ia}} \sum_j g_j e^{-\beta E_{jb}}}{q_a q_b} = \frac{g_i e^{-\beta E_{ia}}}{q_a} \tag{2-18}$$

The result in Eq. (2-18) is just the usual Boltzmann result for the a set alone. Hence if separability holds, the probability that a molecule will be in a certain energy level with respect to one degree of freedom is entirely independent of the energy it has in other degrees of freedom.

It is instructive to derive a relation between the heat capacity C_V and the mean square fluctuation of the energy in the canonical ensemble. From Eq. (1-6) the mean square fluctuation in the energy is

$$\overline{(\delta U)^2} = \overline{U^2} - (\bar{U})^2 \tag{2-19}$$

From Eqs. (1-60), (1-61), and (1-62) the probability that a member of the canonical ensemble is in energy state U_K is

$$P_K = \frac{e^{-\beta U_K}}{Q_N(V,\,T)} \tag{2-20}$$

such that

$$\bar{U} \equiv \text{the thermodynamic energy}$$

$$= \sum_K P_K U_K = \frac{1}{Q_N(V,\,T)} \sum_K U_K e^{-\beta U_K} \tag{2-21.1}$$

$$\overline{U^2} = \sum_K P_K U_K^2 = \frac{1}{Q_N(V,\,T)} \sum_K U_K^2 e^{-\beta U_K} \tag{2-21.2}$$

We may rewrite Eq. (2-21.2) as follows

$$\frac{1}{Q_N(V,\,T)} \sum_K U_K^2 e^{-\beta U_K} = -\frac{1}{Q_N(V,\,T)} \left(\frac{\partial}{\partial \beta}\right)_V \sum_K U_K e^{-\beta U_K}$$

$$= \frac{-1}{Q_N(V,\,T)} \left(\frac{\partial}{\partial \beta}\right)_V [Q_N(V,\,T)\bar{U}] \qquad \text{using Eq. (2-21.1)}$$

$$= -\left(\frac{\partial \bar{U}}{\partial \beta}\right)_V - \bar{U}\left(\frac{\partial}{\partial \beta}\right)_V \ln Q_N(V,\,T)$$

$$= kT^2 \left(\frac{\partial \bar{U}}{\partial T}\right)_V + (\bar{U})^2 \tag{2-22}$$

The last line of Eq. (2-22) follows from

$$\frac{\partial}{\partial \beta} = \left(\frac{\partial \beta}{\partial T}\right)^{-1} \left(\frac{\partial}{\partial T}\right) = -kT^2 \frac{\partial}{\partial T}$$

and use of the thermodynamic relation of Eq. (2-6) in Eq. (2-5.1). Since $C_V = (\partial \bar{U}/\partial T)_V$, substituting Eq. (2-22) into Eq. (2-19) gives us the result

$$\overline{(\delta U)^2} = k T^2 C_V. \tag{2-23.1}$$

It is usual to compare the square root of the mean square fluctuation to the average of the quantity itself (here $\bar{U}$):

$$\frac{[\overline{(\delta U)^2}]^{1/2}}{\bar{U}} = \frac{(k T^2 C_V)^{1/2}}{\bar{U}} \tag{2-23.2}$$

To get an order of magnitude estimate of the ratio we can use the values of $\bar{U}$ and C_V for an ideal monatomic gas, which are $\frac{3}{2} NkT$ and $\frac{3}{2} Nk$, respectively. Hence the ratio goes as $1/\sqrt{N}$ and the relative deviations from the mean will be extremely small for typical macroscopic systems of large N.

Although we treat the classical ideal gas in this chapter, quantum mechanics will provide us with the energy level systems and their degeneracies for use in computing most of the partition functions. If the spacing of the energy levels is small compared to kT we may expect the quantum mechanical results to approach those of classical physics, using the phase space integration method introduced in Sec. 8 of Chap. 1. In Table 2-1 we give a short list of temperatures for which kT is equal to $\Delta\varepsilon$ of various sizes, using units common to experimental spectroscopy:

TABLE 2-1   Temperature at which given $\Delta\varepsilon = kT$

$\Delta\varepsilon$	T/K	λ^*/cm
1 cm^{-1}	1.44	1.000
100 cm^{-1}	143.9	0.010
207.2 cm^{-1}	298.2	4.83×10^{-3}
0.0257 eV	298.2	4.83×10^{-3}
500 cm^{-1}	719.4	2.00×10^{-3}
1000 cm^{-1}	1439	1.00×10^{-3}
0.50 eV	5802	2.48×10^{-4}
1.0 eV	11,604	1.24×10^{-4}
5.0 eV	58,020	2.48×10^{-5}

* Wavelength $\lambda = hc/\Delta\varepsilon$

$$1 \text{ cm}^{-1} = \frac{hc}{1 \text{ cm}} = 1.9865 \times 10^{-23} \text{ joule (J)}$$

$$1 \text{ electron volt (eV)} = 1.6022 \times 10^{-19} \text{ J}$$

2-2
Monatomic Gases

Atoms have no rotational or vibrational degrees of freedom and thus we have

$$q = q_{trans}q_{el}q_{nu} \tag{2-24}$$

with

$$q_{el} = g_{e_0}e^{-\beta E_{e_0}}\left[1 + \frac{g_{e_1}}{g_{e_0}}\exp\left[-\beta(E_{e_1} - E_{e_0})\right] + \frac{g_{e_2}}{g_{e_0}}\exp\left[-\beta(E_{e_2} - E_{e_0})\right] + \ldots\right] \tag{2-25}$$

The usual convention is to take the zero of energy for each atom in its ground electronic, nuclear, and translational states. Thus for atoms $E_{e_0} = 0$, and since the separation of electronic states is generally of the order of an electron volt or more, all higher terms in Eq. (2-25) are negligible at ordinary temperatures (see Table 2-1) and we can put

$$q_{el} \cong g_{e_0} \tag{2-26}$$

The ground state degeneracy g_{e_0} is $2J_0 + 1$, where J_0 is the total angular momentum quantum number for the lowest term of the ground state electron configuration of the atom. Some exceptions to this rule are oxygen atoms (see Example 2-1) and halogen atoms, which have two atomic states, $^2P_{3/2}$ and $^2P_{1/2}$ separated by a rather small energy $\Delta = 0.050, 0.11, 0.46,$ and 0.94 eV for F, Cl, Br, and I, respectively. Thus for halogen atoms

$$q_{el} = 4(1 + \tfrac{1}{2}e^{-\beta\Delta})$$

Since nuclear energy levels are separated by millions of electron volts, only the ground nuclear level is ever of significance and we have

$$q_{nu} = g_{n_0} = 2I + 1 \tag{2-27}$$

where I is the nuclear spin quantum number of the ground state of the nucleus of the atom. By convention one usually omits this q_{nu} factor (i.e., sets it equal to unity even if I is nonzero) since the nuclear state is not altered in any chemical process and does not contribute to any thermodynamic changes.

Example 2-1

The following atomic state energy data are given for the oxygen atom

Electron Configuration	Term	g	Energy cm^{-1}
$1s^2 2s^2 2p^4$	$^3P_2$	5	0
	$^3P_1$	3	158.5
	$^3P_0$	1	226.5
	$^1D_2$	5	15,867.7
	$^1S_0$	1	33,792.4

Compute q_{el}, U_{el}, and $(C_V)_{el}$ at $T = 0$, 100, 300, 1000, 10,000 K for a gas of oxygen atoms.

Solution:

We use Eqs. (2-10) through (2-13), which on conversion of units, for example:

$$\frac{158.5 \text{ cm}^{-1}}{k} = \frac{(158.5 \text{ cm}^{-1})(1.9865 \times 10^{-23} \text{ J} \cdot \text{cm}^{-1})}{(1.38066 \times 10^{-23} \text{ J/K})} = 228.05 \text{ K}$$

read

$$q_{el} = 5 + 3e^{-228.05/T} + e^{-325.89/T} + 5e^{-22,831/T} + e^{-48,621/T}$$

$$q'_{el} = 3\left(\frac{228.05}{T}\right)e^{-228.05/T} + \left(\frac{325.89}{T}\right)e^{-325.89/T}$$

$$+ 5\left(\frac{22,831}{T}\right)e^{-22,831/T} + \left(\frac{48,621}{T}\right)e^{-48,621/T}$$

$$q''_{el} = 3\left(\frac{228.05}{T}\right)^2 e^{-228.05/T} + \left(\frac{325.89}{T}\right)^2 e^{-325.89/T}$$

$$+ 5\left(\frac{22,831}{T}\right)^2 e^{-22,831/T} + \left(\frac{48,621}{T}\right)^2 e^{-48,621/T}$$

$$\frac{U_{el}}{Nk} = T\frac{q'_{el}}{q_{el}}; \qquad \frac{(C_V)_{el}}{Nk} = \frac{q''_{el}}{q_e} - \left(\frac{q'_{el}}{q_e}\right)^2$$

Note that by the calculus

$$\lim_{X \to \infty} Xe^{-X} = \lim_{X \to \infty} X^2 e^{-X} = 0$$

Results are:

T/K	q_{el}	q'_{el}	q''_{el}	$\dfrac{U_{el}}{Nk} \Big/ K$	$\dfrac{(C_V)_{el}}{Nk}$
0	5.0	0	0	0	0
100	5.3451	0.82467	2.0032	15.43	0.3510
300	6.7402	1.4329	1.2088	64.77	0.1341
1000	8.1101	0.77990	0.20087	96.16	0.0155
10,000	9.4179	1.3000	2.8430	1380.	0.2828

NOTE: The qualitative aspects of these results should have been predictable from a simple examination of the given data. The q_{el}, which counts electronic states (not levels) available between zero and kT, rises from 5 to 9 as kT becomes large compared to the spacing of the lower set of levels. It will reach a value of 15 at temperatures well above 50,000 K. The U_{el} always increases with temperature but $(C_V)_{el}$ will exhibit maxima in regions where U_{el} changes most rapidly. For U_{el} changing very slowly (in this case between 1000 and 5000 K) the $(C_V)_{el}$ will be very small.

For monatomic gases most thermodynamic properties (of general symbol f) are given by

$$f = f_{trans} + f_{el} \tag{2-28}$$

with f_{el} given by Eqs. (2-5) to (2-9) using $q_{int} = q_{el}$, and q_{el} given by Eq. (2-25) or (2-26) as appropriate for the given temperature.

2-3
Diatomic Gases

For diatomic molecules the internal partition function is given by

$$q_{int} = q_{rot} q_{vib} q_{el} q_{nu} \tag{2-29}$$

We will discuss each in turn, beginning with the nuclear partition function, which is merely the product of the ground state nuclear degeneracies for the two atoms:

$$q_{nu} = (2I_1 + 1)(2I_2 + 1) \tag{2-30}$$

By convention this is generally omitted from the q_{int} expression. (An exception to this is discussed in Appendix 2.A.)

If $-D_e$ represents the minimum of the potential energy for the stable molecular ground state relative to the separated atoms in their ground electronic states, which we take as our energy zero, then E_{e_0}, the ground electronic state energy, is $-D_e$. Thus the electronic partition function takes the form

$$q_{el} = g_{e_0}e^{+D_e\beta} + g_{e_1}e^{-\beta E_1} \tag{2-31}$$

and generally the energy separation $(E_1 + D_e)$ of these levels is so large that only the first term needs to be kept.

For diatomic molecules there is only one normal mode of vibration, which if treated as a harmonic oscillator in quantum mechanics has a set of nondegenerate quantized states given by

$$\varepsilon_n = (n + \tfrac{1}{2})h\nu \qquad n = 0, 1, 2, \ldots \tag{2-32}$$

with

$$\nu = \frac{1}{2\pi}\sqrt{k/\mu} \tag{2-33}$$

ν being the frequency of the mode, k the force constant, and μ the reduced mass:

$$\mu = \frac{m_1 m_2}{m_1 + m_2} \tag{2-34}$$

The ε_n are measured relative to $-D_e$ as their zero. These energy conventions are explicitly diagramed in Fig. 2-1. The quantity $D_0 = D_e - \tfrac{1}{2}h\nu$, which is the energy difference between the lowest vibrational state and the dissociated (into ground state atoms) molecule, can generally be obtained from spectroscopic data. The vibrational partition function is then by exact summation

$$q_{vib} = \sum_{n=0}^{\infty} e^{-\beta\varepsilon_n} = e^{-\beta h\nu/2} \sum_{n=0}^{\infty} (e^{-\beta h\nu})^n$$

$$= \frac{e^{-\beta h\nu/2}}{1 - e^{-\beta h\nu}} \tag{2-35}$$

This expression is usually written in terms of a characteristic vibrational temperature θ_v:

$$\theta_v = \frac{h\nu}{k} = \frac{hc}{k}\bar{\nu} = (1.4388 \text{ K—cm})\bar{\nu} \tag{2-36}$$

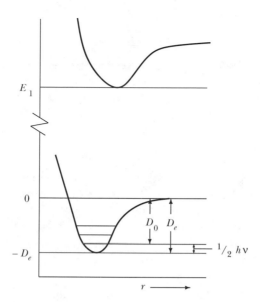

FIGURE 2-1 The ground and first excited electronic states for a diatomic molecule as function of internuclear separation r, showing the quantities D_e, D_0, and E_1.

with $\bar{\nu}$ the wave number in cm^{-1} from spectroscopy to give:

$$q_{vib} = \frac{e^{-\theta_v/2T}}{1 - e^{-\theta_v/T}} \tag{2-37}$$

From Eqs. (2-6), (2-7), and (2-8) we find the following additive vibrational contributions to U, S, and C_V:

$$U_{vib} = Nk\left(\frac{\theta_v}{2} + \frac{\theta_v}{e^{\theta_v/T} - 1}\right) \tag{2-38}$$

$$\frac{S_{vib}}{Nk} = \frac{\theta_v/T}{e^{\theta_v/T} - 1} - \ln\left(1 - e^{-\theta_v/T}\right) \tag{2-39}$$

$$\frac{(C_V)_{vib}}{Nk} = \left(\frac{\theta_v}{T}\right)^2 \frac{e^{\theta_v/T}}{(e^{\theta_v/T} - 1)^2} \tag{2-40}$$

Exercise 2-3

Derive Eqs. (2-38), (2-39), and (2-40).

Exercise 2-4

Show that $U_{vib} \to NkT$ and $(C_V)_{vib} \to Nk$ in accordance with the classical principle of equipartition of energy for an oscillator at high temperature, $(\theta_v/T) \ll 1$.
Use $e^X = 1 + X + (X^2/2) + \dots$ for $X \ll 1$. Also use $1/(1+X) = 1 - X$ for $X \ll 1$.

In Fig. 2-2 $(C_V)_{vib}/Nk$ is plotted against T/θ_v and is seen to approach its classical value asymptotically for $T > \theta_v$. An interesting quantity is the fraction N_n/N of the molecules in the nth vibrational state:

$$\frac{N_n}{N} = \frac{e^{-(n+1/2)\theta_v/T}}{q_{vib}} = e^{-n\theta_v/T}\left(1 - e^{-\theta_v/T}\right) \tag{2-41}$$

This is plotted in Fig. 2-3 for several θ_v/T values. Room temperature is typically (see Table 2-2) $\theta_v/T \approx 10$. For such a ratio essentially all the molecules are in the ground vibrational state and make only a very small contribution to the total heat capacity; that is, the vibration is "frozen" out.

For a rigid rotator we have from quantum mechanics that the levels have a degeneracy of $g_J = 2J + 1$ and allowed energies

$$E_J = J(J + 1)hBc \qquad J = 0, 1, 2, \dots \tag{2-42}$$

in which B, in cm^{-1}, is

$$B = \frac{h}{8\pi^2 cI}$$

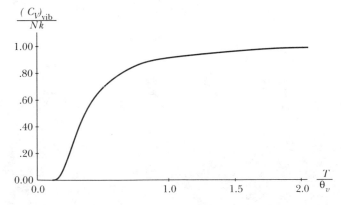

FIGURE 2-2 The vibrational contribution to C_V for one vibrational mode as a function of T/θ_v.

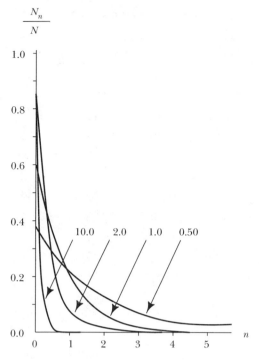

FIGURE 2-3 The fraction of the molecules N_n/N in the nth vibrational state as a function of n for several θ_v/T values.

with I, the one moment of inertia, given in terms of the reduced mass μ and equilibrium internuclear separation r_e as

$$I = \mu r_e^2$$

In terms of a characteristic rotational temperature

$$\theta_r = \frac{hc}{k} B = \frac{h^2}{8\pi^2 Ik} \tag{2-43}$$

the rotational partition function is

$$q_{rot} = \frac{1}{\sigma} \sum_{J=0}^{\infty} (2J + 1) \exp\left[-J(J + 1)\theta_r/T\right] \tag{2-44}$$

In Eq. (2-44) we have included a division by the symmetry number σ, which is 1 for heteronuclear molecules and 2 for homonuclear molecules. The reason for this is a fascinating quantum mechanical symmetry restriction on the number of allowed rotational levels when the nuclei on the ends of the diatomic molecule are identical, such that interchange of them is a symmetry operation for the molecule. This is discussed further in Appendix 2.A.

TABLE 2-2 Spectroscopic Constants for Some Diatomic Molecules

Molecule	D_0/eV	$\bar{v}/\text{cm}^{-1}$	B/cm^{-1}	θ_v/K	θ_r/K
H_2	4.4780	4400.39	60.864	6331	87.57
HD	4.5137	3811.92	45.638	5485	65.66
D_2	4.5562	3118.46	30.442	4487	43.80
N_2	9.760	2358.07	1.9987	3393	2.876
O_2	5.080	1556.30	1.4457	2239	2.080
CO	11.091	2169.82	1.9313	3122	2.779
NO	6.507	1904.12	1.7048	2740	2.453
HF	5.859	4138.32	20.956	5954	30.15
$H^{35}Cl$	4.433	2990.95	10.5934	4303	15.24
HBr	3.759	2648.98	8.4649	3811	12.18
HI	3.057	2308.09	6.5108	3321	9.37
F_2	1.60	891.8	0.8828	1283	1.270
$^{35}Cl_2$	2.4795	559.71	0.2441	805.3	0.351
$^{79}Br\,^{81}Br$	1.9704	323.3	0.08110	465.2	0.117
I_2	1.5424	214.52	0.03739	308.7	0.0538

SOURCE: *American Institute of Physics Handbook* 3d ed. McGraw-Hill, New York, 1972.

The sum in Eq. (2-44) cannot be done exactly, but for $T < 1.4\theta_r$ the first four terms

$$q_{rot} = \frac{1}{\sigma}[1 + 3e^{-2\theta_r/T} + 5e^{-6\theta_r/T} + 7e^{-12\theta_r/T}] \qquad (2\text{-}45)$$

are already sufficient to give the sum to 0.1 percent. For $T > \theta_r$ (and as Table 2-2 shows, almost any practical temperature will be far above θ_r except for diatomics with hydrogen isotopes) we may use a high-temperature approximation. This means replacement of the sum by an integral $[y = J(J + 1)]$

$$q_{rot} \cong \frac{1}{\sigma}\int_0^\infty (2J + 1)\,\exp\,[-J(J + 1)\theta_r/T]\,dJ$$

$$= \frac{1}{\sigma}\int_0^\infty \exp\,(-y\theta_r/T)\,dy = \frac{T}{\sigma\theta_r} \qquad (2\text{-}46)$$

For more precise work one can show [see ref. 4] by use of the Euler-Maclaurin sum formula (see Appendix II) that a better high-temperature approximation is

$$q_{rot} \cong \frac{T}{\sigma\theta_r}\left[1 + \frac{\theta_r}{3T} + \frac{1}{15}\left(\frac{\theta_r}{T}\right)^2 + \frac{4}{315}\left(\frac{\theta_r}{T}\right)^3 + \ldots\right] \tag{2-47}$$

Use of Eq. (2-47) gives the following additive rotational contributions to U, S, and C_V from Eqs. (2-6), (2-7), and (2-8)

$$U_{rot} = NkT\left[1 - \frac{\theta_r}{3T} - \frac{1}{45}\left(\frac{\theta_r}{T}\right)^2 + \ldots\right] \tag{2-48}$$

$$S_{rot} = Nk\ln\left(\frac{T}{\sigma\theta_r}\right) + Nk\left[1 - \frac{1}{90}\left(\frac{\theta_r}{T}\right)^2 + \ldots\right] \tag{2-49}$$

$$\frac{(C_V)_{rot}}{Nk} = 1 + \frac{1}{45}\left(\frac{\theta_r}{T}\right)^2 + \ldots \tag{2-50}$$

Usually we can neglect the terms in powers of (θ_r/T) when using these equations, and we see that they give the classical equipartition theorem results in this limit. As discussed in Problem 2-4, the simple high-temperature result in Eq. (2-46) is identical to that found by using the semiclassical partition function for a rigid rotor. This is true because the small spacing of the rotational levels with respect to kT at ordinary temperatures washes out the quantum effects and $(C_V)_{rot}$ attains in practice its classical temperature-independent form at very low temperature.

Notice that because the degeneracy always increases with J for any temperature greater than θ_r, the rotational level of maximum occupancy will have a J value greater than zero. This topic will be treated in several of the problems.

For a heteronuclear diatomic (e.g., HD) Eq. (2-45) can be used in Eq. (2-13) to calculate the rise (and slight fall) of $(C_V)_{rot}/Nk$ to its classical value. This is shown in Fig. 2-4. There is a maximum $(C_V)_{rot}/Nk = 1.10$ at $T \cong (0.80)\theta_r$. This has been

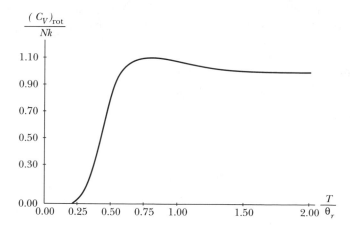

FIGURE 2-4 The rotational contribution to C_V for a heteronuclear diatomic molecule as function of T/θ_r.

observed experimentally for HD. For the homonuclear diatomics H_2 and D_2 the treatment in Appendix 2.A must be used.

Exercise 2-5

Since a rotational C_V cannot be measured directly, describe how the actual C_V data of HD should be treated to obtain experimental points to compare with Fig. 2-4.

2-4
Polyatomic Gases

For polyatomic molecules the q_{int} is given by Eq. (2-29), and by convention the nuclear partition function is set equal to unity. The electronic partition function will be given by Eq. (2-31) and in this text we shall retain only the first term. Thus we shall use

$$q_{el} = g_{e_0} e^{+\beta D_e} \tag{2-51.1}$$

$$U_{el} = -ND_e \tag{2-51.2}$$

$$(C_V)_{el} = 0 \tag{2-51.3}$$

$$S_{el} = Nk \ln g_{e_0} \tag{2-51.4}$$

in which $-D_e$ represents the depth of binding in the ground electronic state of the molecule relative to all the separated atoms of the molecule at rest in their ground electronic states. The value of D_e cannot be obtained from spectroscopy and must be found from some thermochemical data, as shown in Sec. 2-5.

For a molecule of s atoms there will be $3s - 6$ vibrational modes for a nonlinear molecule and $3s - 5$ for a linear molecule (see Sec. 1-8). Thus in the harmonic oscillator approximation the vibrational partition function becomes a product

$$q_{vib} = \prod_{i=1}^{\substack{3s-6 \\ \text{or } 3s-5}} \frac{\exp\left(-\theta_{v_i}/2T\right)}{1 - \exp\left(-\theta_{v_i}/T\right)} \tag{2-52}$$

with every factor of the form given in Eq. (2-37) and the θ_{v_i} obtained from the fundamental vibrational frequencies deduced from spectroscopy. Some experimental data are given in Table 2-3. Thus the vibrational contributions to U, S, and C_V become a sum of terms $(3s - 6)$ or $(3s - 5)$ like those in Eqs. (2-38), (2-39), and (2-40), respectively.

TABLE 2-3 Spectroscopic Constants for Some Polyatomic Molecules

Molecule	Constant, cm^{-1}			Fundamental Vibrations,* cm^{-1}
	A	B_e (linear) B (top)	C	
Linear Molecules				
CO_2	—	0.3906	—	667.3(2), 1388.3, 2349.3
N_2O	—	0.4182[†]	—	588.8(2), 1285, 2223.5
CS_2	—	0.1092[†]	—	396.7(2), 656.5, 1523
HCN	—	1.4878	—	712.1(2), 2089, 3312
C_2H_2	—	1.1838	—	611.8(2), 729.1(2), 1973.8, 3287, 3373.7
Spherical Top Molecules				
CH_4	—	5.252	—	1306.2(3), 1526(2), 2914.2, 3020.3(3)
SF_6	—	—	—	363(3), 524(3), 614(3), 644(2), 775, 965(3)
UF_6	—	0.0554	—	130(3), 200(3), 200(3), 511(2), 640(3), 656
Symmetric Top Molecules				
NH_3	6.31	9.941	—	950, 1627.5(2), 3337, 3414(2)
$^{11}BF_3$	0.178	0.355	—	480.4(2), 691.3, 888, 1445.9(2)
CH_3F	5.10	0.8496	—	1048.2, 1195.5(2), 1471.1(2), 1475.3, 2964.5, 2982.2(2)
CH_3Cl	5.10	0.49	—	731.1, 1015(2), 1354.9, 1454.6(2), 2966.2, 3041.8(2)
CH_3Br	5.08	0.31	—	611, 952(2), 1305.1, 1445.3(2), 2972, 3055.9(2)
C_6H_6	0.0955	0.192	—	400(2), 606.4(2), 671, 685, 849.7(2), 985(2), 992.5, 1010, 1016, 1035(2), 1170, 1178(2), 1298, 1485(2), 1595(2), 1693, 3048(2), 3060, 3062, 3080(2)
Asymmetric Top Molecules				
H_2O	27.33	14.58	9.50	1595, 3651.7, 3755.8
H_2S	10.393	9.040	4.723	1290, 2610.8, 2684
H_2O_2	10.056	0.859	0.791	870, 1370, 1408, 1435, 2869, 3417
C_2H_4	4.867	0.996	0.827	825, 943, 949.2, 995, 1050, 1342.4, 1443.5, 1623.3, 2989.5, 3019.3, 3105.5, 3272.3

Most data from: G. Herzberg. *Infrared and Raman Spectra of Polyatomic Molecules.* Van Nostrand, Princeton, NJ, 1962.
* Degeneracies of normal frequencies are given in parentheses following the frequencies.
† Rotational constant for ground vibrational state.

The rotational partition functions for polyatomic molecules in this text will only be given in the classical approximation, which is accurate for all molecules above 100 K (and only fails for a few, such as CH_4, below 100 K). Most polyatomic molecules are asymmetric tops (with three different principal moments of inertia) for which there is no closed form quantum mechanical result for the rotational energy levels in any case. For linear molecules the relations are the same as for diatomics

$$q_{rot} = \frac{T}{\sigma\theta_r} \tag{2-53}$$

$$\theta_r = \frac{hc}{k} B = \frac{h^2}{8\pi^2 kI} \tag{2-54}$$

with the moment of inertia now a function of two or more internuclear distances. For nonlinear molecules, integration of the classical partition function—details of which we omit [see refs. 1–5]—gives

$$q_{\text{rot}} = \frac{\sqrt{\pi}}{\sigma}\left(\frac{T^3}{\theta_A\theta_B\theta_C}\right)^{1/2} \tag{2-55}$$

in which each θ is a rotational temperature associated with the three principal moments of inertia I_A, I_B, and I_C, for which the constants from spectroscopy (in cm^{-1}) are denoted A, B, and C, respectively. The generalized symmetry number σ must be introduced to take account of certain quantum mechanical symmetry restrictions, which in a classical sense can be thought of as avoiding overcounting indistinguishable configurations achieved via rotation. One counts it as the number of different ways in which the molecule can reach indistinguishable spatial orientations by rotation about its center of mass. It depends on the point group of the molecule and is unity for completely unsymmetrical molecules. It is straightforward to determine the symmetry number; for example, $\sigma = 3$ for NH_3 and $\sigma = 12$ for CH_4.*

The rotational functions then have their classical form

	Linear	Nonlinear	
	$U_{rot} = NkT$	$= \frac{3}{2} NkT$	(2-56.1)
	$S_{rot} = Nk \ln\left(\frac{Te}{\sigma\theta_r}\right)$	$= Nk \ln \frac{\sqrt{\pi}}{\sigma}\left(\frac{T^3 e^3}{\theta_A\theta_B\theta_C}\right)^{1/2}$	(2-56.2)
	$(C_V)_{rot} = Nk$	$= \frac{3}{2} Nk$	(2-56.3)

in which e is the base of natural logarithms.

* Consider the pyramidal shape of NH_3 with three rotations of 120° reaching indistinguishable orientations; similarly, for tetrahedral CH_4 we have 3 × 4, where 4 is the number of ways the H atom that is fixed under 120° rotations can be chosen.

Details concerning more complex situations, such as anharmonic vibrations and nonseparability of rotational and vibrational motions, will be left to more advanced works [note especially references 1, 2, 6]. We shall only mention as a warning to the reader that with single-bonded hydrocarbons, of which the proto-type is ethane, one of the "vibrational" modes is a rotation of one methyl group with respect to the other. This is called an internal rotation and it will be hindered by some potential energy barrier V_0. If $kT < V_0$ then this motion will be roughly vibrationlike, whereas for $kT > V_0$ it will approach that of a free rotation. Hence the C_V contribution of this mode will first rise toward Nk with temperature increase and then at still higher temperatures it will fall off toward $\frac{1}{2}Nk$, characteristic of a free rotation. This a most striking case of the nonseparability of vibrational and rotational modes.

The methods presented here will accurately calculate thermodynamic proper-ties for many molecules. We may compare especially heat capacity and entropy results (termed spectroscopic) with experimental calorimetric data. These latter for the entropy are generally corrected to the ideal gas state, accounting for real gas imperfections by using the relation*

$$S_{id} = S_{cal} + \frac{27RT_c^3 P}{32 T^3 P_c} \tag{2-57}$$

in which T_c is the critical temperature and P and P_c are the actual gas pressure and critical pressure, respectively.

■ ──

Example 2-2

Calculate from the data in Table 2-3 for $CH_3Cl(g)$ at 298.15 K and 1 atm the molar entropy and molar constant volume heat capacity.

Solution:

From Exercise 1-10

$$\frac{(S_m)_{trans}}{R} = \frac{5}{2} \ln (298.15) + \frac{3}{2} \ln (50.488) - \ln 1 - 1.165$$

$$= 18.962$$

$$(S_m)_{trans} = 157.7 \; \text{J} \cdot \text{K}^{-1} \cdot \text{mol}^{-1}$$

─────────────────

* Note for example I. M. Klotz and R. M. Rosenberg. *Chemical Thermodynamics*, 4th ed. Benjamin/Cummings, 1986, p. 235.

From Table 2-3

$$B = 0.49 \text{ cm}^{-1} \; \theta_B = \left(\frac{hc}{k} \Big/ K \cdot cm\right) B = 0.705 \text{ K}$$

$$A = 5.10 \text{ cm}^{-1} \; \theta_A = 7.33 \text{ K}$$

For this symmetric top molecule $\sigma = 3$ and symbol B refers to the moment of inertia that occurs twice with the same value.

From Eq. (2-55)

$$q_{rot} = \frac{\sqrt{\pi}}{3}\left(\frac{T^3}{\theta_A \theta_B^2}\right)^{1/2} = 1.59 \times 10^3$$

meaning there are available some 1600 rotational states at this temperature, and from Eq. (2-56.2)

$$\frac{(S_m)_{rot}}{R} = \ln\left(q_{rot}e^{3/2}\right) = 8.87$$

$$(S_m)_{rot} = 74 \text{ J} \cdot \text{K}^{-1} \cdot \text{mol}^{-1} \qquad \text{(only two significant figures)}$$

From the data on the nine vibrational modes for this molecule, and directly calculating the harmonic oscillator thermodynamic functions from Eqs. (2-39) and (2-40), we have

$\bar{v}/\text{cm}^{-1}$	θ_v/K	$\dfrac{\theta_v}{298.15}$	$\dfrac{S_{vib}}{Nk}$	$\dfrac{(C_V)_{vib}}{Nk}$
2966	4268	14.31	0.0000	0.0001
1355	1949	6.54	0.0113	0.0618
731.1	1052	3.528	0.1367	0.3879
3042(2)	4377(2)	14.68(2)	0.0000	(0.0001)(2)
1455(2)	2093(2)	7.020(2)	(0.0073)(2)	(0.0441)(2)
1015(2)	1460(2)	4.897(2)	(0.0444)(2)	(0.1818)(2)
		Total = 0.2514		0.9018

The numbers in parentheses in the first three columns give the degeneracy of the mode and are thus multiplicative factors in the last two columns. The three modes of lowest frequency contribute 90 percent of the vibrational entropy contribution and 83 percent of the vibrational C_V contribution. Classically, $(C_V)_{vib}/Nk$ would be 9.0. The vibrational results are

$$(S_m)_{vib} = 2.1 \text{ J} \cdot \text{K}^{-1} \cdot \text{mol}^{-1}$$

$$(C_V)_{vib} = 7.50 \text{ J} \cdot \text{K}^{-1} \cdot \text{mol}^{-1}$$

Since the electronic ground state of CH_3Cl is nondegenerate, from Eq. (2-51.4) the electronic entropy contribution is zero.

The total molar entropy is thus

$$S_m = 157.7 + 74 + 2.1 = 234 \text{ J} \cdot \text{K}^{-1} \cdot \text{mol}^{-1}$$

This is in excellent agreement with the experimental entropy value [ref. 7] corrected as in Eq. (2-57) to the ideal gas state, which is $234.5 \text{ J} \cdot \text{K}^{-1} \cdot \text{mol}^{-1}$.

The translational and rotational C_V values have their classical values and thus the total $(C_V)_m$ is

$$(C_V)_m = \tfrac{3}{2}R + \tfrac{3}{2}R + 0.902R = 32.44 \text{ J} \cdot \text{K}^{-1} \cdot \text{mol}^{-1}$$

$$(C_P)_m = (C_V)_m + R = 40.76 \text{ J} \cdot \text{K}^{-1} \cdot \text{mol}^{-1}$$

This last is in precise agreement with the experimental result [7] of $40.8 \text{ J} \cdot \text{K}^{-1} \cdot \text{mol}^{-1}$.

In Table (2-4) we compare the (corrected) calorimetric and spectroscopic values for S_m of a number of gases at 298.15 K and 1 atm. Asterisks mark those cases for which there is a significant discrepancy. In all these cases $(S_m)_{spec} > (S_m)_{cal}$ and the difference is usually referred to as the residual entropy. These residual entropies can be explained in the following way. The spectroscopic results are based on an assumed entropy of zero at absolute zero. The calorimetric entropy starts

TABLE 2-4 Comparison of the Calorimetric and Spectroscopic Statistical Values of Standard Entropy at 298.15 K, 1 atm

Substance	$S^0_{m,cal}$ $\text{J} \cdot \text{mol}^{-1} \text{K}^{-1}$	$S^0_{m,spec}$	Substance	$S^0_{m,cal}$ $\text{J} \cdot \text{mol}^{-1} \text{K}^{-1}$	$S^0_{m,spec}$
Ne	146.5	146.2	H_2O*	185.3	188.7
Kr	163.9	164.0	PH_3	210.7	210.7
CO*	193.4	197.9	CH_3Cl	234.1	234.5
O_2	205.4	205.2	CH_3D*	190.2	201.2
CO_2	213.8	213.7	D_2O*	192.0	195.2
H_2S	205.4	205.7	C_6H_6	269.7	269.5
N_2O*	215.2	220.0	NO*	208.0	211.0
Cl_2	223.1	223.1	C_2H_4	219.7	219.5
N_2	191.4	191.5	CH_4	186.0	186.1

* Case with significant discrepancy.

with measurements on a real solid (via extrapolation to absolute zero), which may have a nonzero entropy frozen in; that is, W, the number of ways of arranging the molecules in the solid at absolute zero, is not equal to unity. Hence the measured entropy difference between the molecules in the disordered solid and the ideal gas state will be less (by the amount $k \ln W$) than that calculated from spectroscopic data. Reasonable estimates have been made for the $W \neq 1$ in all cases with residual entropies. The molecules of CO, for example, have a very small dipole moment and when they crystallize they tend to line up at random rather than in the most energetically favorable way. The resultant crystal is then made up of approximately equal numbers of "up" and "down" dipoles with $W = 2^N$ and a $(\Delta S)_m$ of $R \ln 2 = 5.8 \, \text{J} \cdot \text{K}^{-1} \cdot \text{mol}^{-1}$, which about matches the residual entropy of $4.5 \, \text{J} \cdot \text{K}^{-1} \cdot \text{mol}^{-1}$. A similar argument can be given for the NO molecule. Solid NO contains N_2O_2 dimers of rectangular shape with different side lengths. The two possible orientations of the rectangle for $(N/2)$ dimers from N monomers suggests $W = 2^{N/2}$, $\Delta S_m = (R/2) \ln 2 = 2.9 \, \text{J} \cdot \text{K}^{-1} \cdot \text{mol}^{-1}$. An ingenious argument by Pauling [1, 8] shows that for ice, $W \approx (\frac{3}{2})^N$.

2-5
Chemical Equilibrium in Mixtures of Reacting Ideal Gases

We consider the general homogeneous ideal gas phase chemical reaction

$$a\text{A} + b\text{B} + \ldots \leftrightarrows l\text{L} + m\text{M} + \ldots \tag{2-58}$$

at equilibrium in a closed volume V and at constant temperature T. The lowercase symbols are stoichiometric coefficients and the capital letters represent the reactants and products. The general thermodynamic condition for chemical equilibrium is

$$\sum_i i\mu_I = \sum_j j\mu_J \tag{2-59}$$

in which the sum on i is for reactants, the sum on j for products, and μ_I is the chemical potential per particle of I. Equation (2-59) holds for any kinds of species. However, we are only prepared here to write from statistical mechanics expressions for μ_I for ideal gases in a mixture in terms of their single-molecule partition functions q_I and particle numbers N_I. From Eq. (1-130) we have

$$\mu_I = -k\text{T} \ln \frac{q_I}{N_I} \tag{2-60}$$

Thus from Eq. (2-59), with subscript e denoting at equilibrium, we have

$$-kT \ln \left(\frac{q_A}{N_A}\right)^a - kT \ln \left(\frac{q_B}{N_B}\right)^b - \ldots = -kT \ln \left(\frac{q_L}{N_L}\right)^l - kT \ln \left(\frac{q_M}{N_M}\right)^m - \ldots$$

which becomes

$$\left(\frac{N_L^l N_M^m \cdots}{N_A^a N_B^b \cdots}\right)_e = \frac{q_L^l q_M^m \cdots}{q_A^a q_B^b \cdots} \tag{2-61}$$

All the q's are functions of T times a factor of the total volume V from the translational partition function factor in each. Hence we note that each q_A/V is a function of T only. Thus introducing $\rho_A = N_A/V$, the number density, we have as a function of T only:

$$\left(\frac{\rho_L^l \rho_M^m \cdots}{\rho_A^a \rho_B^b \cdots}\right)_e = K(T) = \frac{\left(\dfrac{q_L}{V}\right)^l \left(\dfrac{q_M}{V}\right)^m \cdots}{\left(\dfrac{q_A}{V}\right)^a \left(\dfrac{q_B}{V}\right)^b \cdots} \tag{2-62}$$

For ideal gases $P_A = \rho_A kT$, and we obtain the conventional thermodynamic equilibrium constant in terms of partial pressures, $K_P(T)$:

$$K_P(T) = \left(\frac{P_L^l P_M^m \cdots}{P_A^a P_B^b \cdots}\right)_e = \frac{(q_L kT/V)^l (q_M kT/V)^m \cdots}{(q_A kT/V)^a (q_B kT/V)^b \cdots} \tag{2-63}$$

Each factor $q_L kT/V$, which has dimensions of a pressure, is to be converted to units of atmospheres to accord with thermodynamic convention* by using $1.01325 \times 10^5 \text{ N/m}^2 = 1$ atm. Recall from thermodynamics that for an ideal gas we have

$$\mu_L = \mu_L^0 + kT \ln (P_L/P_L^0) \tag{2-64}$$

in which P_L^0 is identically 1 atm, P_L is measured in atmospheres, and μ_L^0 is the standard chemical potential (when $P_L = 1$ atm). Thus, using Eq. (2-60) and the ideal gas relation, $P_L V = N_L kT$; we find

$$\mu_L^0 = -kT \ln \left(\frac{q_L kT}{V P_L^0}\right) \tag{2-65}$$

reminding us again that to match the standard tabulations of thermodynamics we must convert $(q_L kT/V)$ to atmosphere units to keep the argument of the logarithm

---

* All modern thermodynamic data tabulations prior to 1982 are based on a standard state pressure, P^0, of 1 atm. However, in 1982 the International Union of Pure and Applied Chemistry (IUPAC) recommended the use of 1 bar (precisely 10^5 N/m^2) as a new standard state pressure. Some tabulations since then, and in particular some computer data base sets, have been adjusted to the 1 bar convention. With the exception of the change in the entropy of an ideal gas, the changes in the data engendered by the change of P^0 are almost always less than the experimental uncertainty in the values. For explicit details of the necessary conversion calculations, consult R. D. Freeman, *J. Chem. Ed.* **62**, 681 (1985).

dimensionless and thus give to μ_L^0 the dimensions of kT. We may now introduce the quantity

$$K_P^0 = \frac{(P_L^0)^l (P_M^0)^m \cdots}{(P_A^0)^a (P_b^0)^b \cdots} \tag{2-66}$$

which has a numerical value of 1 but dimensions equal to that of K_P, and obtain from Eq. (2-63) the well-known thermodynamic relation

$$\ln (K_P/K_P^0) = \frac{-\Delta\mu^0}{kT} \tag{2-67}$$

where

$$\Delta\mu^0 = l\mu_L^0 + m\,\mu_M^0 + \ldots - (a\mu_A^0 + b\mu_B^0 + \ldots) \tag{2-68}$$

On a molar basis we multiply the numerator and denominator of Eq. (2-67) by Avogadro's number to obtain

$$\ln\left(\frac{K_P}{K_P^0}\right) = \frac{-\Delta G^0}{RT} \tag{2-69}$$

in which ΔG^0 is the standard Gibbs free energy change for the reaction as balanced in Eq. (2-58).

Exercise 2-6

Derive Eq. (2-67) from Eq. (2-63) and the defining equations for K_P^0 and μ_L^0.

It is now a straightforward matter to calculate equilibrium constants for reactions of molecules with known D_e values. We illustrate this with an example and then discuss reactions of polyatomic molecules for which some thermochemical data are also needed.

Example 2-3

Calculate K_P for the dissociation of O_2 into O atoms at 4000 K. If the total pressure is 1.0 atm, what fraction of O_2 molecules will be dissociated? Use data from Table 2-2 and Example 2-1. Note that for O_2 the degeneracy of the ground electronic state is $g_{e_0} = 3$.

Solution:

From Eq. (1-116) for the translational particular function

$$\frac{q_{trans}}{V} = \lambda^{-3} = (1.879 \times 10^{26})(MT)^{3/2} \cdot m^{-3}$$

with M in g/mol

For O_2 molecules	For O atoms
$\dfrac{q_{trans}}{V} = 8.605 \times 10^{33} \; m^{-3}$	$\dfrac{q_{trans}}{V} = 3.042 \times 10^{33} \; m^{-3}$
$q_{rot} = \dfrac{T}{2(2.080)} = 961.5$	$q_{el} = 5 + 3e^{-0.05701} + e^{-0.08147}$
$D_0 = D_e - (\frac{1}{2})hv = 5.080 \; eV$	$+ 5e^{-5.708} + e^{-12.16}$
	$= 8.759 \; (\text{Example 2-1})$

From Eqs. (2-31) and Eq (2-35) we have

$$q_{el}q_{vib} = \frac{3e^{D_e/kT}e^{-hv/2kT}}{1 - e^{-hv/kT}}$$

$$= \frac{3e^{D_0/kT}}{(1 - e^{-\theta_v/T})}$$

with

$$D_0/k = 5.893 \times 10^4 \; K$$
$$\theta_v = 2.239 \times 10^3 \; K$$

$q_{el}q_{vib} = 1.751 \times 10^7$ This large value arises because of the way we have set our energy zero.

$kT = k(4000) = 5.523 \times 10^{-20} \; J$

Reaction is $O_2 \rightarrow 2O$

$$K_P = \frac{(q_0/V)^2(kT)}{(q_{0_2}/V)} = \frac{(3.042 \times 10^{33})^2(8.759)^2(5.523 \times 10^{-20})}{(8.605 \times 10^{33})(961.5)(1.751 \times 10^7)}$$

$$= 2.707 \times 10^5 \; N/m^2 = 2.672 \; atm$$

Using SI units, K_P will always come out as N/m^2 to a power (here unity). Conversion to atmosphere units can be done at the end.

If α is the fraction of O_2 dissociated at equilibrium at total pressure P, then for every mole O_2 at the start we have at equilibrium $(1 - \alpha)$ moles O_2 and 2α moles O, thus $(1 + \alpha)$ total moles:

$$K_P = \frac{(2\alpha P/1 + \alpha)^2}{(1 - \alpha/1 + \alpha)P} = \frac{4\alpha^2 P}{1 - \alpha^2}$$

$$\alpha^2 = \frac{K_P}{4P + K_P} = \frac{2.672}{6.672} = 0.400 \qquad \text{for } P = 1.0 \text{ atm}$$

$$\alpha = 0.633$$

The O_2 would be 63.3 percent dissociated at 4000 K and 1 atm total pressure. In dissociation reactions of this type the dimer is favored by the energy difference of the species and the monomer by the entropy increase (square of the huge q_{trans}/V factor in the numerator) and the large electronic degeneracy of the atoms in this case.

One cannot experimentally smash polyatomic molecules into their constituent atoms in their ground states to obtain unambiguous D_e values (it is hard enough to get them for diatomics). Therefore an indirect approach using some thermodynamic data is needed. We may on a molar basis write down equations for the enthalpy function,

$$H_m = U_m + PV_m = U_m + RT \tag{2-70}$$

from our preceeding derivations for U_m. By convention we shall append a superscript zero on these functions, indicating they are standard state ($P = 1$ atm) values, but the reader will recall that for ideal gases H (and U) are independent of pressure. We find, assuming that the rotational contributions are fully classical and there are no contributions from excited electronic states,

For monatomic gases:

$$H_m^0 = \frac{5}{2} RT \tag{2-71.1}$$

For diatomic gases:

$$H_m^0 = \frac{7}{2} RT - N_A D_e + R\left(\frac{\theta_v}{2} + \frac{\theta_v}{e^{\theta_v/T} - 1}\right) \tag{2-71.2}$$

For linear polyatomic:

$$H_m^0 = \frac{7}{2} RT - N_A D_e + \sum_{i=1}^{3s-5} R\left(\frac{\theta_{vi}}{2} + \frac{\theta_{vi}}{e^{\theta_{vi}/T} - 1}\right) \tag{2-71.3}$$

For nonlinear polyatomic:

$$H_m^0 = 4RT - N_A D_e + \sum_{i=1}^{3s-6} R\left(\frac{\theta_{v_i}}{2} + \frac{\theta_{v_i}}{e^{\theta_{v_i}/T} - 1}\right) \tag{2-71.4}$$

All of Eqs. (2-71) are of the form

$$H_m^0 = H_{m_0}^0 + (C_P')_m T + \sum_i \frac{R\theta_{v_i}}{e^{\theta_{v_i}/T} - 1} \tag{2-72}$$

in which

$$H_{m_0}^0 = -N_A D_e + \sum_i \frac{R}{2}\theta_{v_i} = U_{m_0}^0 \tag{2-73}$$

is the enthalpy (and energy) per mole of the molecules at absolute zero and $(C_P')_m$ is the molar constant pressure heat capacity contributed classically from the translational and rotational degrees of freedom only.

For a reaction we may write schematically

$$\Delta H_{m,rx}^0 = \Delta H_{m_0}^0 + T\Delta(C_P')_m + (RT)\Delta \sum_i \frac{\theta_{v_i}/T}{e^{\theta_{v_i}/T} - 1} \tag{2-74}$$

with all the D_e values and zero point energy terms in the $\Delta H_{m_0}^0$ term. If we denote by q_L' the single-molecule partition function for L with the terms $e^{\beta D_e} \prod_i e^{-\theta_{v_i}/2T}$ divided out, Eq. (2-63) becomes

$$K_P(T) = \frac{(q_L' kT/V)^l (q_M' kT/V)^m \cdots e^{-\Delta H_{m_0}^0/RT}}{(q_A' kT/V)^a (q_B' kT/v)^b \cdots} \tag{2-75}$$

Exercise 2-7

Verify the validity of Eqs. (2-71) and (2-75). Start from Eq. (2-70) and the expressions derived for U_m for the various molecule types.

The $\Delta H_{m_0}^0$ may be determined from Eq. (2-74) using at a particular temperature a value of $\Delta H_{m,rx}^0$ from thermochemical measurement. This is so even though different enthalpy zero references are used in the thermochemical tabulations (enthalpy of the atoms in the most stable form of the element at 298.15 K and 1 atm pressure is taken as zero) and in our statistical calculation (enthalpy of the atoms at absolute zero and 1 atm pressure in their ground states and in the state of aggregation that is the most stable form at 298.15 K) because in taking differences of enthalpy for the same number and types of atoms (in different molecules) between products and reactants it is immaterial what arbitrary zero is chosen for any atom.

■

Example 2-4

For the industrially important water-gas reaction

$$CO_2(g) + H_2(g) \rightarrow CO(g) + H_2O(g)$$

for which at 298.15 K $\Delta H^0_{m,rx} = 41.17$ kJ, the following molecular parameters are obtained from Tables 2-2 and 2-3.

Molecule	θ_v/K	θ_r/K
CO_2	1899, 3380, 960.1(2)	0.5620
H_2	6331	87.49
H_2O	5254, 2295, 5405	39.33, 20.98, 13.67
CO	3083	2.779

Calculate $\Delta H^0_{m_0}$ and obtain K_P for this reaction as a function of temperature. Calculate K_P at 300, 600, 900, and 1200 K.

Solution:

Evaluate for each molecule at 298.15 K the terms in the sums on the right-hand side of Eq. (2-74) to obtain

$$41.17 \times 10^3 = \Delta H^0_{m_0} + T(4R + 7/2R - 7/2R - 7/2R)$$
$$+ RT(0.00350 + 0.00033 - 0.27908 - 0)$$
$$= \Delta H^0_{m_0} + (0.22475)RT$$

$$\Delta H^0_{m_0} = 40.61 \times 10^3 \text{ J}, \qquad \frac{\Delta H^0_{m_0}}{R} = 4884 \text{ K}$$

The rotational partition functions are

$$H_2O_{\sigma=2} \qquad q_{rot} = \frac{\sqrt{\pi}}{2}\left(\frac{T^3}{11,280}\right)^{1/2} = (8.344 \times 10^{-3})\,T^{3/2}$$

from Eq. (2-55) and

$$CO_{\sigma=1} \qquad q_{rot} = \frac{T}{\theta_r} = (0.3599)\,T$$

$$H_{2\,\sigma=2} \qquad q_{rot} = \frac{T}{2\theta_r} = (5.715 \times 10^{-3})\,T$$

$$CO_{2\,\sigma=2} \qquad q_{rot} = (0.8997)\,T$$

$$K_P = \frac{(q'/V)_{CO}(q'/V)_{H_2O}e^{-\Delta H_{m_0}^0/RT}}{(q'/V)_{CO_2}(q'/V)_{H_2}}$$

In this example K_P is dimensionless, but in general K_P will have dimensions and the ratio of K_P to K_P^0 must always be used in the expression for ΔG^0, Eq. (2-69).

$$K_P = \frac{(28.01)^{3/2}(18.02)^{3/2}}{(44.01)^{3/2}(2.016)^{3/2}} \frac{(3.003 \times 10^{-3})\,T^{5/2}}{(5.085 \times 10^{-3})\,T^2} (e^{-4884/T})\,K_{vib}$$

$$= (8.014)\,T^{1/2}(e^{-4884/T})\,K_{vib}$$

$$K_{vib} = \frac{(1 - e^{-6331/T})(1 - e^{-960.1/T})^2(1 - e^{-1899/T})(1 - e^{-3880/T})}{(1 - e^{-3083/T})(1 - e^{-5254/T})(1 - e^{-5404/T})(1 - e^{-2295/T})}$$

T/K	K_{vib}	K_P, calc.	K_P, exper.
300	0.9185	1.08×10^{-5}	—
600	0.6252	$3.58 \times 10^{-?}$	—
900	0.4158	0.440	0.46
1200	0.2933	1.39	1.37

The calculated values are probably more accurate than the experimental ones.

■

For more systematic calculations and to compare with standard thermodynamic tables we express the standard chemical potential on a molar basis, $\mu_m^0 \equiv G_m^0$ from Eq. (2-65), as

$$\mu_m^0 = G_m^0 = -RT \ln\left(\frac{qkT}{VP^0}\right) = -RT \ln\left(\frac{q'kT}{VP^0} e^{D_e/kT} \prod_i e^{-\theta_{vi}/2T}\right)$$

$$= -RT \ln\left(\frac{q'kT}{VP^0}\right) - N_A D_e + \sum_i R\frac{\theta_{vi}}{2}$$

$$= -RT \ln\left(\frac{q'kT}{VP^0}\right) + H_{m_0}^0 \qquad (2-76)$$

Then the usually tabulated function

$$-\left(\frac{G_m^0 - H_{m0}^0}{T}\right) = R \ln\left(\frac{q'kT}{VP^0}\right) \tag{2-77}$$

is independent of any choice of zero enthalpy state. Recall that $q'kT/V$ is to be expressed in units of atmospheres and $P^0 \equiv 1$ atm. If values of the left-hand side of Eq. (2-77) are calculated or interpolated from a table, K_P for a reaction is obtained from

$$R \ln\left(\frac{K_P}{K_P^0}\right) = \Delta\left[-\left(\frac{G_m^0 - H_{m0}^0}{T}\right)\right] - \frac{\Delta H_{m0}^0}{T} \tag{2-78}$$

with ΔH_{m0}^0 calculated from Eq. (2-74).

2-6
Negative Absolute Temperatures

It is an amusing fact that since 1951 [9] negative absolute temperatures have been observed in laboratory experiments. We may get an idea of the meaning of a negative absolute temperature by examining the thermodynamic relation

$$\frac{1}{T} = \left(\frac{\partial S}{\partial U}\right)_X \tag{2-79}$$

where X stands for an extensive variable pertinent to the system under study. As long as the entropy increases with increase of system energy, the temperature is positive. The entropy goes to a maximum at infinite temperature, for which generally the energy also becomes infinite—meaning such a situation is unattainable. However, if a finite amount of energy could drive the system to infinite temperature and a maximum in the entropy, it is conceivable that with further energy input the entropy would decrease and we would have a negative temperature system. This also suggests that negative temperatures are hotter (that is, beyond infinity) than infinite temperature! This is verifiable by the following phenomenological argument. If we have an isolated pair of systems, one at a negative absolute temperature T_n and one at a positive temperature T_p, which exchange a quantity of heat Q irreversibly (since T_n and T_p differ by a finite amount), the overall ΔS must be greater than zero by the Second Law and is

$$\Delta S = \frac{Q}{T_n} + \frac{(-Q)}{T_p} > 0$$

or

$$Q\left[\frac{-1}{|T_n|} - \frac{1}{T_p}\right] > 0$$

meaning Q must be negative and it is the system at T_n that gives off heat spontaneously and is thus the hotter one.

An assembly of magnetic moments arising from localized nuclei of nonzero spin (the positions of ^{7}Li nuclei in a LiF lattice, for example) can exhibit negative absolute temperatures. The moments can interact with each other but are isolated from the rest of the universe in the sense that the spin-spin relaxation time is very short compared to any other relaxation time, in particular that of spin-lattice interaction. In practice the spin-spin relaxation time is of the order of 10^{-5} to 10^{-6} seconds while the spin-lattice relaxation time is of the order of 1 to 10 minutes. Only if we have this operative spin-spin interaction while the spin system is decoupled from other energy level systems can we discuss a spin temperature differing from the temperature of the lattice and any other energy levels. The interaction with the lattice corresponds to leakage through vacuum bottle walls in ordinary calorimetry. Assume that the nuclei have spin $\frac{1}{2}$ such that only two magnetic energy levels exist in the presence of an external magnetic field B. The more highly populated state will be that for which the spin magnetic moment is aligned with the field. At infinite temperature both states would be equally populated. If the direction of the external field B is reversed in a time interval much less than the spin-spin relaxation time, the more populated state will be the excited state and the entropy of the spins will increase as they reorient with the new field direction *and* decrease in energy. The derivative in Eq. (2-79) will be negative and we will have a negative temperature system.

We may demonstrate all these properties for a two-state case of N localized nuclei, N_1 in state one with $E_1 = 0$ and N_2 in state two with $E_2 = \varepsilon$, as follows

$$q = 1 + e^{-\varepsilon/kT} \tag{2-80.1}$$

$$Q_N = q^N \tag{2-80.2}$$

$$A = -kT \ln Q_N \tag{2-80.3}$$

$$S = -\left(\frac{\partial A}{\partial T}\right)_V = Nk \ln\left(1 + e^{-\varepsilon/kT}\right) + \left(\frac{N\varepsilon}{T}\right)\frac{e^{-\varepsilon/kT}}{1 + e^{-\varepsilon/kT}} \tag{2-80.4}$$

$$U = A + TS = \frac{N\varepsilon e^{-\varepsilon/kT}}{1 + e^{-\varepsilon/kT}} = N_2\varepsilon \tag{2-80.5}$$

where the last equality must follow since only particles in state two contribute to the energy. Therefore the mole fractions of nuclei in the two states are

$$X_2 = \frac{N_2}{N} = \frac{e^{-\varepsilon/kT}}{1 + e^{-\varepsilon/kT}} = \frac{U}{N\varepsilon} \tag{2-81}$$

$$X_1 = 1 - X_2 = \frac{1}{1 + e^{-\varepsilon/kT}} = 1 - \frac{U}{N\varepsilon} \tag{2-82}$$

Solve Eq. (2-81) explicitly for ε/kT to find

$$\varepsilon/kT = -\ln\left(\frac{U/N\varepsilon}{1 - U/N\varepsilon}\right) \tag{2-83}$$

and substitute into Eq. (2-80.4) to obtain

$$\frac{S}{Nk} = \ln\left(\frac{1}{X_1}\right) - X_2 \ln\frac{(U/N\varepsilon)}{(1 - U/N\varepsilon)} \tag{2-84.1}$$

$$= -X_1 \ln X_1 - X_2 \ln X_2 \tag{2-84.2}$$

$$= -\left[\left(1 - \frac{U}{N\varepsilon}\right)\ln\left(1 - \frac{U}{N\varepsilon}\right) + \left(\frac{U}{N\varepsilon}\right)\ln\left(\frac{U}{N\varepsilon}\right)\right] \tag{2-84.3}$$

Equation (2-84.2) is just of the form of an entropy of mixing, as we could have anticipated in this two-state case, and from Eq. (2-84.3) we obtain

$$\left(\frac{\partial S}{\partial U}\right) = \frac{1}{T} = -\frac{k}{\varepsilon} \ln\frac{(U/N\varepsilon)}{(1 - U/N\varepsilon)} = -\frac{k}{\varepsilon} \ln\left(\frac{X_2}{X_1}\right) \tag{2-85}$$

For $X_2 > X_1$, which means more nuclei in the higher energy state than in the lower state, the absolute temperature of the spin system is negative. In Fig. 2-5 S/Nk is plotted against $U/N\varepsilon$. There is a single maximum in S at $T = \pm\infty$ where $U = N\varepsilon/2$, corresponding to equal occupancy of the two states. For energies greater than this amount $X_2 > X_1$ and the entropy decreases as the energy increases, corresponding to the negative temperature region (and excess of practicles in the upper state).

Another important insight can be gained by considering the heat capacity

$$\frac{C}{Nk} = \left(\frac{\partial U/Nk}{\partial T}\right) = \left(\frac{1}{T^*}\right)^2 \frac{e^{-1/T^*}}{(1 + e^{-1/T^*})^2}$$

$$= \left(\frac{1}{T^*}\right)^2 \frac{e^{+1/T^*}}{(1 + e^{+1/T^*})^2} \tag{2-86}$$

where $T^* = kT/\varepsilon$. This heat capacity is zero at $T = \pm 0$ and also at the single point $T = \pm\infty$. This property of zero heat capacity at infinite temperature is crucial for the existence of an assembly with negative temperatures and is only realized if there is a finite upper bound to the energy spectrum (just ε in our example). If there were not a finite upper bound to the allowed energy states, no finite energy input could drive the assembly into the state of infinite temperature and beyond. As Eq. (2-86) shows, the heat capacity has the same positive value at $\pm T^*$. In Fig. 2-6

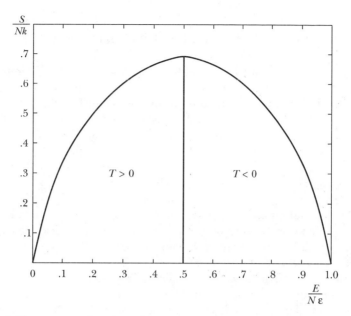

FIGURE 2-5 Entropy as a function of energy for a two-state spin system (separated by energy ε) of N nuclei. (FROM: Charles E. Hecht. *J. Chem. Ed.* **44**, 124 [1967].)

we plot C/Nk against $1/T^*$. References [10, 11] may be consulted for more details on negative absolute temperatures.

In summary, negative absolute temperature systems are transient but observable. They are exceptional cases. Nevertheless, as in many other areas of science the special, even paradoxical, situation serves particularly well in illuminating what we mean by the general concept.

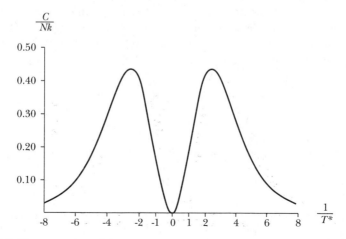

FIGURE 2-6 Heat capacity of the two-state spin system as a function of $1/T^* = \varepsilon/kT$.

2-7
Blackbody Radiation: Ideal Boson (Photon) Gas

We consider an evacuated space of fixed volume V and fixed temperature T. The atoms of the material making up the walls of the space will continually emit and absorb radiation such that at equilibrium the volume is filled with the radiation. We define $\rho(v, T)$ to be the energy per unit volume carried by frequencies between v and $v + dv$ of this radiation (it has dimension energy × time × length^{-3}). That $\rho(v, T)$ is a universal function of v and T and independent of the nature of the material making up the walls is a consequence of the Second Law of thermodynamics. For imagine a second evacuated space at the same temperature with different material walls and suppose there are windows in the walls of both spaces transparent only to radiation and frequencies between v and $v + dv$. Then if the energy density is greater in one cavity than in the other, the one will lose energy and its temperature will fall while the other will gain energy and its temperature will rise—which is an impossible spontaneous process.

Since all radiation in a vacuum travels with the same speed c, the energy arriving per unit time for each v on unit area of any body is directly proportional to $\rho(v, T)$. A blackbody is defined as one which absorbs all the radiation that reaches it, and at equilibrium it will radiate at precisely the same rate as it absorbs, thus at a rate proportional to $\rho(v, T)$. Hence $\rho(v, T)$ is often termed a distribution function for blackbody radiation and can be determined by calculating the rate of emission of such a body. However no real body is truly black at all frequencies and since $\rho(v, T)$ represents the density of radiation in equilibrium with *any* body at T, it is an inherent property of the radiation field itself and may be more simply derived from that point of view. In turn, the radiation field can be treated as waves or as particles, that is, photons. We shall take the photon viewpoint here. (In Chap. 4 we treat the analogous problem of sound waves in a crystal from a wave viewpoint.) Our isolated system of radiation inside walls has fixed U and V. The radiation is considered to be a gas of photons. Because photons absorbed by the walls at one frequency may be emitted at another frequency and in different numbers (for energy conservation), the total number $\mathcal{N}$ in the system is not fixed. This means the chemical potential μ of the photons is zero at all temperatures (since no constraint fixing $\mathcal{N}$ is required), and from Eq. (1-146)

$$\bar{n}_i = \frac{1}{e^{\varepsilon_i/kT} - 1} \tag{2-86}$$

where $\bar{n}_i$ is the occupation number of the ith energy state that maximizes the entropy for the photon gas with energy from the Planck relation

$$\varepsilon_i = h v_i \tag{2-87}$$

Strictly speaking, photons do not interact with each other and so no approximation is made by using ideal gas relations for them. They are particles of spin 1 and they are bosons. They are relativistic particles of zero rest mass. We use for the number of single photon states in phase space (see Sec. 1-8) with momenta between p and $p + dp$,

$$\omega(p)dp = \frac{V4\pi p^2 \, dp}{h^3} \tag{2-88}$$

In Eq. (2-88) we have taken advantage of the spherical symmetry of the non-interacting particle case in using for the volume element in momentum space, $dp_x \, dp_y \, dp_z = 4\pi p^2 \, dp$.

Now two things must be done with Eq. (2-88) to turn it into a counting formula, $\omega'(v) \, dv$, for allowed quantum states with frequencies between v and $v + dv$. Use the de Broglie relation* relating p and v for photons

$$p = h/\lambda = \frac{hv}{c}$$

and multiply by 2, which is the effective spin degeneracy for photons:

$$\omega'(v) \, dv = \frac{8\pi V v^2 \, dv}{c^3} \tag{2-89}$$

From classical physics this factor of 2 arises because the electric field (and magnetic field) vectors are in a plane perpendicular to the direction of propagation of the electromagnetic wave and have two components. Any radiation can be described as made up of a certain contribution that is circularly polarized in a left-handed sense (with electric field vector going around clockwise, viewing the radiation head on), and another contribution circularly polarized in a right-handed sense (electric field vector going around counterclockwise). In quantum physics the photons of spin 1 can only have angular momentum projections along the axis of their propagation of $+(1)\hbar$ for right circular polarization and $(-1)\hbar$ for left circular polarization. They do not have a projection of zero (and thus a spin degeneracy of 3 as might have been expected) because they have a zero rest mass and cannot stand still. This is discussed further by Feynman in reference [12], Secs. 11-4 and 17-4.

Putting together our results and replacing the sum by an integral, using Eq. (2-89), we find for the energy of radiation in volume V:

$$U = \sum_i \varepsilon_i \bar{n}_i = \sum_i \frac{hv_i}{e^{hv_i/kT} - 1} = \int_0^\infty \frac{hv\omega'(v) \, dv}{e^{hv/kT} - 1} \tag{2-90.1}$$

$$= V \int \frac{8\pi h v^3 \, dv}{c^3(e^{hv/kT} - 1)} \tag{2-90.2}$$

$$\equiv V \int \rho(v, \, T) \, dv \tag{2-90.3}$$

---

* The general relativistic formula for the total energy ε of a particle of rest mass m_0 is $\varepsilon = (m_0^2 c^4 + p^2 c^2)^{1/2}$. Thus for photons of zero rest mass, $\varepsilon = pc$. Equation (2-88) becomes equivalent to Eq. (1-153) if we deal with nonrelativistic independent particles with total energy $\varepsilon = p^2/2m$.

Thus we identify

$$\rho(v, T) = \frac{8\pi h v^3}{c^3 (e^{hv/kT} - 1)} \tag{2-91}$$

which is the famous blackbody radiation distribution law first obtained by Planck in 1901 in a different way. We may integrate Eq. (2-90.2) by rewriting the integrand as

$$\frac{v^3}{e^{hv/kT} - 1} = \frac{v^3 e^{-hv/kT}}{1 - e^{-hv/kT}} = v^3 \sum_{n=1}^{\infty} (e^{-hv/kT})^n$$

to obtain, using $x = hv/kT$,

$$\frac{U}{V} = \frac{8\pi (kT)^4}{c^3 h^3} \sum_{n=1}^{\infty} \int_0^{\infty} x^3 e^{-nx} \, dx = \frac{48\pi}{c^3 h^3} (kT)^4 \zeta(4) \tag{2-92}$$

where*

$$\zeta(4) = \sum_{n=1}^{\infty} \frac{1}{n^4} = \frac{\pi^4}{90} \tag{2-93}$$

and so finally

$$\frac{U}{V} = \frac{8\pi^5}{15} \frac{(kT)^4}{(ch)^3} \tag{2-94}$$

An interesting quantity we can now calculate is the energy incident per unit area per unit time on a surface due to the radiation. An argument from the kinetic theory of particles (see Sec. 5-1) shows that the number of particles striking unit area of surface per unit time is $(N/V)(\bar{v}/4)$ where $\bar{v}$ is their average speed. Here we have photons all with speed c. The number per unit volume of frequency v is $\rho(v, T)/hv$, and they carry energy hv such that the incident energy for frequencies v to $v + dv$ is

$$e(v, T) = \frac{c}{4} \rho(v, T) \tag{2-95}$$

For a blackbody the function $e(v, T)$ is also its emissivity of energy radiated per unit area per unit time. The total energy flux radiated by a blackbody inte-

---

* $\zeta(s) \equiv \sum_{n=1}^{\infty} n^{-s}$ are known sums called the Riemann zeta functions; see Appendix II.

grating over all frequencies is, from comparison to Eq. (2-90.3),

$$e(T) = \int_0^\infty e(v,\ T)\ dv = \frac{c}{4}\left(\frac{U}{V}\right) = \sigma T^4 \tag{2-96}$$

Equation (2-96) is the Stefan-Boltzmann radiation law, and from Eq. (2-94)

$$\sigma = \frac{2\pi^5 k^4}{15c^2 h^3} = 5.6702 \times 10^{-8}\ \text{J/m}^2\cdot\text{s}\cdot\text{K}^4 \tag{2-97}$$

It may be assumed that a blackbody at temperature T will exhibit this emissivity whether or not radiation is incident on it; that is, whether or not there is radiative equilibrium as long as some process inside the blackbody keeps it at temperature T.

To obtain the frequency for maximum emission we write from Eqs. (2-91) and (2-95)

$$e(v,\ T) = \frac{2\pi}{(ch)^2}(kT)^3 f(x) \tag{2-98}$$

where

$$f(x) = \frac{x^3}{e^x - 1} \tag{2-99}$$

and maximize $f(x)$ to find numerically

$$x_{max} = \frac{h v_{max}}{kT} = 2.8214 \tag{2-100}$$

with

$$f(x_{max}) = 1.4214 \tag{2-101}$$

Exercise 2-8

Use the Newton-Raphson method of Appendix II to obtain numerically the results in Eqs. (2-100) and (2-101).

To obtain the wavelength of maximum emission for a blackbody, which is *not* found from c/v_{max} because unit wavelength interval corresponds to a larger interval in v at small λ than at large λ, we define

$$\rho(\lambda, T) = \rho(v, T)\left|\frac{dv}{d\lambda}\right| = \frac{8\pi ch}{\lambda^5}\frac{1}{e^{hc/\lambda kT} - 1} \tag{2-102}$$

having used $v = c/\lambda$, and then

$$e(\lambda, T) = \frac{c}{4}\rho(\lambda, T) = \frac{2\pi c^2 h}{\lambda^5(e^{hc/\lambda kT} - 1)} \tag{2-103}$$

The maximum in $e(\lambda, T)$ occurs at

$$\lambda_{max}T = \frac{hc/k}{4.9651} = 2.8978 \times 10^{-3} \text{ mK} \tag{2-104}$$

Exercise 2-9

Use Eq. (2-104) to calculate the lowest temperature at which the maximum emissivity first occurs in the visible region of the spectrum ($\lambda = 770$ to 390 nm). What is the predominant color at this temperature? Explain how and why the color of a heated body will change as λ_{max} moves through and out of the visible range into the ultraviolet with temperature increase.

Example 2-5

Describe how Fig. 2-7 from Hoyle [13] may be plotted from explicit equations. It plots on a logarithmic scale the intensity of energy radiation of a blackbody per unit area per unit time relative to the maximum radiation of a blackbody at 3 K against logarithmic scales in v and λ.

Solution:

The plot is of the integrand of Eq. (2-96)

$$e(v, T) \, dv = \frac{2\pi}{(ch)^2}(kT)^3 f(x) \, dv = \frac{2\pi}{c^2 h^3}(kT)^4 f(x) \, dx$$

$$x = hv/kT = \frac{v}{(2.084)(10^{10})T} = \frac{1.4388}{\lambda T}$$

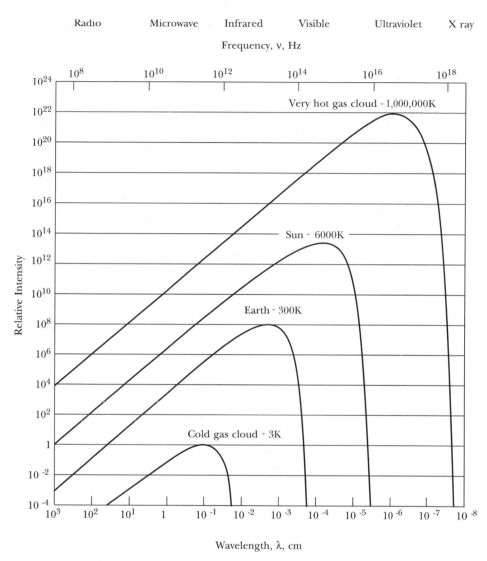

FIGURE 2-7 The intensity per unit area per unit time of energy radiation of a blackbody relative to the maximum of a blackbody at 3 K, plotted against frequency and wavelength; all scales are logarithmic. (FROM: Fred Hoyle. *Astronomy and Cosmology.* W. H. Freeman and Company, San Francisco, 1975, p. 670.)

with λ in cm to accord with the units used in Fig. 2-7,

$$f(x) = \frac{x^3}{e^x - 1}$$

We normalize by taking the ratio to the maximum emission of a blackbody at 3 K where, from Eq. (2-101), $f(x) = 1.4214$. Hence on the ordinate for given T we plot

$$\log \left[\left(\frac{T}{3} \right)^4 \frac{x^3}{e^x - 1} \frac{1}{1.4214} \right]$$

and on the abscissas $\log \lambda/(1 \text{ cm})$ bottom or $\log \nu/(1 \text{ s}^{-1})$ top.

We may now derive other thermodynamic properties of a radiation field. From Eq. (1-136) we have

$$\frac{PV}{kT} = -\sum_i \ln \left(1 - e^{-h\nu_i/kT} \right)$$

$$= -\int_0^\infty \omega'(\nu) \, d\nu \, \ln \left(1 - e^{-h\nu/kT} \right)$$

$$= -\frac{8\pi V}{c^3} \int_0^\infty \nu^2 \, d\nu \, \ln \left(1 - e^{-h\nu/kT} \right) \tag{2-105}$$

The integral in Eq. (2-105) may be integrated by parts using

$$\int u \, dv = vu - \int v \, du$$

with $u = \ln \left(1 - e^{-h\nu/kT} \right)$ and $v = \nu^3/3$ to find (since the function product vanishes at both limits):

$$PV = \frac{8\pi Vh}{3c^3} \int \frac{\nu^3 d\nu}{e^{h\nu/kT} - 1} = \frac{U}{3} \tag{2-106}$$

by comparison with Eq. (2-90.2). Thus radiation exerts a pressure given by

$$P = \frac{4\sigma}{3c} T^4 \tag{2-107}$$

This pressure is negligible under usual terrestrial conditions, but in stars radiation pressure can be more significant than ordinary gas pressure.

Exercise 2-10

Calculate the temperature for which the radiation pressure will be equal to 1 atmosphere.

We obtain the heat capacity of a radiation field from:

$$C_V = \left(\frac{\partial U}{\partial T}\right)_V = \frac{16}{c}\,\sigma V T^3 \tag{2-108}$$

Since $\mu = 0$ we have $G = \mathcal{N}\mu = 0$, and so we find the entropy from:

$$G = U - TS + PV = 0$$

$$S = \frac{U + PV}{T} = \frac{4}{3}\left(\frac{U}{T}\right) = \frac{16}{3}\frac{\sigma}{c}\,T^3 V = \frac{C_V}{3} \tag{2-109}$$

The entropy goes to zero at $T = 0$, as we would expect.

Appendix 2A
The Homonuclear Diatomic Molecule Gas at Low Temperature

The wave function of a molecule in the approximation we use in this chapter is a simple product

$$\psi = \psi_{trans}\psi_{rot}\psi_{vib}\psi_{el}\psi_{NS} \tag{2.A-1}$$

of translational, rotational, vibrational, electronic, and nuclear spin wave functions. If a diatomic molecule is homonuclear, that is, if both atomic nuclei are of the same isotope with the same nuclear spin quantum number I, then the operation of interchanging these two nuclei, leaving the electrons fixed, is a symmetry operation. If the isotope is of even mass number an even number of fermions (the neutrons and protons inside the nucleus) will have been exchanged when the nuclei were exchanged and so the total wave function must not change sign. If the isotope is of odd mass number an odd number of fermions will have been exchanged and the total wave function must change sign.

The translational and vibrational wave functions are completely unaffected by nuclear interchanges and do not change sign. The ψ_{rot} follows the rule of quantum mechanics that for even rotational quantum number J there is *no* sign change but for odd rotational quantum number J there *is* a sign change. For a sign change in ψ_{el} we consider the inversion of the nuclei as first a complete inversion of all electrons and nuclei (which always leaves ψ_{el} unchanged) followed by a reinversion of the electrons only. For the reinversion there will only be a sign change if the electronic state is of odd parity with subscript u ("ungerade," odd) on the term symbol (see any spectroscopy text) or if it is in a Σ_g^- state in distinction to Σ_g^+. Almost all diatomic molecules in their ground electronic states have even parity and their electronic wave functions do not change sign. This is our general assumption in what follows. The (inevitable) exceptional case is $^{16}O_2$ with ground state $^3\Sigma_g^-$, for which the rules that follow must be reversed.

Finally, we must examine the symmetry properties of the ψ_{NS}. Since each nucleus may have a spin oriented in $(2I + 1)$ ways there are a total of $(2I + 1)^2$ functions for a pair of nuclei in a diatomic molecule. Since there are

$$\frac{(2I + 1)(2I + 1 - 1)}{2} = (2I + 1)(I)$$

ways in which the two nuclei can have *different* orientations there will be that number of ways to form the antisymmetric functions:

$$\psi_{NS} = \psi(a, M_{I'}) - \psi(b, M_{I''})$$

which change sign when nucleus a with spin projection $M_{I'}$ is interchanged with nucleus b with projection $M_{I''}$, and also the same number of ways to form the symmetric functions

$$\psi_{NS} = \psi(a, M_{I'}) + \psi(b, M_{I''})$$

In addition we can form $(2I + 1)$ additional symmetric functions of the form

$$\psi(a, M_{I'})\psi(b, M_{I'})$$

where $M_{I'}$ takes $(2I + 1)$ values and both nuclei have the *same* spin orientation. Thus, of the $(2I + 1)^2$ functions

$$(2I + 1)(I)$$

change sign on inversion of nuclei, and

$$(2I + 1)(I) + (2I + 1) = (2I + 1)(I + 1)$$

do not change sign.

Assuming the electronic ground state is symmetric, we have the following rules:

For nuclei of even mass number, e.g., D_2, $^{14}N_2$, which have I equal to zero or integers, the total ψ must not change sign and

- Odd J rotational states go only with $(2I + 1)I$ antisymmetric spin functions

- Even J rotational states go only with $(2I + 1)(I + 1)$ symmetric spin functions

- If I is equal to zero no odd J levels are occupied

For nuclei of odd mass number, e.g., H_2, $^{35}Cl_2$, etc., which have I always half an odd integer (never zero), the total ψ must change sign and

- Odd J rotational states occur with $(2I + 1)(I + 1)$ symmetric spin functions

- Even J rotational states occur with $(2I + 1)(I)$ antisymmetric spin functions

By convention the more abundant high-temperature form, whether with odd mass nuclei (thus the odd J molecules) or with even nuclei (thus the even J molecules), is called the ortho- (o) form while the less abundant form is called the para- (p) form.

The practical application of these rules to the calculation of low-temperature thermodynamic properties of homonuclear diatomic molecules is that the partition functions for rotation and nuclear spin cannot be separated but must be treated as a single entity, which for odd mass nuclei is:

$$q_{rot}q_{nu} = (2I + 1)(I + 1)B + (2I + 1)(I)A \tag{2.A-2}$$

$$A = \sum_{J=0,2,4\ldots}^{\infty} (2J + 1) \exp\left[-J(J + 1)\theta_r/T\right] \tag{2.A-3}$$

$$B = \sum_{J=1,3,5}^{\infty} (2J + 1) \exp\left[-J(J + 1)\theta_r/T\right] \tag{2.A-4}$$

At high temperatures both of the sums A and B become equal to $T/2\theta_r$, and so at high temperature

$$q_{rot}q_{nu} \rightarrow \frac{T}{2\theta_r}\left[(2I + 1)(I + 1) + (2I + 1)(I)\right] = \frac{T}{2\theta_r}(2I + 1)^2$$

thus justifying the introduction of the symmetry number ($\sigma = 2$) in Eq. (2-44) and in accord with the high-temperature q_{nu} given in Eq. (2-30)

Exercise 2-11

Show that both sums A and B when converted to integrals give the result $T/2\theta_r$. Hint: First change summation (later integration) variable J to one that takes on all integer values, namely, use $y = J/2$ in A and $z = (J - 1)/2$ in B.

As far as practical thermodynamic experiments are concerned only H_2 and D_2, with their large θ_r values, can be examined. For H_2 with $I = \frac{1}{2}$

$$q_{rot}q_{nu} = 3B + A \tag{2.A-5}$$

and from this the heat capacity function for the equilibrium mixture of o-H_2 and p-H_2 as shown in Fig. 2-8 can be computed. For the separate species one may use

$$(q_{rot}q_{nu})_o = 3B \tag{2.A-6}$$

$$(q_{rot}q_{nu})_p = A \tag{2.A-7}$$

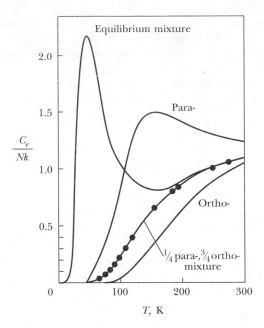

T, K

FIGURE 2-8 The rotational and nuclear contribution to molar heat capacity for $o\text{-}H_2$, $p\text{-}H_2$, an equilibrium mixture of $o\text{-}$ and $p\text{-}H_2$, a metastable 25 percent $p\text{-}H_2$ mixture, and the experimental data. (FROM: K. F. Bonhoeffer and P. Harteck. $\mathcal{Z}$. *Physikal. Chem.* **4B**, p. 113 [1929], in D. McQuarrie. *Statistical Mechanics.* Harper and Row, 1976.)

to obtain the rotational-nuclear heat capacity curves labeled *Ortho-* and *Para*, respectively, in Fig. 2-8. For the equilibrium between $o\text{-}H_2$ and $p\text{-}H_2$ we have from Eq. (2-61), since all factors except the differing rotational-nuclear contributions cancel,

$$K = \left(\frac{N_o}{N_p}\right)_e = \frac{3B}{A} \qquad (2.\text{A-}8)$$

From this the percent *para-* at equilibrium is given by

$$(\%p)_e = \frac{100}{1 + K} \qquad (2.\text{A-}9)$$

and this function is plotted in Fig. 2-9. At equilibrium at 0 K all the molecules will be *para-* because only $p\text{-}H_2$ can reach the $\mathcal{J} = 0$ rotational level. At high temperature $B = A$ and $K = 3$ such that the *para-* will be 25 percent. In the absence of a catalyst (a paramagnetic material is needed) to promote conversion of $o \rightarrow p$, hydrogen will remain a metastable mixture of 25 percent *para-* when it is cooled below room temperature; this is the material for which the experimental points are given in Fig. 2-8. We may calculate this curve using

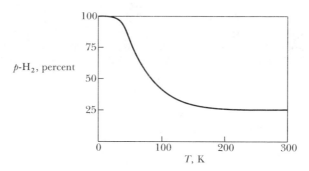

FIGURE 2-9 The percentage of p-H_2 in an equilibrium mixture of p- and o-H_2 as a function of temperature. (FROM: D. McQuarrie. *Statistical Mechanics*. Harper and Row, 1976.)

$$C_V = \tfrac{3}{4}(C_V)_o + \tfrac{1}{4}(C_V)_p \qquad (2.\text{A-}10)$$

In the presence of a catalyst it has been possible to observe the curve for the equilibrium mixture experimentally. With use of a catalyst it is also possible to prepare 100 percent p-H_2 and upon removing it the curve for pure p-H_2 has been observed. To prepare pure o-H_2 is more difficult but has been accomplished by selectively adsorbing it from the metastable mixture on Al_2O_3 at 20 K. [14] At this temperature only the o-H_2 may rotate (it exists in the $J = 1$ level) and its preferential adsorption is probably associated with its being able to continue a hindered rotation on the surface of the Al_2O_3. These fascinating results for H_2 were of great significance in demonstrating the reality of quantum mechanical effects on a macroscopic scale and they are explored further in a number of the problems at the end of this chapter.

References and Recommended Reading

[1] N. Davidson. *Statistical Mechanics*. McGraw-Hill, New York, 1962.

[2] A. Münster. *Statistical Thermodynamics*, vol. 1. Springer, Berlin, 1969.

[3] J. E. Mayer. *Equilibrium Statistical Mechanics*. Pergamon, Oxford, 1968.

[4] T. M. Reed and K. E. Gubbins. *Applied Statistical Mechanics*. McGraw-Hill, New York, 1973.

[5] D. A. McQuarrie, *Statistical Mechanics*. Harper and Row, New York, 1976.

[6] J. E. Mayer and M. G. Mayer. *Statistical Mechanics*, 2d ed. Wiley, New York, 1977).

[7] D. R. Stull, E. F. Westrum, Jr., and G. C. Sinke. *The Chemical Thermodynamics of Organic Compounds*. Wiley, New York, 1969.

[8] L. Pauling. "The residual entropy of ice." *J. Am. Chem. Soc.* **57**, 2680 (1935).

[9] E. M. Purcell and R. V. Pound. *Phys. Rev.* **81**, 279 (1951). First report on observation of a system with a negative absolute temperature.

[10] C. E. Hecht. *J. Chem. Education* **44**, 124 (1967).

[11] N. F. Ramsey. *Phys. Rev.* **103**, 20 (1956). Negative absolute temperatures.

[12] R. P. Feynman. *Lectures on Physics.* vol. III. Addison Wesley, Reading, MA, 1965. Sec. 11-4, 17-4. On polarization states of a photon.

[13] F. Hoyle. *Astronomy and Cosmology.* W. H. Freeman Co., San Francisco, 1975. Blackbody radiation.

[14] C. M. Cunningham, D. S. Chapin, and H. L. Johnston. *J. Am. Chem. Soc.* **80**, 2382 (1958). Preparation of nearly pure *ortho*-H_2.

Problems

1. Assuming only two nondegenerate levels of a set are significantly occupied and that they differ in energy by $300 \text{ kJ} \cdot \text{mol}^{-1}$, at what temperature will the excited level contain 10 percent of the molecules?

2. For HBr at 1000 K calculate the ratio of the number of molecules in the $J = 5$ rotational level of the $n = 2$ vibrational state to the number of molecules in the $J = 2$ rotational level of the $n = 1$ vibrational state.

3. Substance X is an ideal gas containing s atoms per molecule. $(C_P)_m$ for $X(g)$ and $N_2(g)$ is experimentally the same at $0°C$, where you may assume the vibrational contributions for both molecules are negligible. The classically predicted (equipartition) value of $(C_P)_m$ for $X(g)$ is approximately $50 \text{ J} \cdot \text{K}^{-1} \cdot \text{mol}^{-1}$ greater than the classically predicted $(C_P)_m$ for N_2. Deduce from this information as much as you can about s and the general structure of the X molecule.

4. The $\varepsilon(p, q)$ expression for a (diatomic) rigid rotator is

$$\varepsilon = \frac{1}{2I} \left(p_\theta^2 + \frac{p_\phi^2}{\sin^2 \theta} \right)$$

where the momenta p_θ and p_ϕ can each range from $-\infty$ to $+\infty$ and angle ϕ varies from 0 to 2π while angle θ varies from 0 to π. Evaluate the semiclassical partition function for the rigid rotator and show that it gives the result in Eq. (2-46) (assuming $\sigma = 1$).

5. From data given in Sec. 2-3 calculate the fraction of halogen atoms in the first excited electronic state (neglecting all other states) at 300, 1000, and 5000 K for each halogen.

6. Assuming the rotational quantum number J is continuous, derive an equation for the J value as a function of temperature for which N_J is a maximum.

7. Show that Eq. (2-47) for the q_{rot} leads to Eqs. (2-48), (2-49), and (2-50) for U_{rot}, S_{rot}, and $(C_V)_{rot}/R$, respectively.

8. Verify that the high-temperature limit of the average energy per particle for a vibrational mode is kT and not $kT + (k\theta_v/2)$.

9. Prove that a change in the zero of the energy scale from which energy levels of a system are calculated has no effect upon the value of the entropy of the system. Hint: Just change all E_i values by an arbitrary additive term $\Delta(+$ or $-)$ and show that terms in Δ cancel from the entropy expression.

10. Take data for ethylene from Table 2-3. The rotational symmetry number is $\sigma = 4$. Calculate at 300 K and 1 atm $(C_P)_m$ and $(S)_m$ for ethylene and compare with the experimental values, which are 43.7 and 219.7 J·K^{-1}·mol^{-1}, respectively.

11. For CO the equilibrium internuclear distance is 1.115×10^{-10} m. Evaluate the rotational partition function at 150 and 298 K. Assuming J is a continuous variable, obtain an expression as a function of T for the J value at which the rotational energy is equal to kT. Evaluate this J value at 150 and 298 K and show that the number of rotational states (not levels) out to $\varepsilon_{rot} = kT$ is essentially equal to the value of the rotational partition function at both temperatures.

12. From the Boltzmann distribution for vibrational energy the ratio of the number in the nth state to that in the zero state is $N_n/N_0 = e^{-n\theta_v/T}$. Show that the fraction of the molecules that have vibrational quantum numbers q or greater is given by

$$\frac{\sum_{n \geq q} N_n}{N} = e^{-q\theta_v/T}$$

13. From the result in Problem 12 and data in Example 2-3 calculate the fraction of O_2 molecules that have vibrational energy equal to or greater than $0.10D_0$ and equal to or greater than $0.20D_0$ at 4000 K, where D_0 is the dissociation energy of O_2 molecules into ground state O atoms. Comment on the paradox(?) that although these fractions are so small, Example 2-3 shows that the O_2 molecules are 63 percent dissociated at 1 atm total pressure and 4000 K.

14. Take data from Table 2-3 for CH_3F. The rotational symmetry number is 3. The ground electronic state is nondegenerate. Calculate at 298.15 K and 1 atm $(C_P)_m$ and $(S)_m$ for CH_3F. The experimental values are 37.5 and 222.8 J·K^{-1}·mol^{-1}, respectively.

15. For a gas phase reaction of association of two monatomic species to form a diatomic molecule.

$$A(g) + A(g) \rightarrow A_2(g)$$

Write the equilibrium constant for this reaction (K_P) in terms of partition functions (assume only one electronic state of degeneracy g_A contributes for atoms and is nondegenerate for the molecules). Examine in particular the M_A (atomic weight of the atom) and T dependence of K_P in the two limits $\theta_v > T$ and $\theta_v < T$. By using typical orders of magnitude of the partition functions at 300 K estimate the minimum order of magnitude of D_0 in eV (for the diatomic) for this reaction to have $K_P \geqq 1$ atm^{-1}.

16. Assuming the temperature is high enough to use the classical results for rotation, obtain K_P as an explicit function of T for the isotopic gas phase exchange reaction

$$D(g) + H_2(g) \rightarrow H(g) + HD(g)$$

Note that for isotopic species it is an excellent approximation to assume that their electron clouds are identical, which means that they have the same D_e values and the same force constants and the same internuclear r_e values.

17. K. A. Gingerich et al. [*J. Chem. Phys.* **64**, 4028 (1976)] report the following data for Pb_2 molecules in the gas phase:

$$r_e = 3.00 \times 10^{-10} \text{ m} \qquad \bar{v} = 119.1 \text{ cm}^{-1} \qquad g_{el} = 3$$

In addition, by direct mass spectrometric sampling they give the equilibrium constant for $Pb_2(g) \rightarrow 2Pb(g)$ at 968.5 K as 0.266 atm. Calculate a D_0 value for $Pb_2(g)$ assuming the $Pb(g)$ atoms have a nondegenerate ground state.

18. At room temperature the equilibrium constant for the reaction

$$^{16}O_2(g) + {}^{18}O_2(g) \rightarrow 2(^{16}O^{18}O)(g)$$

may be accurately estimated to be _____. (Put in a numerical value on the basis of a simple calculation.) Why is it incorrect to estimate the equilibrium constant at room temperature for

$$D_2(g) + H_2(g) \rightarrow 2HD(g)$$

to have the same value as that for the oxygen isotope reaction?

19. For $O(g)$ calculate $H_m^0(298.15) - H_{m0}^0$. Use data from Example 1.

20. Calculate $-(G_m^0 - H_{m0}^0)/T$ for $H_2O(g)$ and $H_2(g)$ at 298.15 K from data in Example 2-4 and compare these results with tabulated thermochemical values for this function.

21. Imagine a pair of idealized isomeric molecules, A and B, such that per mole $D_{0A} - D_{0B} = 8000$ J·mol^{-1}. Assume both A and B have a simplified set of nondegenerate energy states (hypothetical rotational vibrational type) extending indefinitely with equal spacing of 4000 J·mol^{-1} for A and 2000 J·mol^{-1} for B. Calculate K_P at 298 K for $A(g) \rightarrow B(g)$. How will K_P change as T increases? Give qualitative reasons for this change.

22. Consider the ionization equilibrium for species A (not necessarily monatomic) $A \rightarrow A^+ + 1e^-$.

For this process take the positive species and electron at rest and infinitely separated as the zero of energy. Explain the reasonable cancellations that occur and treat only one electronic level of degeneracy, g_{A+} and g_A, respectively, for the heavy species. The electron degeneracy is 2. Show that

$$\log_{10}(K_P/\text{atm}) = \frac{-C_1\varepsilon}{T} + \frac{5}{2}\log_{10}(T/\text{K}) + \log_{10}\left(\frac{2g_{A+}}{g_A}\right) - C_2$$

where ε is the ionization potential (energy) in electron volts, and evaluate constants C_1 and C_2. This equation is called the Saha equation.

23. With reference to the ionization equilibrium in Problem 22, show that it is reasonable to use only ground electronic levels as we have done, even when there is a large amount of ionization. To do this calculate numerically the T/θ_{ion}, where $\theta_{ion} = \varepsilon/k$, at which the number densities of A^+, A, and e^- are all equal and large (say 10^{19} cm^{-3}). Take $g_{A+} = g_A = 1$ and $\varepsilon = 10.0$ eV as a typical ionization energy.

24. Use the Saha equation (Problem 22) to estimate the temperatures at which gaseous atomic hydrogen in a system at constant total pressure of 1.00×10^{-5} atm would reach 50 percent ionization and 99 percent ionization into electrons and protons. The ionization potential for hydrogen is 13.60 eV.

25. From the data in Table 2-2 calculate $-(G_m^0 - H_{m0}^0)/T$ for HCl(g) at 298.15 and 500 K.

26. From the data in Table 2-3 calculate $-(G_m^0 - H_{m0}^0)/T$ for acetylene (C_2H_2 linear) at 298.15 and 500 K.

27. For vinyl chloride (H_2CCHCl), D. Kivelson et al. [*J. Chem. Phys.* **32**, 205 (1960)] deduce the product of the three moments of inertia to be 3.208×10^{-136} kg^{3}m^6. For the wave numbers of the vibrational modes, C. Gullikson and J. Nielsen [*J. Mol. Spectrosc.* **1**, 158 (1957)] list the following, in units of cm^{-1}:

3121	1369	896
3086	1279	720
3030	1030	620
1608	941	395

The rotational symmetry number is unity.

Calculate $-(G_m^0 - H_{m0}^0)/T$ for this gaseous molecule at 298.15 and 500 K.

28. For the industrially important gas phase synthesis of vinyl chloride from acetylene and HCl,

$$C_2H_2(g) + HCl(g) \rightarrow H_2CCHCl(g)$$

at 298.15 K the $\Delta H_{m,rx}^0 = -9.929 \times 10^4$ J. Use Eqs. (2-74) and (2-78) and the results in Problems 25–27 to calculate K_P for this reaction at 298.15 and 500 K.

29. For a system that can exist in only two levels separated by energy ε (like an electronic q_{el} for a close-lying doublet), measuring from the ground state as zero, use g_0 as the degeneracy of the ground state and g_1 as the degeneracy of the excited state. Derive expressions for the partition function and for A, U, S, and C_V. Describe how the functions will vary from $T = 0$ to $T = \infty$. In particular, for C_V show that there will be a maximum at $T = T_m$ determined by

$$(X - 2) = (X + 2)(g_1/g_0)e^{-X}$$

where $X = \varepsilon/kT_m$. Solve this equation numerically for X and for $(C_V)_m/R$ at $T = T_m$ for $(g_1/g_0) = 1$ and $(g_1/g_0) = 2$.

30. The energy release of nuclear bombs is expressed in kilotons of TNT equivalent (1 kiloton = 4.2×10^{12} joules). Within one microsecond of explosion the core region (treat it as a sphere of radius 10 cm) emits about 75 percent of the energy release as primary thermal radiation (the other 25 percent is in kinetic energy of weapon debris). For weapons of 13 kilotons (Hiroshima bomb) and 2 megatons estimate the effective temperature of the core region and the λ_{max} of the primary radiation and the photon energy in keV associated with λ_{max}. This primary radiation (mainly X rays) is quickly degraded in the distance of a few feet into local heating of the air molecules and formation of the fireball, which then will reradiate in the UV and visible such that overall at large distances from the detonation point only about 35 percent of the energy of the initial explosion arrives as secondary thermal energy. Using these assumptions estimate for the above two weapons the thermal energy in $J \cdot m^{-2}$ and $cal \cdot cm^{-2}$ that eventually (in a few seconds) reach distances of 0.5 and 1.0 miles from the detonation point. (Your estimates will be 50 to 100 percent too high because of neglect of attenuation of thermal energy due to absorption and scattering in the atmosphere.) For more detail on the effects of nuclear weapons see L. Sartori, *Physics Today*, March 1983, p. 32.

31. Consider the integral of Eq. (2-96):

$$e(T) = \int_0^\infty \left[\frac{2\pi}{(ch)^2} (kT)^3 f(x) \right] d\left(\frac{kt}{h} \right) x$$

Plot the bracket term divided by its maximum value vs. frequency in units of kT/h to obtain a universal curve (of value unity at its maximum) as a function of reduced frequency (from 0 to 10). On a copier make three copies of your graph. Your four graphs will now be used for stars of temperatures 2500 K, 5800 K (the sun), 10,000 K, and 30,000 K, respectively. For each graph state the units of the ordinate in $J \cdot m^{-2}$ and of the frequency in s^{-1}. On each graph shade differently the areas under the curve corresponding to the ultraviolet, visible ($\lambda = 770$ to 390 nm), and infrared regions. For each graph give the total energy flux of the star (Stefan-Boltzmann law) and estimate the fraction of the flux that is in the visible region.

32. Although N (photon number) is not conserved, at equilibrium in a radiation field for the number set given by Eq. (2-86), there will be an average number density $\bar{N}/V$ of photons present that is a function of T only. Derive this function. You will need to use the function $\zeta(3) = \sum_{n=1}^{\infty} n^{-3} = 1.202$. Show that in terms of this $\bar{N}/V$ function $P = 0.900(\bar{N}/V)kT$.

33. By differentiating $[z(\partial/\partial z)_{V,T}]\bar{N}$, with $\bar{N}$ the average number of particles present in a replica of the grand canonical ensemble as given in Eq. (1-141), show that the mean square fluctuation in N is given by

$$\overline{N^2} - (\bar{N})^2 = \left(z\frac{\partial}{\partial z}\right)_{V,T} \bar{N} = kT\left(\frac{\partial \bar{N}}{\partial \mu}\right)_{V,T}$$

where $z = e^{\mu/kT}$.

34. Obtain an expression for the mean square fluctuation of the number of photons in an enclosure of volume V and at a temperature T. Use the relation proved in Problem 33 and use $\bar{N}$ for photons from Eq. (1-142). After differentiating, set $z = 1$ since photons have $\mu = 0$. Show that your result is proportional to $\bar{N}$ such that

$$\left(\frac{\overline{N^2} - (\bar{N})^2}{(\bar{N})^2}\right)^{1/2} \propto \frac{1}{\bar{N}^{1/2}}$$

and is negligible. Refer to the equation for $\bar{N}$ in Problem 32.

35. At a temperature of 1.0×10^5 K, if a monatomic ideal gas is present at a particle density of 1.0×10^{17} cm^{-3} in a volume of 1 m^3 along with equilibrium radiation, calculate the particle pressure, the radiation pressure, and the C_V contributions of the particles and of the radiation.

36. For $^{16}O_2$ the nuclear spin of each O is zero, and since the electronic ground state does change sign on inversion of the nuclei, all the *even* J levels are unoccupied. How can one determine from a rotational Raman spectrum of a molecule that all lines of one type are missing? (If both types are present with alternating intensity the situation is obvious.) Review the selection rules for Raman spectroscopy and consider the ratio of the separation of the first rotational lines *across* the exciting line to the common separation of the lines on either side of the exciting line for the three cases: no levels missing; all even levels missing; all odd levels missing. Consult a pure rotational Raman spectrum of O_2 (Fig. 17.16, W. J. Moore, *Physical Chemistry*, 4th ed., Prentice Hall [1972], or Fig. 37, G. Herzberg. *Spectra of Diatomic Molecules*, 2d ed., Van Nostrand [1950], p. 64) and check that all even levels are missing.

37. Discuss D_2 (each nucleus has $I = 1$) qualitatively. How will the percent p-D_2 vary at equilibrium between 0 K and room temperature? Express the experimental $(C_V)_m$ for D_2 (without paramagnetic catalyst) in terms of $(C_V)_m$ of p-D_2 and o-D_2.

38. For pure o-H_2 and for pure p-H_2 obtain expressions for U_m and $(C_V)_m$ due to rotational nuclear energy levels that are accurate for $\theta_r/T > 1$. Show that in this temperature range ΔH for p-$H_2 \rightarrow o$-H_2 is essentially temperature independent.

39. Extend your results for U_m for pure p-H_2 and pure o-H_2 in Problem 38 to calculate S_m for pure p-H_2 and pure o-H_2. Then calculate the *complete* S_m for a metastable mixture of H_2 at 100 K and 1 atm pressure. You will need to include the translational contribution (vibrational contribution is negligible) and an entropy of mixing term. Use $\theta_r = 85.40$ K for H_2.

40. From Eq. (2.A-5) obtain an expression for the heat capacity of the equilibrium p-H_2-o-H_2 mixture that is accurate for $\theta_r/T \geqq 1$ (rotational-nuclear contribution only). To do so keep three or four terms in each sum and do not expand away the denominators. Show with a little numerical calculation that this heat capacity has a maximum at $\theta_r/T = 1.75$.

41. Give a qualitative account based on thermodynamic principles for the fact that $(C_V)_{rot,nu}$ for the equilibrium p-H_2-o-H_2 mixture is so large [indeed has a maximum value of $(2.07)R$] in a temperature range for which the respective *ortho*- and *para*-C_V values are negligible.

Repeat your qualitative analysis of the above situation based on the simple statistical mechanical model of Problem 29.

42. It is interesting to consider the ionization products for pure H_2O and pure D_2O. The known results are as follows:

	10°C	25°C
$K_W(H_2O)$	292×10^{-17}	1008×10^{-17}
$K_D(D_2O)$	54×10^{-17}	195×10^{-17}

We can calculate the ratio of the constants for the following two gas phase reactions

$$H_2O \rightleftharpoons H^+ + OH \qquad K_{(1)}$$
$$D_2O \rightleftharpoons D^+ + OD^- \qquad K_{(2)}$$

as a function of T. If we do so we can in rough approximation hope that this ratio is the same as the ratio of the ionization product constants. Assuming K_W known, calculate K_D and compare with the above experimental results. In doing so we implicitly neglect the somewhat different vapor pressures of the two pure liquids and also any differences in the hydration energies for the different ionic species. Nevertheless we should be able to predict that K_D is less than K_W and obtain a result of the right order of magnitude. In order to calculate the ratio of $K_{(1)}$ to $K_{(2)}$ certain geometric and spectral information is necessary. The moments of inertia of a planar molecule such as H_2O are easily expressed in terms of the bond length a and the angle α given below.

These moments of inertia are:

$$I_{(1)} = \frac{2m_0 m_H a^2 \cos^2 \alpha}{(m_0 + 2m_H)}$$

$$I_{(2)} = 2m_H a^2 \sin^2 \alpha$$

$$I_{(3)} = I_{(1)} + I_{(2)} = \frac{2m_H m_0 a^2 + 4m_H^2 a^2 \sin^2 \alpha}{(m_0 + 2m_H)}$$

Similar expressions hold for D_2O, for which we assume the a and α are identical to those in H_2O. The angle α has been found to be $52°30'$, so that the bond angle in water is $105°$ and the bond length is 0.957×10^{-8} cm. The fundamental frequencies for H_2O and D_2O have been directly observed as follows (in cm^{-1})

H_2O	D_2O
1595	1179
3652	2666
3756	2789

The frequency of vibration in the OH$^-$ ion may be taken as 3680 cm^{-1} and that in the OD$^-$ ion may be found from the former result by using the usual approximation of equal force constants in isotopic molecules and ions.

Calculate the ratio $K_{(1)}/K_{(2)}$ as a function of T and estimate the K_D values at 10°C and 25°C. Also estimate the mean heat of ionization for pure liquid D_2O. At what temperature would the ratio of gas phase constants be unity?

43. The principle of detailed balance asserts that in thermal equilibrium a process and its inverse occur with equal rates. Use the principle of energy conservation, $\varepsilon_1 + \varepsilon_2 = \varepsilon_3 + \varepsilon_4$, to deduce the form of the detailed balance requirement in terms of the mean occupation numbers $\bar{n}_i$ of the states for ideal fermions and for ideal bosons for equilibrium exchange (or scattering) of particles 1 and 2 with particles 3 and 4. Interpret your results.

44. a. For thermal equilibrium in the reaction of two photons (1 and 2) to produce an electron-positron pair and its reverse (pair annihilation), show that the chemical potential of the electron, μ_-, and of the positron, μ_+, must both be zero. Recall that charge conservation must hold.

   b. Write down the form of the detailed balance restriction in this equilibrium in terms of $\bar{n}_1, \bar{n}_2, \bar{n}_+$, and $\bar{n}_-$. Treat the electrons as ideal fermions but with total energy $E = \varepsilon + m_e c^2$, where the rest mass energy is added to the kinetic energy ε and m_e is the electron rest mass. (Note Problem 43.)

   c. Estimate the temperature above which a large number of electron positron pairs will exist at equilibrium by setting the average energy per photon (from Problem 32) equal to $m_e c^2$. (If the two photons do not have an energy of $2m_e c^2$ no pair production can occur in any individual reaction.)

45. From astronomical observations the average density of matter in the universe as a whole at the present time is (to within a factor of 10) about 4×10^{-27} kg·m^{-3}. The universe is also currently bathed in a uniformly distributed blackbody radiation of effective temperature 2.7 K [13]. This radiation is believed to be a relic of the cosmological big bang at the birth of the universe which gave rise to its expansion.

   a. Assuming all matter is in the form of protons, calculate the current number density and the current energy density of matter.

   b. Repeat these calculations for the current number and energy densities of radiation in the universe. (Note Problem 32.)

   c. By comparison of the results in **a** and **b** we note that the energy of the universe is currently mainly found in matter. However in the past, when the volume of the universe was smaller (assume the volume V goes as r^3, where r is the radius of the universe) and the temperature much higher, the energy was mainly in the form of radiation after the pair production discussed in Problem 44 died out. At a certain time and temperature T_C in the past, the energy densities of matter and of radiation were equal. The temperature T_C can be estimated if we assume that the total mass of matter is constant back to this time (no pair production) and that the radiation undergoes a reversible adiabatic change on the way back (meaning its entropy remains fixed). Estimate T_C and the common energy density at that temperature for matter and radiation. Notice that we cannot estimate how long ago T_C was the overall temperature of the universe because we are only using thermodynamic criteria here. For more information on this problem see P. T. Landsberg. *Thermodynamics and Statistical Mechanics.* Oxford, (1978). Chap. 13.

3

Statistical Thermodynamics of Interacting Systems: Imperfect Gases, Liquids, Liquid Solutions

In this chapter we deal on a molecular basis with the real world of interacting substances in the gaseous and liquid states. It is appropriate to treat these states in a single chapter since we know from thermodynamics that by changing temperature and pressure it is possible to pass from any point in the P, T plane assigned as gaseous to any point in the liquid region of the plane without undergoing a phase transition. By contrast, a crystalline solid is always separated from one or the other fluid (to use a common term for the gaseous and liquid states) phases by a first-order phase transition. Crystalline solids (treated in Chap. 4) are characterized by an endlessly (periodically) repeated common structure in space. Atomic positions in a solid can thus be labeled in principle and are called "localized" in the terminology of Sec. 1-4. Although in an ideal gas there is no structure, when interactions between nonlocalized (fluid) molecules become important a structure develops. It is not a structure common to the entire volume of the fluid, but rather a structure relative to any particular molecule chosen as the origin of coordinates. This fluid structure is described by the distribution functions discussed in Sec. 3-6. In that section we also derive relationships between the distribution functions and the fluid equation of state and other fluid thermodynamic properties.

Since fluid structure depends on the law of force between molecules, which in turn arises from molecular electrical properties, we treat these topics in Sec. 3-2 and 3-1, respectively.

Before treating liquids (high density fluids) we treat imperfect gases in Sec. 3-3 and 3-4 essentially as ideal gases perturbed by molecular interactions.

In Sec. 3-5 we present a schematic introduction to computer simulation studies of molecules interacting with (usually) defined model intermolecular force laws. These studies provide results that are essential in evaluating the reliability of the approximate distribution function and perturbation theory methods for liquids discussed in Sec. 3-6 and 3-7.

In Sec. 3-8 and Appendices 3.A and 3.B we discuss critical state phenomena primarily with application to the liquid-gas example but with an introduction to the great power and generality of the modern approach to this subject.

Another approach to the liquid state treats it as a perturbed solid lattice. This approach is followed in Sec. 3-9 with application to solutions of nonelectrolytes.

A standard treatment of the Debye-Hückel theory for solutions of electrolytes (enlivened by use of ion-ion radial distribution functions) is given in Sec. 3-10.

3-1

Electrical Properties of Molecules

Molecular interactions in real systems result from the electrical properties of molecules, primarily their permanent dipole moments (if any) and their polarizabilities, which are a measure of the ease with which dipoles may be induced in the electron clouds of molecules. In our presentation we shall emphasize qualitative meaning and methods of experimentally determining these molecular properties, leaving to standard references (e.g., [1–3]) the details of electrostatic theory.

Consider a parallel-plate charged capacitor in a vacuum, Fig. 3-1a, with charge density $+\sigma$ (in $C \cdot m^{-2}$) on one plate and $-\sigma$ on the other plate. The electric field, E_0, due to these charges is perpendicular to the plates and is given by

$$E_0 = \frac{\sigma}{\varepsilon_0} \tag{3-1}$$

where ε_0 is the permittivity of vacuum. The capacitance, C, is defined as the ratio of the charge on one of the plates to the potential difference, $E_0 d$, where d is the

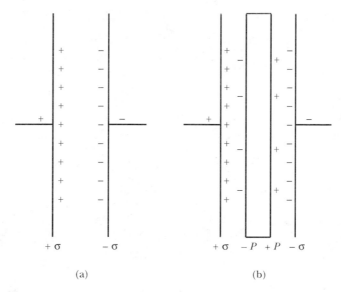

FIGURE 3-1 a. Charged capacitor. b. Charged capacitor with dielectric.

separation of the plates in meters. With S the surface area of the plate:

$$C_0 = \frac{\sigma S}{E_0 d} = \frac{\varepsilon_0 S}{d} \tag{3-2}$$

If the vacuum is replaced with some dielectric (i.e., nonconducting) material the capacitance is always observed to increase and may be written

$$C = C_0 D \tag{3-3}$$

where D (>1) is the relative permittivity of the dielectric, more commonly called the dielectric constant of the medium. This increase occurs because the field E polarizes the dielectric, that is, causes dipoles in the dielectric (permanent or induced) to rotate preferentially, such that their negative ends tend to point toward the positively charged plate and their positive ends toward the negatively charged plate, and hence induces a surface charge density of magnitude $\mp P$ on the surface of the dielectric at the positive and negative plates, respectively, Fig. 3-1b. P is called the polarization of the dielectric, and for electrically isotropic media such as liquids and gases (but not generally solids) it is also the dipole moment per unit volume of the dielectric. This follows because the dielectric will have a charge of

$+PS$ separated by distance d from a charge of $-PS$ in a volume Sd such that

$$\frac{\text{dipole moment}}{\text{volume}} = \frac{PSd}{Sd} = P \tag{3-4}$$

For Eq. (3-3) to hold with the same charge on the plates, the macroscopic electric field E inside the dielectric must be E_0/D, but we may also consider that the effect of the dielectric is to change the charge density from σ to σ-P; thus

$$E = \frac{E_0}{D} = \frac{\sigma}{\varepsilon_0 D} = \frac{\sigma - P}{\varepsilon_0} \tag{3-5}$$

and eliminating σ gives

$$P = \varepsilon_0(D - 1)E \tag{3-6}$$

A molecule inside the dielectric feels, in addition to the macroscopic net field on the plates given by Eq. (3-5), a further contribution generated by the polarization of its neighbors. Assuming the molecule to be in a spherical cavity in a homogeneous medium, this contribution can be shown to be $4\pi/3(P/4\pi\varepsilon_0)$. Therefore the total field, E^*, at the molecule in this approximation becomes

$$E^* = E + \frac{1}{3}\frac{P}{\varepsilon_0} \tag{3-7}$$

This field will distort the electron cloud of the molecule, assumed here to have no permanent dipole moment ($\mu = 0$), and induce a dipole moment m_i that is proportional to and in the direction of E^*:

$$m_i = \alpha E^* \tag{3-8}$$

where α (units of $C \cdot m^2 \cdot V^{-1}$) is called the polarizability of the molecule. Since P is the dipole moment per unit volume we have

$$P = m_i \rho \tag{3-9}$$

where $\rho = N/V$ is the number density of molecules in the dielectric. Then using Eqs. (3-6) and (3-7) we may solve for m_i:

$$m_i = \frac{3\varepsilon_0(D - 1)E^*}{\rho(D - 2)} = \alpha E^* \tag{3-10}$$

Since according to classical electrostatics the polarizability α of a perfectly conducting sphere in vacuum may be shown to be equal to $4\pi\varepsilon_0 r^3$, where r is the radius of the sphere, it has become customary to call $\alpha/3\varepsilon_0 = \frac{4}{3}\pi r^3$ the polarizability volume of a molecule and to call $N_A\alpha/3\varepsilon_0$ the molar polarization P_m (units of volume per mole). We find from Eq. (3-10), using ρ' as the mass density $(\rho M/N_A)$ with M the molar mass of the molecules,

$$P_m = \frac{N_A\alpha}{3\varepsilon_0} = \frac{M}{\rho'}\frac{(D-1)}{(D+2)} \tag{3-11}$$

This is the Clausius-Mossotti equation and holds for nonpolar molecules. For such molecules P_m is independent of temperature (α, being a molecular property, is always independent of temperature) and Eq. (3-11) may be solved for D:

$$D = \frac{1+2s}{1-s} \tag{3-12}$$

where $s = P_m\rho'/M$ is temperature dependent and less than unity. For gases s is very tiny and D is close to unity.

For molecules that possess a permanent dipole moment μ, their mean dipole moment in the absence of an applied field must be zero if the dielectric is a gas or liquid (may be zero in a solid if free rotation is possible) because their rotational motion averages the net effect to zero. In the presence of a field, orientation of their dipoles along the field direction will be energetically preferred. We shall use a classical statistical calculation to determine the average orientational dipole moment $\bar{m}_o$ along the field direction. Taking the z axis of spherical coordinates along the field direction, we may show from electrostatics that the extra separable energy of a dipole that makes an angle θ with the field of strength E^* is

$$\varepsilon = -\mu E^* \cos\theta$$

and of course the component of μ along the E^* direction will be $\mu\cos\theta$. From our previous discussion of the classical phase integral in Sec. 1-8 and 1-9 we have

$$\begin{aligned}
\bar{m}_o &= \frac{\int_0^{2\pi} d\phi \int_0^{\pi} \mu\cos\theta\, e^{-\varepsilon/kT} \sin\theta\, d\theta}{\int_0^{2\pi} d\phi \int_0^{\pi} e^{-\varepsilon/kT} \sin\theta\, d\theta} \\[2mm]
&= \frac{\mu\int_0^{\pi} \cos\theta\, e^{E^*\mu\cos\theta/kT} \sin\theta\, d\theta}{\int_0^{\pi} e^{E^*\mu\cos\theta/kT} \sin\theta\, d\theta}
\end{aligned} \tag{3-13}$$

Introduce $y = \mu E^*/kT$ and note that the numerator is the derivative of the denominator, that is,

$$\bar{m}_o = \mu L(y) \tag{3-14}$$

where $L(y)$, known as the Langevin function, is

$$L(y) = \frac{(d/dy) \int_0^\pi e^{y \cos \theta} \sin \theta \, d\theta}{\int_0^\pi e^{y \cos \theta} \sin \theta \, d\theta} = \frac{(d/dy) \int_{-1}^{+1} e^{yp} \, dp}{\int_{-1}^{+1} e^{yp} \, dp}$$

$$= \frac{(d/dy)[(e^y - e^{-y})/y]}{(e^y - e^{-y})/y} = \frac{e^y + e^{-y}}{e^y - e^{-y}} - \frac{1}{y} \tag{3-15}$$

in which we use a change of variable, $p = \cos \theta$. For small y we may expand $L(y)$ to obtain

$$L(y) = \frac{y}{3} - \frac{y^3}{45} + \dots \tag{3-16}$$

Under practical experimental conditions y is very small (e.g., μ values are generally less than 10^{-29} C·m and E^* values are generally less than 10^7 V·m^{-1}, such that even at 150 K $y < 0.05$) and only the leading term in Eq. (3-16) need be retained. Therefore, for a polar molecule in the dielectric the net dipole moment along the field direction is a sum of the induced and orientational contributions:

$$m = m_i + \bar{m}_o = \alpha E^* + \frac{\mu^2 E^*}{3kT} \tag{3-17}$$

Exercise 3-1

Check that the expansion of Eq. (3-15) gives Eq. (3-16) as the first two terms.

The polarization is now Eq. (3-9) with m in place of m_i:

$$P = m\rho = \rho E^* \left(\alpha + \frac{\mu^2}{3kT} \right) \tag{3-18}$$

and the molar polarization remains defined by the right-hand side of Eq. (3-11) but now consists of two terms:

$$P_m \equiv \frac{M}{\rho'} \frac{(D-1)}{(D+2)} = \frac{N_A \alpha}{3\varepsilon_0} + \frac{N_A \mu^2}{9\varepsilon_0 kT} \tag{3-19}$$

Equation (3-19) is called the Debye equation, and it shows that P_m is temperature dependent. In fact, by experimentally measuring D as a function of T and ρ' [by using Eq. (3-3) and measuring capacitance with and without the dielectric sample] a plot of P_m against $1/T$ should give a straight line from which α and μ may be determined.

Example 3-1

The following values have been reported for the dielectric constant of BrF_5 vapor at 1.00 atm pressure [M. Rogers et al., *J. Am. Chem. Soc.* **78**, 44 (1956)].

T/K	D	T/K	D
345.6	1.006320	402.4	1.004910
362.6	1.005824	417.2	1.004603
374.9	1.005525	430.8	1.004378
388.9	1.005180		

Assuming ideal gas behavior for density values, calculate and plot appropriate quantities to determine α and μ for BrF_5.

Solution:

Since we assume an ideal gas at 1.00 atm,

$$\frac{M}{\rho'} = (0.08206)10^{-3} \, T/m^3 \cdot mol^{-1}$$

From Eq. (3-19)

$$\frac{3\varepsilon_0}{N_A} \frac{M}{\rho'} \frac{(D-1)}{(D+2)} = \alpha + \frac{\mu^2}{3kT}$$

$$\varepsilon_0 = 8.8542 \times 10^{-12} \, C/V \cdot m$$

$$\frac{3\varepsilon_0}{N_A} \frac{M}{\rho'} \frac{(D-1)}{(D+2)} = (3.6196 \times 10^{-39}) \, T\left(\frac{D-1}{D+2}\right) C \cdot m^2/V$$

Draw up the table

T/K	$10^3(1/T)/K^{-1}$	D	$\left(\dfrac{D-1}{D+2}\right)10^3$	$10^{39}\left[\dfrac{3\varepsilon_0}{N_A}\dfrac{M}{\rho'}\dfrac{(D-1)}{(D+2)}\right]$ $C \cdot m^2/V$
345.6	2.894	1.006320	2.102	2.629
362.6	2.758	1.005824	1.938	2.544
374.9	2.667	1.005525	1.838	2.494
388.9	2.571	1.005180	1.724	2.427
402.4	2.485	1.004910	1.634	2.380
417.2	2.397	1.004603	1.532	2.313
430.8	2.321	1.004378	1.457	2.272

The data are plotted in Fig. 3-2 to illustrate the good linearity of the results. Using the least squares routine of a pocket calculator on these data,

$$\text{slope} = \frac{\mu^2}{3k} = 6.275 \times 10^{-37}\ C \cdot m^2 \cdot V^{-1} \cdot K$$

$$\mu = 5.10 \times 10^{-30}\ C \cdot m$$

Much of the older data on dipole moments are given in Debye units, D,

$$1\ D = 10^{-18}\ \text{statC} \cdot cm = 3.33564 \times 10^{-30}\ C \cdot m$$

$$\mu = 1.53\ D$$

$$\text{intercept} = \alpha = 8.15 \times 10^{-40}\ C \cdot m^2/V$$

$$\frac{\alpha}{4\pi\varepsilon_0} = 7.32 \times 10^{-30}\ m^3$$

It must be pointed out that the Debye equation is only accurate for gas phase dielectrics or, with simple adaptation, to dilute solutions of polar molecules in a nonpolar solvent. It cannot be used for pure condensed phases (solids or liquids) of polar molecules. Absurd results can appear, such as $(D-1)/(D+2)$ calculated from Eq. (3-19) being greater than unity, which is physically impossible. The reason

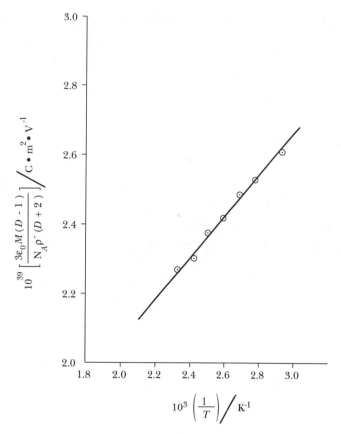

FIGURE 3-2 Debye equation plot for BrF$_5$ vapor at 1.00 atm pressure for the labeled conditions.

for this failure is the inadequate treatment of the actual dipole-dipole interactions between the molecules in dense polar phases, which in the present simple theory are only implicitly introduced by the second term in Eq. (3-7) for the local field E^*. Further discussion of this matter may be found in reference [4].

In Table 3-1 we list polarizabilities, $\alpha/4\pi\varepsilon_0 = r^3$ values, dipole moments, and liquid boiling points (at 1 atm) for a large number of species. The larger and looser a molecule's electron cloud, the more easily it can be deformed by an external field. Hence we should expect that $(\alpha/4\pi\varepsilon_0)^{1/3}$ should roughly parallel molecular size. This is shown to be true in Table 3-2, which compares this quantity with the van der Waals effective hard sphere radius $\sigma/2$ of the molecule from the relation $b = 4N_A[4/3\pi(\sigma/2)^3]$. Except for He and H$_2$, which have only two electrons, r is about 85 percent of $\sigma/2$.

So far we have implicitly assumed that the experimental technique of measuring D involves measurement of capacitance in an electrical circuit with and

TABLE 3-1   Dipole Moments μ, Polarizabilities, α, Liquid Boiling Points BP

	$10^{30}\ \mu/\text{C}\cdot\text{m}$	$10^{40}\ \alpha/\text{C}\cdot\text{m}^2\cdot\text{V}^{-1}$	$10^{30}\left(\dfrac{\alpha}{4\pi\varepsilon_0}\right)/\text{m}^3$	BP/K
		Inorganic Species		
He	0	0.23	0.21	4.2
Ne	0	0.45	0.40	27.2
Ar	0	1.81	1.63	87.5
Kr	0	2.76	2.48	120.9
Xe	0	4.45	4.00	166.1
H_2O	6.14	1.66	1.49	373.2
HF	6.37	0.91	0.80	292.7
HCl	3.44	2.93	2.63	188.3
HBr	2.64	4.02	3.61	206.2
HI	1.27	6.06	5.45	237.8
SO_2	5.37	4.14	3.72	263
NH_3	4.90	2.51	2.26	239.8
CO	0.33	2.17	1.95	81.7
HCN	9.77	2.88	2.59	299
CO_2	0	2.95	2.65	195 (sub)
H_2	0	0.88	0.79	20.3
N_2	0	1.96	1.76	77.4
O_2	0	1.78	1.60	90.2
Cl_2	0	5.13	4.61	238.6
		Organic Species		
CH_4	0	2.89	2.60	112
C_6H_6	0	11.5	10.3	353
CCl_4	0	11.7	10.5	350
C_2H_6	0	4.97	4.47	184.6
C_3H_8	0	7.00	6.29	231.0
$(CH_3)_3CH$	0	9.10	8.14	263.0
$(CH_3)_2C = CH_2$	1.6	9.22	8.29	267
C_2H_4O	6.27	5.8	5.2	283.9
$(CH_3)_2O$	4.30	5.74	5.16	249.6
$(CH_3)_3N$	2.0	9.22	8.29	276.7
$p\text{-}C_6H_4(NO_2)_2$	0	20.5	18.4	572 (sub)

SOURCE: *Landolt-Börnstein Tabellen* (Tables), vol. I. Springer, Berlin, 1951. Part 3, pp. 386–517.

TABLE 3-1 (*continued*)

	$10^{30}\ \mu/\text{C}\cdot\text{m}$	$10^{40}\ \alpha/\text{C}\cdot\text{m}^2\cdot\text{V}^{-1}$	$10^{30}\left(\dfrac{\alpha}{4\pi\varepsilon_0}\right)/\text{m}^3$	BP/K
$m\text{-}C_6H_4(NO_2)_2$	12.6	21	19	576
$o\text{-}C_6H_4(NO_2)_2$	20.0	21	19	592
CH_3Cl	6.20	5.07	4.56	249.0
CH_2Cl_2	5.27	7.21	6.48	313
$CHCl_3$	3.40	9.16	8.23	334.9
C_6H_5Cl	5.67	13.6	12.25	405
$p\text{-}C_6H_4Cl_2$	0	16.1	14.47	446
$m\text{-}C_6H_4Cl_2$	5.57	15.8	14.23	445
$o\text{-}C_6H_4Cl_2$	8.64	15.8	14.17	454
CH_3OH	5.70	3.59	3.23	338.2
$(CH_3)_2C{=}O$	9.47	7.04	6.33	329.7

without the dielectric. The circuit is usually an alternating current circuit with frequency of electrical field reversal very low compared to any frequency $(\Delta E/h)$ characteristic of energy excitations of the molecules of the dielectric. If the frequency could be raised to the microwave region $(10^{12}\ \text{s}^{-1})$, polar molecules could not align themselves quickly enough to give rise to a net orientation polarization $P_o \equiv \mathcal{N}_A\mu^2/9\varepsilon_0 kT$, and so this term is suppressed in Eq. (3-19); that is, it cannot

TABLE 3-2 Comparisons of Molecular Size from van der Waals b and Polarizability Value α. $\sigma/2 = (3b/16\pi N_A)^{1/3}$

Molecule	$10^6\ b/\text{m}^3\cdot\text{mol}^{-1}$	$(\sigma/2)/\text{pm}$	$(\alpha/4\pi\varepsilon_0)^{1/3}/\text{pm}$
He	23.7	133	59
H_2	26.6	138	92
O_2	31.8	147	117
N_2	39.1	157	121
HCl	40.8	159	138
CH_4	42.8	162	138
Xe	51.0	172	159
Cl_2	56.2	177	166
$(CH_3)_2C{=}O$	99.4	214	185
CCl_4	138.3	239	219

physically contribute under the conditions of measurement. The remaining term can be written as a sum

$$\frac{\mathcal{N}_A \alpha}{3\varepsilon_0} = P_a + P_{el} \tag{3-20}$$

of atomic polarization P_a due to motions of the atomic nuclei (generally a very small term) and electronic polarization P_{el} due to electron response to the alternating electric field. For frequencies in the infrared $(10^{13}–10^{14} \text{ s}^{-1})$ the atoms will no longer be able to follow the field changes and P_a will also be strictly zero (recall that vibrational transitions are in the infrared). At still higher frequencies (visible and ultraviolet), only P_{el} will remain. The schematic variation of molar polarization with frequency is shown in Fig. 3-3. The type of measurement at such high frequencies cannot be with an alternating current circuit, for the only practical source of such rapidly alternating electric field vectors is simply electromagnetic radiation of the required frequency. After all, such radiation consists precisely of alternating electric field vectors! The quantity measured is the index of refraction,

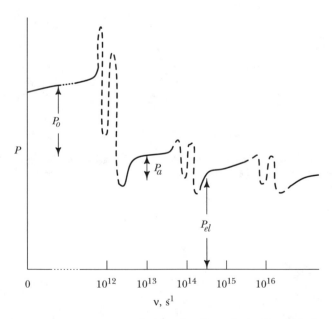

FIGURE 3-3 Representation of the variation of molar polarization with frequency of electric field reversal. As frequency increases, orientational polarization P_o and atomic polarization P_a are successively washed out until only electronic polarization P_{el} remains. Anomalous dispersion (maxima and minima) is a quantum effect that occurs when the frequency matches particular rotational, vibrational, or electronic transitions of the molecules of the dielectric.

SOURCE: N. Davidson. *Statistical Mechanics*. McGraw Hill, New York, 1962.

n_r, of the dielectric for such radiation. As Maxwell first derived, $D(v)$, the dielectric constant at frequency v, is given by

$$D(v) = n_r^2 \qquad (3\text{-}21)$$

At such high frequencies, with the orientation effect washed out the $D(v)$ values will not be large and Eq. (3-19) in the form

$$\frac{M}{\rho'}\left(\frac{n_r^2 - 1}{n_r^2 + 2}\right) = \frac{N_A \alpha}{3\varepsilon_0} = P_a + P_{el} \qquad (3\text{-}22)$$

will be accurate for condensed phases as well as gaseous phases of any type of molecule. The index of refraction is the ratio of speed of light in vacuum to that in the dielectric medium. The speed in the medium is slower as the wavelength is shorter (the frequency is higher) because higher frequency induces a higher frequency of oscillation among the electrons of the dielectric. This is the basis of the dispersion of light by a glass prism: n_r for blue light is larger than that of red light and blue rays are bent more than are red. For a detailed physical explanation of this phenomenon the treatment by Feynman [5] should be studied.

The quantity R_m, the molar refractivity,

$$R_m = \frac{M}{\rho'}\left(\frac{n_r^2 - 1}{n_r^2 + 2}\right) \qquad (3\text{-}23)$$

using a fixed wavelength is nearly independent of temperature, pressure, and state of aggregation of the molecules. Using visible radiation (the only kind for which n_r can be conveniently measured), R_m is strictly equal to P_{el}. To good accuracy atoms and particular bonding arrangements (e.g., π electrons of double and triple bonds) make constant, additive contributions to the molar refractivity of any molecule of which they are a part. Table 3-3 lists such contributions of atomic groups deduced from an analysis of many organic compounds using yellow sodium light ($\lambda = 589$ nm).

Example 3-2

Use data in Table 3-3 to estimate n_r of liquid methyl phenylamine $C_6H_5NHCH_3$ at 20°C ($\rho' = 0.9891$ g/cm^3). Compare with the experimental value (using $\lambda = 589$ mm), $n_r = 1.5684$.

Solution:

From Table 3-3:

$$R_m = 25.463 + 3.550 + 2.591 + 4(1.028) = 35.716 \text{ cm}^3/\text{mol}$$

TABLE 3-3   Molar Refractivities at $\lambda = 589$ nm

Group	$R_m/\text{cm}^3 \cdot \text{mol}^{-1}$	Group	$R_m/\text{cm}^3 \cdot \text{mol}^{-1}$
C	2.591	$C_{10}H_7$	43.00
H	1.028	$>$S	7.729
$=$O	2.122	$=$S	7.921
$>$O	1.643	C$\equiv$N	5.459
OH	2.553	N (primary aliphatic)	2.376
F	0.81		
Cl	5.844	N (secondary aliphatic)	2.582
Br	8.741	N (aromatic)	3.550
I	13.954	Added ethylenic double bond contribution	1.575
C_6H_5	25.463	Added acetylenic triple bond contribution	1.977

SOURCE: *Handbook of Chemistry and Physics*, 56th ed. Chemical Rubber Pub. Co., 1975.

Solving Eq. (3-23):

$$n_r^2 = \frac{1 + 2r}{1 - r} \qquad r = \frac{R_m}{M/\rho'} = \frac{35.716}{107.2/0.9891} = 0.32954$$

$$n_r^2 = \frac{1.6591}{0.67046} = 2.4746$$

$$n_r = 1.573$$

This method is quite accurate. Note also that n_r will have a small temperature dependence because of the temperature dependence of ρ'.

3-2

Intermolecular Forces

In dealing with statistical thermodynamics in the last chapter we neglected molecular interactions. However these forces can only be ignored in systems of low density, low pressure, and high temperature. To discuss imperfect gases and liquids in this chapter we must include intermolecular forces. The sources of these are the electrical properties of molecules described in the preceding section. We shall in fact focus our attention on the average potential energy of pairwise molecular interaction $\varepsilon(r)$, assumed to depend only on molecule-molecule separation r (any angle dependence of the potential energy is assumed to have been averaged over as described below). This function is related to intermolecular force f by the equation

$$f = \frac{-d}{dr}\,\varepsilon(r) \tag{3-24}$$

We will discuss first the comparatively long range attractive forces between neutral molecules. These are called van der Waals forces and are of three kinds: dipole-dipole, dipole-induced dipole, and induced dipole-induced dipole or dispersion forces. The actual calculation of these forces is complex and lengthy and best left to advanced works [e.g., 6, 7]. Here we merely quote the results for each $\varepsilon(r)$ function and qualitatively justify them.

The potential energy for a pair of molecules with permanent electric dipoles μ_1, μ_2 arises from the orientation-dependent attraction and repulsion that the dipoles exert upon each other. If the two dipoles were oriented completely at random with respect to each other this potential energy would average to zero, because repulsive orientations would occur as frequently as attractive ones. However the Boltzmann distribution slightly favors the lower energy attractive orientations. A calculation analogous to that of Eq. (3-13) results in

$$\varepsilon_{or}(r) = \frac{-2\mu_1^2\mu_2^2}{3kTr^6(4\pi\varepsilon_0)^2} \tag{3-25}$$

The competition between favorable orientation and randomizing thermal motion is reflected in the factor of kT in the denominator, which causes $\varepsilon_{or}(r)$ to decrease as the temperature rises.

The presence of a polar molecule with dipole moment μ_1 near a second molecule (which may itself be polar or nonpolar) will induce a transient dipole in this molecule, which must by electrostatics be aligned so as to attract the dipole of the first molecule. The magnitude of the transient dipole will be proportional to α_2. Since the interaction is always favorable, no temperature dependence appears in the result, which for a single pair is

$$\varepsilon_{in}(r) = \frac{-\mu_1^2\alpha_2}{(4\pi\varepsilon_0)^2 r^6} - \frac{\mu_2^2\alpha_1}{(4\pi\varepsilon_0)^2 r^6} \tag{3-26}$$

The second term accounts for the transient induced dipole in molecule 1 if molecule 2 is also polar.

Finally, there is a force between two molecules even if both are nonpolar. It arises from the same factors that cause light of different wavelengths to have different speeds inside a dielectric (light dispersion as mentioned in Sec. 3-1), and so this force is termed a dispersion force. A small instantaneous dipole is always present in any electron distribution, controlled as it is by a probabilistic wave function. This dipole induces another dipole in any nearby electron cloud, which in turn interacts attractively with the original instantaneous dipole. Averaging these effects in a quantum mechanical calculation (first done by F. London) gives the attractive potential energy for a pair of like molecules to be approximately

$$\varepsilon_d(r) = -\frac{3}{4}\frac{h\nu\alpha^2}{(4\pi\varepsilon_0)^2 r^6} \tag{3-27}$$

where ν is a characteristic frequency of oscillation of the charge distribution of the molecule. A usual further approximation is to put $h\nu = I$, where I is the ionization potential of the molecule. If the two molecules differ,

$$\varepsilon_d(r) = \frac{-3I_1 I_2 \alpha_1 \alpha_2}{2(I_1 + I_2)(4\pi\varepsilon_0)^2 r^6} \tag{3-28}$$

Exercise 3-2

Show that the expressions for ε_{or}, ε_{in}, and ε_d are all of energy dimensions.

Generally, the dispersion energy, which increases as the square of the electron cloud volume of the molecule, contributes more to the potential energy of interaction than the orientation and induction energies, even for polar molecules. This is illustrated by the data in Table 3-4, where only the very polar NH_3, H_2O, and HCN molecules are exceptions.

Exercise 3-3

Verify some of the numerical entries in Table 3-4 from the data in Table 3-1 and reference to ionization potential data. (Generally all molecule ionization potentials are of order 10 eV.)

At very small molecular separations the attractive effects will be negligible compared to the repulsion engendered by the incipient overlap of electron clouds. This may be represented by a repulsive energy term, ε_r. No precise theoretical equation

TABLE 3-4 Contribution to the Intermolecular Potential Energy of Two Like Gaseous Molecules at 25°C

Substance	$-10^{79}\, \varepsilon r^6/\mathrm{J}\cdot\mathrm{m}^6$		
	Orientation (Dipole-Dipole)	Dipole-Induced Dipole	Dispersion
He	0	0	1.3
Ar	0	0	50.1
Xe	0	0	233
CCl_4	0	0	1524
C_3H_8	0	0	528
C_6H_6	0	0	1186
CO	0.0016	0.038	64.0
HCl	18.3	5.60	106
HBr	6.35	4.53	182
HI	0.34	1.58	370
NH_3	75.4	9.74	62.4
CH_2Cl_2	101	32.3	573
SO_2	109	19.3	205
H_2O	186	10.1	33.7
HCN	1192	44.4	111

has been worked out for this term. A useful empirical form that is commonly used is

$$\varepsilon_r(r) = \frac{B}{r^n} \tag{3-29}$$

where B is a positive constant and n is an integer that may range from 9 to 15. For convenience we will take $n = 12$ (since r^{-12} is the square of r^{-6}), and adding all contributions to the intermolecular potential energy of a pair of like molecules we find

$$\varepsilon(r) = \varepsilon_r(r) + \varepsilon_{or}(r) + \varepsilon_{in}(r) + \varepsilon_d(r)$$

$$= \frac{B}{r^{12}} - \frac{A}{r^6} \tag{3-30}$$

where

$$A = \frac{2\mu^4}{3(4\pi\varepsilon_0)^2 kT} + \frac{2\alpha\mu^2}{(4\pi\varepsilon_0)^2} + \frac{3I\alpha^2}{4(4\pi\varepsilon_0)^2} \tag{3-31}$$

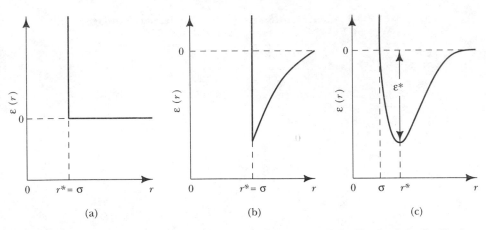

FIGURE 3-4 Model intermolecular potential energy functions. a. Hard sphere. b. Hard sphere with attractive tail. c. Lennard-Jones.

It is convenient to rewrite Eq. (3-30) in terms of two natural parameters: ε^*, the depth of the curve at its minimum, and r^*, the molecular separation at the minimum as shown in Fig. 3-4. Using the conditions

$$\left(\frac{\partial \varepsilon}{\partial r}\right)_{r=r^*} = 0 \quad \text{and} \quad \varepsilon(r^*) = -\varepsilon^*$$

the parameters A and B may be expressed in terms of r^* and ε^*:

$$A = 2\varepsilon^*(r^*)^6 \tag{3-32.1}$$

$$B = \varepsilon^*(r^*)^{12} \tag{3-32.2}$$

such that Eq. (3-30) becomes

$$\varepsilon(r) = \varepsilon^*\left[\left(\frac{r^*}{r}\right)^{12} - 2\left(\frac{r^*}{r}\right)^6\right] \tag{3-33}$$

If we further define σ (roughly a collision diameter of the molecules) to be the finite separation at which $\varepsilon(\sigma) = 0$, we find

$$r^* = 2^{1/6}\sigma \tag{3-34}$$

and Eq. (3-33) becomes

$$\varepsilon(r) = 4\varepsilon^*\left[\left(\frac{\sigma}{r}\right)^{12} - \left(\frac{\sigma}{r}\right)^6\right] \tag{3-35}$$

The form of Eq. (3-33) or (3-35) is termed the Lennard-Jones (LJ) potential, after J. E. Lennard-Jones, who first chose $n = 12$ in Eq. (3-29).

Exercise 3-4

Derive Eqs. (3-32) from the two conditions specifying the minimum of the intermolecular potential energy curve.

Figure 3-4 shows the LJ potential and two other model potentials: that for hard spheres [$\varepsilon(r) = \infty$ for $r \leq \sigma$, $\varepsilon(r) = 0$ for $r > \sigma$], and that for hard spheres with an attractive tail. These latter will be discussed in Secs. 3-4 and 3-7 and are much less realistic than the LJ potential.

The constants ε and σ obtained by fitting P, V, T data yielding second virial coefficients (see Sec. 3-4) or from gas viscosity data (see Sec. 5-4) for a number of gases are listed in Table 3-5 and illustrated in Fig. 3-5. Although the LJ potential is most realistic for nonpolar molecules with spherical electron distributions, even for such molecules the LJ function does not represent the true shape of the intermolecular potential. Recent monographs [7, 8] review more precise approximations for this function.

TABLE 3-5   Lennard-Jones Potential Parameters

Substance	$(\varepsilon^*/k)/K$	$10^{10}\,\sigma/m$	Substance	$(\varepsilon^*/k)/K$	$10^{10}\,\sigma/m$
He	10.22	2.56	CO_2	189	4.49
H_2	37.00	2.93	CH_4	148.2	3.82
Ne	35.60	2.75	CCl_4	327	5.88
Ar	122	3.40	Cl_2	357	4.12
Kr	171	3.60	Br_2	520	4.27
Xe	221	4.10	I_2	550	4.98
N_2	95.05	3.70	HCl	360	3.30
O_2	118	3.46	HI	324	4.12
CO	100.2	3.76	CH_2Cl_2	406	4.76

SOURCE: J. O. Hirschfelder, C. F. Curtiss, and R. B. Bird. *The Molecular Theory of Gases and Liquids.* Wiley, New York, 1954.

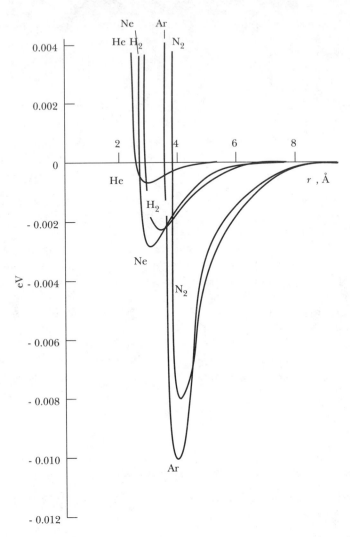

FIGURE 3-5 Lennard-Jones potential energy functions for several specific molecules. Source: W. J. Moore. *Physical Chemistry*, 4th ed. Prentice-Hall, 1972.

The attractive or cohesive energy $\varepsilon_c(r)$ of one pair of molecules is given by the second term of Eq. (3-30):

$$\varepsilon_c(r) = -\frac{A}{r^6} \tag{3-36}$$

It is interesting to estimate the average cohesive energy between all the molecules in a gas or liquid. We may do this approximately by assuming a random distribution of molecules about any one molecule such that with $\rho = N/V$ the number of molecules in a spherical shell at distance r from a particular molecule is

$$dN = \rho \, dV = 4\pi r^2 \rho \, dr$$

and then integrating from the smallest r value, which is σ, to some macroscopic distance R characterizing the size of the system, so as to include all the $(N-1)$ other molecules to obtain the interaction energy of one pair averaged over all radial separations:

$$\overline{(\varepsilon_c)_1} = \frac{1}{(N-1)} \int \varepsilon_c(r) \, dN = \frac{\rho}{(N-1)} \int_{r=\sigma}^{r=R} \frac{-A}{r^6} 4\pi r^2 \, dr$$

$$= -\frac{4\pi A \rho}{(N-1)} \left(\frac{-1}{3R^3} + \frac{1}{3\sigma^3} \right) = \frac{-4\pi \rho A}{3\sigma^3 (N-1)} \tag{3-37}$$

The term in R^{-3} is negligible compared to that of σ^{-3}. The total number of pairs in the system is $N(N-1)/2$, and thus we find for the total cohesive energy

$$U_c = \frac{N(N-1)}{2} \overline{(\varepsilon_{c_1})} = \frac{-2}{3} \frac{\pi N \rho A}{\sigma^3} \tag{3-38}$$

which on a molar basis becomes

$$U_{c_m} = \frac{-2\pi N_A^2 A}{3\sigma^3 V_m} \tag{3-39}$$

If we deal with a liquid of small molar volume compared with the molar volume of the equilibrium gas phase such that U_{c_m} in the gas phase is negligible, we can approximate ΔU_{vap} by

$$U_{vap_m} = |U_{c_m}| = \frac{2\pi N_A^2 A}{3\sigma^3 V_m} \tag{3-40}$$

Example 3-3

Estimate ΔH_{vap_m} for liquid CCl_4 at 25°C, $\rho' = 1.59$ g/cm^3, using molecular parameters of CCl_4 and using LJ parameters of CCl_4. Comment on the discrepancies between them and with the experimental result, $\Delta H_{vap_m} = 32.8$ kJ·mol^{-1}.

Solution:

From molecular parameters (Table 3-4) for CCl_4

$$A = 1.52 \times 10^{-76} \, \text{J·m}^6$$

From Eq. (3-32.1)

$$A = 4\varepsilon^*\sigma^6 = 4(327)(1.38 \times 10^{-23})(5.88 \times 10^{-10})^6$$

$$= 7.46 \times 10^{-76} \text{ J} \cdot \text{m}^6 \qquad \text{(Table 3-5)}$$

Here we note that A from LJ parameters is almost five times that from Eq. (3-31). In general, A from LJ parameters is always at least twice that calculated from Eq. (3-31) and this is a reflection of the failure of the LJ model potential to be consistent with the more fundamental molecular properties (α and μ). However, reasonable estimates are obtainable if the collision diameter σ in Eq. (3-40) for molecular properties is chosen as $2(\alpha/4\pi\varepsilon_0)^{1/3}$ and as the LJ σ when A is obtained from Eq. (3-32.1). Thus from molecular properties, $V_m = 153.8/1.59 = 96.4 \text{ cm}^3/\text{mol} = 96.4 \times 10^{-6} \text{ m}^3/\text{mol}$

$$|U_{cm}| = \frac{2\pi}{3} \frac{\mathcal{N}_A^2(1.52 \times 10^{-76})}{(96.4 \times 10^{-6})(8)(10.5 \times 10^{-30})}$$

$$= 14.3 \times 10^3 \text{ J} \cdot \text{mol}^{-1}$$

$$\Delta H_{vap_m} = |U_{cm}| + RT = 16.8 \times 10^3 \text{ J} \cdot \text{mol}^{-1}$$

From effective LJ parameters:

$$|U_{cm}| = \frac{2\pi}{3} \frac{\mathcal{N}_A^2(7.46 \times 10^{-76})}{(96.4 \times 10^{-6})(5.88 \times 10^{-10})^3}$$

$$= 28.9 \times 10^3 \text{ J} \cdot \text{mol}^{-1}$$

$$\Delta H_{vap_m} = 31.4 \times 10^3 \text{ J} \cdot \text{mol}^{-1}$$

The almost precise match of the LJ estimate with the experimental result ($32.8 \text{ kJ} \cdot \text{mol}^{-1}$) is fortuitous.

The strength of intermolecular attraction, and its relation to liquid phase normal boiling points dependent only on molecular polarizability and dipole moment (see Problems 6 and 7), as so far presented fails completely for molecules that have an exposed hydrogen atom bonded to the very small (thus highly electronegative) N, O, or F atoms. Molecular association via hydrogen bonds then occurs and leads to abnormally high boiling points for such substances as alcohols, amines, and HF and H_2O as compared to compounds of comparable polarizability or to the congeners in higher periods (HCl and H_2S, respectively) for the last two. Hydrogen bonds can hold complex structures together with definite relative orientations. The well-known Watson-Crick [9] model for the structure of deoxyribonucleic acid (DNA) depends on hydrogen bonds.

Hydrogen bonding is not fully understood theoretically. Although it is primarily an electrostatic interaction, there is a significant amount of covalent character to it due to the sharing of a lone pair of electrons on the N, O, or F atom with the hydrogen. From experiment we know that the magnitude of hydrogen bond energies (generally in the range of $10–50$ kJ·mol^{-1}) is greater than van der Waals interaction energies (0.4 to 8 kJ·mol^{-1}) but substantially less than normal covalent bond energies ($120–1000$ kJ·mol^{-1}).

3-3
Imperfect Gases: Canonical Partition Function

As early as 1901 experimentalists such as K. Onnes in the Netherlands began to represent experimental data on real gases in a series expansion in the number density $\rho = N/V$:

$$P/kT = \rho + B_2(T)\rho^2 + B_3(T)\rho^3 + \ldots \tag{3-41}$$

This is now called the virial series, and the virial coefficients B_i are functions of temperature only. This technique implicitly contained the insight that successive deviations from the ideal gas law may be treated as interactions of molecular pairs, of molecular triplets, and so on. This will be confirmed in our derivation of this series using statistical mechanics. Indeed, the virial expansion is a Taylor series expansion about the $\rho = 0$, ideal gas state:

$$P/\rho kT = 1 + \left[\frac{\partial(P/\rho kT)}{\partial\rho}\right]_{T,\rho=0}\rho + \frac{1}{2!}\left(\frac{\partial^2(P/\rho kT)}{\partial\rho^2}\right)_{T,\rho=0}\rho^2 + \ldots \tag{3-42}$$

such that

$$B_n(T) = \frac{1}{(n-1)!}\left[\frac{\partial^{n-1}(P/\rho kT)}{\partial\rho^{n-1}}\right]_{T,\rho=0} \tag{3-43}$$

This shows that the $B_n(T)$ are properties of a gas extrapolated to the limit of zero density and also makes clear that experimentally it is difficult to accurately obtain B_n values for $n > 2$.

Using the classical canonical partition function, we shall first indicate a way to obtain the start of the virial series; then in Sec. 3-4 we shall give a more general prescription using the grand canonical partition function. We shall neglect any coupling between interacting center of mass motion and the internal vibrations and rotations of a molecule. This is accurate for the vibrations since vibrational energy differences are generally comparable to or larger than the depth of the attractive well in the potential energy function (ε^* in Eq. 3-35). It is less accurate for rotational motion, but for spherically symmetric molecules the effects average out to zero. The electronic levels of a molecule may be substantially affected by inter-molecular forces, but because at ordinary temperatures the population of the excited electronic states is usually negligible, it is still a good approximation to neglect the effect of intermolecular forces on the internal electronic partition function as well.

By a straightforward adaptation of Eqs. (1-181) and (1-185) we write the classical canonical partition function as

$$Q_N^{cl} = \frac{q_{int}^N Z_N}{\lambda^{3N} N!}$$
(3-44)

where q_{int} given by Eq. (2-15) is the single molecule internal partition function (a function of temperature only) and Z_N is the configuration integral given by Eq. (1-182):

$$Z_N = \int \cdots \int \exp\left(\frac{-V_N}{kT}\right) d\bar{r}_1 \ldots d\bar{r}_N$$
(3-45)

By $\int d\bar{r}_i$ we mean a threefold volume integration over the position of the ith particle relative to some origin.

We will make the further assumption that the interaction energy of the N particles, V_N, can be written as a sum of pairwise potentials each only dependent on the radial separation r_{ij} of molecules i and j—that is, on $r_{ij} = |\bar{r}_j - \bar{r}_i|$. Advanced works [6–8] treat cases where one or both of these assumptions are not made.

$$V_N = \sum\sum_{1 \leq i < j \leq N} \varepsilon(r_{ij})$$
(3-46)

The limits on the sums in Eq. (3-46) result from the fact that there are N ways to select the first molecule of a pair and $N - 1$ ways to select the second. The order of choosing the terms does not matter and the total number of pairs is $N(N - 1)/2$, and the i and j must always differ. Introducing Eq. (3-46) into Eq. (3-45) results in a repeated product expression in the integrand:

$$Z_N = \int \cdots \int d\bar{r}_1 \cdots d\bar{r}_N \prod_{1 \leq i < j \leq N} (e^{-\varepsilon(r_{ij})/kT})$$
(3-47)

It is useful to introduce the Mayer cluster function [10], defined as

$$f_{ij}(r_{ij}) = (e^{-\varepsilon(r_{ij})/kT} - 1)$$
(3-48)

This function is illustrated and compared with the $\varepsilon(r_{ij})$ function in Fig. 3-6. It is negative for very small r_{ij}, quickly rises to a positive maximum, and is zero for r_{ij} larger than about twice the molecular diameter. It is a very short range function.

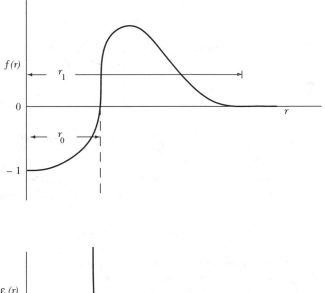

FIGURE 3-6 A Mayer cluster function plotted versus the intermolecular separation r correlated with the intermolecular potential energy function versus r. r_0 is a measure of the hard core diameter, r_1 is a measure of the range of the potential energy function.

Thus Eq. (3-47) becomes

$$Z_N = \int \ldots \int d\bar{r}_1 \ldots d\bar{r}_N \prod_{1 \le i < j \le N} (1 + f_{ij}) \tag{3-49}$$

Expansion of the repeated product will give schematically

$$Z_N = \int \ldots \int d\bar{r}_1 \ldots$$

$$d\bar{r}_N \left(1 + \sum_{1 \le i < j \le N} \sum f_{ij} + \sum\sum f_{ij} f_{i'j'} + \ldots \right) \tag{3-50}$$

Exercise 3-4

Work out in detail the full repeated product in Eq. (3-49) for $N = 3$ and $N = 4$.

The leading term (unity) in the bracket gives V^N when integrated and will be seen to yield the ideal gas result. We will assume that retaining only the terms in the first double sum will be sufficient to obtain the first-order corrections to the ideal gas law. Since the molecular coordinates are all integrated over, each of the $N(N-1)/2 \simeq N^2/2$ terms f_{ij} in the sum retained gives the same result:

$$V^{N-2} \int d\bar{r}_1 \, d\bar{r}_2 f_{12} = V^{N-2} \, V \int d\bar{r}_{12} f_{12} \tag{3-51}$$

In Eq. (3-51) we make use of the important cluster property of the f_{ij}, namely that it vanishes unless i and j are within a few molecular diameters of each other, and so we may choose one molecule as origin and measure the position of the other molecule relative to the one chosen. We put $d\bar{r}_1 \, d\bar{r}_2 = d\bar{r}_1 \, d\bar{r}_{12}$, and since molecule 1 may move through the entire gas volume, the integral over $\int d\bar{r}_1$ yields a factor V. The $d\bar{r}_{12}$ represents a threefold integration over relative coordinates with $d\bar{r}_{12} = 4\pi r_{12}^2 \, dr_{12}$ for angle-independent potentials, with r_{12} varying from 0 to ∞ (since f_{12} vanishes for r_{12} beyond 1 nm, details of the actual upper limit on r_{12} are irrelevant and ∞ may be used).

If we define

$$b_2 = \frac{1}{2} \int f_{12} \, d\bar{r}_{12} = 2\pi \int_0^\infty f_{12} r_{12}^2 \, dr_{12} \tag{3-52}$$

then we have from Eq. (3-50) in this approximation

$$Z_N = V^N + \frac{N^2}{2} V^{N-1}(2b_2) = V^N \left(1 + \frac{N^2 b_2}{V}\right) \tag{3-53}$$

We hope $N^2 b_2/V$ is small compared to unity (it is zero for ideal gases).

In the canonical ensemble we have

$$A = -kT \ln Q_N^{cl}, \qquad P = -\left(\frac{\partial A}{\partial V}\right)_T$$

such that from Eq. (3-44)

$$P = kT \left(\frac{\partial}{\partial V}\right)_T \ln \left(\frac{q_{int}^N}{\lambda^{3N} N!}\right) + kT \left(\frac{\partial}{\partial V}\right)_T \ln Z_N$$

and the first term is zero since the logarithm is a function of T only. Thus

$$P = kT\left(\frac{\partial}{\partial V}\right)_T \ln V^N + kT\left(\frac{\partial}{\partial V}\right)_T \ln\left(1 + \frac{N^2 b_2}{V}\right) \qquad (3\text{-}54.1)$$

$$= \frac{NkT}{V} - \frac{N^2 b_2}{V^2} \frac{kT}{1 + \left(\frac{N^2 b_2}{V}\right)} \qquad (3\text{-}54.2)$$

$$= \frac{NkT}{V} - \frac{N^2 b_2 kT}{V^2}\left(1 - \frac{N^2 b_2}{V}\right) \qquad (3\text{-}54.3)$$

If $N^2 b_2/V$ is small compared to unity it may be discarded, and on comparison of Eq. (3-41) and Eq. (3-54.3) we may identify B_2:

$$B_2(T) = -b_2 \qquad (3\text{-}55)$$

The result in Eq. (3-55) is correct but the derivation presented here is wrong because $N^2 b_2/V$ is not small. It is huge.

Exercise 3-5

If b_2 is of the order of the volume of a single molecule (note Eq. 3-76, below) estimate the magnitude of $N^2 b_2/V$ for a gas at STP.

The terms in the further sums in Eq. (3-50) should not have been discarded and must be treated properly to derive Eq. (3-55) correctly. This has been done by Mayer and others [10, 11] but is too complex to present here. The only brief *and* correct derivation involves the use of the grand canonical partition function and is given in the next section. Much of our presentation in this section will be found useful there as well.

3-4
Imperfect Gases: Grand Canonical Partition Function

From Eqs. (1-63) and (1-64) the grand canonical partition function is

$$\mathcal{Q}(z, V, T) = e^{PV/kT} = \sum_{N=0}^{\infty} z^N Q_N(V, T) \qquad (3\text{-}56)$$

where

$$z = e^{\mu/kT} \tag{3-57}$$

To be consistent with the meaning of a zero particle case, $Q_0(V, T) = 1$. The one-particle partition function $Q_1(V, T)$ is, from Chap. 2, given by

$$Q_1(V, T) = \frac{V}{\lambda^3} q_{int} \tag{3-58}$$

for classical particles but we do *not* need to assume classical particles here. We may define

$$\mathcal{Z}_N = \frac{N! Q_N(V, T) V^N}{Q_1^N} \tag{3-59}$$

If we do deal with classical particles, use of Eq. (3-58) gives

$$Q_N^{cl} = \frac{(q_{int})^N \mathcal{Z}_N}{\lambda^{3N} N!} \tag{3-60}$$

which by comparison with Eq. (3-44) shows that $\mathcal{Z}_N$ is the configuration integral given by Eq. (3-45). Now insert Eq. (3-59) into Eq. (3-56) to obtain

$$e^{PV/kT} = \mathcal{Q}(z, V, T) = \sum_{N=0}^{\infty} \frac{z^N Q_1^N \mathcal{Z}_N}{V^N N!} \tag{3-61.1}$$

$$= 1 + \sum_{N=1}^{\infty} \zeta^N \frac{\mathcal{Z}_N}{N!} \tag{3-61.2}$$

where

$$\zeta = \frac{z Q_1}{V} \tag{3-62.1}$$

and for classical particles

$$\zeta = \frac{z}{\lambda^3} q_{int} \tag{3-62.2}$$

We will find below that with these definitions, as the number density $\rho \to 0$ then $\zeta \to \rho$. Hence we will assume that the pressure can be expanded in a power series in ζ according to

$$P/kT = \sum_{j=1}^{\infty} b_j(T)\zeta^j \qquad (3\text{-}63)$$

and work backward to determine the initially unknown $b_j(T)$ in terms of Z_N. We insert Eq. (3-63) into Eq. (3-61.1) and begin to expand the exponential using

$$e^x = 1 + x + \frac{x^2}{2} + \frac{x^3}{6} + \ldots$$

$$e^{PV/kT} = 1 + \zeta Z_1 + \zeta^2 \frac{Z_2}{2} + \zeta^3 \left(\frac{Z_3}{6}\right) = \exp\left(V \sum_j b_j \zeta^j\right)$$

$$= 1 + V(b_1\zeta + b_2\zeta^2 + b_3\zeta^3 + \ldots)$$

$$+ \frac{1}{2} V^2(b_1\zeta + b_2\zeta^2 + \ldots)^2$$

$$+ \frac{1}{6} V^3(b_1\zeta + \ldots)^3 \qquad (3\text{-}64.1)$$

which becomes

$$1 + \zeta Z_1 + \frac{1}{2}\zeta^2 Z_2 + \frac{1}{6}\zeta^3 Z_3 = 1 + V b_1\zeta + \zeta^2\left(b_2 V + b_1^2 \frac{V^2}{2}\right)$$

$$+ \zeta^3\left(b_3 V + b_1 b_2 V^2 + b_1^3 \frac{V^3}{6}\right) \qquad (3\text{-}64.2)$$

Equating coefficients of like powers of ζ on both sides of the equality we find

$$Z_1 = V b_1 \qquad (3\text{-}65.1)$$

$$\frac{Z_2}{2} = V b_2 + \frac{b_1^2 V^2}{2} \qquad (3\text{-}65.2)$$

$$\frac{Z_3}{6} = V b_3 + V^2 b_1 b_2 + \frac{b_1^3 V^3}{6} \qquad (3\text{-}65.3)$$

Solving these consecutively for the b's gives

$$V b_1 = Z_1 \qquad (3\text{-}66.1)$$

$$2 V b_2 = Z_2 - Z_1^2 \qquad (3\text{-}66.2)$$

$$6 V b_3 = Z_3 - 3 Z_1 Z_2 + 2 Z_1^3 \qquad (3\text{-}66.3)$$

Now use the relation for N in the grand canonical ensemble from Eq. (1-141):

$$N = \left(z\frac{\partial}{\partial z}\right)_{V,T} \ln \mathcal{Q} = \left(\zeta\frac{\partial}{\partial \zeta}\right)_{V,T} \ln \mathcal{Q} = \left(\zeta\frac{\partial}{\partial \zeta}\right)_{V,T} \left(\frac{PV}{kT}\right)$$

$$= \left(\zeta\frac{\partial}{\partial \zeta}\right)_{V,T} V \sum_{j=1} b_j\zeta^j$$

$$= V \sum_{j=1} jb_j\zeta^j \tag{3-67}$$

From Eq. (3-66.1), since $\mathcal{Z}_1 = V$ we note that $b_1 = 1$, and Eq. (3-67) becomes

$$\rho = \frac{N}{V} = \zeta + 2b_2\zeta^2 + 3b_3\zeta^3 + \dots \tag{3-68}$$

We can invert this series by writing

$$\zeta = \rho + a_2\rho^2 + a_3\rho^3 + \dots \tag{3.69}$$

and put this into Eq. (3-68):

$$\rho = \rho + a_2\rho^2 + a_3\rho^3 + 2b_2(\rho^2 + 2a_2\rho^3 + \dots) + 3b_3\rho^3 + \dots$$

and by equating coefficients of like powers of ρ on both sides we determine the a's:

$$a_2 = -2b_2 \tag{3-70.1}$$

$$a_3 = 8b_2^2 - 3b_3 \tag{3-70.2}$$

Now finally we use Eq. (3-70) in Eq. (3-69), which gives us ζ as a function of ρ, and we substitute this into Eq. (3-63) to obtain the pressure as a power series in ρ—the virial series!

$$\frac{P}{kT} = \zeta + b_2\zeta^2 + b_3\zeta^3 + \dots$$

$$= \rho + a_2\rho^2 + a_3\rho^3 + b_2(\rho^2 + 2a_2\rho^3 + \dots) + b_3\rho^3 + \dots$$

$$= \rho - b_2\rho^2 + \rho^3(4b_2^2 - 2b_3) + \dots \tag{3-71}$$

By comparison of Eq. (3-71) and Eq. (3-41) we can read off two virial coefficients:

$$B_2 = -b_2 \tag{3-72}$$

$$B_3 = 4b_2^2 - 2b_3 \tag{3-73}$$

The technique used here is similar to that used in Sec. 1-7 to derive the virial coefficients for ideal bosons and ideal fermions. The reader must admit that very simple algebra has derived these very general and important results. (Of course to go on to obtain $B_4, B_5, \ldots$ would be tedious but not difficult.) Consider that we have expressed B_2 and B_3 in terms of quantities from Eq. (3-66) that depend on completely general two- and three-body partition functions (note Eq. 3-59) since we have not had to make any assumption as to the type of particle interaction. These expressions are correct for the completely quantum mechanical case as well as any classical case. However, even the two-body interacting quantum mechanical Z_2 function is complicated* and cannot be discussed here. In Example 3-4 we specialize our results to the classical case using the approximate potential function of Eq. (3-46).

■ ──

Example 3-4

Obtain the expressions for B_2 and B_3 for classical particles interacting via a sum of pair potentials that depend on particle-particle radial separation only. Use Mayer f_{ij} cluster functions to simplify the integrals.

Solution:

$$Z_1 = \int d\bar{r}_1 = V; \qquad Z_2 = \int d\bar{r}_1\, d\bar{r}_2 \exp\left(\frac{-\varepsilon_{12}}{kT}\right)$$

from Eq. (3-66.2)

$$b_2 = \frac{1}{2V}\left(Z_2 - Z_1^2\right)$$

write

$$Z_1^2 = \int d\bar{r}_1 \int d\bar{r}_2 = V^2$$

then

$$b_2 = \frac{1}{2V}\iint d\bar{r}_1\, d\bar{r}_2 \left[\exp\left(\frac{-\varepsilon_{12}}{kT}\right) - 1\right]$$

────────────

* K. Huang. *Statistical Mechanics*, 2d ed. Wiley, New York, 1987. Chapter 10, Section 3.

use the method of Eq. (3-51) and the definition

$$f_{12} = \exp\left(\frac{-\varepsilon_{12}}{kT}\right) - 1$$

$$b_2 = \frac{1}{2V} \iint d\bar{r}_1 \, d\bar{r}_{12} \, f_{12} = \frac{1}{2} \int d\bar{r}_{12} f_{12}$$

$$B_2 = -b_2 = -\frac{1}{2} \int d\bar{r}_{12} f_{12}$$

This is identical to Eq. (3-55).

$$Z_3 = \iiint d\bar{r}_1 \, d\bar{r}_2 \, d\bar{r}_3 \exp\left[-\frac{1}{kT}(\varepsilon_{12} + \varepsilon_{13} + \varepsilon_{23})\right]$$

Rewrite

$$\exp\left[-\frac{1}{kT}(\varepsilon_{12} + \varepsilon_{13} + \varepsilon_{23})\right] = \left[\exp\left(\frac{-\varepsilon_{12}}{kT}\right)\right]\left[\exp\left(\frac{-\varepsilon_{13}}{kT}\right)\right]\left[\exp\left(\frac{-\varepsilon_{23}}{kT}\right)\right]$$

$$= (f_{12} + 1)(f_{13} + 1)(f_{23} + 1)$$

$$= 1 + f_{12} + f_{13} + f_{23} + f_{12}f_{13}$$

$$+ f_{13}f_{23} + f_{12}f_{23} + f_{12}f_{13}f_{23}$$

From Eq. (3-66.3)

$$b_3 = \frac{1}{6V} Z_3 - \frac{1}{2V} Z_1 Z_2 + \frac{1}{3V} Z_1^3$$

use

$$Z_1 = \int d\bar{r}_3 \qquad Z_1^3 = V^3 = \int d\bar{r}_1 \, d\bar{r}_2 \, d\bar{r}_3$$

From Eq. (3-73)

$$B_3 = -2b_3 + 4b_2^2$$

$$= -\frac{2}{6V} \iiint d\bar{r}_1 \, d\bar{r}_2 \, d\bar{r}_3 (\underline{f_{12}f_{13}f_{23}} + f_{12}f_{23} + f_{13}f_{23}$$

$$+ f_{12}f_{13} + f_{12} + f_{13} + f_{23} + 1)$$

$$+ \frac{1}{V} \int d\bar{r}_3 \int d\bar{r}_1 \, d\bar{r}_2 (1 + f_{12}) - \frac{2}{3V} \int d\bar{r}_1 \, d\bar{r}_2 \, d\bar{r}_3$$

$$+ 4\left(\frac{1}{2V} \int d\bar{r}_1 \, d\bar{r}_2 \, f_{12}\right)\left(\frac{1}{2V} \int d\bar{r}_1 \, d\bar{r}_2 \, f_{12}\right)$$

In each integral go to coordinates relative to one molecule and integrate over the position of this molecule to give a factor of V to cancel all V's in the denominators. For example,

$$-\frac{2}{6V} \iiint d\bar{r}_1 \, d\bar{r}_2 \, d\bar{r}_3 \, f_{12}f_{23} = -\frac{2V}{6V} \int d\bar{r}_{12} \, d\bar{r}_{23} \, f_{12}f_{23}$$

$$= -\frac{1}{3} \int d\bar{r}_{12} \, f_{12} \int d\bar{r}_{23} \, f_{23}$$

and

$$\frac{1}{V} \int d\bar{r}_3 \, d\bar{r}_2 \, d\bar{r}_1 \, f_{12} = \int d\bar{r}_1 \, d\bar{r}_2 \, f_{12} = V \int d\bar{r}_{12} \, f_{12}$$

Every term in B_3 has dimensions of (volume/particle)^2. All the terms cancel except the one term underlined. Note, for example, $\int d\bar{r}_{12} \, f_{12} = \int d\bar{r}_{23} \, f_{23}$, and so on.

The final result is

$$B_3 = -\frac{2}{6V} \iiint d\bar{r}_1 \, d\bar{r}_2 \, d\bar{r}_3 \, f_{12}f_{13}f_{23} = -\frac{1}{3} \iint d\bar{r}_{12} \, d\bar{r}_{23} \, f_{12}f_{23}f_{13}$$

Note that in the above integrand $|\bar{r}_{13}|$ in the f_{13} function will generally depend on all six coordinates (four angles and the lengths r_{12} and r_{23}) involved in the sixfold integration, and the integral will be hard to evaluate.

For reference we state our above two results as numbered equations:

$$B_2 = \frac{1}{2} \int d\bar{r}_{12} \, f_{12} = -2\pi \int r_{12}^2 f_{12} \, dr_{12} \tag{3-74}$$

$$B_3 = -\frac{1}{3} \iint d\bar{r}_{12} \, d\bar{r}_{23} \, f_{12}f_{23}f_{13} \tag{3-75}$$

For any intermolecular potential use of Eq. (3-48) in Eq. (3-74) permits the easy evaluation of B_2. For the hard sphere potential (see Fig. 3-4) the integral is split into two parts

$$(B_2)_{HS} = -2\pi \int_0^\sigma (e^{-\infty/kT} - 1)r_{12}^2 \, dr_{12} - 2\pi \int_\sigma^\infty (e^{-0/kT} - 1)r_{12}^2 \, dr_{12}$$

$$= +\frac{2}{3}\pi\sigma^3 \equiv b_0 \tag{3-76}$$

This and all virial coefficients of the hard sphere potential are independent of temperature. For the Lennard-Jones potential given by Eq. (3-35)

$$B_2 = -2\pi \int_0^\infty r_{12}^2 \, dr_{12} \left(\exp\left\{ -\frac{4\varepsilon^*}{kT} \left[(\sigma/r_{12})^{12} - (\sigma/r_{12})^6 \right] \right\} - 1 \right)$$

and with definition

$$T^* = \frac{kT}{\varepsilon^*} \tag{3-77}$$

and new variable $X = r_{12}/\sigma$ this becomes

$$B_2^* \equiv \frac{(B_2)_{LJ}}{\tfrac{2}{3}\pi\sigma^3} = 3 \int_0^\infty X^2 \, dX \left[1 - \exp\left(-\frac{4}{T^*} X^{-12} + \frac{4}{T^*} X^{-6} \right) \right] \tag{3-78}$$

in which B_2^* is a function only of T^*. The integral cannot be done analytically but is easy to do numerically and extensive tables are available [6]. It is negative for small T^* where $\varepsilon^* > kT$ and attractive effects dominate, passes through zero (the Boyle temperature) at $T^* = 3.42$, and has a maximum at $T^* = 24$, beyond which it slightly decreases. This function is shown in Fig. 3-7. Also shown in Fig. 3-7 are data for the inert gas second virial coefficients [12] reduced by their LJ parameters[†] given in Table 3-5; that is, the experimental B_2 values are divided by $\tfrac{2}{3}\pi\sigma^3$ and plotted against kT/ε^*. We see that except for helium the LJ potential gives a generally good representation of these virial coefficients although there are deviations at low temperatures. The failure for helium is because of quantum effects (see Problem 1-14). Below 40 K helium gas does not follow Maxwell-Boltzmann statistics. The deviations for the other inert gases are due to the approximate nature of the LJ potential itself.

The fact that $(B_2)_{LJ}$ reduced by σ^3 comes out as a universal function of T^* is a special case of a general principle of corresponding states that holds in classical physics whenever it is possible to assume that the internal partition function, q_{int}, is independent of intermolecular forces (as discussed in Sec. 3-3) and that the intermolecular potential energy is given as a sum of pairwise terms in the form

$$\varepsilon_{ij} = \varepsilon^* \phi\left(\frac{r_{ij}}{\sigma} \right) \tag{3-79}$$

[†] Actually these parameters themselves have been obtained from second virial coefficient data. For two temperatures calculate the ratio $B_2(T_2)/B_2(T_1)$ from experiment and by trial and error refine a guess as to ε^* that gives T_2^* and T_1^* such that from interpolation in the tabulation of B_2^* versus T^* the ratio $B_2^*(T_2^*)/B_2^*(T_1^*)$ = experimental ratio. This gives ε^*. Then

$$\left(\frac{2\pi\sigma^3}{3} \right) = \frac{B_2(T_1)}{B_2^*(T_1^*)} = \frac{B_2(T_2)}{B_2^*(T_2^*)}$$

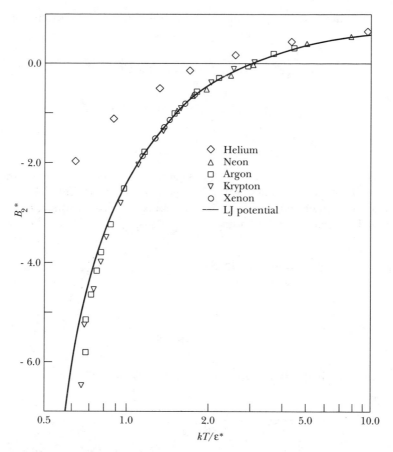

FIGURE 3-7 Second virial coefficients for the inert gases compared in reduced units with the predictions of the Lennard-Jones potential $B_2^* = (B_2)_{LJ}/\frac{2}{3}\pi\sigma^3$.
Source: R. O. Watts and I. McGee. *Liquid State Chemical Physics*. Wiley-Interscience, New York, 1976.

where ϕ is the same function for all substances that conform to the principle. Different conformal substances will of course have different ε^* and σ values. Then we may add to the reduced temperature, Eq. (3-77), the reduced volume,

$$V^* = \frac{V}{N\sigma^3} \tag{3-80}$$

and the reduced pressure,

$$P^* = \frac{P\sigma^3}{\varepsilon^*} \tag{3-81}$$

For all conformal substances the configuration integral may be written using $r'_{1x} = r_{1x}/\sigma$, etc., as

$$Z_N = \int \ldots \int d\bar{r}_1 \ldots d\bar{r}_N \exp(-V_N/kT)$$

$$= \int \ldots \int \sigma^{3N} d\bar{r}'_1 \ldots d\bar{r}'_N \exp\left\{-\sum_{i,j}\left[\frac{\phi(r'_{ij})}{T^*}\right]\right\}$$

$$= (\sigma^3)^N F(T^*, V^*, N) \tag{3-82}$$

since the (volume)^N dimensionality of Z_N is contained in the $(\sigma^3)^N$ factor and function F must be of dimensionless variables only. Since the Helmholtz free energy A is an extensive thermodynamic variable, it must be proportional to N times a function of V/N, that is, of V^* and

$$A = -kT \ln Q_N = -kT \ln \frac{q_{int}^N Z_N}{N! \lambda^{3N}}$$

shows that for this to be the case

$$F(T^*, V^*, N) = f^N(T^*, V^*)$$

Thus we find for the pressure (using $V = N\sigma^3 V^*$, $kT = \varepsilon T^*$)

$$P = -\left(\frac{\partial A}{\partial V}\right)_T = -\left(\frac{\partial}{\partial V}\right)_T \ln Z_N$$

$$= \frac{kT}{N\sigma^3} \frac{\partial}{\partial V^*} \ln (\sigma^3 f)^N = \left(\frac{kT}{\sigma^3}\right) \frac{(\partial/\partial V^*)(\sigma^3 f)}{\sigma^3 f}$$

$$= \frac{\varepsilon T^*}{\sigma^3}\left(\frac{\partial}{\partial V^*}\right)_{T^*} \ln f(V^*, T^*)$$

or

$$P^* = T^*\left(\frac{\partial}{\partial V^*}\right)_{T^*} \ln f(V^*, T^*) \tag{3-83}$$

This last is an equation for the reduced pressure as a universal function of the reduced temperature and reduced volume. In particular this implies that at the critical point the reduced critical properties will be the same for all gases following the same two-parameter potential. In Table 3-6 we illustrate the accuracy of the principle of corresponding states for a number of nearly spherical nonpolar molecules not only for the critical region, but also as to the behavior of B_2 and for

TABLE 3-6 Principle of Corresponding States Illustrated for Several Nearly Spherical Nonpolar Molecules

	T_c/K	P_c/atm	$\dfrac{P_c V_c}{Nk T_c}$	$\dfrac{kT_c}{\varepsilon^*}$	$\dfrac{P_c\sigma^3}{\varepsilon^*}$	$\dfrac{V_c}{N\sigma^3}$	T_B/T_c	T_s/T_c
^{4}He	5.2	2.26	0.300	0.52	0.027	5.78	—	—
H_2	33.3	12.8	0.304	0.92	0.064	4.37	—	—
Ne	44.5	26.9	0.307	1.25	0.115	3.33	2.74	0.563
Ar	150.7	48.4	0.292	1.26	0.116	3.17	2.73	0.577
Xe	289.8	57.9	0.290	1.31	0.132	2.88	2.65	0.580
N_2	126.1	33.5	0.292	1.33	0.131	2.96	2.59	0.588
O_2	154.5	49.7	0.292	1.31	0.142	2.69	2.63	0.583
CH_4	190.7	45.8	0.289	1.29	0.126	2.96	2.67	0.581
CO	133.0	34.5	0.294	1.33	0.134	2.92	2.60	0.593

Critical properties reduced by LJ potential parameters:

T_B = Boyle temperature (temperature at which $B_2 = 0$)
T_s = Temperature for which liquid vapor pressure is equal to $(P_c/50)$

SOURCES: J. O. Hirschfelder, C. F. Curtiss, and R. B. Bird. *Molecular Theory of Gases and Liquids*. Wiley, New York, 1954. E. A. Guggenheim. *Thermodynamics*, 4th ed. North Holland, Amsterdam, 1959.

reduced liquid vapor pressure. Helium and hydrogen do not follow the principle because of quantum mechanical deviations. Extension of the corresponding states principle to include nonspherical and polar molecules by introducing a third parameter (in addition to ε^* and σ) has been effected by Pitzer [13]. In summary this principle permits estimation of P, V, T behavior of a substance if its LJ parameters are known. Conversely, for conformal substances one can estimate their LJ parameters by

$$\frac{\varepsilon^*}{k} \simeq \frac{T_c}{1.3} \qquad \sigma \simeq \left(\frac{kT_c}{10P_c}\right)^{1/3} \tag{3-84}$$

Exercise 3-6

Obtain Eqs. (3-84) from data in Table 3-6 and use them to estimate ε^*/k and σ for Cl_2 ($T_c = 417$ K, $P_c = 76.1$ atm). Compare with Table 3-5.

To compute corrections to the thermodynamic functions due to gas imperfection it is most convenient to use

$$G = N\mu$$

$$dG = N \, d\mu = -S \, dT + V \, dP$$

and at constant T

$$d\mu = \frac{1}{\rho} dP = \frac{kT}{\rho} \left(\frac{dP/kT}{d\rho} \right)_T d\rho \qquad (3\text{-}85)$$

$$\rho = \frac{N}{V}$$

from the virial series, Eq. (3-41)

$$\frac{P}{kT} = \rho + \sum_{n=2}^{\infty} B_n \rho^n$$

$$\frac{1}{\rho} \left(\frac{dP/kT}{d\rho} \right)_T = \frac{1}{\rho} \left(1 + \sum_{n \geq 2} n B_n \rho^{n-1} \right) = \frac{1}{\rho} + \sum_{n \geq 2} n B_n \rho^{n-2}$$

Integrating Eq. (3-85) at constant temperature

$$\frac{\mu}{kT} = \int d\rho \left(\frac{1}{\rho} + \sum_{n \geq 2} n B_n \rho^{n-2} \right)$$

$$= \text{const} + \ln \rho + \sum_{n \geq 2} \left(\frac{n}{n-1} \right) B_n \rho^{n-1} \qquad (3\text{-}86)$$

The constant of integration in the above is actually a function of temperature, which we may identify by noting from Eq. (3-69) that as ρ becomes very small $\zeta \to \rho$, and by Eq. (3-62.2)

$$\ln \zeta \equiv \ln z + \ln (q_{int}/\lambda^3) \equiv \frac{\mu}{kT} + \ln \left(\frac{q_{int}}{\lambda^3} \right) \simeq \ln \rho \qquad \text{for small } \rho$$

or

$$\frac{\mu}{kT} \to \ln \left(\frac{\lambda^3}{q_{int}} \right) + \ln \rho \qquad \text{for small } \rho$$

$$\therefore \quad \text{const} = \ln \left(\frac{\lambda^3}{q_{int}} \right)$$

and Eq. (3-86) becomes

$$\frac{\mu}{kT} = \ln\left(\frac{\rho\lambda^3}{q_{int}}\right) + \sum_{n\geq 2}\left(\frac{n}{n-1}\right)B_n\rho^{n-1} \tag{3-87}$$

where the first term is just the ideal gas result (compare Eq. (1-130)). From Eq. (3-87) we obtain A by integrating

$$\left(\frac{\partial A}{\partial N}\right)_{T,V} = \mu$$

$$A = \int_0^N \mu\, dN = -NkT + NkT\ln\left(\frac{\rho\lambda^3}{q_{int}}\right) + NkT\sum_{n\geq 2}\frac{B_n}{(n-1)}\rho^{n-1} \tag{3-88.1}$$

$$A = A_{id} + NkT\sum_{n\geq 2}\left(\frac{B_n}{n-1}\right)\rho^{n-1} \tag{3-88.2}$$

where A_{id} is the ideal gas result. All other thermodynamic functions may be derived from Eq. (3-88.2) and are discussed in Problems 13 and 14.

■

Exercise 3-7

Obtain Eq. (3-88.1) by integration of Eq. (3-87) at constant T, V using $\rho = N/V$.

■

In practical terms, to measure or to calculate B_n for $n \geq 4$ is prohibitively difficult (except for certain simple model potentials such as hard spheres). However, much theoretical work has been done in examining the virial series formally for large n. At one time it was conjectured that the phenomenon of gaseous condensation to the liquid state occurs at the density and temperature for which the virial series diverges in a mathematical sense. This is probably not true but many unanswered questions remain to be investigated. It is certainly true that the critical behavior of a real gas cannot be described accurately by use of the virial series with only two or three virial coefficients retained. We will adopt a more global (non-series expansion) viewpoint using distribution functions to treat liquids in Sec. 3-6. To finish this section we discuss the higher virial coefficients and an accurate equation of state for hard sphere molecules since this will have an application to liquid state theory as well. B_3 for hard spheres may be calculated exactly from Eq. (3-75), and B_4 through B_7 may be calculated using numerical methods (and

TABLE 3-7   Hard Sphere Molecule Virial Coefficients

$B_2 = b_0 = 2/3\pi\sigma^3$	$B_5 = (0.1100 \pm 0.0005)b_0^4$
$B_3 = 5/8b_0^2$	$B_6 = (0.0386 \pm 0.0004)b_0^5$
$B_4 = 0.2869b_0^3$	$B_7 = (0.0138 \pm 0.0004)b_0^6$

SOURCE: F. H. Ree and W. G. Hoover. *J. Chem. Phys.* **46**, 4181 (1967).

computers) to estimate the values of the many integrals involved. Results are given in Table 3-7.

Computer simulations of the hard sphere system (to be discussed in Sec. 3-5) can obtain essentially exact numerical values for $P/\rho k T$ and other thermodynamic properties of the hard sphere system as a function of density. In 1969 Carnahan and Starling [14] suggested the following simple form for the hard sphere equation of state:

$$\frac{P}{\rho k T} = \frac{1 + \eta + \eta^2 - \eta^3}{(1 - \eta)^3} \tag{3-89}$$

where

$$\eta = \frac{N b_0}{4 V} = \frac{\sqrt{2\pi} V_0}{6 V} \qquad V_0 = \frac{N\sigma^3}{\sqrt{2}} \tag{3-90}$$

and V_0 is the close-packed volume of a set of N hard spheres of diameter σ. Note that the largest possible value for η is 0.7405. (A way to obtain this equation is discussed in Problem 12). This equation has been found to match the computer simulation results almost exactly and may be used as a practically exact equation of state for fluid hard spheres. As is expected for a purely repulsive potential, there can be no condensation to liquid and no liquid-gas critical point for hard spheres. Like ordinary gases above their critical temperature, hard spheres exhibit only a single fluid phase without division into gaseous and liquid regions. The reader will note that because the hard sphere virial coefficients are independent of temperature there is only one curve of $P/\rho k T$ (from Eq. 3-89) as a function of density [$\eta = (b_0/4)\rho$], in distinction to the different isotherms shown for $P/\rho k T$ by ordinary molecules with attractive forces. For hard spheres $P/\rho k T$ always increases with increase of density and it is always ≥ 1. Hard spheres do show in computer simulations* a freezing or solidification transition (not predicted by Eq. 3-89) at a fluid $\eta = 0.494$ ($V/V_0 = 1.50$) to a solid form with $\eta = 0.545$ ($V/V_0 = 1.36$).

---

* W. G. Hoover and F. H. Ree. *J. Chem. Phys.* **49**, 3609 (1968).

Example 3-5

B. J. Alder and T. E. Wainwright [*J. Chem. Phys.* **33**, 1439 (1960)] report the following computer simulation results for hard spheres:

V/V_0	10.0	2.36	1.77	1.50	1.41
$P/\rho kT$	1.36	4.29	7.73	12.5	15.0

Calculate the results for $P/\rho kT$ given by the virial series through B_7 (Table 3-7) and by the Carnahan-Starling (CS) equation at these V/V_0 values.

Solution:

Use $b_0 = 4\eta(V/N)$ and rewrite the virial series in terms of powers of η using the data in Table 3-7.

$$\frac{P}{\rho kT} = 1 + B_2\left(\frac{N}{V}\right) + B_3\left(\frac{N}{V}\right)^2 + B_4\left(\frac{N}{V}\right)^3 + B_5\left(\frac{N}{V}\right)^4 + B_6\left(\frac{N}{V}\right)^5 + B_7\left(\frac{N}{V}\right)^6$$

$$= 1 + 4\eta + 10\eta^2 + 18.36\eta^3 + 28.16\eta^4 + 39.5\eta^5 + 56.5\eta^6$$

The CS equation is Eq. (3-89).

Draw up the following table.

		$P/\rho kT$		Virial
V/V_0	η	Exact	CS	Series
10.0	0.07405	1.36	1.36	1.36
2.36	0.314	4.29	4.28	4.26
1.77	0.419	7.73	7.75	7.47
1.50	0.494	12.5	12.5	11.3
1.41	0.524	15.0	15.3	13.3

The seven-term virial series becomes inadequate as the density increases (beyond $\eta = 0.3$).

3-5
Numerical Statistical Mechanics: Monte Carlo and
Molecular Dynamics Techniques

The liquid state is intermediate between the gaseous and solid states. It has neither the regular long-range structural order of solids nor the large intermolecular separation characteristic of gases. Thus it is more difficult to treat theoretically, for neither the perturbation on the ideal gas (virial series) nor the perturbation of the independent harmonic oscillator model (see Chap. 4) for solids is realistic for liquids. Progress in describing liquids in recent years has been greatly dependent on the development of methods of numerical statistical mechanics using the ever more powerful computers that have become so widely available since 1960.

We shall limit our discussion of these methods to those dealing with classical particles because, although ways of treating explicitly quantum mechanical systems are available, they are all much more complex than those for classical cases because of the need to take into account the wave function symmetry requirements of Bose or Fermi statistics (see Sec. 1-4).

The numerical methods calculate the properties of finite N (usually 100 to 1000) particle systems with *given* intermolecular potential energy and provide much more detail about the microscopic behavior than any conceivable experiment could obtain. These methods are called computer simulations and give practically exact results for the model intermolecular potential energy chosen. One can add and subtract terms in the potential energy and learn how this affects the thermodynamic properties. Although this technique is sometimes used to generate a precise intermolecular potential energy function for a specific molecular liquid [8], the more usual application is to develop the properties of a hypothetical exact LJ molecule system or an exact hard sphere model system that can serve as an unperturbed liquid model, which in turn can be used as a basis for more general approximate analytic theories of liquids. Furthermore, these different approximate liquid theories can be checked as to their accuracy by comparing their results for a given potential to those from computer simulations using this same model potential. Thus one gets around the added ambiguity in comparing theoretical results with ordinary experiments on real liquids that arises because one does not know the precise intermolecular potential energy for real liquids.

There are two techniques used in computer simulations, the Monte Carlo (MC) method and the molecular dynamics (MD) method. The MC method introduced by Metropolis et al.* provides data on equilibrium properties only, whereas the MD technique pioneered by Alder and his coworkers at the University of California† gives equilibrium properties and data on the approach to equilibrium and on transport properties of gases and liquids (see Chaps. 5 and 6).

---

* N. Metropolis, A. W. Rosenbluth, M. N. Rosenbluth, A. H. Teller, and E. Teller. *J. Chem. Phys.* **21**, 1087 (1953).
† B. J. Alder and T. E. Wainwright. *J. Chem. Phys.* **27**, 1208 (1957); *J. Chem. Phys.* **31**, 459 (1959).

In the Monte Carlo method (for three-dimensional problems in the canonical ensemble) a cubic box of volume V encloses the N particles at a given temperature in any configuration allowed by the potential energy, V_N, chosen (i.e., hard cores cannot overlap). A particle selected at random is moved in a random way (but within prescribed limits) to generate a new particle configuration. If this move takes the particle outside the cube into a surrounding or image cube, application of periodic boundary conditions means that a new particle is inserted at the corresponding position in the original cube (see Fig. 3-8). Whether the new configuration is accepted depends on the Boltzmann factor exp $(-V_N/kT)$ before and after the move (the computer program calculates V_N after every move) according to rules (involving random number generation by the computer program) which ensure that configurations that are accepted will occur with a frequency proportional to their Boltzmann factors. This is absolutely essential because configurations chosen completely at random would generally have such high V_N values as to be of negligible significance for real systems and they also should not be significant in any simulation. If the new configuration is not accepted, the old configuration is counted again. Then another particle is randomly chosen to be moved. Consecutive configurations differ very little (if at all), and indeed the program is often most efficient if V_N is stored for calculating averages only after every particle on average is made to attempt one move. Classical phase space is thus directly sampled and the averages of V_N or of the total energy, U, become arithmetic averages over their values in the 10^5–10^7 configurations sampled. This is so because, as mentioned above, these configurations can be generated with frequencies *proportional* to their Boltzmann factors. Other system mechanical properties, such as the heat capacity, can be calculated from Eq. (2-23.1) and the pressure from Eq. (3-99). This last requires that during the simulation information be stored on the pair distribution function (see Sec. 3-6). The constant of proportionality (in the probability of occurrence of configurations) that involves the configuration integral, Eq. (3-45), is not known. Hence it is considerably more difficult to calculate the inherently statistical properties of a system, such as its entropy or Helmholtz free energy.

The moves in the Monte Carlo method are artificial rather than dynamical. In the molecular dynamics method (at the cost of more complex programming and more lengthy computer runs) it is possible to solve Newton's classical laws of motion (replacing the differential equations by finite-difference equations with minimum time increments of about 10^{-14} s) for systems involving $100 < N < 1000$ particles and follow their complete history for real-time intervals up to 10^{-11} or 10^{-10} s. (The computer time needed to follow through even 10^{-11} s will be of the order of hours on the fastest current machines.) The programs take account of all collisions and the momentum change in each collision (according to the chosen form of intermolecular potential). The pressure may be calculated from the total rate of change of momentum in real time after the arbitrary initial conditions have evolved to equilibrium. All sorts of details of the approach to equilibrium may be examined as well as transport of momentum, kinetic energy, and particles across an arbitrary plane in the equilibrium fluid. Although the 10^{-11} s real-time limit precludes study of any hydrodynamic effects in fluids with relaxation times longer than 10^{-11} s, equilibrium is reached in much shorter periods. In the case of hard spheres (all started off with the *same* kinetic energy but with velocities in random directions and with the same periodic boundary conditions as in the MC method) equilibrium is reached after only two to four collisions per particle, depending on the density, thus in much less than 10^{-12} s. For molecules with attractive tails in their intermolecular potential somewhat more than four collisions per particle are needed to reach equilibrium.

Two extensive review articles should be consulted for more details on these simulation methods [15]. Roughly, the errors in simulation results, associated with hav-

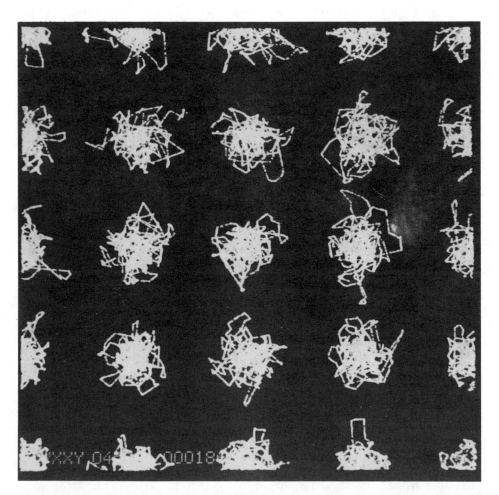

FIGURE 3-8 The traces made up of a succession of positions of a 32 particle hard sphere system at $V/V_0 = 1.525$ in a cubic box projected onto a plane. The periodic boundary conditions allow the same particle to make traces on opposite edges of the figure. If the system were close packed and truly stationary only 16 positions would be visible since in a planar projection one set of particles is right on top of the other set. The figure is for 3000 collisions in the middle of the initial solid "observation" period. SOURCE: T. W. Wainwright and B. J. Alder, *Nuovo Cimento* (*Supplement*) **9**, Ser. 10, 116 (1958).

ing to use finite N, are neglected terms of order $1/N$. Thus if more than 100 particles are used results are generally good to better than 1 percent. There is one serious difficulty in artificially keeping a constant number of molecules in the cube: critical phenomena that involve very large fluctuations in number density (see Sec. 3-8) cannot be treated accurately. Also, first-order phase transitions in general are cumbersome to treat, because they involve equilibrium between two phases of different density and for small N systems in the vicinity of a phase transition only one or the other phase will be "observed" in the simulation at a given time. For example, in the first MD study of 32 hard spheres started off in a face-centered cubic configuration at $V/V_0 = 1.5$, for the first 93,000 collisions (between all 32 particles) a solid phase was observed as shown in Fig. 3-8 for a subset of 3000 of these collisions. Then between 93,000 and 203,000 collisions a fluid phase (at a different pressure) was observed, as shown in Fig. 3-9. Of course in the years since

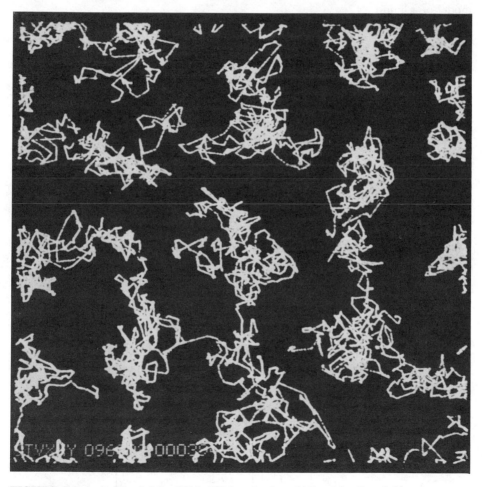

FIGURE 3-9 As in Fig. 3-8 for 3000 collisions in the middle of the first fluid "observation" period.

the introduction of these simulation techniques, many improvements have been devised [16] such that phase transitions and even critical phenomena can now be adequately treated. Even though we have these excellent numerical procedures available in principle, it is still important to develop approximate analytical theories since numerical results are only valid for the specific case or cases solved.

3-6

Theory of Liquids: Distribution Functions

For liquids the density is generally too large to fruitfully use density expansions such as those used in dealing with imperfect gases (e.g., Eqs. 3-42 and 3-69). Before we examine some of the modern methods we present an ultrasimplified free-volume model of a liquid.

Example 3-6

Treat a model liquid of N_l *independent* classical nonlocalized particles in a total volume $V = N_l v_f$, where v_f is the free volume per molecule in the liquid. Take the energy of each particle as

$$\varepsilon = \frac{1}{2m} (p_x^2 + p_y^2 + p_z^2) - \chi$$

where $-\chi$ is a *constant* average potential energy due to interaction with the other molecules (depth of the potential well in which the particles of the liquid are held). Treat the equilibrium gas phase of N_g molecules as ideal, and by equating chemical potentials derive an equation for the vapor pressure of the liquid. Obtain an equation for the entropy difference of the two phases and give a thermodynamic meaning to $N_A \chi$.

Solution:

For independent nonlocalized molecules in the liquid

$$Q_N^{c\ell} = \frac{1}{N_l!} (q^{c\ell})^{N_l}$$

where from Eq. (1-185) the one-particle partition function is

$$q^{c\ell} = \frac{q_{int}}{h^3} \int \ldots \int \exp\left[-\frac{1}{kT} \left(\frac{p_x^2}{2m} + \frac{p_y^2}{2m} + \frac{p_z^2}{2m} \right) \right]$$
$$\times e^{+\chi/kT} \, dp_x \, dp_y \, dp_z \, dx \, dy \, dz$$

The volume integration for the liquid will give the full free volume, $V = N_l v_f$, the momentum integrals are standard, and the *constant* factor $e^{\chi/kT}$ comes outside the integral signs

$$q^{cl} = \frac{N_l v_f e^{+\chi/kT} q_{int}}{\lambda^3}$$

where

$$\lambda = \frac{h}{\sqrt{2\pi m k T}}$$

Using the Stirling approximation for $\ln N_l!$

$$A_l = -kT \ln Q_N^{cl} = -N_l \chi - N_l kT \ln \left(\frac{e v_f q_{int}}{\lambda^3} \right)$$

where e = base of natural logs. Taking A_l as a function of N_l (using $v_f = V/N_l$) at constant V, T

$$\mu_l = \left(\frac{\partial A_l}{\partial N_l} \right)_{V,T} = kT - \chi - kT \ln \left(\frac{e v_f q_{int}}{\lambda^3} \right)$$

$$= -\chi - kT \ln \left(\frac{v_f q_{int}}{\lambda^3} \right)$$

Similarly, for the ideal gas of N_g molecules in volume V_g

$$P = \frac{N_g kT}{V_g}$$

and from Eq. (2-5)

$$A_g = -N_g kT \ln \left(\frac{e v_g q_{int}}{\lambda^3} \right)$$

with $v_g = V_g/N_g$. Thus

$$\mu_g = \left(\frac{\partial A_g}{\partial N_g} \right)_{V,T} = -kT \ln \left(v_g \frac{q_{int}}{\lambda^3} \right)$$

at equilibrium $\mu_g = \mu_l$, and assuming q_{int} is the same for liquid and gaseous states,

$$kT \ln \frac{v_g}{v_f} = \chi$$

$$\frac{v_g}{v_f} = e^{+\chi/kT} \qquad\qquad \text{Eq. (A)}$$

and

$$v_g = \frac{kT}{P}$$

and the vapor pressure is given by

$$P = \frac{kT}{v_f} e^{-\chi/kT} = \frac{RT}{N_A v_f} \exp\left(-N_A\chi/RT\right)$$

From the entropy expressions

$$S_l = -\left(\frac{\partial A_l}{\partial T}\right)_V \qquad S_g = -\left(\frac{\partial A_g}{\partial T}\right)_V$$

taking differences on a molar basis

$$S_{g_m} - S_{l_m} = R \ln \frac{v_g}{v_f} = \frac{R\chi}{kT} = \frac{N_A\chi}{T}$$

which identifies $N_A\chi = \Delta H_{v_m}$ (the molar enthalpy of vaporization). Thus the vapor pressure equation reads

$$P = \frac{RT}{N_A v_f} \exp\left(-\Delta H_{v_m}/RT\right)$$

Since we took χ to be a constant, this model gives a temperature-independent enthalpy of vaporization but the form of the vapor pressure equation is similar to that from the Clausius-Clapeyron equation [see Appendix I, Eq. (I-43)].

If we use Trouton's rule for nonpolar liquids at their normal boiling points, $\Delta H_{v_m}/T_b \equiv 88\ \mathrm{J \cdot K^{-1} \cdot mol^{-1}}$, Eq. (A) gives at T_b

$$\ln \frac{N_A v_g}{N_A v_f} = 10.6$$

$$\frac{N_A v_g}{N_A v_f} = 4.0 \times 10^4$$

Since the molar volume in the gas phase at 1 atm is in the range 25,000 cm³ ($T_b = 300$ K) to 33,000 cm³ ($T_b = 400$ K), $N_A v_f$ is generally of order less than 1 cm³/mol, that is, less than 1 percent of a typical liquid molar volume (which is usually 80–100 cm³/mol at 1 atm). An argument of Hirschfelder [6, p. 280] illustrated in Fig. 3-10 estimates the free volume of a liquid by assuming a simple cubic packing [molecules on average separated by $(V_m/N_A)^{1/3}$] with all molecules of diameter σ momentarily frozen except for one wanderer (shaded) as

$$v_f^{1/3} = 2(V_m/N_A)^{1/3} - 2\sigma$$

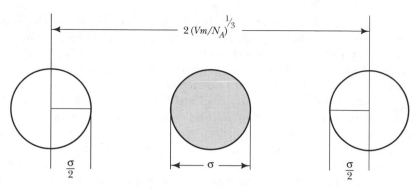

FIGURE 3-10 Illustration of method of estimating the free volume per molecule, v_f, in a liquid. Shaded molecule can move while others are momentarily frozen in a simple cubic array.

that is,

$$N_A v_f = 8[(V_m)^{1/3} - 0.782(b)^{1/3}]^3$$

where the van der Waals constant $b = \frac{2}{3}\pi N_A \sigma^3$. Applying this to CCl_4 at 25°C ($V_m = 96.4$ cm^3/mol, $b = 138$ cm^3/mol, $N_A v_f = 1.3$ cm^3/mol, $P/\text{atm} = 63 T \exp(-\Delta H_{v_m}/RT)$. This gives quantitatively a very poor representation of the vapor pressure of CCl_4, but of course the model and the estimate of $N_A v_f$ is very rough.

■

There is an interesting added point about Example 3-6 that deserves comment. Should we have treated the molecules in the liquid as nonlocalized? If we had been doing an analogous model calculation for a solid we would have localized each molecule in its own tiny free volume v_f such that each q^{el} would have been divided by N_l and we would also *not* have divided by $N_l!$ (see Secs. 1-4 and 1-8) in the Q_N^{el} expression. Thus we would have had the factor (per mole) of $(N_A!/(N_A)^{N_A})$ times the Q_N^{el} we used. By Stirling's approximation this factor is equal to e^{-N_A}. Tracing this through $S = -(\partial A/\partial T)_V = +(\partial/\partial T)(kT \ln Q_N)$ gives a new term $-N_A k = -R$ per mole in the entropy. Thus if we had treated the liquid molecules as localized, their molar entropy would have been less by this amount, which is called the *communal* entropy (the entropy gained when the molecules can move in the common full volume). Certainly the full communal entropy is gained by a substance "somewhere" between the solid and gaseous states. In Example 3-6 we have implicitly assumed that the liquid has this full entropy. In the past the assumption of the acquisition of the full communal entropy upon melting was usually made in free volume and cell theories of liquids [6]. This has been definitely shown to be incorrect from computer simulation studies for hard spheres. When the entropy increase due to the increase of volume per molecule upon melting is subtracted from the total entropy increase of melting the result is less than $0.2R$. Hard spheres have no other means of entropy increase on melting except *partial* delocalization and molar volume increase. Real polyatomic substances also have rotational and vibrational contributions to the entropy of melting and so generally have a total molar entropy of melting larger than R. This does *not* mean that they have gained the full communal entropy at the melting point. In fact some monatomic substances, such as the alkali metals, have a molar entropy of melting less than R, clearly showing that their communal entropy increase at melting is much less than R. It may be concluded that all real substances only gradually acquire their communal entropy and may only fully do so upon vaporization.

Exercise 3-8

Show that $N_A!/(N_A)^{N_A} = e^{-N_A}$. Use the Stirling approximation.

We turn now to a more formal treatment of classical fluids. In analogy to the treatment of classical statistical mechanics in Sec. 8 of Chap. 1 [see Eq. (1-191)] we may write for the fraction of the molecules with positions at $\bar{r}_1 \ldots \bar{r}_N$ (or equivalently for the probability that molecule 1 is in $d\bar{r}_1$ at $\bar{r}_1$, molecule 2 is in $d\bar{r}_2$ at $\bar{r}_2$, etc.):

$$\frac{dN}{N} = \frac{\exp\left(-V_N/kT\right) d\bar{r}_1 \ldots d\bar{r}_N}{Z_N} \tag{3-91}$$

having integrated out over all momenta of center of mass motion and over all internal degrees of freedom. Z_N is the configuration integral of Eq. (3-45). The corresponding distribution function with respect to molecular positions is by definition

$$f^{(N)}(\bar{r}_1 \ldots \bar{r}_N) = \frac{dN}{N \, d\bar{r}_1 \ldots d\bar{r}_N} = \frac{\exp\left(-V_N/kT\right)}{Z_N} \tag{3-92}$$

The reduced n particle distribution function for molecule 1 to be at $\bar{r}_1$, molecule 2 to be at $\bar{r}_2 \ldots$ molecule n to be at $\bar{r}_n$ is found as usual by integrating over the positions of the other molecules in which we are *not* interested:

$$f^{(n)}(\bar{r}_1 \ldots \bar{r}_n) = \frac{dN}{N \, d\bar{r}_1 \ldots d\bar{r}_n} = \frac{\int \ldots \int \exp\left(-V_N/kT\right) d\bar{r}_{n+1} \ldots d\bar{r}_N}{Z_N} \tag{3-93}$$

It proves convenient, as we will see below, to multiply Eq. (3-93) by the number of ways to pick out n particles from a set of N, that is, by N ways to pick the first, times $(N-1)$ ways to pick the second, ... times $(N-n+1)$ ways to pick the nth, or by $N!/(N-n)!$ ways in all. Then we will have a distribution function for *any* molecule to be at $\bar{r}_1$ and *any* second molecule to be at $\bar{r}_2$, and so on. In this way we define dimensionless particle correlation functions $g^{(n)}(\bar{r}_1 \ldots \bar{r}_n)$ as

$$\frac{N!}{(N-n)!}\, f^{(n)}(\bar{r}_1 \ldots \bar{r}_n) \equiv \rho^n g^{(n)}(\bar{r}_1 \ldots \bar{r}_n)$$

$$= \frac{N!}{(N-n)!}\, \frac{\int \cdots \int \exp\left(-V_N/kT\right) d\bar{r}_{n+1} \ldots d\bar{r}_N}{Z_N} \tag{3-94}$$

where $\rho = N/V$ is the number density.

Note the normalization requirements that follow from Eq. (3-94) and the fact that from Eq. (3-93) if $f^{(n)}$ is integrated over $d\bar{r}_1 \ldots d\bar{r}_n$ we must get unity:

$$\int \rho^n g^{(n)}(\bar{r}_1 \ldots \bar{r}_n)\, d\bar{r}_n = (N-n+1)\rho^{n-1}g^{(n-1)}(\bar{r}_1 \ldots \bar{r}_{n-1}) \tag{3-95.1}$$

$$\int \cdots \int \rho^n g^{(n)}(\bar{r}_1 \ldots \bar{r}_n)\, d\bar{r}_1 \ldots d\bar{r}_n = \frac{N!}{(N-n)!} \tag{3-95.2}$$

The simplest of these functions is $g^{(1)}(\bar{r}_1)$. The quantity $\rho g^{(1)}(\bar{r}_1)\, d\bar{r}_1$ is the probability that *any* one molecule will be found at $\bar{r}_1$. In a fluid all points inside the full volume V are equivalent, and so $g^{(1)}(\bar{r}_1)$ must be independent of $\bar{r}_1$. Thus, using Eq. (3-95.2) with $n = 1$,

$$\int \rho g^{(1)}(\bar{r}_1)\, d\bar{r}_1 = N$$

$$g^{(1)}\rho \int d\bar{r}_1 = g^{(1)}\rho V = N$$

$$\therefore \quad g^{(1)} = 1 \tag{3-96}$$

This result does not hold for crystals since their lattice structure makes $g^{(1)}(\bar{r}_1)$ a periodic function of $\bar{r}_1$ with maxima at the lattice sites.

Consider next $n = 2$ in Eq. (3-95.2) and go to relative coordinates as in Example 3-4:

$$\iint \rho^2 g^{(2)}(\bar{r}_1\bar{r}_2)\, d\bar{r}_1\, d\bar{r}_2 = \rho^2 \int d\bar{r}_1 \int 4\pi r_{12}^2\, dr_{12}\, g^{(2)}(r_{12})$$

$$= \rho^2 V \int 4\pi r_{12}^2\, dr_{12}\, g^{(2)}(r_{12})$$

$$= N\rho \int 4\pi r_{12}^2\, dr_{12}\, g^{(2)}(r_{12})$$

$$= N(N-1)$$

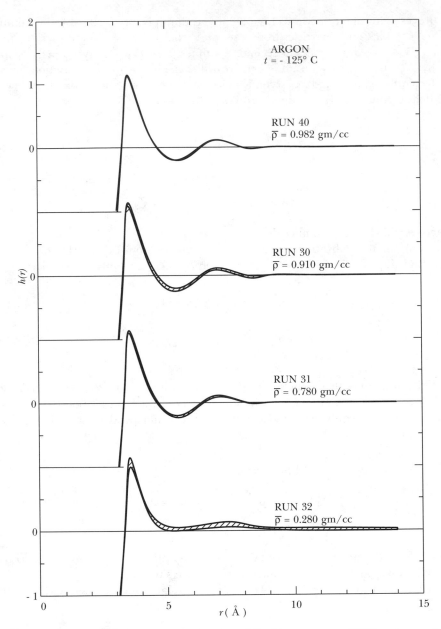

FIGURE 3-11 The radial distribution function of liquid argon at $-125°C$ and several densities $[g^{(2)}(r) - 1]$ versus r.

SOURCE: Reference [17].

Thus we have

$$\int_0^\infty \rho g^{(2)}(r_{12}) 4\pi r_{12}^2 \, dr_{12} = \mathcal{N} - 1 \qquad (3\text{-}97)$$

which shows the following very important property of $g^{(2)}(r)$: The average number of molecules at a distance between r and $r + dr$ from any given molecule is given by $\rho g^{(2)}(r) 4\pi r^2 \, dr$, and integrating over all r must count all the other molecules, $(\mathcal{N} - 1)$ in number, in the system. Compare this with the method used in Eq. (3-37), in which we assumed a completely random arrangement of molecules about a given one and took $g^{(2)}(r) = 1$. The function $g^{(2)}(r)$ is called the radial distribution function and accounts in a concise way for nonrandom molecular pair arrangements due to intermolecular forces. It is a function of density and temperature as well as of radial separation r. It can be obtained experimentally for any fluid from X-ray or neutron scattering data [17]. Some experimental curves for argon are shown in Fig. 3-11, and in Fig. 3-12 some curves for a model LJ fluid are shown as obtained from a molecular dynamics simulation study. The qualitative nature

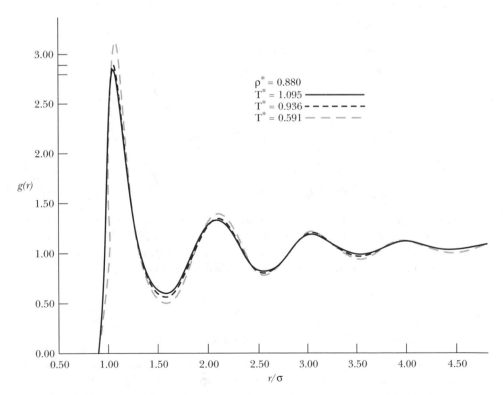

FIGURE 3-12 The radial distribution function of a Lennard-Jones model fluid from a molecular dynamics simulation study. $T^* = kT/\varepsilon^*$, $\rho^* = \rho\sigma^3$.
SOURCE: Reference [18].

of the $g^{(2)}(r)$ function is easily understood: Strong repulsion at short distances causes it to be zero, from which it rises to a first and highest maximum corresponding to a set of fluctuating first nearest neighbors at a distance corresponding to the minimum in the intermolecular potential energy. Further oscillations are more noticeable if the temperature is low or the density is high. For large r it must approach unity because the effect of the given molecule about which r is measured becomes negligible.

If V_N is a sum of pair potentials, all the thermodynamic properties of a fluid may be written in terms of $g^{(2)}(\bar{r})$. The energy is easiest to express, since if we take any molecule as central the total intermolecular potential energy between this one and all the other fluid molecules between r and $r + dr$ will be $\varepsilon(r)[\rho g^{(2)}(r)]4\pi r^2 \, dr$ such that by integrating and multiplying by $N/2$ we have*

$$U = U_{id} + \frac{N\rho}{2} \int_0^\infty \varepsilon(r)g^{(2)}(r, \rho, T)4\pi r^2 \, dr \qquad (3\text{-}98)$$

where $\varepsilon(r)$ is the intermolecular potential energy function and U_{id} is the ideal gas energy due to center of mass motion and internal degrees of freedom as worked out in Chap. 2. Using the formal definition of $g^{(2)}$ in Eq. (3-94) it may be shown [18] that the pressure is given by

$$\frac{P}{\rho kT} = 1 - \frac{\rho}{6kT} \int_0^\infty r \frac{d\varepsilon(r)}{dr} g^{(2)}(r, \rho, T)4\pi r^2 \, dr \qquad (3\text{-}99)$$

Finally, in a way similar to the derivation of Eq. (3-87) we can show

$$\frac{\mu}{kT} = \ln\left(\frac{\lambda^3 \rho}{q_{int}}\right) - \frac{\rho}{6kT} \int_0^\infty r \frac{d\varepsilon(r)}{dr} g^{(2)}(r, \rho, T)4\pi r^2 \, dr$$

$$- \int_0^\rho \frac{d\rho}{6kT} \int_0^\infty r \frac{d\varepsilon(r)}{dr} g^{(2)}(r, \rho, T)4\pi r^2 \, dr \qquad (3\text{-}100)$$

We note that although the pressure and energy can be evaluated if we know $g^{(2)}(r, \rho, T)$ at one particular density and temperature, Eq. (3-100) shows that to evaluate the chemical potential one needs to know $g^{(2)}(r, \rho, T)$ as a function of density in the entire range 0 to ρ at the specified temperature.

■

Example 3-7

Equation (3-99) is a general fluid equation of state and holds in the gas phase as well as in the liquid phase. Consider a density expansion for the radial distribution

---

* Division by 2 is necessary so as not to count all pairs twice over.

function:

$$g^{(2)}(r, \rho, T) = g_0^{(2)}(r, T) + \rho g_1^{(2)}(r, T) + \rho^2 g_2^{(2)}(r, T) + \dots$$

By comparison to Eq. (3-41) for the virial series, express the virial coefficients in terms of the $g_i^{(2)}(r, T)$. By comparison to the result for $B_2(T)$ in Problem 3-11, determine $g_0^{(2)}(r, T)$.

Solution:

By use of Eqs. (3-41) and (3-99) we have

$$\frac{P}{\rho k T} = 1 + B_2(T)\rho + B_3(T)\rho^2 + \dots$$

$$= 1 - \frac{\rho}{6kT} \int_0^\infty r \frac{d\varepsilon(r)}{dr} g_0^{(2)}(r, T) 4\pi r^2 \, dr$$

$$- \frac{\rho^2}{6kT} \int_0^\infty r \frac{d\varepsilon(r)}{dr} g_1^{(2)}(r, T) 4\pi r^2 \, dr + \dots$$

Thus

$$B_2(T) = -\frac{1}{6kT} \int_0^\infty r \frac{d\varepsilon(r)}{dr} g_0^{(2)}(r, T) 4\pi r^2 \, dr$$

$$B_3(T) = -\frac{1}{6kT} \int_0^\infty r \frac{d\varepsilon(r)}{dr} g_1^{(2)}(r, T) 4\pi r^2 \, dr$$

and so on.

Since from Problem 3-11 we have

$$B_2(T) = -\frac{2\pi}{3kT} \int_0^\infty \frac{d\varepsilon(r)}{dr} \exp\left[-\varepsilon(r)/kT\right] r^3 \, dr$$

we identify

$$g_0^{(2)}(r, T) = \exp\left[-\varepsilon(r)/kT\right]$$

Comment: Note that this result—that $g^{(2)}(r, \rho, T) \rightarrow \exp\left[-\varepsilon(r)/kT\right]$—holds only at low density. At higher density it can be shown that $\varepsilon(r)$ should be replaced by a function $\omega^{(2)}(r)$ called the potential energy of mean force, which embodies the averaged effects of the simultaneous interactions with the $(\mathcal{N} - 2)$ other molecules that affect any two-body interaction in a dense system. In fact it is easy to show that the next term is

$$g_1^{(2)}(r, T) = \exp\left[-\varepsilon(r)/kT\right] \int d\bar{r}_3 f_{23} f_{13}$$

where the integral over the Mayer cluster functions is also a function of r_{12}.

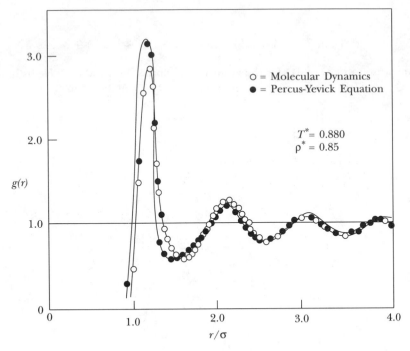

FIGURE 3-13 The radial distribution function from the Percus-Yevick integral equation for the Lennard-Jones model fluid compared with that from a molecular dynamics study. SOURCE: F. Mandel, R. J. Bearman, and M. Y. Bearman. *J. Chem. Phys.* **52**, 3315 (1970).

Since the late 1950's a number of integral equations have been derived that relate the $g^{(2)}(r_{12})$ function to integrals over $d\bar{r}_3$ involving integrands as a function of the potential functions $\varepsilon(r_{13})$, $\varepsilon(r_{23})$, and of $g^{(2)}(r_{23})$ and $g^{(2)}(r_{13})$. A concise presentation of these theories is given by Barker and Henderson [19]. These equations must be solved numerically in a self-consistent manner. This is another large-scale task for computers. These theories are very successful in deriving the structure of simple liquids, that is, in obtaining the $g^{(2)}(r_{12})$ function as shown in Fig. 3-13 (the Percus-Yevick equation is one of the integral equation theories). However, in comparison to the precise molecular dynamics results there are discrepancies with regard to the height of the first peak of $g^{(2)}(r)$ and with respect to the position and shape of the first minimum. These errors can give rise to appreciable errors in the computed thermodynamic properties, especially for the pressure. Furthermore, since these theories generate approximate $g^{(2)}(r)$ functions, different thermodynamic routes for computing a given property—all of which would give the same result with an exact $g^{(2)}(r)$ function—give different results. This is illustrated in Fig. 3-14. Improvement of these theories is an active field of current research.

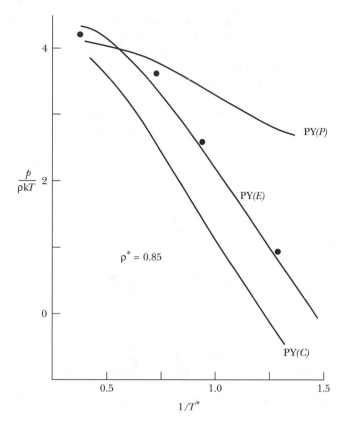

FIGURE 3-14 Results for the pressure from the Percus-Yevick (PY) theory as calculated by three different sets of thermodynamic equations, P, E, C. Solid dots give computer simulation results. An exact theory would give only one curve coincident with the simulation results.
SOURCE: J. A. Barker and D. Henderson. *Ann. Rev. Phys. Chem.* **23**, 439 (1972).

3-7

Theory of Liquids: Augmented van der Waals Equation

In recent years much work has been done on perturbation theories for the liquid state and these have turned out to be quite accurate. What is equally important is that they are much easier to understand intuitively than are the integral equation theories. In fact first-order perturbation theory is merely an augmented van der Waals equation. Perturbation theories can be applied to any system that is in some sense slightly perturbed from some reference system whose properties are known. The hard sphere system is usually used as the reference system since it is physically realistic for high-density systems such as liquids. In addition, there is now available from computer simulations an accurate $g^{(2)}(r)$ function and an accurate equation of state for hard spheres. In this technique thermodynamic properties are always calculated from the Helmholtz free energy, as will be shown below, and the actual $g^{(2)}(r)$ function of the system need not be used, although it can be computed as an interesting quantity in its own right.

We use a subscript zero to denote the properties of the unperturbed system and a prime on V_N and on $\varepsilon(r)$ to denote the additional terms in the intermolecular potential energy due to the perturbing potential (which in the simplest case is merely an attractive tail outside the hard core). The ratio of the classical partition functions of the perturbed (i.e., of the actual system) to the unperturbed system by Eq. (3-44) is, with $\beta = 1/kT$,

$$\frac{Q}{Q_0} = \frac{Z}{Z_0} = \frac{1}{Z_0} \int \ldots \int \exp - \beta(V_{N_0} + V_N') \, d\bar{r}_1 \ldots d\bar{r}_N \qquad (3\text{-}101)$$

We expand

$$\exp\left(-\beta V_N'\right) = 1 - \beta V_N' + \frac{\beta^2}{2}(V_N')^2 + \ldots \qquad (3\text{-}102)$$

and use

$$V_N' = \sum_{i<j=2}^{N} \varepsilon'(r_{ij}) \qquad (3\text{-}103)$$

Collecting all like integrals we obtain, keeping only terms through β^2:

$$\frac{Q}{Q_0} = 1 - \beta \frac{N(N-1)}{2} \int \ldots \int \frac{e^{-\beta V_{N_0}}}{Z_0} \varepsilon'(r_{12}) \, d\bar{r}_1 \, d\bar{r}_2 \ldots d\bar{r}_N$$

$$+ \frac{\beta^2}{2} \left\{ \frac{1}{4} \frac{N!}{(N-4)!} \int \ldots \int \frac{e^{-\beta V_{N_0}}}{Z_0} \varepsilon'(r_{12})\varepsilon'(r_{34}) \, d\bar{r}_1 \ldots d\bar{r}_N \right.$$

$$+ \frac{N!}{(N-3)!} \int \ldots \int \frac{e^{-\beta V_{N_0}}}{Z_0} \varepsilon'(r_{12})\varepsilon'(r_{23}) \, d\bar{r}_1 \ldots d\bar{r}_N$$

$$+ \frac{1}{2} \frac{N!}{(N-2)!} \int \ldots \int \frac{e^{-\beta V_{N_0}}}{Z_0} [\varepsilon'(r_{12})]^2 \, d\bar{r}_1 \ldots d\bar{r}_N \bigg\} + \ldots \qquad (3\text{-}104)$$

The terms in β^2 have the correct coefficients to count up correctly the $[N(N-1)/2]^2$ pairs of pairs coming from the $(V_N')^2$ term in Eq. (3-102) when Eq. (3-103) is used for V_N'. They count respectively the cases where all indices on r_{ij} and $r_{i'j'}$ differ, where one index is the same, and where both indices are the same, that is:

$$\frac{N!}{4(N-4)!} + \frac{N!}{(N-3)!} + \frac{N!}{2(N-2)!} = \left[\frac{N(N-1)}{2}\right]^2$$

as may be easily checked.

Using the general definition in Eq. (3-94) of the $g^{(n)}$ functions, we can rewrite Eq. (3-104) in terms of the distribution functions of the *unperturbed* system with subscript zero.

$$\frac{Q}{Q_0} = 1 - \frac{\beta}{2}\, \rho^2 \iint g_0^{(2)}(r_{12})\varepsilon'(r_{12})\, d\bar{r}_1\, d\bar{r}_2$$

$$+ \frac{\beta^2}{2}\left\{ \frac{\rho^4}{4} \int \cdots \int g_0^{(4)}(1, 2, 3, 4)\varepsilon'(12)\varepsilon'(34)\, d\bar{r}_1 \cdots d\bar{r}_4 \right.$$

$$+ \rho^3 \int \cdots \int g_0^{(3)}(1, 2, 3)\varepsilon'(12)\varepsilon'(23)\, d\bar{r}_1 \cdots d\bar{r}_3$$

$$\left. + \rho^2 \iint g_0^{(2)}(1, 2)[\varepsilon'(12)]^2\, d\bar{r}_1\, d\bar{r}_2 \right\} + \cdots \tag{3-105}$$

Note in particular that the first set of integrations is now only a two-particle form and will be simple to do. Since the Helmholtz free energy is $A = -kT \ln Q$, we have also

$$\frac{Q}{Q_0} = \exp - \beta(A - A_0) \tag{3-106}$$

such that if we introduce an expansion with coefficients ω_n

$$-\beta(A - A_0) = \sum_{n=1}^{\infty} \frac{\omega_n}{n!}\, (-\beta)^n \tag{3-107}$$

$$A = A_0 + \omega_1 - \frac{\omega_2}{2!}\,\beta + \cdots \tag{3-108}$$

Then, introducing another series,

$$\frac{Q}{Q_0} \equiv \sum_{k=0}^{\infty} \frac{(-\beta)^k}{k!}\, \alpha_k \tag{3-109}$$

comparison of this with Eqs. (3-102) and (3-104) shows

$$\alpha_0 = 1$$

$$\alpha_1 = \text{coeff of } -\beta \text{ in } Q/Q_0 = \langle V'_N \rangle_0 \tag{3-110}$$

$$\alpha_2 = \text{coeff of } +\beta^2/2 \text{ in } Q/Q_0 = \langle (V'_N)^2 \rangle_0$$

We may interpret $\langle V'_N \rangle_0$ as the average of the perturbing potential over the unperturbed system

$$\langle V'_N \rangle_0 = \frac{\rho^2}{2} \iint g_0^{(2)}(r_{12})\varepsilon'(r_{12})\, d\bar{r}_1\, d\bar{r}_2 \tag{3-111}$$

Similarly, $\langle (V'_N)^2 \rangle_0$ is the average of the square of the perturbing potential over the unperturbed system.

Inserting Eq. (3-109) and Eq. (3-107) into Eq. (3-106) and expanding the exponential,

$$\sum_{k=0}^{\infty} \frac{(-\beta)^k}{k!} \alpha_k = \exp\left(-\beta\omega_1 + \frac{\beta^2}{2}\omega_2 + \ldots \right)$$

$$= 1 - \beta\omega_1 + \frac{\beta^2}{2}\omega_2 + \frac{1}{2}\left(\beta^2\omega_1^2 + \ldots \right) + \ldots$$

we easily identify the ω_i of Eq. (3-108) in terms of the α_i of Eq. (3-110)

$$\omega_1 = \alpha_1 = \langle V'_N \rangle_0$$
$$\omega_2 = \alpha_2 - (\alpha_1)^2 = \langle (V'_N)^2 \rangle_0 - [\langle V'_N \rangle_0]^2 \qquad (3\text{-}112)$$

In perturbation theory thermodynamic properties are calculated from A of Eq. (3-108) with the coefficients given by Eq. (3-112). With hard spheres as the reference system good results are obtained merely by keeping the first-order correction [only the ω_1 term in Eq. (3-108)]. This is because the liquid structure is mainly determined at high and moderate density by the geometric requirements of packing the hard cores. The effect of the perturbing attractive potential is to provide a nearly uniform background potential in which the molecules move, which is taken into account by the ω_1 term. Notice that ω_2 is of the form of a fluctuation in V'_N, being the difference between the average of the square and the square of the average. This fluctuation is bound to be small at high density where interparticle separations can vary only slightly. This theory is expected to be less accurate at lower densities characteristic of the critical point. This is true, but computer simulations discussed by Alder* show that ω_2 is practically volume independent, such that the equation of state for the pressure (obtained as the volume derivative of A) is still accurately given by the first-order term such that critical properties will also be accurate.

Of course keeping only the ω_1 term is the van der Waals idea. The equation of state becomes [recall that $g_0^{(2)}$ is a function of ρ, i.e., of V but not of temperature]:

$$P = -\left(\frac{\partial A}{\partial V} \right)_T = P_0 - \frac{\partial}{\partial V}\left[\frac{\rho^2}{2} \iint g_0^{(2)}(\rho, r)\varepsilon'(r)\, d\bar{r}_1 d\bar{r}_2 \right]$$

$$P = P_0 - \frac{\partial}{\partial V}\left[\frac{\rho^2 V}{2} \int g_0^{(2)}(\rho, r)\varepsilon'(r) 4\pi r^2\, dr \right] \qquad (3\text{-}113)$$

This is the augmented van der Waals equation of state. To obtain the ordinary van der Waals equation, write P_0 for hard spheres just keeping the term in $B_2 =$

---

*B. J. Alder, W. E. Alley, and M. Rigby. *Physica* **73**, 143 (1974).

$2/3\pi\sigma^3 = b_0 = b/N_A$, where b is the van der Waals constant. Then to this order,

$$P_0 = \frac{NkT}{V}\left(1 + b_0\frac{N}{V}\right) \cong \frac{NkT}{V(1 - b_0 N/V)}$$

In addition take $g_0^{(2)}$ as density independent

$$\begin{aligned} g_0^{(2)} &= 0 && \text{for } r < \sigma \\ &= 1 && \text{for } r > \sigma \end{aligned}$$

and define

$$a = -2\pi \int_\sigma^\infty \varepsilon'(r)r^2\, dr \tag{3-114}$$

such that $N_A^2 a$ is the van der Waals constant. Thus

$$P = P_0 + \frac{\partial}{\partial V}\left(\frac{aN^2}{V}\right) = \frac{NkT}{V(1 - b_0 N/V)} - \frac{aN^2}{V^2} \tag{3-115}$$

which is the van der Waals equation! Note that a as defined in Eq. (3-114) will be positive since the $\varepsilon'(r)$ function is attractive, that is, negative outside of $r = \sigma$.

The augmented equation improves on the ordinary van der Waals theory by writing the equation of state as

$$\frac{PV_m}{RT} = \frac{P_0 V_m}{RT} - \frac{2\pi\sqrt{2}}{T^*X}\left(I - X\frac{\partial}{\partial X}I\right) \tag{3-116}$$

where $P_0 V_m/RT$ for hard spheres is given in the virial series form of Example 3-5 or better by Eq. (3-89), with $X = V/V_0$, and the integral I is dimensionless and defined by

$$I = \frac{-1}{\sigma^3\varepsilon^*}\int_0^\infty g_0^{(2)}(r, \rho)\varepsilon'(r)r^2\, dr \tag{3-117}$$

with ε^* some characteristic energy depth of the attractive tail function $\varepsilon'(r)$, and $T^* = kT/\varepsilon^*$.

Exercise 3-9

Derive Eq. (3-116) from Eq. (3-113) using the definition of I in Eq. (3-117).

The main improvement over van der Waals lies in the use of the correct hard sphere equation for $P_0 V_m / RT$. The function $[I - X(\partial/\partial X)I]$ is only moderately density dependent (in van der Waals, I is independent of density). In a study of the square well potential

$$\varepsilon'(r) = -\varepsilon^* \qquad \sigma \le r \le s\sigma$$

$$\varepsilon'(r) = 0 \qquad r > s\sigma \tag{3-118}$$

added to a hard core for $s = 1.5$, Alder and Hecht [20] found that I changed only about 20 percent over the entire liquid (and solid) region below the X value of the critical point ($X_c = 4.2$), and again only another 20 percent from $X = 4.2$ to infinite dilution ($X = \infty$). The $g_0^{(2)}(r, \rho)$ function for hard spheres needed to numerically integrate Eq. (3-117) is now easily available from computer simulation studies. In fact, a program that computes this function at any r and ρ due to D. Henderson is reproduced in Appendix D of reference [18].

Interesting qualitative results concerning the dependence of the critical properties of a fluid on the shape of the attractive potential tail [20] can be obtained by using the augmented van der Waals equation and the critical point conditions

$$\left(\frac{\partial P}{\partial V}\right)_{T_c} = \left(\frac{\partial^2 P}{\partial V^2}\right)_{T_c} = 0 \tag{3-119}$$

Different shapes of tail were "normalized" to have the same volume integral of the potential beyond the hard core as that of an LJ potential:

$$\frac{-2\pi}{\sigma^3 \varepsilon^*} \int_\sigma^\infty \varepsilon'(r) r^2 \, dr = \text{const} = \frac{-2\pi}{\sigma^3 \varepsilon^*} \int_\sigma^\infty 4\varepsilon^* \left[\left(\frac{\sigma}{r}\right)^{12} - \left(\frac{\sigma}{r}\right)^6 \right] r^2 \, dr$$

$$= \frac{16\pi}{9} \tag{3-120}$$

For a square well to 1.5σ of constant depth ε_s, for example,

$$\frac{-2\pi}{\sigma^3 \varepsilon^*} \int_\sigma^{1.5\sigma} (-\varepsilon_s) r^2 \, dr = \frac{19}{12} \pi \left(\frac{\varepsilon_s}{\varepsilon^*}\right) = \frac{16\pi}{9}$$

$$\varepsilon_s = 1.1\varepsilon^*$$

This has been rounded to $\varepsilon_s \simeq \varepsilon^*$ for square wells to 1.5σ in Table 3-8. Other attractive potentials considered in Table 3-8 besides the LJ and various widths of square wells are the Sutherland potential:

$$\varepsilon'(r) = -\varepsilon_s \frac{\sigma^6}{r^6} \qquad \text{for } r > \sigma \tag{3-121}$$

and the Kac potential:

$$\varepsilon'(r) = -\alpha \gamma^3 e^{-\gamma r} \qquad \text{for } r > \sigma \tag{3-122}$$

TABLE 3-8 Liquid-Gas Critical Parameters from Augmented van der Waals Equation for Various Attractive Tails.[a] (All cases have a hard core of diameter σ, close packed volume $V_0 = N\sigma^3/\sqrt{2}$)

Tail	Range[b] in Units of σ	Maximum[c] Potential Depth	$X_c = (V_c/V_0)$	$\dfrac{kT_c}{\varepsilon^*}$	$\left(\dfrac{PV_m}{RT}\right)_c$
Sutherland	0.395	$2.67\varepsilon^*$	3.25	1.80	0.44
Square well to 1.5 σ	0.500	ε^*	4.49	1.41	0.44
Exact MD results for square well to 1.5σ	0.500	ε^*	4.15 ± 0.2	1.26 ± 0.01	0.28 ± 0.01
Lennard-Jones	0.586	ε^*	4.72	1.35	0.41
Square well to 1.8σ	0.800	$0.50\varepsilon^*$	5.54	1.10	0.40
Square well to 2.0σ	1.000	$0.34\varepsilon^*$	5.81	0.98	0.38
Kac potential[d]	∞	0	5.66	1.01	0.36

[a] GENERAL SOURCE: B. J. Alder and C. E. Hecht. *J. Chem. Phys.* **50**, 2032 (1969).
[b] For square wells the range is taken as their width. For infinitely extended tails the range is taken as $R\sigma - \sigma$, where $R\sigma$ is the distance beyond σ for which the volume integral of the attractive tail falls to $(1/e)$ the volume integral out from $R = \sigma$. e = base of natural logs.
[c] Expressed in terms of the maximum depth, ε^*, of the LJ potential. All tails are normalized to have a volume integral of attractive tail equal to that of the LJ potential.
[d] For discussion of the Kac potential see Problems 3-20 and 3-21, and J. Lebowitz, S. Baer, and G. Stell. *Phys. Rev.* **141**, 198 (1966).

with α a constant and $\gamma = 1/r_1$, with r_1 the range of the potential and the limit $r_1 \to \infty$ taken such that this is an infinitesimally weak attraction of infinite range that results in a finite value for the integral I (see Problems 3-21 and 3-22). It has been proved rigorously* that for the Kac potential the augmented van der Waals equation is exact (in the sense that all $\omega_i = 0$ in Eq. (3-108), for $i > 1$). The physical reason for this is that for such a potential the potential energy will not vary from one configuration of particles to another. As the range of the different attractive potentials (defined in Table 3-8) in units of the hard core σ increases, the $X_c = (V_c/V_0)$ value also increases. Long-range attraction causes the critical number density to be low, short-range attraction causes the critical density to be high. The reduced critical temperature becomes larger as the depth of the potential increases, and conversely the temperature can be made lower without the possibility of liquid

---

* P. C. Hemmer, M. Kac, and G. E. Uhlenbeck. *J. Math. Phys.* **5**, 60 (1964).

phase formation as the well becomes shallower. Notice that for the wide and shallow square well to 2.0σ, the critical parameters are close to those of the van der Waals limit, that is, of the Kac potential. The inaccuracy of the augmented van der Waals equation is reflected in the comparison of the square well to 1.5σ results with those from a molecular dynamics calculation for this case, also shown in Table 3-8. Some of this inaccuracy is due to the fact that in reference [20] the $g_0^{(2)}(r, \rho)$ used for hard spheres was from early computer simulations and does not represent the better form now (post 1972) available.

■

Exercise 3-10

Show that the range of the LJ potential as defined in Table 3-8 is $(R\sigma - \sigma)$ where $R > 1$ is determined by

$$\frac{-2\pi}{\sigma^3 \varepsilon^*} \int_{R\sigma}^{\infty} 4\varepsilon^* \left[\left(\frac{\sigma}{r} \right)^{12} - \left(\frac{\sigma}{r} \right)^6 \right] r^2 \, dr = \frac{16\pi}{9e}$$

where e is the base of natural logs. Note Eq. (3-120). Calculate R numerically.

■

Table 3-9 presents experimental V_c/V_0 values for a number of real substances. From the correlation suggested by Table 3-8, the range of the attractive tails is seen to increase from argon to the alkali metals to water.

TABLE 3-9   Experimental V_c/V_0 Values for Some Real Substances

Substance	$10^{10}\,\sigma/m$	T_c/K	V_c/V_0
Ar	3.4	151	4.4
Cs	5.2[a]	2057	5.2[c]
Rb	4.7	2093	5.6
K	4.4	2223	5.8
Na	3.6	2573	5.8
H_2O	2.76[b]	647	6.2[d]

[a] J. C. Slater. *J. Chem. Phys.* **41**, 3199 (1964).
[b] J. S. Rowlinson. *Trans. Faraday Soc.* **47**, 120 (1951).
[c] B. S. Swanson et al. *J. Chem. Phys.* **44**, 4229 (1966).
[d] J. Levelt-Sengers. *Physica* **73**, 73 (1974).

Perturbation theory calculations have been extended in an approximate way to second order with many refinements [19]. An example of some such results for the equation of state is shown in Fig. 3-15.

To complete our study of simple liquids a few remarks are in order about the solid-liquid boundary. This boundary never ends in a critical point since it is not possible to pass gradually from a liquid to a solid, which must either *have* or *not*

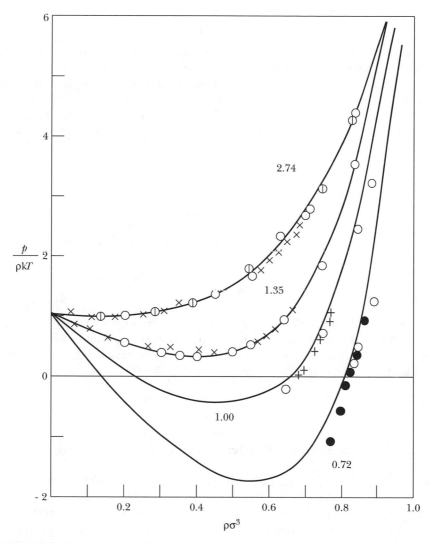

FIGURE 3-15 The equation of state for the Lennard-Jones potential from (approximate) second-order perturbation theory. The curves are labeled with T^* values. The points are from computer simulations.
SOURCE: J. A. Barker and D. Henderson. *J. Chem. Phys.* **47**, 4714 (1967).

have certain symmetry elements (which cannot ever be partially present). Most substances melt upon expanding about 30 percent in volume from their close-packed (or 0 K) form. As mentioned at the end of Sec. 3-4, this is even observed for hard spheres with no attractive interactions! Hard spheres have no specific liquid phase; their triple point may be said to coincide with their critical point. To properly describe melting it has been found necessary to focus attention on the exponent n in the repulsive term of the intermolecular potential, rewritten from Eq. (3-29) as

$$\varepsilon_r(r) = \varepsilon_0 \left(\frac{\sigma}{r}\right)^n \tag{3-123}$$

It is also necessary to include an attractive tail. Metals (especially alkali metals) exhibit a soft repulsion, $n = 4$ or 5, while solidified inert gases have a much harder repulsion, $n \simeq 15$. This different choice of exponents can be shown* to reproduce the very different melting properties of these two classes of substances: Metals have a very large liquid range, with a ratio of their critical point temperature to their triple point temperature greater than 6, while this ratio is only ~ 1.8 for the liquified gases.

Exercise 3-11

Show that the choice $n = \infty$ in Eq. (3-123) is equivalent to a hard sphere repulsion.

Finally, the reader must take note that the theories we have presented are only accurate for simple liquids, that is, those with nonpolar and almost spherical molecules. Much work has been done [19] on complicated liquids, which are after all the usual type encountered in chemistry. Much work remains to be done.

3-8
Critical Phenomena: Critical Exponents and Scaling Laws

In the preceding two sections we briefly treated some aspects of phase transitions and the liquid-gas critical point. The existence of a phase transition implies the instability of one phase relative to some other phase. An ordinary critical point is the state at which two phases become identical, having been driven into homo-

* H. Matsuda and Y. Hiwatari, *Prog. Theor. Phys.* (Japan) **47**, 741 (1972).

geneity by continuous qualitative changes in both phases. It is a particularly delicate and strange state due to the fact that so many physical properties become zero or infinite when it is reached. In this section we present a short review of the critical state, including a description of some of the profound and fascinating new ideas [21–28] introduced to describe it. In doing so we will present many purely thermodynamic relations.

There are several different types of system that show critical phenomena besides the liquid-gas example. One is a system of partially miscible liquids in which two liquid phases of different composition become identical at a critical point, T_c. The second-order transition in ferromagnets is another example. Below their critical (or Curie) temperature ferromagnetic materials can develop a spontaneous magnetization even in the absence of an external magnetic field. Conversely, as ferromagnets reach the Curie temperature from below, their net magnetization (as vector sum of contributions from two or more different directions) vanishes. Another example of second-order transition critical points is the ordering temperature T_c in some homogeneous crystals, such as β-brass (Cu-Zn), below which one crystal sublattice is predominantly occupied by one species and the other sublattice is predominantly occupied by the other species. Also, the lambda transition in liquid ^{4}He and the transition of many metals and alloys to the superconducting state (with zero resistance to electron flow) exhibit many features in common with ordinary critical point phenomena.

One can study theoretically (and even experimentally for certain adsorbed materials) critical phenomena in two dimensions. A particularly important statistical mechanical model is the two-dimensional Ising model for ferromagnetic interaction, which treats single spins located on lattice sites. Each spin can only be up or down, as for a single unpaired electron in a paramagnetic atom. A nearest-neighbor-only interaction is assumed, which gives a lower energy to a pair if their spins are parallel than if they are antiparallel. In zero external magnetic field this model develops a spontaneous magnetization below a critical temperature. Furthermore, all the details of its thermodynamics have been worked out exactly [21]. In what follows we shall refer to some of these exact results and also to known approximate results for the three-dimensional Ising model.

This rich variety of critical behavior can be classified and understood from a common viewpoint. We shall describe this after study of the gas-liquid case.

We use the term "classical" to mean standard, in that it is assumed that the Helmholtz free energy and the pressure are analytic at the critical point in the mathematical sense of being expandable in a Taylor series about their critical point values. This turns out to be equivalent physically to assuming that the state of any particular molecule of the system is determined by the average properties (such as net magnetization or average density) of the system as a whole [28]. Assuming this is the same as saying that all particles contribute equally to the force on every particle or that the forces have infinite range. [This is what made the augmented van der Waals theory exact for the Kac potential of Eq. (3-122) but of course real molecules do not interact with a Kac potential.] Classical theories are thus also called mean field theories. They neglect fluctuations in local density or in local magnetization.

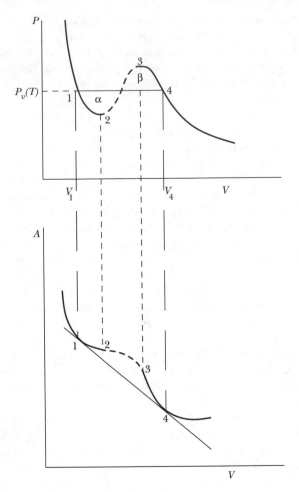

FIGURE 3-16 Correlated subcritical $(T_1 < T_c)$ isotherms in the P-V and A-V planes from any approximate classical theory of the equation of state. The Maxwell construction yields the vapor pressure $P_v(T_1)$ of the two coexisting phases at V_1 and V_4 in the P-V plane and the common tangent $(\partial A/\partial V)_T = -P_{vap}(T_1)$ in the A-V plane.

In Fig. 3-16 we show correlated isotherms from any classical equation of state in the P-V and A-V planes for a subcritical temperature. Since it is always required for stability that $(\partial P/\partial V)_T \leq 0$, we note that both curves show nonphysical portions (2-3) where $(\partial P/\partial V)_T > 0$ or where $(\partial^2 A/\partial V^2)_T < 0$. Maxwell first showed that because of the equality of the chemical potentials of liquid (l) and gas (g) the equilibrium vapor pressure is found by drawing the horizontal tie line 1-4 in the P-V plane such that area α = area β (see Problem 3-26). This corresponds to drawing the common tangent 1-4 in the A-V plane. That is, for the two-phase equilibrium we require,

$$T_l = T_g \tag{3-124.1}$$

$$\left(\frac{\partial A_m}{\partial V_m}\right)_{T_l} = -P_l = -P_g = \left(\frac{\partial A_m}{\partial V_m}\right)_{T_g} \tag{3-124.2}$$

$$\mu_l = G_{m_l} = A_{m_l} - V_{m_l}\left(\frac{\partial A_m}{\partial V_m}\right)_{T_l}$$

$$= \mu_g = A_{m_g} - V_{m_g}\left(\frac{\partial A_m}{\partial V_m}\right)_{T_g} \tag{3-124.3}$$

such that the common (negative) tangent is

$$\left(\frac{\partial A_m}{\partial V_m}\right)_T = \frac{A_{m_l} - A_{m_g}}{V_{m_l} - V_{m_g}} = -P_v.$$

The 1-2 and 3-4 sections of the curves in Fig. 3-16 correspond in principle to observable metastable states of superheated liquid and supercooled gas, respectively. In Fig. 3-17 we show a series of isotherms in the P-V plane for a classical equation

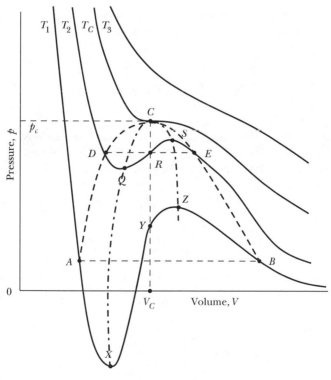

FIGURE 3-17 Representation of the isotherms of a classical equation of state leading to a critical point. Also shown are coexistence curve —————, and spinodal curve —·—·—.
SOURCE: Reference [6].

of state and sketch the saturation or coexistence curve and the curve of limit of stability, called the spinodal curve. Note that the isothermal compressibility

$$\beta_T = -\frac{1}{V}\left(\frac{\partial V}{\partial P}\right)_T$$

always becomes infinite on the spinodal curve and that this curve coincides with the saturation curve at the critical point.

Referring again to Fig. 3-16, the critical point is the case for which the tangent line 1-4 is vanishingly small such that all points 1, 2, 3, 4 coincide along with the point between 2 and 3 at which $(\partial P/\partial V)_T$ is a maximum and $(\partial^2 P/\partial V^2)_T = 0$. Thus at the critical point

$$(V_m)_{l_c} = (V_m)_{g_c}$$

$$P_c > 0 \qquad \left(\frac{\partial P}{\partial V}\right)_{T_c} = 0 \qquad \left(\frac{\partial^2 P}{\partial V^2}\right)_{T_c} = 0 \qquad\qquad (3\text{-}125.1)$$

By further requirement for thermodynamic stability [25], some odd derivative of P with respect to V must be the first nonvanishing (negative) derivative. We usually take this as

$$\left(\frac{\partial^3 P}{\partial V^3}\right)_{T_c} < 0 \qquad\qquad (3\text{-}125.2)$$

Before we examine the implications of the classical model we must discuss critical point exponents, which are the exponents of the leading terms in powers of $\pm t$ by which a property goes to zero or to infinity as the critical temperature is approached, where

$$t = \frac{T - T_c}{T_c} \qquad\qquad (3\text{-}126)$$

We will use a subscript $+(-)$ to denote above T_c (below T_c) values of exponents. Generally, for a positive function $f(t)$ the critical point exponent $\lambda_\pm$ of $f(t)$ is defined by

$$\lambda_\pm = \lim_{t \to 0_\pm} \frac{\ln f(t)}{\ln(\pm t)} \qquad\qquad (3\text{-}127)$$

This λ gives the rate of approach of f to zero if λ is positive or to infinity if λ is negative. Notice that below T_c, $f(t)$ need not be as simple as $B_0(-t)^{\lambda_-}$ but will generally be like $B_0(-t)^{\lambda_-}[1 + B_1(-t)^x + \ldots]$ with $x > 0$. Then application of

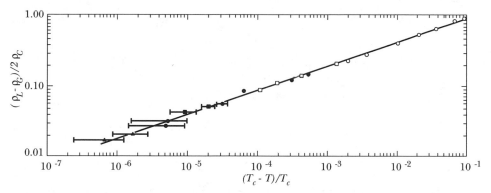

FIGURE 3-18 The coexistence curve for CO_2. Exponent β is deduced to be 0.335 ± 0.020. SOURCE: P. Heller. *Reports on Progress in Physics* **30**(II), 731 (1967).

l'Hospital's rule to Eq. (3-127) will still give λ_-

$$\lambda_- = \lim_{t \to 0_-} \frac{d/dt \ln f(t)}{d/dt \ln (-t)} = \frac{d \ln f(t)}{d \ln (-t)}$$

Furthermore, this shows that a plot of $\ln f(t)$ versus $\ln (-t)$ will give λ_- from the slope of the straight line part if we are near enough to T_c. Two examples of such plots are shown in Figs. 3-18 and 3-19. They represent data on the coexistence curves of two entirely different systems: pure CO_2 and liquid mixtures of CCl_4-C_7F_{14} very close to their respective critical temperatures. The lines are practically coincident, and more significantly, they yield the same value for the exponent β [see Eq. (3-128)].

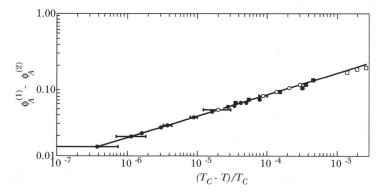

FIGURE 3-19 The coexistence curve of liquid mixtures of CCl_4-C_7F_{14}, $\phi_A^{(1)} =$ concentration of CCl_4 in molecules/cm^3 in phase 1 divided by its number density in pure CCl_4. Exponent β is deduced to be 0.340 ± 0.015. SOURCE: P. Heller. *Reports on Progress in Physics* **30**(II), 731 (1967).

Exercise 3-12

Show that if $f(t) = -a_- \ln(-t) + b_-$ for $T < T_c$, and $f(t) = -a_+ \ln t + b_+$ for $T > T_c$, where both a_+ and a_- are > 0, then $\lambda_+ = \lambda_- = 0$. The heat capacity, C_V, of the two-dimensional Ising model is of this form.

The following functional forms suggested from experimental results serve to define the most important critical point exponents. Experimental *conditions* that determine the path followed in taking the limit must be specified.

The order parameter of the transition—which is a quantity that is nonzero below T_c and zero above T_c and thus only defined below T_c—is defined as the density difference between coexisting liquid and gaseous phases:

$$\frac{\rho_l(T) - \rho_g(T)}{2\rho_c} = B_0(-t)^\beta(1 + \ldots) \qquad (3\text{-}128)$$

Condition: along the saturation curve for $T < T_c$

This defines the exponent β. Division by $2\rho_c$ is included so that we deal with dimensionless quantities and so that the constant B_0 will change only slightly from system to system (it is of order unity).

The compressibility, β_T, being infinite at $T = T_c$, we define exponents γ_- and γ_+ by (the negative signs in these definitions make them both positive)

$$\beta_T/(P_c^{-1}) = \frac{\rho_c^2}{\rho^2}\Gamma_-(-t)^{-\gamma_-}(1 + \ldots) \qquad (3\text{-}129.1)$$

Condition: $T < T_c$, $\rho = \rho_l$ or ρ_g on the saturation
curve

and

$$\beta_T/(P_c^{-1}) = \Gamma_+(t)^{-\gamma_+}(1 + \ldots) \qquad (3\text{-}129.2)$$

Condition: $T > T_c$, $\rho = \rho_c$

The experimental heat capacity at constant volume C_V seems to become infinite at T_c if $\rho = \rho_c$. For $\rho < \rho_c$ there is a finite discontinuity at a $T < T_c$ corresponding physically to the disappearance of the last drop of liquid (the two-phase system having the higher C_V—see Appendix 3.B). For $\rho > \rho_c$ there is a corresponding discontinuity at the disappearance of the last bubble of vapor. The heat capacity exponents α_+, α_- are defined by:

$$\frac{C_V}{Nk} = A_-(-t)^{-\alpha_-}(1 + \ldots) + B_- \tag{3-130.1}$$

Condition: $T < T_c$ in a two-phase system
with $\rho = \rho_c$ overall

and

$$\frac{C_V}{Nk} = A_+(t)^{-\alpha_+}(1 + \ldots) + B_+ \tag{3-130.2}$$

Condition: $T > T_c, \rho = \rho_c$

Exercise 3-13

If $\alpha_+ = \alpha_- = 0$ in Eq. (3-130), show that the prediction of these equations is a finite discontinuity in C_V going from the two-phase to the one-phase system at $T = T_c$ and $\rho = \rho_c$. This is the behavior of C_V for classical theories. It is easy to show experimentally that this prediction is false. If $\alpha_+ = \alpha_-$ is small (less than 0.25) and positive, why is it hard to distinguish experimentally from the logarithmic singularity suggested in Exercise 3-12?

The one exponent δ describes the variation of $P - P_c$ with $\rho - \rho_c$ along the critical isotherm as one takes the limit $\rho \to \rho_c$:

$$\frac{P - P_c}{P_c} = D\left|\frac{\rho - \rho_c}{\rho_c}\right|^{\delta} \operatorname{sgn}(\rho - \rho_c) \tag{3-131}$$

where sgn (I) means sign of I, that is, a factor of $+1$ or -1 only.

The other important critical exponents refer to the behavior of the total pair correlation function $h(r)$ in the critical region, where

$$h(r) \equiv g^{(2)}(r) - 1 \tag{3-132}$$

and $g^{(2)}(r)$ is the radial distribution function defined in Sec. 3-6. We know that for large r, $g^{(2)}(r) \to 1$ and so $h(r) \to 0$. The expected behavior [23] of $h(r)$ in the one-phase region for $T > T_c$ is

$$h(r) = f(r)e^{-r/\zeta} \tag{3-133}$$

where $f(r)$ is slowly varying compared to the exponential and ζ is the range of order or *correlation length* in the fluid. It is a measure of the *greatest distance* from a given particle at which the presence of this particle can still cause nonrandom effects in particle number density. Except near the critical point, ζ is of the order of only a few molecular diameters. However as the temperature gets near to T_c the correlation length will increase to macroscopic size if $\rho = \rho_c$ (indeed for practical purposes we can say it becomes infinite). The correlations take the form of density or local number of particle fluctuations. High-density and low-density regions will appear in the fluid on a larger and larger scale as $T \to T_c$ until macroscopic volumes of different overall densities exist in the fluid, leading to the two-phase situation as T falls below T_c. Although the maximum size of these density fluctuations increases, the smaller fluctuations do not disappear [28]. Inside the high-density region there will be smaller regions of low density containing within themselves still smaller pockets of high density, and so on. These density fluctuations scatter visible light very strongly as soon as ζ becomes as large as the wavelength, $\sim 10^{-6}$ m. This is the phenomenon of critical opalescence, which persists even when the temperature becomes even closer to T_c and $\zeta \gg 10^{-6}$ m.

We can directly relate the divergence (to infinity) of the density fluctuations to the divergence of the isothermal compressibility. In Problem 33 of Chap. 2 we proved that

$$\overline{\mathcal{N}^2} - (\overline{N})^2 = kT\left(\frac{\partial \overline{N}}{\partial \mu}\right)_{V,T} \tag{3-134}$$

We rewrite

$$\beta_T = -\frac{1}{V}\left(\frac{\partial V}{\partial P}\right)_{T,\overline{N}} = \frac{1}{\rho}\left(\frac{\partial \rho}{\partial P}\right)_{T,\overline{N}} \tag{3-135}$$

and note that since $\overline{N}/V = \rho$ is a function of P and T only it does not matter if we differentiate ρ at constant $\overline{N}$ or at constant V. Thus,

$$\beta_T = \frac{V}{\overline{N}}\left[\frac{\partial}{\partial P}\left(\frac{\overline{N}}{V}\right)\right]_{T,V} = \frac{1}{\overline{N}}\left(\frac{\partial \overline{N}}{\partial P}\right)_{T,V} = \frac{\rho}{\overline{N}\rho}\left(\frac{\partial \overline{N}}{\partial P}\right)_{T,V} \tag{3-136}$$

Then from the fundamental relations of the grand canonical ensemble in Sec. 7 of Chap. 1:

$$PV = kT \ln \mathcal{Q}$$

$$\overline{N} = kT\left(\frac{\partial}{\partial \mu}\right)_{V,T} \ln \mathcal{Q} = \sum_N \frac{N z^N Q^N}{\mathcal{Q}}$$

$$z = e^{\mu/kT}$$

we have

$$\left(\frac{\partial P}{\partial \mu}\right)_{T,V} = \frac{\bar{N}}{V} = \rho \tag{3-137}$$

Inserting Eq. (3-137) into Eq. (3-136) and using Eq. (3-134), we find

$$\bar{N}kT\rho\beta_T = \overline{(N - \bar{N})^2} \tag{3-138}$$

which shows that β_T goes to infinity at the critical state because the fluctuations go to infinity, which in turn depends on the correlation length ζ going to infinity. Of course, strictly to go to infinity the system must be of infinite size, but the finite size effects are too minor to be experimentally significant. No instrument can measure an infinite value in any case.

The critical exponents v_- and v_+ are defined by

$$\zeta = \zeta_{0_-}(-t)^{-v_-} \qquad T < T_c$$

$$\rho = \rho_l \text{ or } \rho_g \text{ on the saturation curve} \tag{3-139.1}$$

and

$$\zeta = \zeta_{0_+}(+t)^{-v_+} \qquad T > T_c, \rho = \rho_c \tag{3-139.2}$$

It is easy (see Problem 3-25) to prove from Eq. (3-138), using the definitions of $g^{(n)}$ in Eqs. (3-94) and (3-95), that

$$kT\rho\beta_T = 1 + \rho \int d\bar{r} h(r) \tag{3-140}$$

For this integral to diverge at $\rho \to \rho_c$ and $T = T_c$, $h(r)$ itself must become long range in the sense that in d dimensions it must decrease more slowly with r (as $r \to \infty$) than $1/r^d$ since $d\bar{r}$ goes as r^d in d dimensions. There is a classical theory [23] for this due to Ornstein and Zernike which shows that as $r \to \infty$ and $T > T_c$,

$$h(r) \to \frac{D_d e^{-r/\zeta}}{r^{d-2}} \tag{3-141}$$

However in two dimensions ($d = 2$) Fisher [23] showed that this form leads to the physically impossible prediction that $h(r)$ increases as r increases. Therefore it is suspect in three dimensions as well. Accordingly, he introduced another exponent η, which must be $0 < \eta < 2$, as follows:

$$h(r) \to \frac{D_d e^{-r/\zeta}}{r^{d-2+\eta}} \qquad \text{as } r \to \infty \tag{3-142.1}$$

$$\rho = \rho_c$$

$$h(r) \to \frac{D_d}{r^{d-2+\eta}} \qquad \text{as } r \to \infty \qquad (3\text{-}142.2)$$

$$\rho = \rho_c, \, T = T_c$$

The classical value of η is zero and experiments on real fluids in three dimensions indicate η is very small (about 0.03). For the Ising model in two dimensions $\eta = \frac{1}{4}$ and is very significant.

Example 3-8

As an example of deriving relationships between the critical point exponents, insert Eq. (3-142.1) into Eq. (3-140) for $d = 3$ and derive $\gamma_+ = (2 - \eta)\nu_+$

Solution:

For the three-dimensional case with $T > T_c$, $\rho = \rho_c$

$$\rho \int d\bar{r} \, h(r) \simeq 4\pi\rho \int_0^\infty r^2 \frac{dr \, D_3 e^{-r/\zeta}}{r^{1-\eta}} = 4\pi D_3 \rho \zeta^{2-\eta} \int_0^\infty x^{1-\eta} e^{-x} \, dx$$

using $x = r/\zeta$. We may assume η is less than 1 and the integral

$$\int_0^\infty x^{1-\eta} e^{-x} \, dx$$

is certainly convergent, being unity if $\eta = 0$.

Thus in the relation of Eq. (3-140) $(kT\rho\beta_T)$ diverges as $t^{-\gamma_+}$ by Eq. (3-129.2), while $1 + \rho \int d\bar{r} h(r)$ diverges as

$$\zeta^{2-\eta} = (t^{-\nu_+})^{2-\eta} \qquad \text{by Eq. (3-139.2)}$$

$$\therefore \quad \gamma_+ = (2 - \eta)\nu_+$$

Comment: For the classical case $\eta = 0$ and $\gamma_+ = 2\nu_+$.

The critical point exponents are important first of all because two strict thermodynamic inequalities among some of them may be proved [21, 23]:

$$\alpha_- + 2\beta + \gamma_- \geq 2 \qquad (3\text{-}143)$$

$$\alpha_- + \beta(1 + \delta) \geq 2 \qquad (3\text{-}144)$$

In addition, modern theories (that seem to be correct experimentally) show that these inequalities are actually satisfied in nature as *equalities* and further that

$$\alpha_- = \alpha_+ \equiv \alpha \tag{3-145.1}$$

$$\nu_- = \nu_+ \equiv \nu \tag{3-145.2}$$

$$\gamma_- = \gamma_+ \equiv \gamma \tag{3-145.3}$$

$$\gamma = (2 - \eta)\nu \tag{3-145.4}$$

Finally, theory gives us an interesting relationship involving the dimensionality d of the system:

$$\nu d = 2 - \alpha \tag{3-146}$$

Therefore, of the original nine critical point exponents only two are independent and the other seven can be calculated once any two are determined.

We return now to investigate the predictions of any classical model for the critical point exponents, which assumes one can expand the Helmholtz free energy or any of its volume derivatives $[P = -(\partial A/\partial V)_T]$ about its value at the critical point. It proves convenient to discuss the expansion of $(\partial P/\partial V)_T$. We define

$$\tau = T - T_c = T_c t \tag{3-147}$$

$$\omega = V_m - V_{m_c} \tag{3-148}$$

$$P_{ij} = \left(\frac{\partial^{i+j} P}{\partial V_m^i \partial T^j} \right)_c \tag{3-149}$$

where the last is a multiple partial derivative evaluated at the critical point ($T = T_c$, $P = P_c$, $V_m = V_{m_c}$) such that

$$P_{11} = \left(\frac{\partial^2 P}{\partial V_m \partial T} \right)_c \qquad P_{20} = \left(\frac{\partial^2 P}{\partial V_m^2} \right)_c$$

and so on. Then by Taylor series expansion near the critical point

$$\left(\frac{\partial P}{\partial V_m} \right)_T = P_{11}\tau + P_{20}\omega + \frac{1}{2!} P_{30}\omega^2 + P_{21}\omega\tau + \frac{1}{2!} P_{12}\tau^2 + \ldots \tag{3-150}$$

By Eq. (3-125.1) $P_{20} = 0$ and by Eq. (3-125.2) P_{30} is negative. Near enough to the critical point we may neglect the terms in $\omega\tau$ and in τ^2 with respect to the terms in τ and ω^2. We retain the term in ω^2 since it may not be (and in fact *is* not) smaller than the term in τ. Hence we use

$$\left(\frac{\partial P}{\partial V_m} \right)_T = P_{11}\tau + \frac{1}{2} P_{30}\omega^2 = \left(\frac{\partial P}{\partial \omega} \right)_T \tag{3-151}$$

Since we know $(\partial P/\partial V_m)_T$ must be negative for all stable states, consideration of the case of τ positive ($T > T_c$), for which all states are stable *and* $\omega = 0$ (on the critical isochore above T_c), shows that P_{11} must be negative. Integration of Eq. (3-151) at constant T gives

$$P = P_{11}\tau\omega + \tfrac{1}{6}P_{30}\omega^3 + f(\tau) \tag{3-152}$$

where the "constant" of integration is a function of temperature, thus of τ:

$$f(\tau) = P_c + \left(\frac{\partial P}{\partial T}\right)_{V_c} \tau + \ldots \tag{3-153}$$

The expression for the pressure is thus of the usual sigmoid (S) type, cubic in the volume, with three real roots for ω for $T < T_c$ in certain pressure ranges, but with only two roots making physical sense. By the Maxwell construction (see Fig. 3-16) that requires ΔG between coexisting liquid and gaseous phases to vanish at constant temperature below T_c, we have the relation

$$\Delta G = 0 = \int_{P_l}^{P_g} V_m \, dP = \int_{P_l}^{P_g} \omega \, dP = \int_{\omega_l}^{\omega_g} \omega \left(\frac{\partial P}{\partial \omega}\right)_T d\omega \tag{3-154}$$

In writing Eq. (3-154) we used $V_m = \omega + V_{m_c}$, and since V_{m_c} is a constant, the $\int_{P_l}^{P_g} V_{m_c} \, dP$ is identically zero since $P_l = P_g$. Inserting Eq. (3-151) into Eq. (3-154) and integrating we have

$$\int_{\omega_l}^{\omega_g} (P_{11}\tau\omega + \tfrac{1}{2}P_{30}\omega^3) \, d\omega = 0$$

$$\frac{P_{11}\tau}{2} (\omega_g^2 - \omega_l^2) + \frac{P_{30}}{8} (\omega_g^4 - \omega_l^4) = 0 \tag{3-155}$$

which has for a solution

$$\omega_l^2 = \omega_g^2$$

The only physically acceptable solution is

$$\omega_l = -\omega_g \tag{3-156}$$

with ω_l being negative since the ω's are differences and the molar volumes of liquid and gas in equilibrium below T_c must differ. Notice that Eq. (3-156) also means

$$V_{m_l} + V_{m_g} = 2V_{m_c}$$

However this is a result of only keeping the leading terms in Eq. (3-150). An extended analysis* of the *classical* case shows that the sum of the coexisting volumes is not a constant but is linear in τ:

$$V_{m_l} + V_{m_g} = 2V_{m_c} + \frac{2}{5}\tau\left(\frac{P_{11}}{P_{30}}\right)\left(3\frac{P_{40}}{P_{30}} - 5\frac{P_{21}}{P_{11}}\right) \tag{3-157}$$

Inserting Eq. (3-156) into Eq. (3-152) for the pressure and equating $P_l = P_g$, we have

$$\begin{aligned}
P_g &= P_{11}\tau\omega_g + \tfrac{1}{6}P_{30}\omega_g^3 + f(\tau) \\
&= P_l = P_{11}\tau\omega_l + \tfrac{1}{6}P_{30}\omega_l^3 + f(\tau)
\end{aligned} \tag{3-158}$$

$$P_{11}\tau\omega_g + \tfrac{1}{6}P_{30}\omega_g^3 = -P_{11}\tau\omega_g - \tfrac{1}{6}P_{30}\omega_g^3$$

$$P_{11}\tau\omega_g + \tfrac{1}{6}P_{30}\omega_g^3 = 0 \tag{3-159}$$

since if something is equal to its negative it must be zero. Thus

$$\omega_g = -\omega_l = \left(\frac{-6P_{11}\tau}{P_{30}}\right)^{1/2} \tag{3-160}$$

Recall that τ is negative below T_c and P_{11} and P_{30} are both negative, so we have the square root of a positive quantity. It is easy to show (see Problem 3-27) from Eq. (3-160) that

$$\rho_l - \rho_g = \frac{2\rho_c}{V_{m_c}}\left(\frac{-6P_{11}\tau}{P_{30}}\right)^{1/2} \tag{3-161}$$

Thus by comparison with Eq. (3-128) the critical point exponent β for classical theories is $\frac{1}{2}$. Note from Eq. (3-152) and Eq. (3-153) for the pressure on the critical isotherm $(P \neq P_c, V \neq V_c, \text{ but } T = T_c)$

$$P - P_c = \tfrac{1}{6}P_{30}\omega^3 \tag{3-162}$$

such that the critical exponent δ for classical theories is 3.

From Eq. (3-152)

$$\beta_T = \frac{-1}{V_m}\left[\left(\frac{\partial P}{\partial V_m}\right)_T\right]^{-1} = \frac{-1}{V_m}\frac{1}{(P_{11}\tau + \tfrac{1}{2}P_{30}\omega^2)} \tag{3-163}$$

---

* R. Barieau. *J. Chem. Phys.* **45**, 3175 (1966).

such that approaching $T = T_c$ from below along the coexistence curve (τ negative) in either liquid or gas, $\omega^2 = -6P_{11}\tau/P_{30}$,

$$\beta_T = \frac{1}{2V_m P_{11}\tau} \qquad (3\text{-}164.1)$$

and $\gamma_- = 1$ for classical theories [compare with Eq. (3-129.1)]. Approaching $T = T_c$ from above along the $\rho = \rho_c$ (or $\omega = 0$) critical isochore with τ positive,

$$\beta_T = \frac{-1}{V_{m_c} P_{11}\tau} \qquad (3\text{-}164.2)$$

and $\gamma_+ = 1$ for classical theories [compare with Eq. (3-129.2)]. If we consider β_T^{-1}, which is zero at the critical point as we move *away* from T_c to lower temperatures, β_T^{-1} increases twice as fast as when we move *away* from $T = T_c$ to higher temperatures.

As mentioned in Exercise 3-13, we can also show that for classical theories the exponents in the C_V relations are $\alpha_+ = \alpha_- = 0$ with a finite discontinuity but no infinity in C_V. Since [note Eq. (I-34)]

$$C_P = C_V - T\frac{[(\partial P/\partial T)_V]^2}{(\partial P/\partial V)_T}$$

C_P diverges like β_T in classical theory because $(\partial P/\partial T)_V$ is finite and continuous at the critical point. The reason for this is that the vapor pressure curve (symbol σ in Fig. 3-20) below $T = T_c$ is continuous with the one-phase critical isochore (P

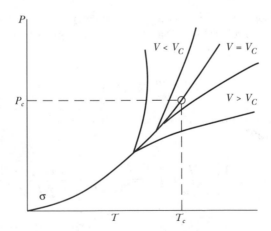

FIGURE 3-20 The vapor pressure curve, σ, for $T \leq T_c$ is continuous with the one-phase critical isochore for $T \geq T_c$. Other isochores intersect the vapor pressure curve in pairs that represent the coexisting volumes at $T = T_\sigma < T_c$.

as function of T only at fixed $V = V_c$) beyond the critical point, as shown in Fig. 3-20. General isochores for $V \neq V_c$ in the P-T plane will intersect in pairs on the vapor pressure curve at temperatures $T_\sigma < T_c$, thus representing the coexisting volumes at particular T_σ points.

Example 3-9

Take the specific classical van der Waals equation of state and using Eqs. (3-152) and (3-153) write the expansion for the pressure in the critical region. Obtain a final expression in fully dimensionless form using

$$a = \frac{9}{8} V_{mc} R T_c \qquad b = \frac{V_{mc}}{3} \qquad \left(\frac{PV_m}{RT}\right)_c = \frac{3}{8}$$

Obtain particular forms of your expression on the critical isotherm and on the coexistence curve [see Eq. (3-160)].

Solution:

$$P = \frac{RT}{V_m - b} - \frac{a}{V_m^2}$$

$$\left(\frac{\partial P}{\partial V_m}\right)_T = \frac{-RT}{(V_m - b)^2} + \frac{2a}{V_m^3} \qquad \left(\frac{\partial^2 P}{\partial V_m \partial T}\right) = \frac{-R}{(V_m - b)^2}$$

$$\left(\frac{\partial^2 P}{\partial V_m^2}\right)_T = \frac{2RT}{(V_m - b)^3} - \frac{6a}{V_m^4} \qquad P_{11} = \left(\frac{\partial^2 P}{\partial V_m \partial T}\right)_c = \frac{-R}{(\frac{2}{3} V_{mc})^2}$$

$$\left(\frac{\partial^3 P}{\partial V_m^3}\right)_T = \frac{-6RT}{(V_m - b)^4} + \frac{24a}{V_m^5}$$

$$P_{30} = \left(\frac{\partial^3 P}{\partial V_m^3}\right)_c = \frac{-6RT_c}{(\frac{2}{3} V_{mc})^4} + \frac{24}{(V_{mc})^5}\left(\frac{9}{8} V_{mc} R T_c\right)$$

$$= \frac{-27}{8}\frac{RT_c}{V_{mc}^4}$$

From Eqs. (3-152) and (3-153)

$$\left(\frac{\partial P}{\partial T}\right)_V = \frac{R}{V_m - b} \qquad P = P_c + \left(\frac{\partial P}{\partial T}\right)_{V_c} \tau + P_{11}\tau\omega + \frac{1}{6}P_{30}\omega^3 + \dots$$

$$\left(\frac{\partial P}{\partial T}\right)_{V_c} = \frac{R}{\frac{2}{3} V_{mc}} \qquad P = P_c + \frac{3}{2}\frac{R\tau}{V_{mc}} - \frac{9}{4}\frac{R\tau\omega}{V_{mc}^2} - \frac{9}{16}\frac{RT_c}{V_{mc}^4}\omega^3 + \dots$$

To obtain a dimensionless equation divide by P_c and use

$$t = \frac{\tau}{T_c} = \frac{T - T_c}{T_c}$$

$$\omega' = \frac{\omega}{V_{m_c}} = \frac{V_m - V_{m_c}}{V_{m_c}}$$

$$\frac{P}{P_c} = 1 + \frac{3}{2}\left(\frac{RT_c}{P_c V_{m_c}}\right) t - \frac{9}{4}\left(\frac{RT_c}{P_c V_{m_c}}\right) t\omega' - \frac{9}{16}\frac{RT_c}{P_c V_{m_c}}(\omega')^3 + \dots$$

Since

$$\frac{RT_c}{P_c V_{m_c}} = \frac{8}{3}$$

we have

$$\frac{P}{P_c} = 1 + 4t - 6t\omega' - \frac{3}{2}(\omega')^3$$

On the critical isotherm $(t = 0)$ near the critical point

$$\frac{P}{P_c} = 1 - \frac{3}{2}(\omega')^3$$

and P may be less than or greater than P_c depending on whether $V_m > V_{m_c}$ or $V_m < V_{m_c}$.

On the coexistence curve we may use Eq. (3-160) to eliminate t since, whether liquid or gas,

$$V_{m_c}^2(\omega')^2 = \omega^2 = \frac{-6P_{11}\tau}{P_{30}} = \frac{-6P_{11}t T_c}{P_{30}} = \frac{-6[-R/(4V_{m_c}^2/9)]t T_c}{-(27/8)(RT_c/V_{m_c}^4)}$$

such that on the coexistence curve with t negative,

$$(\omega')^2 = -4t$$

$$\frac{P_v}{P_c} = 1 - (\omega')^2 - 6\left[\frac{-(\omega')^2}{4}\right]\omega' - \frac{3}{2}(\omega')^3$$

$$= 1 - (\omega')^2 \qquad P_v = \text{vapor pressure}$$

which is a symmetric parabola and P_v can only be $\leq P_c$. The symmetry of this parabola is lost if higher order terms in the expansion are retained.

TABLE 3-10   Critical Exponents from Various Theoretical Models

Exponent	Exact 2D Ising Model ($n = 1$)	3D Models		Classical (Mean Field) Theory (any D)
		Ising* ($n = 1$)	Heisenberg ($n = 3$)	
α	0 (logarithmic)	0.110	-0.115	0 (discontinuous)
β	$\frac{1}{8}$	0.325	0.365	$\frac{1}{2}$
γ	$\frac{7}{4}$	1.240	1.387	1
δ	15	4.82	4.80	3
v	1	0.630	0.705	$\frac{1}{2}$
η	$\frac{1}{4}$	0.03	0.03	0

* It is now accepted that if all the data used in the determination of the *experimental* critical exponents for pure fluids and for fluid-fluid mixtures are for $|t| < 10^{-3}$ then the *experimental* results are all consistent with the 3D Ising theoretical exponents.
SOURCE: J. Levelt-Sengers, R. Hocken, and J. V. Sengers. *Physics Today* **30**, 42 (Dec. 1977).

All the exponent results for classical theory are collected in Table 3-10. The classical results are all wrong as regards experiment on the gas-liquid critical point (and all critical points). Experimental results show a universal (same for all substances) set of exponents that are neither integers nor probably even the ratio of integers! If the experiments make use of data for $|t| < 10^{-3}$ the gas-liquid experimental results are consistent (and believed to coincide) with the approximate exponents of the 3D Ising model also given in Table 3-10. If all or most of the data are for $|t| > 10^{-3}$, correction terms beyond the leading term will give an effective exponent that differs from the true leading term exponent. Most experiments prior to 1976 fall into this category (see Figs. 3-18 and 3-19).

An important scaling law hypothesis first formulated by Widom* resulted in the derivation of the relations among exponents given in Eqs. (3-143–3-145) and suggested a form for the equation of state in the critical region. This is discussed in Appendix 3.A.

Even more remarkable has been the work of K. Wilson (building on some earlier intuitive ideas of Kadanoff) that has shown how to calculate these exponents from first principles [28]. For this achievement Wilson was awarded the Nobel Prize in physics for 1982. The fundamental idea is that because the correlation length goes to infinity at the critical point, all details of molecular interaction such as the force laws between the molecules (or between the spins in a magnetic case) are irrelevant. Examination of Table 3-10 suggests what *is* relevant—first of all spatial dimensionality d, and second, another geometric property, the dimensionality n of the order parameter. This second dimensionality counts the number of ways

* B. Widom. *J. Chem. Phys.* **43**, 3892, 3898 (1965).

that the two phases that are to become identical at the critical point can differ below the critical point. For the gas-liquid case and the fluid-fluid case $n = 1$, since there is a *single* density or concentration difference between the phases that are to become coincident. In the Ising model $n = 1$ also, because the spins can only be up or down with respect to *one* axis. In other more realistic magnetic cases—such as the Heisenberg spin model, in which a spin has components along all d spatial dimensions—$n = d$ (i.e., $n = 3$ for the 3D Heisenberg model, for which the exponents are also given in Table 3-10). Theory and experiment show that all systems with the same n and d should have the same set of critical exponents.[†] This is known as the principle of universality.

The relation in Eq. (3-146) involving the dimensionality is particularly interesting. It can only be satisfied by the mean field or classical theories at one $d = d^*$, namely $d^* = 4$. The methods of Wilson, called renormalization group theory, show that in four and more dimensions ordinary critical points always follow classical theory and thus Eq. (3-146) no longer holds for $d > 4$. It is intuitively reasonable that as the number of dimensions increases (with accompanying increase of nearest neighbors of a given particle), the mean field concept should become more accurate. Indeed it has been known since the 1930's that in one dimension no phase transitions are possible whatsoever, because with such low connectivity fluctuations will always wash out any incipient phase differentiation.

A crucial idea in Wilson's method [28] is that one builds up to treating long-range correlation effects one step at a time by averaging first over fluctuations on an atomic scale and then repeating this averaging for longer and longer scales of length. This is termed renormalization. Many other complex problems, such as turbulent hydrodynamic flow of fluids and the binding of elementary particles, may also be usefully treated by renormalization techniques.

3-9

Elementary Statistical Mechanical Theory for Solutions of Nonelectrolytes

In this section we present a treatment of a molecular model that is the statistical mechanical basis for those solutions termed *strictly regular*, for which the excess molar entropy of mixing, $\Delta S_{m,mxg}^E$ is zero.

[†] This is strictly true only for forces that fall off more rapidly than $1/r^3$. For very long range forces (reminiscent of the Kac potential) exponents close to the *classical* values are observed. This is the case for solutions of metals such as sodium dissolved in liquid ammonia separating into two solutions of different metal concentration below a critical temperature. Some electron delocalization in these solutions probably gives rise to very long range interactions (M. J. Sienko. *J. Chem. Phys.* **53**, 566 [1970]).

The model is called the quasi-lattice model and makes the following assumptions [29]:

1. Quasi-lattice structure. The molecules of the liquid are assumed to have motions mainly restricted to cells defined by a lattice passing through the cell centers about which the molecules oscillate. Each cell center is surrounded by z nearest neighbor cells and their oscillating molecules. This assumption works best for solid solutions, but it is true that in liquids the number of nearest neighbors does have a fairly well defined *average* value. Furthermore, the available volume per molecule in a liquid near the melting point is generally only about 10 percent greater than that in the solid state.

2. Extensive factorization of the *classical* canonical partition function for N_a molecules of type a, N_b of type b . . .

$$Q = Q_{conf}Q_{int} \tag{3-165}$$

where by a generalization of Eq. (3-44) the configurational partition function for the mixture is

$$Q_{conf} = \frac{1}{N_a!N_b! \dots} \int \dots \int \exp -\left(\frac{V_{N_a,N_b \dots}}{kT}\right)(d\bar{r}_a)^{N_a}(d\bar{r}_b)^{N_b} \dots \tag{3-166}$$

and the internal partition function is

$$Q_{int} = (\lambda_a^{-3N_a})(\lambda_b^{-3N_b})(\dots)(q_{int_a})^{N_a}(q_{int_b})^{N_b} \dots \tag{3-167}$$

It is assumed that the internal partition functions q_{int_a} . . . do not change upon mixing. This is a poor approximation for molecules with angular dependent forces and for those that can form hydrogen bonds. Thus the Helmholtz free energy can be written

$$A = A_{conf} + A_{int}$$

with

$$A_{conf} = -kT \ln Q_{conf} \tag{3-168}$$

We further factor Q_{conf} into

$$Q_{conf} = Q_{latt}Q_{vib} \tag{3-169}$$

assuming V_{N_a,N_b} can be split into a potential energy contribution from each molecule at rest on its equilibrium site on the lattice, giving rise to the Q_{latt} factor, plus a contribution due to vibrations about their sites, giving the Q_{vib} factor. The Q_{vib} factor will be discussed in Chap. 4. Here it is assumed to be unaffected by mixing and will cancel in the calculation of mixing properties in what follows.

3. The lattice is taken to be rigid and we do not take into account changes of intermolecular distances due to changes in composition. This means that in this model the excess molar volume of mixing is taken to be zero.

$$\Delta V^E_{m,mxg} = 0 \qquad (3\text{-}170)$$

We also do not take into account thermal expansion of the lattice. In many real solutions, particularly those composed of molecules with different sizes, changes of lattice distances with composition change must be taken into account to properly match experimental data.

4. Each molecule is assumed to occupy one lattice site. This assumption must be relaxed to treat even simple models of polymer solutions.

With these four assumptions we now treat a binary solution of $\mathcal{N}_a$ molecules of type a and $\mathcal{N}_b$ of type b on a lattice of $\mathcal{N} = \mathcal{N}_a + \mathcal{N}_b$ sites each with a coordination (nearest neighbor) number z. There are $\frac{1}{2}z\mathcal{N}$ pairs of nearest neighbors of three kinds, aa, bb, ab. By obvious normalization of pair numbers,

$$\begin{aligned} z\mathcal{N}_a &= 2\mathcal{N}_{aa} + \mathcal{N}_{ab} \\ z\mathcal{N}_b &= 2\mathcal{N}_{bb} + \mathcal{N}_{ab} \end{aligned} \qquad (3\text{-}171)$$

such that

$$z(\mathcal{N}_a + \mathcal{N}_b) = z\mathcal{N} = 2(\mathcal{N}_{aa} + \mathcal{N}_{bb} + \mathcal{N}_{ab})$$

which is twice the total number of pairs, as it must be.

If s = nearest neighbor distance on the lattice, every pair contributes a quantity $\varepsilon_{ij} \equiv \varepsilon_{ij}(s)$ to the lattice potential energy where $\varepsilon_{ij}(r)$ is of the form given in Eq. (3-79). We neglect interactions of nonnearest neighbors for which $r > s$. Thus the lattice energy, U_{latt}, is

$$\begin{aligned} U_{latt} &= \mathcal{N}_{aa}\varepsilon_{aa} + \mathcal{N}_{bb}\varepsilon_{bb} + \mathcal{N}_{ab}\varepsilon_{ab} \\ &= \frac{z}{2}\mathcal{N}_a\varepsilon_{aa} + \frac{z}{2}\mathcal{N}_b\varepsilon_{bb} + \mathcal{N}_{ab}\left(\varepsilon_{ab} - \frac{1}{2}\varepsilon_{aa} - \frac{1}{2}\varepsilon_{bb}\right) \end{aligned} \qquad (3\text{-}172)$$

having used Eq. (3-171). Since s is assumed not to change on mixing and $(z/2)\mathcal{N}_i\varepsilon_{ii} = $ lattice energy of pure $i \equiv U_{ii}$, we have

$$U_{latt} = U_{aa} + U_{bb} + \mathcal{N}_{ab}\omega \qquad (3\text{-}173)$$

where

$$\omega = \varepsilon_{ab} - \frac{1}{2}(\varepsilon_{aa} + \varepsilon_{bb}) \qquad (3\text{-}174)$$

is the energy change on mixing per ab pair formed in the solution.

If we denote $g \equiv g(N_a, N_b, N_{ab})$ = number of arrangements of N_a type a and N_b type b, such that there are N_{ab} type ab pairs on the lattice, the lattice configuration partition function may be written as

$$Q_{latt} = \sum_{N_{ab}} ge^{-U_{latt}/kT}$$

$$= e^{-(U_{aa}+U_{bb})/kT} \sum_{N_{ab}} ge^{-N_{ab}\omega/kT} \tag{3-175}$$

Thus the contribution of the lattice partition function to the free energy of the solution is

$$-kT \ln Q_{latt} = U_{aa} + U_{bb} - kT \ln \left(\sum_{N_{ab}} ge^{-N_{ab}\omega/kT} \right) \tag{3-176}$$

and by the assumptions enumerated above for constant pressure mixing we find

$$\Delta A_{mxg} = \Delta G_{mxg} = -kT \ln \left(\sum_{N_{ab}} ge^{-N_{ab}\omega/kT} \right) \tag{3-177}$$

Exercise 3-14

Go through the steps of writing the Helmholtz free energy for pure a, for pure b, and for the mixture. Take the difference and derive Eq. (3-177).

There is no exact formula in three-dimensional space for the function g itself. The problem we are confronted with here is in fact equivalent to the Ising model discussed in Sec. 3-8. However we do know that the *sum*

$$\sum_{N_{ab}} g(N_a, N_b, N_{ab}) = \text{total number of ways to arrange } N_a \text{ and } N_b \text{ on } N \text{ sites}$$

$$= \frac{(N_a + N_b)!}{N_a!N_b!} \tag{3-178}$$

Thus if $\omega = 0$ in Eq. (3-177), we have

$$\Delta G_{mxg} = -kT \ln \sum_{N_{ab}} g = -kT \ln \left(\frac{N!}{N_a!N_b!} \right)$$

$$= -kTN(\ln N - X_a \ln N_a - X_b \ln N_b)$$

$$= nRT(X_a \ln X_a + X_b \ln X_b)$$

$$= \Delta G_{mxg}^{id} \tag{3-179}$$

which is the change in Gibbs function on mixing for an ideal solution. In obtaining Eq. (3-179) the Stirling approximation was used and X_i = mole fraction of i. The vanishing of ω is a sufficient condition to obtain an ideal solution (*not* an ideal gas, since $\varepsilon_{ij} \neq 0$) for our crude lattice model. Real solutions may very well not be ideal even if $\omega = 0$.

We shall evaluate the sum in Eq. (3-177) in the Bragg-Williams or random mixing approximation. This assumes that g achieves its maximum value when the particles are distributed randomly on the lattice such that

$$\text{Probability of finding molecule } a \text{ on any site} = \frac{N_a}{N_a + N_b}$$

$$\text{Probability of finding molecule } b \text{ on any site} = \frac{N_b}{N_a + N_b}$$

$$\text{Probability of finding pair } a\text{-}b = \frac{2N_a N_b}{(N_a + N_b)^2}$$

Then the most probable or average value of N_{ab} (written $\bar{N}_{ab}$) will be given by this last probability times the total number of pairs:

$$\bar{N}_{ab} = \frac{z}{2}(N_a + N_b) \frac{2N_a N_b}{(N_a + N_b)^2} = \frac{zN_a N_b}{(N_a + N_b)} \tag{3-180}$$

Now recall the trick (see Chap. 1) of replacing a sum by its single largest term when particle numbers are huge and one is taking a logarithm, as in Eq. (3-177). If we were to replace the full sum in Eq. (3-178) with its single largest term, that term should be $(N_a + N_b)!/N_a!N_b!$ exactly! Hence in replacing the sum in Eq. (3-177) with its single largest term, we assume the g has this value and we use $\bar{N}_{ab}$ in the exponential factor of the largest term:

$$\Delta G_{mxg} = -kT \ln \left[g(N_a, N_b, \bar{N}_{ab}) e^{-\bar{N}_{ab}\omega/kT} \right]$$

$$= nRT(X_a \ln X_a + X_b \ln X_b) + \frac{z\omega N_a N_b}{(N_a + N_b)} \tag{3-181}$$

Hence the excess molar Gibbs function of mixing for our model in the random mixing approximation is:

$$\Delta G^E_{m,mxg} = \frac{\Delta G_{mxg} - \Delta G^{id}_{mxg}}{n} = z\omega' X_a X_b \tag{3-182}$$

with

$$\omega' = N_A \omega \tag{3-183}$$

being on a molar basis.

Also,

$$\Delta H^E_{m,mxg} = -T^2 \frac{\partial}{\partial T}\left(\frac{\Delta G^E_{m,mxg}}{T}\right)_P = z\omega' X_a X_b \qquad (3\text{-}184)$$

since ω' depends only on the intermolecular spacing s, which we assumed was temperature independent. Thus we also have

$$\Delta S^E_{m,mxg} = 0 \qquad (3\text{-}185)$$

The entire excess free energy in this model has a purely energetic interpretation. This is in strong disagreement with many real solutions for which $T\Delta S^E_{mxg}$ and ΔH^E_{mxg} are of the same sign and of about equal magnitude such that ΔG^E_{mxg} is not very large and may in fact have an opposite sign to that of ΔH^E_{mxg}. Nevertheless, this model exhibits liquid-liquid phase immiscibility and an upper critical solution temperature T_c if ω' is positive (note Problems 41 and 42), which is the case when two ab pairs are at a higher lattice energy than the sum of aa and bb pair energies. These critical point results are

$$X_{ac} = X_{bc} = \frac{1}{2} \qquad (3\text{-}186)$$

$$T_c = \frac{z\omega'}{2R} \qquad (3\text{-}187)$$

It is interesting to note that T_c drops when z decreases. The Bragg-Williams approximation is really a mean field theory in the sense discussed at the beginning and end of Sec. 3-8. Solution of the exact one-dimensional model analogous to this model in three dimensions shows that there is no phase transition ($T_c = 0$) in one dimension.

The general subject of solution theories for nonelectrolytes is a very large area of current research [30, 31]. The computationally more difficult theories are extensions of those involving distribution functions and perturbation methods introduced for pure liquids in Sec. 3-6 and 3-7. A good descriptive survey of some of this work is provided by Berry et al. [32].

3-10
Statistical Mechanical Theory for Solutions of Electrolytes: Debye-Hückel Model

Solutions of electrolytes are greatly different from solutions of nonelectrolytes because of the long-range Coulombic force ($f \propto 1/r^2$). As long ago as 1923 P. Debye and E. Hückel introduced a primitive model to describe the properties of electrolyte solutions. Their theory is straightforward and is based on a very clever combination of microscopic and macroscopic concepts. We will use it first to obtain an equation for individual ion activities and then combine these to obtain the mean ionic activity coefficient $\gamma_\pm$ and other thermodynamic properties.

Recall from any physical chemistry text that the chemical potential of a single ionic species in solution is given by

$$\mu_j = \mu_j^0(T) + RT \ln\left(\frac{m_j}{m_j^0}\right) + RT \ln\left(\frac{\gamma_j}{\gamma_j^0}\right)$$

$$= \mu_j^{id} + \mu_j^E = \mu_j^{id} + \mu_j^{el} \tag{3-188}$$

where m_j is the molality of ion j in the solution, γ_j is its corresponding activity coefficient, and $\mu_j^0(T)$ is the chemical potential of ion j in the hypothetical standard state of unit activity with $m_j^0 = 1$, $\gamma_j^0 = 1$. μ_j^{id} is the chemical potential of ion j in an ideal solution for which $\gamma_j = 1$. μ_j^E is the excess chemical potential of ion j in the solution and it is responsible for the deviation of ion j in the actual solution from ideal behavior. In the Debye-Hückel theory it is assumed that nonideal behavior arises only from ionic interactions and so $\mu_j^E = \mu_j^{el}$.

The Debye-Hückel model was based on the following assumptions:

1. The interaction between ions is completely determined by Coulombic forces, all other interparticle forces are neglected as minor compared with the Coulombic. (We shall however also include a repulsive hard sphere core of diameter σ per ion.)

2. The dielectric constant of the solution is equal to that of the solvent treated as a continuous and structureless dielectric medium. This incorrectly neglects the effect of the ions on the dielectric constant of the medium near them.

3. The ions are regarded as spherical and unpolarizable charges producing spherically symmetric electric fields.

4. The solution is sufficiently dilute that at the average interionic separation the Coulombic interaction potential energy (see below) is small compared to kT.

5. Strong electrolytes are completely dissociated at all concentrations.

There are chiefly two competing factors operative in ionic solutions: the Coulombic force, which tends to arrange the ions into an organized structure, and the thermal motion of ions and solvent molecules, which tends to disrupt any organized structure. The result is a compromise in which negative ions predominate as the nearest neighbors of any chosen positive ion and vice versa. Any central ion will be surrounded by a group of ions called its ionic atmosphere, which will have a net charge of opposite sign to that of the central ion. We will make this idea explicit in Eq. (3-203), which follows.

In our development we use

$$\mathcal{N}_j = \text{ionic concentration of type } j \text{ in units of ions per unit volume}$$

$$\mathcal{N}_{ji}(r) = \text{average concentration of type } i \text{ ions at distance } r \text{ from ion } j$$

The potential energy of a test charge q_i at distance r from a charge q_j in a medium of dielectric constant D, assuming only the two charges present, is $(q_i q_j / 4\pi\varepsilon_0 Dr)$ because the negative gradient of this potential energy (in a spherically symmetric field) gives us Coulomb's force law:

$$f = -\frac{\partial}{\partial r}\left(\frac{q_j q_i}{4\pi\varepsilon_0 Dr}\right) = \frac{q_j q_i}{4\pi\varepsilon_0 Dr^2}$$

Thus the electrical potential Φ'_j of the single ion j at the radial distance r from j (which by definition is the ratio of the potential energy of the test charge i divided by the charge of i) is

$$\Phi'_j = \frac{q_j}{4\pi\varepsilon_0 Dr} \tag{3-189}$$

In a real electrolyte we use $\Phi_j(r)$ to represent the electrical potential due to ion j *and* its atmosphere (which arises from *all* the other ions in the medium) and $\rho_j(r)$ to represent the charge density at a distance r from the jth ion. These must be related self-consistently by Poisson's differential equation from electrostatics:

$$\nabla^2 \Phi_j(r) = \frac{-\rho_j(r)}{\varepsilon_0 D} \tag{3-190}$$

where $\rho_j(r)$ is given by a sum over all ionic types i (including type j), both positive and negative, that are in the solution at distance r from j:

$$\rho_j(r) = \sum_i \mathcal{N}_{ji}(r) q_i \tag{3-191}$$

We obtain $\mathcal{N}_{ji}(r)$ in terms of the radial distribution function for ions i and j (see Sec. 3-6)

$$\mathcal{N}_{ji}(r) = \mathcal{N}_i g_{ji}^{(2)}(r) \tag{3-192}$$

and make the approximation of Example 3-7 (only accurate for low ionic density)

$$g_{ji}^{(2)}(r) = \exp\left[-\omega_{ji}(r)/kT\right] \tag{3-193}$$

where ω_{ji} is the potential energy of ion i in the field of ion j

$$\omega_{ji}(r) = \Phi_j(r) q_i \tag{3-194}$$

This ω_{ji} also has the meaning of the work done in charging the ith ion up to charge q_i in the potential Φ_j.

In spherical polar coordinates the ∇^2 operator of Eq. (3-190) acting on a function of r only is

$$\nabla^2\Phi_j(r) = \frac{1}{r^2}\frac{d}{dr}\left(r^2\frac{d}{dr}\right)\Phi_j(r)$$

such that using Eqs. (3-191–3-194) we have

$$\frac{1}{r^2}\frac{d}{dr}\left(r^2\frac{d}{dr}\right)\Phi_j(r) = \frac{-1}{\varepsilon_0 D}\sum_i \mathcal{N}_i q_i \exp\left[-\Phi_j(r)q_i/kT\right] \tag{3-195}$$

This is a complicated differential equation that can only be solved numerically, giving little insight into the structure and properties of electrolyte solutions. However Debye and Hückel made it into a linear differential equation by expanding

$$\exp\left[-\Phi_j(r)q_i/kT\right] = 1 - \Phi_j(r)q_i/kT + \ldots \tag{3-196}$$

This is justified if $\Phi_j(r)q_i < kT$, which is correct only if even the nearest neighbors of the j th ion are at large r values—which again means only for dilute solutions. Use Eq. (3-196) in Eq. (3-195) and note that the first term sums to zero, since by electrical neutrality

$$\sum_i \mathcal{N}_i q_i = 0 \tag{3-197}$$

We further define

$$q_i^2 = z_i^2 e^2 \tag{3-198}$$

where z_i is the charge, positive or negative, of the ion i in units of the magnitude of the charge on the electron e, and

$$\kappa^2 = \frac{e^2}{\varepsilon_0 DkT}\sum_i \mathcal{N}_i z_i^2 \tag{3-199}$$

Then Eq. (3-195), which we must solve, is

$$\nabla^2\Phi_j(r) = \frac{1}{r^2}\frac{d}{dr}\left[\frac{r^2\,d\Phi_j(r)}{dr}\right] = \kappa^2\Phi_j(r) \tag{3-200}$$

Substitute $\Phi_j(r) = \Upsilon(r)/r$ into Eq. (3-200) to turn it into

$$\frac{d^2\Upsilon(r)}{dr^2} = \kappa^2\Upsilon(r)$$

for which the easily checked general solution is

$$Y(r) = C_1 e^{-\kappa r} + C_2 e^{+\kappa r} \quad \text{or} \quad \Phi_j(r) = \frac{C_1 e^{-\kappa r}}{r} + \frac{C_2 e^{+\kappa r}}{r}$$

with C_1 and C_2 constants to be fixed by two physical boundary conditions of our problem. Since $\Phi_j(r)$ must go to zero as $r \to \infty$, the constant C_2 must be zero. The physically correct solution is thus

$$\Phi_j(r) = C_1 \frac{e^{-\kappa r}}{r} \tag{3-201}$$

We next determine C_1. By Eq. (3-190) and Eq. (3-200) the charge density

$$\rho_j(r) = -\varepsilon_0 D \nabla^2 \Phi_j(r) = \frac{-\varepsilon_0 D \kappa^2 C_1 e^{-\kappa r}}{r} \tag{3-202}$$

If we take the hard sphere diameter of ion j to be σ (no two ions can get closer than σ if we assume all the ions have the same diameter) and then if we count up all the charges in the solution on ions existing at $r \geq \sigma$, we must by electrical neutrality obtain a charge $(-z_j e)$ that is the opposite of the charge on ion j:

$$\int_\sigma^\infty \rho_j(r) 4\pi r^2 \, dr = \int_\sigma^\infty \left(\frac{-\varepsilon_0 D \kappa^2 C_1 e^{-\kappa r}}{r} \right) 4\pi r^2 \, dr = -z_j e \tag{3-203}$$

Doing the integral and solving for C_1 we find

$$C_1 = \frac{z_j e \, e^{+\kappa \sigma}}{4\pi \varepsilon_0 D(1 + \kappa \sigma)} \tag{3-204}$$

Substituting into Eq. (3-201) we have for the total electrical potential at distance $r \geq \sigma$ from a finite-size ion j

$$\Phi_j(r) = \frac{z_j e \, e^{-\kappa(r-\sigma)}}{4\pi \varepsilon_0 D(1 + \kappa \sigma) r} \tag{3-205}$$

It is a general principle of electrostatics that any potential is a sum of potentials arising from all sources in a medium. We have already stated that $\Phi_j(r)$ is a sum of the potential in Eq. (3-189) due to itself and of the potential of its complete ionic atmosphere, Φ_j^{atm}. Thus by difference

$$\Phi_j^{atm}(r) = \frac{z_j e \, e^{-\kappa(r-\sigma)}}{4\pi \varepsilon_0 D(1 + \kappa \sigma) r} - \frac{z_j e}{4\pi \varepsilon_0 D r} \tag{3-206}$$

Furthermore, since no ions can penetrate the region $r < \sigma$ we may take

$$\Phi_j^{atm}(r = \sigma) \equiv \Phi_j^* = \frac{z_j e}{4\pi\varepsilon_0 D\sigma}\left(\frac{1}{1 + \kappa\sigma} - 1\right) = \frac{-z_j e\kappa}{4\pi\varepsilon_0 D(1 + \kappa\sigma)} \quad (3\text{-}207)$$

as the additional amount of electrical potential characteristic of each ion due to the effects of all the other ions.

We imagine charging up a single jth ion from charge zero to $z_j e$ in the potential Φ_j^* and equate the reversible work needed to do this to the change in chemical potential *per ion* due to the interaction with all the other ions [33]. By Eq. (3-188) this change in chemical potential is $kT \ln (\gamma_j)$. We introduce a charging parameter λ that may vary from 0 to 1 such that $q_j = z_j e\lambda$ and $dq_j = z_j e\, d\lambda$. The potential Φ_j^* will also depend on λ:

$$\Phi_j^*(\lambda) = \frac{-z_j e\lambda\kappa}{4\pi\varepsilon_0 D(1 + \kappa\sigma)}$$

$$\text{Work} = kT \ln (\gamma_j) = \int_0^{z_j e} \Phi_j^*(\lambda)\, dq_j$$

$$= \int_0^1 \frac{-z_j e\lambda\kappa(z_j e\, d\lambda)}{4\pi\varepsilon_0 D(1 + \kappa\sigma)}$$

$$= \frac{-z_j^2 e^2 \kappa}{8\pi\varepsilon_0 D(1 + \kappa\sigma)} \quad (3\text{-}208)$$

This is a change in Helmholtz free energy that must be equal to the *negative* of the total work done *by* the system in a reversible process. From Eq. (3-208), which is the negative of a positive quantity, the work done *by* the system must be positive. This is correct because the potential energy of the system must *decrease* as a new charge and its oppositely charged atmosphere separated roughly by the distance $(1/\kappa)$ form inside the system. The last sentence gives a physical meaning to $(1/\kappa)$ that follows from the charge density of Eq. (3-202), in that the probability of finding charge about the jth ion is $4\pi r^2|\rho_j(r)|$ and this probability has a maximum at $r = 1/\kappa$. We may say $(1/\kappa)$ is the most probable thickness of the ionic atmosphere of any ion. Another physical meaning comes from using Eq. (3-193) with Eq. (3-194) and Eq. (3-205):

$$g_{ji}^{(2)}(r) = \exp\left[-z_i e\Phi_j(r)/kT\right] \simeq 1 - \frac{z_i e\Phi_j(r)}{kT} + \frac{[z_i e\Phi_j(r)]^2}{2(kT)^2} + \cdots$$

$$= 1 - \frac{z_i z_j e^2\, e^{-\kappa r}(e^{+\kappa\sigma})}{4\pi\varepsilon_0 D(1 + \kappa\sigma)rkT} + \cdots \quad (3\text{-}209)$$

For $r > 1/\kappa$ this radial distribution function will rapidly fall off to unity, meaning the correlation between two ions only extends over a distance of about $1/\kappa$. For this reason $1/\kappa$ is called the Debye *screening* length. Because of the formation of oppositely charged atmospheres to any ion the long-range Coulomb potential is screened and ionic correlations are not so very long range in actual fact.

Example 3-10

Express $1/\kappa$ in meters for 1-1 and 2-1 electrolytes in water at 25°C in terms of the molar concentration c of electrolyte. Also express $1/\kappa$ in terms of the ionic strength I of a solution

$$I = \frac{1}{2} \sum_i m_i z_i^2 \qquad (3\text{-}210)$$

$D_{H_2O}(25°C) = 78.5$.

Solution:

Let N_i be the number of ions per liter of solution

$$c_i = \frac{n_i}{V} = \text{molar concentration of } i$$

From Eq. (3-199)

$$\kappa^2 = \frac{e^2 N_A}{\varepsilon_0 DkT} \sum_i c_i z_i^2 \qquad \varepsilon_0 = 8.8542 \times 10^{-12} \text{ C/V·m}$$

$$\frac{1}{\kappa} = \left[\frac{\varepsilon_0 DkT(10^{-3})}{e^2 N_A \left(\sum_i c_i z_i^2 \right)} \right]^{1/2} \qquad \text{in meters since 1 liter} = 10^{-3} \text{ m}^3$$

$$= \frac{2.812 \times 10^{-12} (DT)^{1/2}}{\left(\sum_i c_i z_i^2 \right)^{1/2}}$$

For water at 25°C $(DT)^{1/2} = 153$

<table>
<tr><td>1-1 electrolyte</td><td>2-1 electrolyte</td></tr>
<tr><td>$\dfrac{1}{\kappa} = \dfrac{3.04 \times 10^{-10} \text{ m}}{\sqrt{c}}$</td><td>$\dfrac{1}{\kappa} = \dfrac{1.76 \times 10^{-10} \text{ m}}{\sqrt{c}}$</td></tr>
</table>

The general relation between c_i and m_i is

$$c_i = \frac{1000 \rho' m_i}{1000 + m_i M_i} \left(1 - \sum_{j \neq i} \frac{c_j M_j}{1000 \rho'} \right) \qquad (3\text{-}211)$$

where ρ' is the solution density in g/cm^3

$$M_i = \text{molecular weight (g/mol) of species } i$$

For dilute solutions for which the Debye-Hückel theory is valid

$$c_i = \rho' m_i \simeq \rho'_1 m_i$$

where ρ'_1 is the pure solvent density in g/cm^3. Hence, introducing the ionic strength,

$$\frac{1}{\kappa} = \left[\frac{\varepsilon_0 D k T (10^{-3})}{2e^2 N_A \rho'_1 I}\right]^{1/2} \tag{3-212}$$

Returning now to the activity coefficient from Eq. (3-208) we have

$$\ln (\gamma_j) = \frac{-z_j^2 e^2 \kappa}{8\pi\varepsilon_0 D k T (1 + \kappa\sigma)} \tag{3-213}$$

Since single ion activity coefficients cannot be measured we will combine them into the mean molal activity coefficient $\gamma_\pm$, which is experimentally measurable. If one molecule of electrolyte produces v ions of which v_j are of the jth kind, and there are p different types of ions from one molecule,

$$\ln (\gamma_\pm) = \frac{1}{v} \sum_{j=1}^{p} v_j \ln (\gamma_j) \tag{3-214}$$

In the particular case that only two types, one positive and one negative, are produced, and using z_+ and z_- *both* as positive integers,

$$\ln (\gamma_\pm) = \frac{1}{v} [v_+ \ln (\gamma_+) + v_- \ln (\gamma_-)]$$

$$= \frac{1}{(v_+ + v_-)} [-v_+ z_+^2 - v_- z_-^2] \frac{e^2 \kappa}{8\pi\varepsilon_0 D k T (1 + \kappa\sigma)} \tag{3-215}$$

This may be rewritten using Eq. (3-212) and conventional logarithms to the base 10 as

$$\log_{10} (\gamma_\pm) = \frac{-z_+ z_- A I^{1/2}}{1 + B\sigma I^{1/2}} \tag{3-216}$$

where

$$A = \frac{1}{2.3026} \left(\frac{2000 e^6 N_A \rho'_1}{64\pi^2 \varepsilon_0^3 (D k T)^3}\right)^{1/2} = \frac{1.8247 \times 10^6 (\rho'_1)^{1/2}}{(DT)^{3/2}} \tag{3-217}$$

and

$$B = \left(\frac{2000e^2 N_A \rho_1'}{\varepsilon_0 DkT}\right)^{1/2} = \frac{5.029 \times 10^{11} (\rho_1')^{1/2}}{(DT)^{1/2}} \qquad (3\text{-}218)$$

in which ρ_1' is the density of the pure solvent in g/cm^3 and all other quantities are in SI units. For water at 25°C $\rho_1' = 1.00$ g/cm^3, $D = 78.5$, $A = 0.509$ $(molal)^{-1/2}$, $B = 3.29 \times 10^9$ $m^{-1} \cdot (molal)^{-1/2}$.

■

Exercise 3-15

Obtain Eq. (3-216) from Eq. (3-215) using Eq. (3-212). Note for electrical neutrality that $v_+ z_+ = v_- z_-$. Also check the numerical coefficients given in Eq. (3-217) and Eq. (3-218).

■

In the original Debye-Hückel theory point ions were assumed and σ taken as zero such that the DH limiting law reads as

$$\log_{10}(\gamma_\pm) = -z_+ z_- A I^{1/2} \qquad (3\text{-}129)$$

The striking result is that this combination of fundamental constants and solvent properties does give the correct limiting slope of a plot of $\log_{10}(\gamma_\pm)$ against $I^{1/2}$, as shown in Fig. 3-21. It is only accurate up to $I \simeq 0.01$. The experimental curves of this function generally pass through a minimum and increase such that $\gamma_\pm$ may even become greater than unity in concentrated solutions. This is due to several factors, such as the ionic hard sphere exclusions, the decrease of free solvent molecules as the ionic concentration increases, and the increased attachment of solvent molecules to the ions in solvation spheres. Some of this is mimicked by Eq. (3-216), which as an expansion in powers of $I^{1/2}$ may be written as

$$\log_{10}(\gamma_\pm) = -z_+ z_- A I^{1/2} + C'I \qquad (3\text{-}220)$$

with

$$C' = z_+ z_- AB\sigma \qquad (3\text{-}221)$$

This form is accurate up to $I \simeq 0.10$. It is better however to fit C' to experimental results for a given electrolyte, particularly to the experimental minimum in the

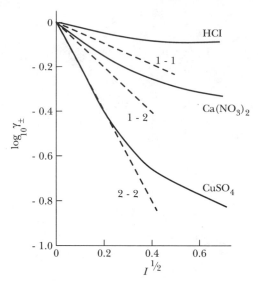

FIGURE 3-21 The $\log_{10} \gamma_{\pm}$ versus the square root of the ionic strength. The Debye-Hückel limiting law is shown as dashed lines for the three different types of electrolyte. The experimental curves generally pass through a minimum and then increase (not shown here).

SOURCE: G. W. Castellan. *Physical Chemistry*, 3d ed. Addison Wesley, 1983.

curve at I'_{ex}. Then

$$C' = \frac{A z_+ z_-}{2(I'_{ex})^{1/2}} \tag{3-222}$$

and Eq. (3-220) will be accurate up to $I \simeq 1.0$.

Much work has been and is being done to extend the theory of electrolytes to more concentrated solutions. The Bjerrum-Fuoss theory of ion association [33] is presented in Problems 51–53. More complex theories are described in references [34–36].

Before closing this section more of the interesting results of the Debye-Hückel theory itself will be examined. Pitzer [37] has pointed out that the structure of a charged hard sphere model is given very accurately by the radial distribution function of Eq. (3-209) in exponential form (and not as a two- or three-term expansion thereof). This function is accurate as gauged by comparison of Monte Carlo model calculations of the $g_{ji}^{(2)}$ appropriate to a 1-1 electrolyte in water at 25°C with a common $\sigma = 4.25 \times 10^{-10}$ m carried out by Card and Valleau.* Note that the Monte Carlo calculation is for a model that only has hard cores and a Coulomb interaction. Real solvent-solute and specific solute-solute interactions are

---

* D. N. Card and J. P. Valleau. *J. Chem. Phys.* **52**, 6232 (1970).

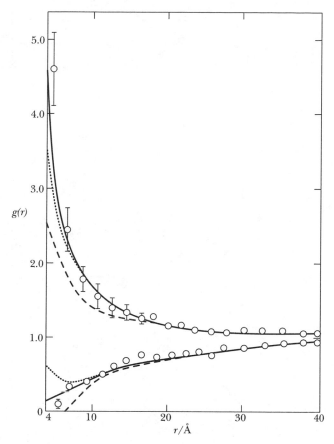

FIGURE 3-22 The radial distribution functions ($g^{(2)}_{+-}$ above, $g^{(2)}_{++} = g^{(2)}_{--}$ below)
appropriate to a 0.00911 M aqueous solution of 1-1 electrolyte with common
$\sigma = 4.25 \times 10^{-10}$ m at 25°C. The points are Monte Carlo calculations, the solid curve
is the exponential Debye-Hückel expression, the dotted and dashed curves are respectively
the three-term and two-term expansions of the exponential.
SOURCE: Reference [37].

absent from the model as from Debye-Huckel theory. Results are shown in Figs.
3-22 and 3-23. Of course all the $g^{(2)}$ functions are zero for $r < \sigma$.

It is useful to consider a hypothetical charging process not just for one ion but
for all the ions simultaneously in an electrolyte solution of one solute (denoted by
subscript 2) and so arrive at an expression for the Helmholtz free energy due to the
electrical charges in the Debye-Hückel approximation. We replace all the charges
by $\lambda z_j e$ where λ can vary from zero to unity. The value of κ during the charging process will be $\lambda \kappa$ since by Eq. (3-199) κ is proportional to the square root of the sum
of the squares of the charges, each of which goes as λ^2. Then for type j ions whose
extra ionic atmosphere potential reaches $-z_j e\kappa/[4\pi\varepsilon_0 D(1 + \kappa\sigma)]$ at the end of the
process we must have an atmospheric potential $-z_j e\kappa\lambda^2/[4\pi\varepsilon_0 D(1 + \kappa\lambda\sigma)]$ during

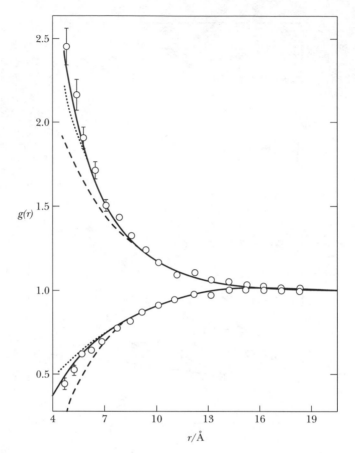

FIGURE 3-23 The radial distribution functions for a 0.425 M aqueous solution. Other details as in Fig. 3-22.
SOURCE: Reference [37].

the process. If one ion j with infinitesimal charge $z_j e \, d\lambda$ is brought up to this potential, the infinitesimal work done on the system will be the product of these terms. Integrating and multiplying by the total number of j type ions, $\mathcal{N}_j V$, (with V the total volume and $\mathcal{N}_j$ the number per unit volume) we find

$$A_j^{el} = -\frac{\mathcal{N}_j V z_j^2 e^2 \kappa}{4\pi\varepsilon_0 D} \int_0^1 \frac{\lambda^2 \, d\lambda}{(1 + \kappa\lambda\sigma)}$$

$$= \frac{-\mathcal{N}_j V z_j^2 e^2 \kappa}{4\pi\varepsilon_0 D} \, s(\kappa\sigma) \tag{3-223}$$

with

$$s(\kappa\sigma) = \frac{\ln\,(1 + \kappa\sigma)}{(\kappa\sigma)^3} + \frac{1}{2\kappa\sigma} - \frac{1}{(\kappa\sigma)^2} \tag{3-224}$$

For a single electrolyte giving only two types of ions the total A^{el} is

$$A^{el} = \sum_{\text{types}} A_j^{el} = A_+^{el} + A_-^{el}$$

$$= \frac{Ve^2\kappa}{4\pi\varepsilon_0 D} s(\kappa\sigma)(N_+ z_+^2 + N_- z_-^2) \tag{3-225}$$

Use N_2 = molecules of electrolyte (originally undissociated) per unit volume, such that

$$N_+ = v_+ N_2$$

$$N_- = v_- N_2$$

$$A^{el} = -\frac{VN_2 e^2\kappa}{4\pi\varepsilon_0 D} s(\kappa\sigma)(v_+ z_+^2 + v_- z_-^2) \tag{3-226}$$

Since κ^2 for this case may be written from Eq. (3-199) as

$$\kappa^2 = \frac{e^2 N_2}{\varepsilon_0 DkT}(v_+ z_+^2 + v_- z_-^2) \tag{3-227}$$

$$A^{el} = -\frac{V\kappa^3 kT}{4\pi} s(\kappa\sigma) \tag{3-228}$$

For the limiting law result, $\kappa\sigma \to 0$,

$$A^{el} = -\frac{V\kappa^3 kT}{12\pi} \tag{3-229}$$

Exercise 3-16

By expansion of $\ln(1 + \kappa\sigma)$ for $\kappa\sigma \to 0$ show that $s(\kappa\sigma) \to 1/3$.

If n_2 = total moles of electrolytic solute, 2,

$$n_2 = \frac{N_2 V}{N_A}$$

and the expression for κ is

$$\kappa = \left(\frac{e^2 n_2 N_A}{V\varepsilon_0 DkT}(v_+ z_+^2 + v_- z_-^2)\right)^{1/2} \tag{3-230}$$

such that

$$\left(\frac{\partial \kappa}{\partial n_2}\right)_{V,T} = \frac{\kappa}{2n_2} \qquad (3\text{-}231)$$

Thus on a molar basis

$$\mu_2^{el} = \left(\frac{\partial A^{el}}{\partial n_2}\right)_{V,T} = \frac{-V}{12\pi} kT(3\kappa^2)\frac{\kappa}{2n_2} = -\frac{V\kappa^3 kT}{8\pi n_2} = \frac{3}{2}\frac{A^{el}}{n_2} \qquad (3\text{-}232)$$

Exercise 3-17

Since $\mu_2^{el} = \nu RT \ln{(\gamma_\pm)}$, obtain Eq. (3-219) from Eq. (3-232) and Eq. (3-229).

$$G^{el} = n_2\mu_2^{el} = \frac{3}{2} A^{el} = \frac{-V\kappa^3 kT}{8\pi} \qquad (3\text{-}233)$$

The reversible charging process that gave us Eq. (3-225) was at constant volume and temperature and represents total work done on the system, not net work done, because with the appearance of charge there is a change in osmotic pressure due to the charging. This is why the result was A^{el} and not G^{el}. The latter is given by Eq. (3-233).

Appendix 3.A
Scaling Theory and the Critical Region

There is no unique way to present scaling theory. We shall follow the method used by Widom [26], who introduced a form for the equation of state of a fluid in the critical region that could accommodate nonclassical critical point exponents. To understand this method we need to state a number of thermodynamic relations in terms of μ_m (the chemical potential per mole) and $\rho_m = 1/V_m$ (the molar density).

From the definition of the Gibbs free energy

$$d\mu_m = -S_m dT + \frac{1}{\rho_m} dP \qquad (3.A\text{-}1)$$

$$\left(\frac{\partial u_m}{\partial T}\right)_P = -S_m \qquad (3.A\text{-}2)$$

and so this derivative is always negative while

$$\left(\frac{\partial \mu_m}{\partial P}\right)_T = V_m = \frac{1}{\rho_m} \tag{3.A-3}$$

is always positive. In addition

$$\left(\frac{\partial \mu_m}{\partial T}\right)_{\rho_m} = -S_m + \frac{1}{\rho_m}\left(\frac{\partial P}{\partial T}\right)_{\rho_m} \tag{3.A-4}$$

and this derivative is always negative, being dominated by the first term (but less negative for the liquid phase than for the gaseous phase). In Fig. 3-24 we show a plot in the $\mu_m - T$ plane analogous to Fig. 3-20 in the $P - T$ plane. The common μ_m of liquid and gas on the saturation (σ) curve for $T \le T_c$ is continuous with the μ_m on the critical isochore ($\rho_m = \rho_{mc}$) of the one-phase system for $T \ge T_c$. Isochores with $\rho_m \ne \rho_c$ meet in pairs on the saturation curve at T_σ values $< T_c$, corresponding to the saturation curve temperatures and their coexisting ρ_m values.

Other important relations are

$$\left(\frac{\partial^2 \mu_m}{\partial T^2}\right)_{\rho_m} = \frac{-C_{Vm}}{T} + \frac{1}{\rho_m}\left(\frac{\partial^2 P}{\partial T^2}\right)_V \tag{3.A-5}$$

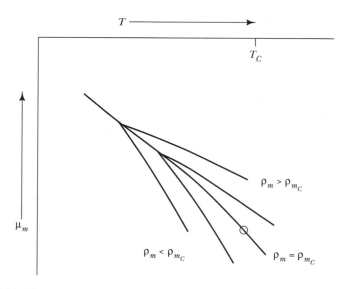

FIGURE 3-24 The common μ_m of liquid and gas on the saturation curve (σ) for $T \le T_c$ is continuous with the μ_m on the critical isochore of the one-phase system for $T \ge T_c$. Isochores with $\rho_m \ne \rho_c$ meet in pairs on the saturation curve at T_σ values $< T_c$.

from Eq. (3.A-4), with the realization that constant ρ_m means constant V, and

$$\left(\frac{\partial \mu_m}{\partial \rho_m}\right)_T = \frac{1}{\rho_m}\left(\frac{\partial \mu_m}{\partial \rho_m}\frac{\partial P}{\partial \mu_m}\right)_T = \frac{1}{\rho_m}\left(\frac{\partial P}{\partial \rho_m}\right)_T = (\rho_m^2 \beta_T)^{-1} \tag{3.A-6}$$

from Eq. (3.A-3) and Eq. (3-135). Since for physically realizable states $\beta_T > 0$, we have $(\partial \mu_m / \partial \rho_m)_T$ always positive.

We shall also need the Helmholtz free energy density, a:

$$a = \frac{A_m}{V_m} = \frac{1}{V_m}(\mu_m - PV_m) = \mu_m \rho_m - P \tag{3.A-7}$$

such that

$$\left(\frac{\partial a}{\partial T}\right)_{\rho_m} = \rho_m \left(\frac{\partial \mu_m}{\partial T}\right)_{\rho_m} - \left(\frac{\partial P}{\partial T}\right)_{\rho_m} = -\rho_m S_m \tag{3.A-8}$$

by use of Eq. (3.A-4). Finally we note that

$$\left(\frac{\partial^2 a}{\partial T^2}\right)_{\rho_m} = -\rho_m \left(\frac{\partial S_m}{\partial T}\right)_{\rho_m} = -\rho_m \left(\frac{C_{V_m}}{T}\right) \tag{3.A-9}$$

One can derive several relations between critical point exponents from the idea that the temperature dependence of all quantities near the critical point arises from their dependence on the correlation length, ζ, which in turn has a temperature dependence given by Eq. (3-139). For example, the fluctuations in density that occur in a fluid are strongly correlated (the high-density regions must balance the low-density regions that appear) but only if the density regions are separated by distances less than ζ. Roughly, we may say that inside volumes of size ζ^d in d-dimensional space the fluctuations are correlated, while fluctuations in different ζ^d-size volumes of the fluid are independent. From Eq. (3.A-9), since C_V varies as $|t|^{-\alpha}$, the free energy density a must vary as $|t|^{2-\alpha}$ because two differentiations with respect to temperature must give $|t|^{-\alpha}$. We may identify the free energy density that gives rise to the singularity in C_V with a common free energy cost of order kT for any fluctuation inside a volume ζ^d; that is,

$$a \propto \frac{kT}{\zeta^d} \qquad \text{which varies as} \qquad \frac{1}{|t|^{-vd}}$$

but because a varies as $|t|^{2-\alpha}$, by the argument above we have

$$vd = 2 - \alpha \tag{3.A-10}$$

which is the relation between exponents given by Eq. (3-146).

As another example, consider the result of dividing Eq. (3-138) by V^2:

$$\overline{(\rho - \bar{\rho})^2} = \frac{\rho^2 k T \beta_T}{V} \qquad (3.A-11)$$

Now if, as is physically reasonable,* the mean square density fluctuation is just of the magnitude of the coexisting liquid-gas density difference

$$\overline{(\rho - \bar{\rho})^2} \sim (\rho_l - \rho_g)^2 \sim (-t)^{2\beta}$$

and if this fluctuation occurs in a volume V of size ζ^d

$$\frac{\rho^2 k T \beta_T}{V} \sim \frac{(-t)^{-\gamma}}{(-t)^{-\nu d}}$$

we obtain

$$2\beta = \nu d - \gamma \qquad (3.A-12)$$

which when combined with Eq. (3.A-10) gives us

$$\alpha + 2\beta + \gamma = 2 \qquad (3.A-13)$$

which is Eq. (3-143) holding as an equality.

We turn now to Widom's formulation of the equation of state. We define

$$\Delta \mu_m = \mu_m(\rho_m, T) - \mu_m(\rho_{m_c}, T) \qquad (3.A-14)$$

$$\Delta \rho_m = \rho_m - \rho_{m_c} \qquad (3.A-15)$$

and assume

a. $\Delta \mu_m$ is an odd function of $\Delta \rho_m$ (i.e., when $\Delta \rho_m$ changes sign $\Delta \mu_m$ also merely changes sign) on isotherms near the critical point.

b. $\mu_m(\rho_m, T)$ is analytic in both variables in any one phase region.

c. $\mu_m(\rho_{m_c}, T)$ is analytic in T even at $T = T_c$.

Notice from Fig. 3-24 that $\Delta \mu_m$ must vanish not only for $\rho_m = \rho_{m_c}$ but also for pairs of ρ_m values at $T = T_\sigma(\rho_m)$ on the coexistence curve. The classical equation

---

* Widom [26] asserts that this idea explains how a homogeneous system upon reaching the coexistence curve "knows" what the density of the newly appearing phase should be. "It shall be that density that the original homogeneous fluid had been *practicing* making in its spontaneous density fluctuations."

that fits the above is (in dimensionless form)

$$\frac{\Delta\mu_m}{(P_c/\rho_{m_c})} = \left(\frac{\Delta\rho_m}{\rho_{m_c}}\right)\left(\frac{T - T_\sigma(\rho_m)}{T_c}\right)\Phi \tag{3.A-16}$$

with Φ a constant. We may rewrite the second factor as

$$\frac{T - T_\sigma(\rho_m)}{T_c} = t + \left(\frac{T_c - T_\sigma(\rho_m)}{T_c}\right) \tag{3.A-17}$$

in which we use $t = (T - T_c)/T_c$ and rewrite Eq. (3-128) using

$$\rho_{m_l} - \rho_{m_g} = (\rho_{m_l} - \rho_{m_c}) - (\rho_{m_g} - \rho_{m_c}) = 2|\Delta\rho_m|$$

and

$$\frac{T_c - T_\sigma(\rho_m)}{T_c} = X_0 \left|\frac{\Delta\rho_m}{\rho_{m_c}}\right|^{1/\beta} \tag{3.A-18}$$

where $X_0 = B_0^{-1/\beta}$ with B_0 the constant used in Eq. (3-128). Widom notes that the factor in Eq. (3.A-17) is a homogeneous function of degree unity in t and in $|\Delta\rho_m/\rho_{m_c}|^{1/\beta}$. Recall (see Appendix II) the following about homogeneous functions:
A function of two variables $f(x, y)$ is said to be homogeneous of degree q if

$$f(\lambda x, \lambda y) = \lambda^q f(x, y) \tag{3.A-19}$$

for any value of the parameter λ. If we make the special choice $\lambda = y^{-1}$, we have

$$f(x/y, 1) = y^{-q} f(x, y) \tag{3.A-20}$$

Hence if $f(x, y)$ is homogeneous of degree q it can also be written as y^q times a function of the ratio x/y.
Widom generalized Eq. (3.A-16) as follows:

$$\frac{\Delta\mu_m}{(P_c/\rho_{m_c})} = \left(\frac{\Delta\rho_m}{\rho_{m_c}}\right)\psi\left(t, \left|\frac{\Delta\rho_m}{\rho_{m_c}}\right|^{1/\beta}\right) \tag{3.A-21}$$

in which ψ is a function homogeneous of degree γ in its two variables. Thus by Eq. (3.A-20)

$$\Delta\mu_m = \left(\frac{P_c}{\rho_{m_c}}\right)\left(\frac{\Delta\rho_m}{\rho_{m_c}}\right)\left|\frac{\Delta\rho_m}{\rho_{m_c}}\right|^{\gamma/\beta} h(x) \tag{3.A-22}$$

where

$$x = \frac{l}{|\Delta\rho_m/\rho_{m_c}|^{1/\beta}} \tag{3.A-23}$$

and

$$\psi(x, 1) \equiv h(x) \tag{3.A-24}$$

Then

$$\Delta\mu_m \propto \text{sgn}\,(\Delta\rho_m)|\Delta\rho_m|^{(\gamma/\beta)+1} \tag{3.A-25}$$

where sgn (I) means the sign of I ($+1$ or -1 only).
  Define

$$\Delta P = P(\rho_m, T) - P(\rho_{m_c}, T) \tag{3.A-26}$$

and from Eq. (3.A-6)

$$\frac{1}{\rho_m}\left(\frac{\partial P}{\partial\rho_m}\right)_T = \frac{1}{(\Delta\rho_m + \rho_{m_c})}\left[\frac{\partial(\Delta P)}{\partial(\Delta\rho_m)}\right]_T = \left[\frac{\partial(\Delta\mu_m)}{\partial(\Delta\rho_m)}\right]_T \tag{3.A-27}$$

which we may integrate along an isotherm

$$\Delta P \propto \int (\Delta\rho_m + \rho_{m_c})d(\Delta\rho_m)[(\Delta\rho_m)^{\gamma/\beta}]$$

$$\propto (\Delta\rho_m)^{(\gamma/\beta)+2} + (\Delta\rho_m)^{(\gamma/\beta)+1}$$

where the second term dominates as $\Delta\rho_m \to 0$ such that the critical point exponent of the critical isotherm, δ, [note Eq. (3-131)] is the same as the critical point exponent of $\Delta\mu_m$, namely

$$\delta = (\gamma/\beta) + 1 \tag{3.A-28}$$

Using Eq. (3.A-28) and Eq. (3.A-13), we find

$$\alpha + \beta(1 + \delta) = 2 \tag{3.A-29}$$

which is Eq. (3-144) holding as an equality.
  We can see that the γ introduced as the degree of homogeneity of ψ is the same as γ in Eq. (3-129) by using Eq. (3.A-6)

$$(\rho_m^2\beta_T)^{-1} = \left[\frac{\partial(\Delta\mu_m)}{\partial(\Delta\rho_m)}\right]_T \propto (\Delta\rho_m)^{\gamma/\beta}$$

since the left-hand side goes as $[|t|^{-\gamma}]^{-1}$ and the right-hand side varies as $[|t|^{\beta}]^{\gamma/\beta}$.

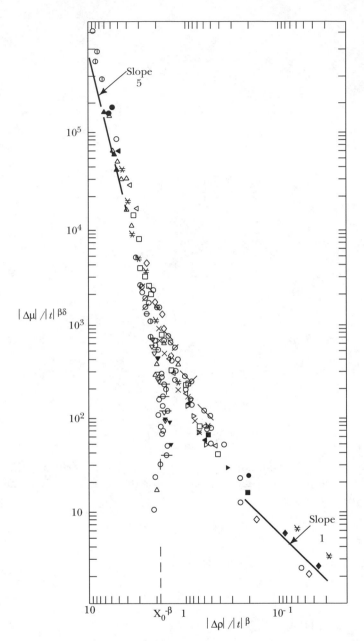

FIGURE 3-25 First test of the scaling hypothesis for the equation of state of fluids in the critical region.

$$z \equiv \frac{|\Delta\mu_m|\, \rho_{m_c}/P_c}{|t|^{\beta\delta}} \qquad \text{versus} \qquad y \equiv \frac{|\Delta\rho_m/\rho_{m_c}|}{|t|^{\beta}} \qquad t = \frac{T - T_c}{T_c}$$

Of the two branches the one on the left is for t negative, the one on the right for t positive. $\beta = 0.35$, $\delta = 5.0$ (current best values for these are $\beta = 0.325$, $\delta = 4.82$).
SOURCE: M. S. Green, M. Vicentini-Missoni, and J. M. H. Levelt-Sengers. *Phys. Rev. Letters* **18**, 1113 (1967).

The accuracy of the scaled equation of state has been definitely confirmed by experiment. [From experiments on ΔP as a function of $\Delta \rho_m$, on isotherms one may obtain $\Delta \mu_m$ by numerical integration of Eq. (3.A-27).] From Eq. (3.A-22) and Eq. (3.A-23) we have

$$\Delta \mu_m = \frac{P_c}{\rho_{m_c}} \left[\text{sgn} \left(\Delta \rho_m \right) \right] \left| \frac{\Delta \rho_m}{\rho_{m_c}} \right|^\delta h(x)$$

$$= \frac{P_c}{\rho_{m_c}} \left[\text{sgn} \left(\Delta \rho_m \right) \right] \frac{|t|^{\beta \delta}}{|x|^{\beta \delta}} h(x) \qquad (3.A-30)$$

Define

$$z = \frac{|\Delta \mu_m| (\rho_{m_c} / P_c)}{|t|^{\beta \delta}} \qquad (3.A-31)$$

and

$$y = |x|^{-\beta} = \frac{|\Delta \rho_m / \rho_{m_c}|}{|t|^\beta} \qquad (3.A-32)$$

such that the equation of state reads

$$z = y^\delta h(x) = y^\delta h[(\text{sgn } x) y^{-1/\beta}] \qquad (3.A-33)$$

This predicts z (the properly *scaled* $|\Delta \mu_m|$ on an isotherm where $|t|$ is known) to be a universal function of y (the properly *scaled* $|\Delta \rho_m|$ on an isotherm). The successful first test of this relation for five different fluids is shown in Fig. 3-25. There are two branches of this function, one for $t, x > 0$ and one for $t, x < 0$, which join asymptotically for y large—in practice for y greater than 4.0, which corresponds to isotherms being close to the critical isotherm on either side of $\rho_m = \rho_{m_c}$. Every isotherm in the $\mu_m - \rho_m$ plane will approach the critical isotherm for large enough $|\Delta \rho_m|$, as shown in Fig. 3-26. In the region of joining, where $|x|$ is very small, one can show that $h(x) \to$ constant such that $\log z$ versus $\log y$ has a slope equal to δ (chosen as 5.0 in Fig. 3-25). On the coexistence curve for $T < T_c$, z vanishes at a y value from Eq. (3.A-18) and Eq. (3.A-31) that can be no smaller than $X_0^{-\beta} \simeq 1.9$, and so the branch with $t < 0$ runs down almost vertically to end at $y \simeq 1.9$. On the other branch for $T > T_c$, z is only zero when $y = 0$, and on a log-log plot the slope can be shown to be unity for small y (in practice for $y < 0.4$). The fact that all the data for five different fluids fall onto the two expected branches of the z versus y curve confirms the scaling hypothesis. Similar confirmation has been found for isotropic ferromagnets when scaled magnetic field is plotted against scaled net magnetization [22]. Only two critical point exponents are independent. All others may be obtained from the two chosen as independent. For further details see reference [22].

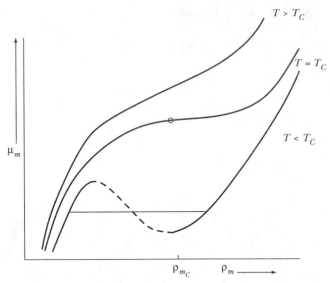

FIGURE 3-26 Plot of isotherms in the μ_m-ρ_m plane. All isotherms for sufficiently large $|\rho_m - \rho_{m_c}|$ will get very close to the critical isotherm. Since for physically realizable states $(\partial \mu_m / \partial \rho_m)_T$ is positive, there is a Maxwell equal area rule in this plane also for drawing the horizontal line of common μ_m of coexisting liquid and gaseous phases for isotherms with $T < T_c$.

Appendix 3.B
Heat Capacity of a Two-Phase Liquid-Gas System

The following development is needed to facilitate working problems on the heat capacity function in the critical region. We treat a two-phase liquid (l)-gas(g) system below T_c. We obtain the heat capacity from the entropy expression

$$S = n_l S_{m_l} + n_g S_{m_g} \tag{3.B-1}$$

with total volume and total number of moles, $n = n_l + n_g$, fixed.

$$dn_l = -dn_g \tag{3.B-2}$$

$$dS = n_l dS_{m_l} + S_{m_l} dn_l + n_g dS_{m_g} + S_{m_g} dn_g$$
$$= n_l dS_{m_l} + n_g dS_{m_g} - (S_{m_g} - S_{m_l}) dn_l \tag{3.B-3}$$

$$\frac{\Delta S}{\Delta V} = \left(\frac{dP}{dT}\right)_\sigma \tag{3.B-4}$$

where subscript σ means on the saturated vapor pressure curve. The C_{V_2}, which is the overall constant volume heat capacity of the two-phase system, is obtained from

$$C_{V_2} dT = T dS \tag{3.B-5}$$

Heating along the saturation curve

$$T \, dS = n_l C_{\sigma_l} \, dT + n_g C_{\sigma_g} \, dT - T \Delta V \left(\frac{dP}{dT} \right)_\sigma \left(\frac{dn_l}{dT} \right)_\sigma dT \qquad (3.\text{B-}6)$$

$$V = n_l V_{m_l} + n_g V_{m_g} = n_l V_{m_l} + (n - n_l) V_{m_g} \qquad (3.\text{B-}7)$$

$$n_l = \frac{n V_{m_g} - V}{V_{m_g} - V_{m_l}} \qquad (3.\text{B-}8)$$

at constant V, n

$$\left(\frac{dn_l}{dT} \right)_\sigma = \frac{n (\partial V_{m_g} / \partial T)_\sigma}{(V_{m_g} - V_{m_l})} - \frac{(n V_{m_g} - V)}{(V_{m_g} - V_{m_l})^2} \left[\left(\frac{\partial V_{m_g}}{\partial T} \right)_\sigma - \left(\frac{\partial V_{m_l}}{\partial T} \right)_\sigma \right] \qquad (3.\text{B-}9)$$

Eliminate n from Eq. (3.B-9) by use of Eq. (3.B-8) to obtain

$$\left(\frac{dn_l}{dT} \right)_\sigma = \frac{1}{V_{m_g} - V_{m_l}} \left[n_g \left(\frac{\partial V_{m_g}}{\partial T} \right)_\sigma + n_l \left(\frac{\partial V_{m_l}}{\partial T} \right)_\sigma \right] \qquad (3.\text{B-}10)$$

Thus by comparison of Eqs. (3.B-6) and (3.B-5), using Eq. (3.B-10) we have

$$C_{V_2} = n_l C_{\sigma_l} + n_g C_{\sigma_g} - T \left(\frac{dP}{dT} \right)_\sigma \left[n_l \left(\frac{\partial V_{m_l}}{\partial T} \right)_\sigma + n_g \left(\frac{\partial V_{m_g}}{\partial T} \right)_\sigma \right] \qquad (3.\text{B-}11)$$

For any one phase in equilibrium, liquid or gas [see Eq. (I-44)],

$$C_\sigma = C_V - T \left(\frac{\partial P}{\partial V_m} \right)_T \left(\frac{\partial V_m}{\partial T} \right)_P \left(\frac{\partial V_m}{\partial T} \right)_\sigma \qquad (3.\text{B-}12)$$

We use

$$\left(\frac{\partial V_m}{\partial T} \right)_\sigma = \left(\frac{\partial V_m}{\partial T} \right)_P + \left(\frac{\partial V_m}{\partial P} \right)_T \left(\frac{dP}{dT} \right)_\sigma \qquad (3.\text{B-}13)$$

to eliminate $(\partial V_m / \partial T)_P$, and substituting C_{σ_l} and C_{σ_g} separately into Eq. (3.B-11) we get C_{V_2} in terms of C_{V_l} and C_{V_g} of the coexisting phases. We introduce the coexisting number densities

$$\left(\frac{\partial V_{m_g}}{\partial T} \right)_\sigma = \frac{-N_A}{\rho_g^2} \left(\frac{\partial \rho_g}{\partial T} \right)_\sigma \qquad (3.\text{B-}14.1)$$

$$\left(\frac{\partial V_{m_l}}{\partial T} \right)_\sigma = \frac{-N_A}{\rho_l^2} \left(\frac{\partial \rho_l}{\partial T} \right)_\sigma \qquad (3.\text{B-}14.2)$$

and the corresponding isothermal compressibilities

$$\beta_{T_i} = \frac{-1}{V_{m_i}} \left(\frac{\partial V_{m_i}}{\partial P} \right)_T \tag{3.B-15}$$

Then, dividing Eq. (3.B-11) by the total number of moles we obtain the final result per mole of the two-phase system, using $X_i = \dfrac{n_i}{n}$ = mole fraction of i:

$$C_{V_2} = X_l C_{V_l} + X_g C_{V_g} + \frac{X_l T N_A}{\rho_l^3 \beta_{T_l}} \left(\frac{\partial \rho_l}{\partial T} \right)_\sigma^2 + \frac{X_g T N_A}{\rho_g^3 \beta_{T_g}} \left(\frac{\partial \rho_g}{\partial T} \right)_\sigma^2 \tag{3.B-16}$$

In any experimental setup, from Eq. (3.B-6) with $V/n = V_m$ fixed, $V/n = V_m = X_g V_{m_g} + X_l V_{m_l}$

$$X_l = \frac{V_{m_g} - V_m}{V_{m_g} - V_{m_l}} = \frac{(1/\rho_g) - (1/\rho)}{(1/\rho_g) - (1/\rho_l)} = \frac{\rho_l}{\rho} \frac{(\rho - \rho_g)}{(\rho_l - \rho_g)} \tag{3.B-17}$$

If $\rho < \rho_c$ as one heats the two-phase system, the meniscus (separating the phases) will fall until the last drop of liquid disappears at a $T < T_c$, $\rho_l = 0$, $X_l = 0$, with a finite discontinuity in C_v (a drop of C_v in going to the one-phase system) from Eq. (3.B-16) of

$$\frac{T N_A}{\rho_g^3 \beta_{T_g}} \left(\frac{\partial \rho_g}{\partial T} \right)_\sigma^2$$

If $\rho > \rho_c$ as one heats the system, the meniscus will rise until the last bubble of gas disappears at a $T < T_c$, $\rho_g = 0$, $X_l = 1$, with a finite discontinuity in C_V of

$$\frac{T N_A}{\rho_l^3 \beta_{T_l}} \left(\frac{\partial \rho_l}{\partial T} \right)_\sigma^2$$

If $\rho = \rho_c$,

$$X_l = \frac{\rho_l(\rho_c - \rho_g)}{\rho_c(\rho_l - \rho_g)}$$

and with heating the meniscus will vanish from the center of the vessel (neglecting gravity effects, which make experiments very difficult because β_T becomes infinite) at $T = T_c$, $X_l = \frac{1}{2}$, $\rho_l = \rho_g = \rho_c$. We use from Eq. (3-128) very near $T = T_c$

$$\frac{\rho_l - \rho_g}{2\rho_c} = \frac{\rho_c - \rho_g}{2\rho_c} + \frac{\rho_l - \rho_c}{2\rho_c} = B_0(-t)^\beta \tag{3.B-18}$$

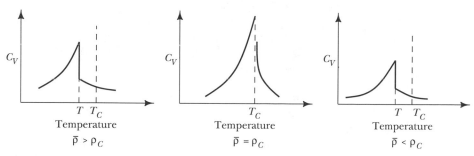

FIGURE 3-27 Schematic real fluid heat capacity (C_V) behavior for various overall number densities.
SOURCE: P. Heller. *Reports on Progress in Physics* **30**(II), 731 (1967).

and

$$\frac{\rho_l - \rho_c}{2\rho_c} = \frac{\rho_c - \rho_g}{2\rho_c} = \frac{B_0}{2}\,(-t)^\beta \qquad (3.\text{B-19})$$

The "discontinuity" is now

$$C_{V_2} - C_{V_1} = \frac{1}{2}\frac{T_c N_A}{\rho_c^3 (\beta_T)_{l_c}}\left(\frac{\partial \rho_{l_c}}{\partial T}\right)_\sigma^2 + \frac{1}{2}\frac{T_c N_A}{\rho_c^3 (\beta_T)_{g_c}}\left(\frac{\partial \rho_{g_c}}{\partial T}\right)_\sigma^2 \qquad (3.\text{B-20})$$

and for real fluids it is infinite because C_{V_2} and C_{V_1} are both infinite at $T = T_c$. For classical fluids the discontinuity is finite even at $T = T_c$. Further discussion is provided in Problems 35–37. Figure 3-27 shows this C_V behavior schematically for a real fluid.

References and Recommended Reading

[1] F. Bitter. *Currents, Fields and Particles*. Technology Press, Cambridge, MA, 1956.

[2] R. J. W. LeFèvre. *Dipole Moments*, 3d ed. Methuen, London, 1953.

[3] C. P. Smyth. *Dielectric Behavior*. McGraw-Hill, New York, 1955.

[4] J. T. Edsall and J. Wyman. *Biophysical Chemistry*. Academic Press, New York, 1958. Chapter 6.

[5] R. P. Feynman. *Lectures on Physics*, vol. I. Addison Wesley, Reading, MA, 1965. Section 31.

[6] J. O. Hirschfelder, C. F. Curtiss, and R. B. Bird. *The Molecular Theory of Gases and Liquids*. Wiley, New York, 1954.

[7] G. C. Maitland, M. Rigby, E. B. Smith, and W. A. Wakeham. *Intermolecular Forces: Their Origin and Determination*. Oxford Univ. Press, Oxford, 1981.

[8] R. O. Watts and I. J. McGee. *Liquid State Chemical Physics*. Wiley-Interscience, New York, 1976.

[9] J. D. Watson and F. H.C. Crick. *Nature* **17**, 964 (1953). Original publication of the double helix structure of DNA.

[10] J. E. Mayer and M. G. Mayer. *Statistical Mechanics*, 2d ed. Wiley, New York, 1977. Chapters 7, 8, 11.

[11] E. Mason and T. Spurling. *The Virial Equation of State*. Pergamon, London, 1969.

[12] J. H. Dymond and E. B. Smith. *The Virial Coefficients of Gases: A Critical Compilation*. Oxford Univ. Press, Oxford, 1969.

[13] K. S. Pitzer. *J. Am. Chem. Soc.* **77**, 3427 (1955). K. S. Pitzer and R. F. Curl, Jr. *J. Am. Chem. Soc.* **79**, 2369 (1957). Extended principle of corresponding states.

[14] N. F. Carnahan and K. E. Starling. *J. Chem. Phys.* **51**, 635 (1969); *J. Chem. Phys* **53**, 600 (1970); *Phys. Rev.* **A1**, 1672 (1970). Practically exact equation of state for hard spheres.

[15] W. W. Wood in *Physics of Simple Liquids*, H. N. V. Temperley, J. S. Rowlinson, and G. S. Rushbrooke, ed. North Holland, Amsterdam, 1968. Chapter 5. F. H. Ree in *Physical Chemistry, An Advanced Treatise*, vol. 8A, H. Eyring, D. Henderson, and W. Jost, eds. Academic Press, New York 1971. Chapter 3. On computer simulation methods for classical particles.

[16] J. P. Valleau and G. M. Torrie in *Statistical Mechanics Part A: Equilibrium Techniques*, B. J. Berne, ed., Plenum, New York, 1977. Chapter 5. K. Binder and D. Stauffer in *Applications of the Monte Carlo Method in Statistical Physics*, K. Binder, ed. Springer, Berlin, 1984. Chapter 1.

[17] C. J. Pings in *Physics of Simple Liquids*, H. N. V. Temperley, J. S. Rowlinson, and G. S. Rushbrooke, eds. North Holland, Amsterdam, 1968. Chapter 10.

[18] D. A. McQuarrie. *Statistical Mechanics*. Harper and Row, New York, 1976. Chapter 13.

[19] J. A. Barker and D. Henderson. *Rev. Modern Physics* **48**, 587 (1976). Review of theories of the liquid state.

[20] B. J. Alder and C. E. Hecht. *J. Chem. Phys.* **50**, 2032 (1969). Augmented van der Waals theory.

[21] H. E. Stanley. *Introduction to Phase Transitions and Critical Phenomena*. Oxford Univ. Press, Oxford, 1971.

[22] J. Stephenson in *Physical Chemistry, An Advanced Treatise*, vol. 8B, H. Eyring, D. Henderson, and W. Jost, eds. Academic Press, New York, 1971. Chapter 10.

[23] M. E. Fisher. *J. Math. Phys.* **5**, 944 (1964).

[24] M. E. Fisher. *Reports on Progress in Physics* **30**, 615 (1967).

[25] J. S. Rowlinson and F. L. Swinton. *Liquids and Liquid Mixtures*, 3d ed. Butterworths, London, 1982. Chapter 3.

[26] B. Widom. *Physica* **73**, 107 (1974).

[27] J. L. Sengers, R. Hocken, and J. V. Sengers. *Physics Today* **30**, 42 (Dec., 1977).

[28] K. G. Wilson. *Scientific American* **241**, 158 (Aug., 1979).

[29] E. A. Guggenheim. *Mixtures*. Oxford Univ. Press, Oxford, 1952.

[30] I. Prigogine. *The Molecular Theory of Solutions*. North Holland, Amsterdam, 1957.

[31] D. Henderson and P. J. Leonard in *Physical Chemistry, An Advanced Treatise*, vol. 8B, H. Eyring, D. Henderson, and W. Jost, eds. Academic Press, New York, 1971. Chapter 7.

[32] R. S. Berry, S. A. Rice, and J. Ross. *Physical Chemistry*. Wiley, New York, 1980. Chapter 25, Sections 9, 10.

[33] J. O'M. Bockris and A. K. Reddy. *Modern Electrochemistry*. Plenum, New York, 1970.

[34] H. L. Friedman. *Ionic Solution Theory*. Wiley, New York, 1962.

[35] P. Resibois. *Electrolyte Theory*. Harper and Row, New York, 1968.

[36] S. Petrucci, ed. *Ionic Interactions*. Academic Press, New York, 1971.

[37] K. S. Pitzer. *Accounts of Chem. Research* **10**, 371 (1977).

Problems

1. K. B. McAlpine and C. P. Smyth, [*J. Chem. Phys.* **3**, 55 (1935)] report for fluorobenzene in the gas phase

T/K	343.6	371.4	414.1	453.2	507.0
$10^6 P_m/m^3/mol$	69.9	66.8	62.5	59.3	55.8

Determine μ and α for this molecule.

2. Show for gases at moderate pressure that their dielectric constant may be written as

$$(D - 1) = \frac{3s}{1 - s} \qquad s = \frac{P}{T}(a + b/T)$$

with P the pressure in atm, $a = (0.02763)(\alpha)10^{40}$, $b = 6.673(\mu \times 10^{30})^2$. Calculate D for water vapor at 140°C and pressures of 1.0 and 10.0 atm.

3. Measurements on C_2H_5Br in the gas phase give $\mu = 6.67 \times 10^{-30}$ C·m and $\alpha = 11.1 \times 10^{-40}$ C·m²·V⁻¹. If one were to heedlessly apply the Debye equation to C_2H_5Br in the liquid phase at 300 K ($\rho' = 1.43$ g/cm³), what would be the calculated D value? The experimental value is $D = 9.3$.

   As shown in reference [4], an improved but still rough equation for polar molecules in the liquid phase is

$$P_m = \frac{M}{\rho'} \frac{(2D + 1)(D - 1)}{9D} = \frac{N_A \alpha}{3\varepsilon_0} + \frac{N_A \mu^2}{9\varepsilon_0 kT}$$

Use this equation and the gas phase data to calculate D of liquid C_2H_5Br at 300 K.

4. Use Table 3-3 to estimate n_r for liquid acetic acid, CH_3COOH, at 25°C ($\rho' = 1.046$ g/cm^3).

5. Use data in Table 3-1 to calculate D for liquid CCl_4 at 20°C ($\rho' = 1.595$ g/cm^3).

6. From data in Table 3-1 compare T_b (normal boiling points) among the following triples:
   a. $(CH_3)_3CH$, $(CH_3)_2C\!=\!CH_2$, $(CH_3)_3N$
   b. $(CH_3)_2CH_2$, $(CH_3)_2O$, C_2H_4O
   c. p-, m-, o-dinitrobenzene
   with reference to μ and α values.

7. Show that Eq. (3-40) implies that ΔU_{vap_m} for nonpolar molecules ($\mu = 0$) depends on $(\alpha/4\pi\varepsilon_0)^n$, where n is somewhat less than 1.0. Also show that T_b (normal boiling point) is given by

$$T_b = \left(\frac{\Delta U_{vap_m}}{\Delta S_{vap_m} - R}\right)$$

such that if ΔS_{vap_m} is constant (Trouton's rule), T_b is proportional to $(\alpha/4\pi\varepsilon_0)^n$. Consider the data in Table 3-1 for Ne, N_2, Ar, O_2, CH_4, Kr, Xe, C_2H_6, and C_3H_8. Let $y = [10^{30}(\alpha/4\pi\varepsilon_0)]^n$ and consider a fit to

$$T_b = sy + b$$

where s and b are respectively the slope and intercept of a least-squares program for a linear equation (pocket calculator). Physically, we should expect b to be zero and we can use this to fix n. Start with a subset of the data for Ne, Ar, Kr, and C_3H_8 to see what guess for n (start at 1.0 and decrease) will bring b close to zero. Then refine this guess on the data for all nine substances. Use your final result to calculate $\alpha/4\pi\varepsilon_0$ for $(CH_3)_3CH$, given $T_b = 263$ K. Also calculate T_b for p-dichlorobenzene and p-dinitrobenzene from their $\alpha/4\pi\varepsilon_0$ values.

8. For a van der Waals gas we have $(\partial U_m/\partial V_m)_T = a/V_m^2$. If we identify U_m with U_{c_m}, the molar cohesive energy, obtain an expression for the van der Waals a. Calculate its value using molecular parameters and using Lennard-Jones parameters for He, Xe, and CCl_4. Compare with the experimental results, which are 0.034, 4.19, and 20.4 atm·L^2·mol^{-2}, respectively.

9. a. Explain briefly why at high temperatures (and explain with respect to what the temperature is high) the virial coefficients of a real gas approach those of a hard sphere gas.
   b. At 423 K, B_2 for He is 18.4×10^{-30} m^3 per atom. Estimate B_3 for He at this temperature and explain your means of estimation.

10. Obtain the second virial coefficient for the following three potential energy functions.

a. Hard spheres

$$\varepsilon(r) = \infty \qquad 0 \leq r < \sigma$$

$$\varepsilon(r) = 0 \qquad r \geq \sigma$$

b. Square well

$$\varepsilon(r) = \infty \qquad 0 < \sigma$$

$$\varepsilon(r) = -\varepsilon^* \qquad \sigma \leq r \leq s\sigma$$

$$\varepsilon(r) = 0 \qquad r > s\sigma$$

c. Hard core with attractive tail

$$\varepsilon(r) = \infty \qquad 0 \leq r < \sigma$$

$$\varepsilon(r) = -\lambda/r^6 \qquad r \geq \sigma$$

Assume $\lambda/\sigma^6 kT$ is small.

d. Compare these virial coefficients with the second virial coefficient obtainable from the van der Waals equation. Use notation $b_0 = \frac{2}{3}\pi\sigma^3$, $\Delta = (e^{\varepsilon^*/kT} - 1)$.

11. Show that the expression for the second virial coefficient of a classical gas

$$B_2 = -2\pi \int_0^\infty (e^{-\varepsilon(r)/kT} - 1) r^2 \, dr$$

can be rewritten as

$$B_2 = \frac{-2\pi}{3kT} \int_0^\infty \frac{d\varepsilon(r)}{dr} e^{-\varepsilon(r)/kT} r^3 \, dr$$

if a certain condition on $\varepsilon(r)$ is satisfied. State the condition. Hint: Consider integration by parts.

12. Take the virial series for hard spheres through B_6 as shown in Example 3-5 and write the coefficients as nearest integers:

$$\frac{P}{\rho kT} = 1 + 4\eta + 10\eta^2 + 18\eta^3 + 28\eta^4 + 40\eta^5$$

$$n = \qquad 2 \qquad 3 \qquad 4 \qquad 5 \qquad 6$$

Try to find a function $f(n)$ of n to fit all these coefficients. Start with $f(n) = an + b$. Then try $f(n) = an^2 + bn + c$ and beyond if necessary to find a, b or a,

b, c such that all the coefficients are reproduced. Then perform the sum

$$\frac{P}{\rho kT} = 1 + \sum_{n=2}^{\infty} f(n)\eta^{n-1}$$

to derive the Carnahan-Starling equation of state for hard spheres (Eq. 3-89). Use geometric series and differentiation of these series to perform the sums exactly.

13. From Eq. (3-88) for A, derive expressions for U and S and C_V of a nonideal gas in each case as an ideal gas part and another part as a series depending on the virial coefficients.

14. For argon the following B_2 data are given:

$t/°C$	25	50	75
$B_2/cm^3 \cdot mol^{-1}$	-15.49	-11.06	-7.14

Use these data to calculate the corrections to S_m, U_m, and C_{V_m} for gas imperfection of argon at 50°C and 1.00 atm and 50°C and 100 atm. You will note that these effects are small. Compare your entropy corrections to those given by Eq. (2-57). Use equations from Problem 13.

15. In Eq. (3-96) we showed $g^{(1)} = 1$ for a fluid. Prove this directly from the defining equation, Eq. (3-94), by examining the ratio of integrals in that equation and going to coordinates relative to particle 1. Why will your proof not hold for crystals?

16. Consider the density expansion for $g^{(2)}$ for hard spheres described in Example 3-7. $g_0^{(2)}(r) = e^{-\varepsilon(r)/kT}$, and it may be shown that

$$g_1^{(2)}(r) = 0 \qquad 0 < r < \sigma$$

$$= \pi\sigma^3 \left[\frac{4}{3} - \frac{r}{\sigma} + \frac{1}{12}\left(\frac{r}{\sigma}\right)^3 \right] \qquad \sigma \leq r \leq 2\sigma$$

$$= 0 \qquad r > 2\sigma$$

Plot the two-term approximate $g^{(2)}(r, \rho)$ function for hard spheres for r/σ from 0 to 3 for $\rho\sigma^3 = 0.880$ and compare your result to Fig. 3-12, which shows the $g^{(2)}(r, \rho, T)$ for the Lennard-Jones potential at $\rho\sigma^3 = 0.88$.

17. A way to estimate the number of nearest neighbors or the first coordination number, N_1, of a given molecule in a fluid is to evaluate

$$N_1 = \int_0^{r_D} 4\pi r^2 \rho g^{(2)}(r) \, dr$$

where r_D is at the first minimum in the function $r^2 g^{(2)}(r)$ (note reference [17], pp. 410–411). Evaluate $\mathcal{N}_1$ for hard spheres at $V/V_0 = 1.60$ and $V/V_0 = 3.00$ from the following molecular dynamics data taken from reference [20]. Integrate numerically using Simpson's rule, which takes an odd number of integrand points (called y_m) at fixed interval h in x and estimates

$$I = \int y \, dx = \frac{h}{3} \left[y_0 + 4(y_1 + y_3 + \ldots + y_{m-1}) \right.$$

$$\left. + 2(y_2 + y_4 + \ldots + y_{m-2}) + y_m \right]$$

	$V/V_0 = 1.60$	$V/V_0 = 3.00$		$V/V_0 = 1.60$	$V/V_0 = 3.00$
$x = r/\sigma$	$g(x)$	$g(x)$	$x = r/\sigma$	$g(x)$	$g(x)$
1.00	4.95	2.07	1.48	0.66	1.02
1.04	3.73	1.92	1.52	0.65	0.99
1.08	2.89	1.78	1.56	0.65	0.97
1.12	2.23	1.65	1.60	0.67	0.95
1.16	1.76	1.53	1.64	0.69	0.94
1.20	1.44	1.43	1.68	0.74	0.92
1.24	1.20	1.35	1.72	0.80	0.92
1.28	1.01	1.28	1.76	0.87	0.92
1.32	0.89	1.21	1.80	0.96	0.92
1.36	0.80	1.16	1.84	1.03	0.92
1.40	0.73	1.10	1.88	1.10	0.93
1.44	0.69	1.06	1.92	1.17	0.95

18. Show that for hard spheres Eq. (3-99) for the pressure may be written

$$\frac{P}{\rho k T} = 1 + \frac{2}{3} \pi \rho \sigma^3 g^{(2)}(\sigma)$$

where $g^{(2)}(\sigma)$ is the hard sphere radial distribution function at $r = \sigma$. Hint: Use the identity

$$g(r) \frac{d\varepsilon}{dr} = -kT g(r) e^{+\varepsilon/kT} \frac{d}{dr} \left(e^{-\varepsilon/kT} \right)$$

and consider the properties of a δ (delta) function.

19. Use the result of the previous problem to obtain an expression for $g^{(2)}(\sigma)$ for hard spheres using the Carnahan-Starling equation in terms of η. Calculate $g^{(2)}(\sigma)$ at $V/V_0 = 1.60$ and $V/V_0 = 3.00$ and compare with the computer results given in Problem 17. What is the largest value that $g^{(2)}(\sigma)$ can have?

20. Consider the equation of state given by perturbation theory for a square well as perturbing potential:

$$\varepsilon'(r) = -\varepsilon^* \qquad \sigma \leq r \leq s\sigma$$

$$\varepsilon'(r) = 0 \qquad r > s\sigma$$

and hard spheres as the unperturbed potential.

a. Work out the second virial coefficient of this equation of state using as expansion of the unperturbed radial distribution function

$$g_0^{(2)}(r, \rho) = e^{-\varepsilon_0(r)/kT}\left(1 + \rho \int d\bar{r}_3 f_{13} f_{23} + \cdots\right)$$

(Show that you do not need to evaluate $\int d\bar{r}_3 f_{13} f_{23}$ to do this.)

b. Compare your answer to the exact second virial coefficient for a square well

$$B_2 = b_0[1 - (s^3 - 1)(e^{\varepsilon^*/kT} - 1)]$$

and show how to obtain your answer to **a** by expanding the exact answer. How is this expansion consistent with thermodynamic perturbation theory?

21. The integral I of Eq. (3-117) may be simply evaluated for the Kac potential of Eq. (3-122) as follows

$$I = \frac{-1}{\sigma^3 \varepsilon^*} \int_0^\infty g_0^{(2)}(r, \rho)(-\alpha\gamma^3)e^{-\gamma r}r^2\, dr$$

$$= \frac{\alpha}{\sigma^3 \varepsilon^*} \int_\sigma^\infty g_0^{(2)}(r, \rho)\gamma^3 e^{-\gamma r}r^2\, dr$$

since $g_0^{(2)}$ for hard spheres is zero for $r < \sigma$. Change variables and use $y = \gamma r/\sigma$

$$I = \frac{\alpha}{\varepsilon^*} \int_\gamma^\infty g_0^{(2)}\left(\frac{\sigma y}{\gamma}\right)e^{-\sigma y}y^2\, dy$$

Now we take limit $\gamma \to 0$, which means $g_0^{(2)}(\infty)$ occurs in the integrand and $g_0^{(2)}(\infty) = 1$, irrespective of the density

$$I = \frac{\alpha}{\varepsilon^*} \int_0^\infty e^{-\sigma y}y^2\, dy = \frac{2\alpha}{\sigma^3 \varepsilon^*}$$

To fit the "normalization" of Eq. (3-120), show that

$$\frac{\alpha}{\sigma^3 \varepsilon^*} = \frac{4}{9} \qquad \alpha \text{ has units of } (\text{length})^3(\text{energy})$$

Write out the augmented van der Waals equation of state with the Kac potential as a power series in $1/X$, where $X = V/V_0$, using data in Table 3-7 through B_6 only.

22. Obtain the critical point parameters of the augmented van der Waals equation with Kac tail derived in Problem 21. Since the critical point is determined by setting

$$\left(\frac{\partial P}{\partial V}\right)_T = \frac{1}{V_0}\left(\frac{\partial P}{\partial X}\right)_T = 0$$

and also

$$\left(\frac{\partial^2 P}{\partial V^2}\right)_T = \frac{1}{V_0^2}\left(\frac{\partial^2 P}{\partial X^2}\right)_T = 0$$

it will be convenient to multiply your result in Problem 21 by $1/X = V_0/V_m$ to obtain a form

$$\frac{PV_0}{RT} = f(X) - \frac{h(X)}{T^*}$$

Then, denoting derivatives with respect to X with primes,

$$f'(X_c) - \frac{h'(X_c)}{T_c^*} = 0$$

$$f''(X_c) - \frac{h''(X_c)}{T_c^*} = 0$$

from which

$$T_c^* = \frac{h'(X_c)}{f'(X_c)}$$

and X_c is found numerically using Newton's method to solve:

$$f''(X_c)h'(X_c) = h''(X_c)f'(X_c)$$

Compare your results to those in Table 3-8.

23. In the Lennard-Jones and Devonshire cell theory of a liquid, the model of Example 3-6 is used with the constant average potential $-\chi$ expressed in terms of LJ potential interactions of a molecule and its 12 first nearest neighbors, assuming a face-centered cubic (FCC) arrangement on the average:

$$-\chi = \frac{12(4\varepsilon^*)}{2}\left[\left(\frac{\sigma}{d}\right)^{12} - \left(\frac{\sigma}{d}\right)^{6}\right]$$

with d = the nearest neighbor distance. (Divide by 2 so as not to count each interaction twice.) The total liquid volume is $N_l v$, where v is the actual volume per molecule (the volume of the FCC unit cell divided by the four atoms per cell).

$$v = \frac{2^{3/2} d^3}{4} = \frac{d^3}{\sqrt{2}}$$

The *free* volume per molecule, v_f, is defined arbitrarily to be of spherical shape

$$v_f = \frac{4}{3} \pi \left(\frac{d - \sigma}{2} \right)^3$$

Introduce also the close-packed volume per molecule

$$v_0 = \frac{\sigma^3}{\sqrt{2}}$$

Show that

$$-\chi = 24\varepsilon^* \left[\left(\frac{v_0}{v} \right)^4 - \left(\frac{v_0}{v} \right)^2 \right]$$

$$v_f = \frac{\pi}{6} \sqrt{2v} \left[1 - \left(\frac{v_0}{v} \right)^{1/3} \right]^3$$

Using equations in Example 3-6, obtain the equation of state for this model as

$$Pv = \frac{kT}{1 - (v_0/v)^{1/3}} + 48\varepsilon^* \left[2 \left(\frac{v_0}{v} \right)^4 - \left(\frac{v_0}{v} \right)^2 \right]$$

This equation is very roughly of the augmented van der Waals form. (Why?) However, it does not give accurate results even for Lennard-Jones potential molecules. Evaluate $N_A v_f$ for CCl_4 at 25°C using data in Example 3-6.

24. First-order perturbation theory for the Helmholtz free energy gives

$$A = A_0 + 2\pi \frac{N^2}{V} \int_0^\infty r^2 \varepsilon' g_0^{(2)}(r, \rho) \, dr$$

where the second term is independent of temperature for hard spheres as the unperturbed system. Take U as the energy of the system given by

$$U = A + TS \qquad S = - \left(\frac{\partial A}{\partial T} \right)$$

and $U_0 = \frac{3}{2}NkT$ since there is no interaction energy among hard spheres. If ε' is a square well to 1.5σ, show how the close-packed value $(V = N\sigma^3/\sqrt{2})$ of the integral

$$\int_1^{1.5} x^2 g_0^{(2)}(x)\, dx \qquad x = r/\sigma$$

can be obtained exactly. (At close packing for any site there are 12 nearest neighbors at distance σ, 6 second nearest neighbors at distance $\sqrt{2}\sigma$, 24 third nearest neighbors at $\sqrt{3}\sigma$, etc.)

25. Use the identity

$$\overline{N^2} - (\bar{N})^2 = \bar{N} + \overline{N(N-1)} - (\bar{N})^2$$

and Eq. (3-95.2) in Eq. (3-138) to prove that

$$kT\rho\beta_T = 1 + \rho \int d\bar{r}\, h(r)$$

You should use an explicit average notation in using Eq. (3-95.2), that is,

$$\int \cdots \int \rho^n g^{(n)}\, d\bar{r}_1 \cdots d\bar{r}_n = \overline{\left[\frac{N!}{(N-n)!}\right]}$$

26. For a subcritical isotherm from a classical theory, such as the one shown in Fig. 3-16, in the P–V plane in order to assure equality of chemical potentials as well as of pressures in the coexisting phases, the vapor pressure line must be drawn such that the areas shown are equal. Prove this rule using $\Delta G = \mu_{m_g} - \mu_{m_l} = \int V_m\, dP = 0$. Hint: Integrate by parts.

27. Derive Eq. (3-161) from Eq. (3-160).

28. How many decades (powers of 10) closer to T_c must t become for the correlation length to increase from 10^{-9} to 10^{-6} m? Critical opalescence would be noticeable at about 10^{-6} m. Repeat the calculation for increase of correlation length from 10^{-6} to 10^{-1} m. Use $v_+ = v_- = v = 0.630$.

29. With reference to Table 3-10, state the chief qualitative difference that would show up experimentally between experiments on the gas-liquid critical point and those on an isotropic ferromagnetic material near its Curie (critical) point. Assume the magnet follows the three-dimensional Heisenberg model.

30. What is the critical exponent for the vanishing of ΔH_{vap_m} at the critical point (in terms of the exponents listed in Table 3-10)? Hint: Use the Clapeyron equation. Derive an expression for ΔH_{vap_m} for a van der Waals fluid near its critical point.

31. Starting from results in Example 3-9, evaluate Eq. (3-157) for the van der Waals fluid and show that

$$\frac{V_{m_l} + V_{m_g}}{2V_{m_c}} = 1 - \frac{18}{5}t$$

32. From Eq. (3-160) for a classical fluid

$$V_{m_g} - V_{m_c} = \left(\frac{-6P_{11}\tau}{P_{30}}\right)^{1/2} = V_{m_c} - V_{m_l}$$

and for a van der Waals fluid this reads

$$V_{m_g} - V_{m_c} = 2V_{m_c}(-t)^{1/2} = V_{m_c} - V_{m_l}$$

Combining this with the result of Problem 31, it is reasonable to put as the separate volume expansions of a van der Waals fluid

$$\frac{V_{m_g}}{V_{m_c}} = 1 + 2(-t)^{1/2} - \frac{18}{5}t$$

$$\frac{V_{m_l}}{V_{m_c}} = 1 - 2(-t)^{1/2} - \frac{18}{5}t$$

Use $\rho_i/\rho_c = V_{m_c}/V_{m_i}$ and obtain the law of rectilinear diameters for the van der Waals fluid

$$\frac{\rho_l + \rho_g}{2\rho_c} = 1 + q(-t)$$

and determine the value of q.

Some comments should be made about this interesting law. It has been derived by classical methods, yet by experiment it holds very precisely in density units for real fluids over a very large temperature range, not just close to T_c. [The analogous relation of Eq. (3-157) in volume units is not accurately followed by real fluids except very close to T_c.] The simple scaling laws (Appendix 3.A) and use of only leading terms in the $h(x)$ function of Eq. (3.A-24) cannot derive this law. The parameter q, of order unity for many real fluids, shows significant variation from one to another. It also shows significant quantum effects and is in fact the only critical state parameter to do so. For ^{4}He q is about 0.1 and for ^{3}He q is very slightly *negative*. These quantum effects are understandable from the point of view that at 0 K, ($t = -1$), where $\rho_g = 0$, quantum liquids should have comparatively small reduced densities because of their very large zero point energy.

33. The spinodal curve exists inside the coexistence region on a P-V plot (see Fig. 3-17). It is where β_T^{-1} goes to zero for $T < T_c$. It is a curve that gives the limit of metastability (superheated liquid or supercooled gas). Using results in Example 3-9, obtain equations for ω_s' as function of t_s and P_s as function of ω_s' for a van der Waals fluid where the subscript s refers to the spinodal curve. Under what conditions can P_s be negative?

34. The Helmholtz free energy per unit volume for a van der Waals fluid is

$$\frac{A_m}{V_m} = \frac{A_{0m}(T)}{V_m} - \frac{RT}{V_m} \ln (V_m - b) - \frac{a}{V_m^2}$$

A_{0m} being a function of T only.

Obtain* an expression for μ_m in terms of $\rho_m = 1/V_m$. Eliminate the van der Waals constants via

$$b = \frac{V_{m_c}}{3} = \frac{1}{3\rho_{m_c}} \qquad a = \frac{9}{8} \frac{RT_c}{\rho_{m_c}}$$

Expand μ_m in a Taylor series about $\rho_m = \rho_{m_c}$, thus in powers of $\Delta\rho_m = \rho_m - \rho_{m_c}$, with coefficients as functions of T only being $(\partial\mu_m/\partial\rho_m)_{\rho=\rho_c}$, etc., through terms in $(\Delta\rho)^4$ and obtain $\Delta\mu_m$ as a function of $\Delta\rho_m$, where $\Delta\mu_m = \mu_m - \mu_m(\rho_{m_c}, T)$. How well does your result fit Widom's form for the equation of state of a real fluid near the critical point? Compare with Eq. (3.A-22).

35. a. What is the exponent of $-t$ for the discontinuity term of Eq. (3.B-20) in terms of β and γ?

   b. If we assume [and this cannot be derived from the purely thermodynamic analysis that gave us Eq. (3.B-20)] that C_{V_1}, which is the heat capacity at T_c of the single critical phase reached from $T < T_c$, diverges with the same exponent found in **a** and not with one more negative, what is the result as to an exponent identity?

   c. What is the more general result as an exponent inequality if the assumption in **b** is not made?

36. Apply Eq. (3-164.1) and results from Problem 32 to a van der Waals fluid and show that at the critical point the discontinuity in C_V of Eq. (3.B-20) is finite and equal to $\frac{9}{2}R$ per mole.

37. To a previously evacuated volume of 30.6 liters, 2.00 moles of water are added at some low temperature. Calculate the temperature t_f at which the final drop

---

* NOTE:

$$\mu = \left(\frac{\partial A}{\partial N}\right)_{T,V} = \left(\frac{\partial(A/V)}{\partial(N/V)}\right)_{T,V}$$

$$N\mu = \left(\frac{\partial(A/V)}{\partial(1/V)}\right)_{T,N} \quad \text{and on a molar basis } \mu_m = \left[\frac{\partial(A_m/V_m)}{\partial\rho_m}\right]_{T,N}$$

of liquid water evaporates. Use Eq. (3.B-16) to calculate the C_V of this system at 25, 50, 75, 100, 110, t_f and $(t_f + 1)°C$. Use the following data

$$\Delta H_{vap_m} = 40.6 \times 10^3 \text{ J·mol}^{-1} \text{ at } 100°C$$

$$P_{vap} = 1.00 \text{ atm at } 100°C$$

$$\left.\begin{array}{l} C_{P_l} \simeq C_{V_l} = 75.3 \text{ J·K}^{-1}\text{·mol}^{-1} \\ C_{P_g} \simeq 4R = 33.3 \text{ J·K}^{-1}\text{·mol}^{-1} \end{array}\right\} \quad \begin{array}{l} \text{assumed temperature} \\ \text{independent} \end{array}$$

$$(\beta_T)_l = 4.6 \times 10^{-5} \text{ atm}^{-1} \qquad \text{assumed constant}$$

$$\frac{1}{V_{m_l}}\left(\frac{\partial V_{m_l}}{\partial T}\right)_P = 5.5 \times 10^{-4} \text{ K}^{-1} \qquad \text{at } 65°C$$

Assume that the gaseous water follows the ideal gas law and show that you are justified in neglecting the term $(TN_A/\rho_l^3\beta_{T_l})(\partial\rho_l/\partial T)_\sigma^2$ in the equation for C_{V_2} under these conditions, which are far from the critical point.

38. Guggenheim [29] has extended the quasi-lattice model of solutions to take account of nonrandom occupation of the lattice sites because of energy effects; that is, if two ab pairs are favorable as to energy over a sum of aa and bb pair energies, then $\bar{N}_{ab}$ should be increased over its random occupation value and vice versa. The changes from that of the strictly regular solution model are small, but $\Delta S_{m,mxg}^E$ is no longer zero. What must be its sign? Qualitatively, what would be a more significant change in the assumptions of the quasi-lattice model in order to deal with a wider class of real solutions?

39. Liquid solutions of ^{3}He and ^{4}He isotopes, which remain liquid all the way to absolute zero under their own vapor pressure, exhibit below 0.8 K liquid-liquid immiscibility, separating into a ^{4}He-rich and a ^{3}He-rich phase. Explain why the quasi-lattice model of solutions cannot account for this and suggest at least two factors that might explain this phenomenon.

40. It was assumed that the number of ways $g(N_a, N_b, N_{ab})$ to put N_a type a and N_b type b on a lattice such that N_{ab} ab nearest neighbor pairs are formed, is a maximum in the random case where $N_{ab} = \bar{N}_{ab} = zN_aN_b/(N_a + N_b)$. An *approximate* formula for g may be written by considering all pairs as independent of each other and treating the ab pairs as made up of *two* types, ab and ba, in equal numbers with fictitious orientation in two ways. (The real reason for taking two "types" is that there are two ways in which an ab pair can be formed on two sites.) Then

$$g \simeq \frac{(\tfrac{1}{2}zN)!}{(N_{aa})!(N_{ab}/2)!(N_{ab}/2)!(N_{bb})!}$$

Show that this approximate g is maximized if $N_{ab} = \bar{N}_{ab}$.

41. Use Eq. (3-181) to write an expression for the Gibbs function of a mixture of n_a moles of a and n_b moles of b in terms of μ_a^* and μ_b^* (the chemical potentials of pure a and b, respectively). Obtain expression for μ_a and μ_b in the strictly regular solution. Also obtain expressions for the respective activity coefficients, activities, and fugacities for species a and b in such a solution. Then by reference to Eq. (I-54) and (I-55) show that the critical properties of this model are given by Eqs. (3-186) and (3-187).

42. Let X_b' and X_b'' denote the mole fractions of component b in the two different liquid solutions that are formed for $T < T_c$ in the strictly regular solution model. By equating activities of components a and of components b in the two solutions, verify that

$$X_b' + X_b'' = 1$$

Then obtain an equation for T/T_c in terms of X_b' or in terms of X_b''. Plot this coexistence curve as X_b' falls from 0.5 to zero (and X_b'' rises from 0.5 to 1.0). The two values, X_b' and X_b'', have the same T/T_c of course.

From data on $T/T_c > 0.90$ calculate the slope of $\ln(X_b'' - X_b')$ versus $\ln(1 - T/T_c)$. Comment on your result.

43. a. Express the Henry's law constant for a component a of a strictly regular solution in terms of $z\omega'$ and f_a^* (the fugacity of pure a at the same temperature and pressure as the solution).

 b. For a constant temperature such that $z\omega'/RT = 8/3$, describe how the total vapor pressure (assuming the gas phase is ideal) of a strictly regular solution varies from that of pure a as component b is added until one gets a final solution almost pure in b. Contrast this behavior with that of a pair of completely immiscible components.

44. a. How, in general, will the value of the constant A in Eq. (3-219) change when solvents other than water are used at $25°C$? How will A change for temperatures lower than $25°C$?

 b. Calculate A for a 80 wt. percent water mixture of water and dioxane for which $D = 60.8$, $\rho' = 1.02 \text{ g/cm}^3$ at $25°C$. Also calculate A for methanol at $25°C$, $D = 31.5$, $\rho_1' = 0.79 \text{ g/cm}^3$.

 c. Calculate A for liquid ammonia at $-33°C$, for which $D = 22.0$, $\rho_1' = 0.69 \text{ g/cm}^3$.

45. Using experimental $\gamma_\pm$ for NaCl in water at $25°C$, deduce a value for σ for NaCl solution in a sensible way.

c/molarity	$\gamma_\pm$
0.0010	0.966
0.0020	0.953
0.0050	0.929
0.010	0.904

The "common" σ value should be considered as the sum of the radii of the positive and negative ions of the solute since ions of like charge will never come into contact in practice.

46. Use the empirical extended Debye-Hückel theory for mean molal activity coefficient, Eq. (3-220), to find the coefficient of the term linear in I for LiCl in water at 25°C, given that $\gamma_\pm$ is a minimum for $I = 0.49$ molal. Calculate this minimum $\gamma_\pm$ and also the nonzero I value for which $\gamma_\pm$ reaches unity in concentrated LiCl solution.

47. Keeping Eq. (3-228) for A^{el}, derive an expression for G^{el}. Then from these obtain an equation for π^{el}, the osmotic pressure of the solute due to charge effects in an electrolyte solution. In the limit $\kappa\sigma \to$ zero, calculate π^{el} for a 1:1 electrolyte (10^{-2} M) and for a 2-1 (10^{-3} M) electrolyte in water at 25°C. What would be the total osmotic pressure of these respective solutions? Give osmotic pressures in atmospheres.

48. By definition, the osmotic coefficient ϕ of a solution is defined by $\phi = \pi/kT \sum_i \mathcal{N}_i$, where π is the osmotic pressure due to the solutes. From Problem 47 obtain an equation for $1 - \phi$ and obtain its limit for $\kappa\sigma \to$ zero. In this limit show that

$$1 - \phi = z_+ z_- \left(\frac{2.3026A}{3}\right) I^{1/2}$$

where A is the constant given by Eq. (3-217).

49. a. What percentage of the balancing charge of the ionic atmosphere of any ion is located beyond $r = 1/\kappa$ from the ion? Express your answer in terms of $\kappa\sigma$ and give the limit when $\kappa\sigma = 0$. Of course to make sense, $1/\kappa > \sigma$.

    b. At what molarity will $1/\kappa = \sigma$ for a 1-1 and for a 2-2 electrolyte? (Assume σ for 1-1 to be 4.0×10^{-10} m, and for 2-2 to be 5.0×10^{-10} m.)

50. Calculate and plot $g^{(2)}_{+-}$ and $g^{(2)}_{++} = g^{(2)}_{--}$ (thus reproducing the solid curves in Fig. 3-22) using the exponential form for $g^{(2)}$ of Eq. (3-209) for the case relevant to Fig. 3-22.

51. The differential number of ions i (*not* number density) $d\mathcal{N}'_i(r)$ at distance r to $r + dr$ from an ion j, from Eq. (3-192) and Eq. (3-194) is

$$d\mathcal{N}'_i(r) = 4\pi r^2 \mathcal{N}_i \exp\left[-\Phi_j(r) z_i e/kT\right] dr \equiv f(r)\, dr$$

For r values not much larger than σ, approximate $\Phi_j(r)$ by

$$\Phi_j(r) = \frac{z_j e}{4\pi\varepsilon_0 D r}$$

Show that if i and j are of opposite charge this differential number and $f(r)$

have a minimum at $r = q$ where

$$q = \frac{|z_j z_i| e^2}{8 \pi \varepsilon_0 D k T}$$

Calculate the value of q in water at $25°C$ and in a solvent with $D = 20$ at $25°C$ for 1-1 and 2-2 electrolytes.

52. With reference to Problem 51 for symmetrical electrolytes (1-1 or 2-2),

$$\bar{N}'_0 = \int_{r=\sigma}^{r=q} dN'_0(r)$$

gives the average number of charges *opposite* to the central ion in the range $\sigma \le r \le q$ and these pairs will be net uncharged. Then $\bar{N}'_0$ may be interpreted* as the fraction of the ions associated, $1 - \alpha$. Only the ion fraction α that are dissociated will follow Debye-Hückel theory. If $q < \sigma$ no change from DH theory results. The physical idea is that for those oppositely charged ions less than $r = q$ away from a central ion the potential energy $|\Phi_j(r) z_i e|$ will be so large compared to kT that they will associate, while for those at $r > q$ away their potential energy is small compared to kT and thermal motion will separate the ions. Show that

$$(1 - \alpha) = 32 \pi N_i q^3 Q(2q/\sigma)$$

$$Q(b) = \int_2^b \frac{e^y}{y^4} \, dy$$

$$N_i = \text{number of } + \text{ ions per unit volume}$$
$$= \text{number of } - \text{ ions per unit volume}$$

53. Calculate and comment on the meaning of $1 - \alpha$ for the four cases specified in Problem 51, each at the three molarities 10^{-4}, 10^{-3}, 10^{-2} M. Assume σ for a 1-1 electrolyte to be 4.0×10^{-10} m and for a 2-2 electrolyte to be 5.0×10^{-10} m. You will find it necessary to use from numerical integration:

b	$Q(b)$
5.7	0.94
7.0	1.42
22.0	1.90×10^4

---

* Bjerrum-Fuoss theory of ion association. See R. Fuoss. *Trans. Faraday Soc.* **30**, 967 (1934); *Chem. Reviews* **17**, 27 (1935). Also note reference [33].

54. a. The electrical Helmholtz free energy in the limiting law form of Eq. (3-229) may be considered the free energy change ΔA^{el} between the electrolyte solution and an otherwise equivalent solution of *uncharged* particles. Show that the corresponding entropy change is

$$\Delta S^{el} = \frac{\Delta A^{el}}{2}\left(\frac{1}{T} + \frac{3}{D}\frac{\partial D}{\partial T}\right)$$

where D is the dielectric constant of the solvent.

b. The temperature dependence of D for water is approximately

$$D = 305.7e^{-T/219}$$

What will be the *sign* of ΔS^{el} for aqueous solution (temperature range 0 to 100°C) in this hypothetical charging process which neglects solvation effects? Suggest a reason for this sign.

55. From Eq. (3-232), on a molar basis

$$\mu_2^{el} = \left(\frac{\partial \Delta A^{el}}{\partial n_2}\right)_{V,T} = \frac{3}{2}\frac{\Delta A^{el}}{n_2}$$

where n_2 is the total moles of electrolytic solute and the total μ_2 is

$$\mu_2 = \mu_2^0 + \nu RT \ln\left(\frac{(m_\pm)_2}{(m_\pm^0)_2}\right) + \mu_2^{el}$$

a. Obtain an equation for the total change in partial molal entropy of solute 2 upon dilution from molality m_b to molality m_a. Use results from Problem 54. Assume that temperature is constant.

b. Calculate numerically the increase in partial molal entropy of NaCl in aqueous solution for dilution from 0.010 to 0.0010 M at 25°C.

4

Statistical Thermodynamics of Solids

4-1
Lattice Partition Function for Ideal Atomic Crystals:
Einstein Model; Debye Approximation

We have already remarked in Sec. 4 of Chap. 1 that the counting formula, Eq. (1-29), does apply to localized identical particles on a crystal lattice. Hence no division by $N!$ is needed to use it. In the case that these localized particles are independent of one another (and we will describe in what sense that comes about in what follows) the set of occupation numbers of the single-particle levels will still be given by Eq. (1-103). This is because the maximization that gave us Eq. (1-103) cannot be affected by not dividing by the *constant* $N!$ However the entropy expression for the crystal lattice (subscript c) will now read

$$\frac{S_c}{k} = N \ln q_c + \frac{U_c}{kT} \qquad (4\text{-}1)$$

in place of the result in Eq. (1-105). Of course the single-particle partition function for the lattice, q_c, is not a sum over the levels of translation in a gas. It is a sum over the levels available to the particle on a lattice (oscillator levels, as we will see). Thus for the chemical potential we have

$$\frac{-\mu_c}{T} = \left(\frac{\partial S_c}{\partial \mathcal{N}}\right)_{U,V} = k \ln q_c \tag{4-2}$$

$$e^{-\mu_c/kT} = q_c \tag{4-3}$$

in place of the result in the nonlocalized gas (g) case from Eq. (1-105)

$$e^{-\mu_g/kT} = \frac{q_g}{\mathcal{N}} \tag{4-4}$$

From Eq. (1-114) we know q_g is proportional to the total volume, so that Eq. (4-4) is an intensive quantity (being $[(V/\mathcal{N})f(T)]$), as it must be. We deduce that q_c cannot be a function of the entire volume V, but must be a function of the volume per particle $V/\mathcal{N}$. This is the fundamental difference between independent localized and independent nonlocalized particles.

Now we turn to the matter of how the particles on the lattice can be treated in first approximation as effectively independent. Physically they are not independent and the potential energy of the lattice can be written as a *sum* over all $\mathcal{N}$ particles of their real nearest neighbor interaction potentials, which for a one-dimensional lattice will be of the form $v(a + x_{i+1} - x_i)$ between the ith and $(i + 1)$th particles, where a is the equilibrium lattice site spacing and the x's are displacements from their respective equilibrium positions. One expands this function in a Taylor series about its equilibrium value:

$$v(a + x_{i+1} - x_i) = v(a) + \left[\frac{dv(x)}{dx}\right]_{x=a} (x_{i+1} - x_i)$$

$$+ \frac{1}{2}\left[\frac{d^2v(x)}{dx^2}\right]_{x=a} (x_{i+1} - x_i)^2 + \ldots \tag{4-5}$$

Since at equilibrium there must be no net forces on any atom, the first derivative term evaluated at $x = a$ will always vanish by definition. Hence we have

$$v(a + x_{i+1} - x_i) = v(a) + \frac{1}{2}\left[\frac{d^2v(x)}{dx^2}\right]_{x=a} (x_{i+1} - x_i)^2 + \ldots \tag{4-6}$$

Now the problem is analogous to finding the normal coordinates of a molecule as long as the higher order *anharmonic* terms like $(x_{i+1} - x_i)^3$ and beyond are neglected. The result of this transformation to normal coordinates in *three* dimensions is that the energy of a lattice configuration becomes the sum of $(3\mathcal{N} - 6)$ squared terms of normal coordinates in the potential energy and $(3\mathcal{N} - 6)$ squared momenta in the kinetic energy plus an additive term $\mathcal{N}\phi(0)/2$ that represents the potential energy of the lattice when all the atoms are at their equilibrium sites. $\phi(0)$ is negative when we take the usual convention that the infinitely separated atoms have zero energy. It represents the interaction of a single reference atom with all the other atoms in the crystal when all are on their equilibrium sites. It will be dominated by the nearest neighbor interactions.

By the classical equipartition principle (see Sec. 1-8) the thermodynamic C_{V_m} of the lattice will be $2(3\mathcal{N}_A - 6)(k/2) \simeq 3\mathcal{N}_A k = 3R$. This is the law of Dulong and Petit.

From a quantum mechanical viewpoint, which must be used because the spacing of the quantized oscillator levels is generally not small compared to kT, we have effectively obtained a set of independent quantized harmonic oscillators. Thus the lattice canonical partition function factors

$$Q_c = \exp\left[-\mathcal{N}\phi(0)/2kT\right] \prod_{i=1}^{3N} q_{vib}(\theta_i) \tag{4-7}$$

with q_{vib} from Eq. (2-37)

$$q_{vib} = \frac{\exp\left(-\theta_{v_i}/2T\right)}{1 - \exp\left(-\theta_{v_i}/T\right)} \tag{4-8}$$

and

$$\theta_{v_i} = h\nu_i/k \tag{4-9}$$

Since $\mathcal{N}$ is the number of atoms in the crystal, $3\mathcal{N} - 6 \simeq 3\mathcal{N}$.

To find the $3\mathcal{N}$ frequencies is an impossible task. However it is convenient and legitimate to introduce a continuous frequency distribution $g(\nu)$ such that $g(\nu)\,d\nu$ is the number of normal modes with frequencies between ν and $\nu + d\nu$. Then

$$-\ln Q_c = \frac{\mathcal{N}\phi(0)}{2kT} - \sum_{i=1}^{3N}\left[\frac{-\theta_{v_i}}{2T} - \ln\left(1 - e^{-\theta v_i/T}\right)\right] \tag{4-10.1}$$

$$= \frac{\mathcal{N}\phi(0)}{2kT} + \int_0^\infty \left(\ln\left(1 - e^{-h\nu/kT}\right) + \frac{h\nu}{2kT}\right)g(\nu)\,d\nu \tag{4-10.2}$$

where we must normalize the distribution so as to maintain the correct count of frequencies

$$\int_0^\infty g(\nu)\,d\nu = 3\mathcal{N} \tag{4-11}$$

The simplest distribution is that all the frequencies are identically the same, ν_E. This was first proposed by Einstein, in 1907, who suggested that to understand the decrease of C_{v_m} of solids to zero as the temperature falls one must quantize the harmonic oscillator motion. Then $g(\nu)/3\mathcal{N}$ is a delta function (see Appendix II) and Eq. (4-10.2) gives

$$-\ln Q_c = \frac{N\phi(0)}{2kT} + \frac{3Nh\nu_E}{2kT} + 3N \ln \left(1 - e^{-h\nu_E/kT} \right) \qquad (4\text{-}12)$$

from which all the thermodynamic functions of this Einstein model solid may be derived:

$$A_c = -kT \ln Q_c = \frac{N\phi(0)}{2} + \frac{3Nh\nu_E}{2} + 3NkT \ln \left(1 - e^{-h\nu_E/kT} \right) \quad (4\text{-}13)$$

$$S_c = -\left(\frac{\partial A_c}{\partial T} \right)_V = 3Nk \left[\frac{\theta_E/T}{(e^{\theta_E/T} - 1)} - \ln \left(1 - e^{-\theta_E/T} \right) \right] \qquad (4\text{-}14)$$

$$U_c = A_c + TS_c = \frac{N\phi(0)}{2} + \frac{3}{2} Nk\theta_E + 3NkT \left(\frac{\theta_E/T}{e^{\theta_E/T} - 1} \right) \qquad (4\text{-}15)$$

$$C_{V,c} = \left(\frac{\partial U_c}{\partial T} \right)_V = 3Nk \left(\frac{\theta_E}{T} \right)^2 \frac{e^{\theta_E/T}}{(e^{\theta_E/T} - 1)^2} \qquad (4\text{-}16)$$

in which θ_E, called the Einstein temperature of the solid, is given by

$$\theta_E = \frac{h\nu_E}{k} \qquad (4\text{-}17)$$

Note that derivatives at constant V and N mean ν_E and $\phi(0)$ are also constant. Both of these quantities depend on the nearest-neighbor distance on the lattice and so in fact are functions of V/N. Thus, as pointed out at the beginning of this section, Q_c for localized particles is a function of V/N. The second term in Eq. (4-15) is the zero point energy and this always decreases the binding energy of the lattice at 0 K [which is the negative of the sum of the first two terms in Eq. (4-15), recalling that $\phi(0)$ itself is negative by our zero of energy convention].

Care must be taken in obtaining the chemical potential. Since the Helmholtz free energy is a function of N, T, and $v = V/N$, we have at constant T

$$dA_c = \left(\frac{\partial A_c}{\partial N} \right)_{v,T} dN + \left(\frac{\partial A_c}{\partial v} \right)_{N,T} dv$$

$$\mu_c = \left(\frac{\partial A_c}{\partial N} \right)_{T,V} = \left(\frac{\partial A_c}{\partial N} \right)_{v,T} + \left(\frac{\partial A_c}{\partial v} \right)_{N,T} \left(\frac{\partial v}{\partial N} \right)_{T,V}$$

$$= \left(\frac{\partial A_c}{\partial N} \right)_{v,T} - \frac{V}{N^2} \left(\frac{\partial A_c}{\partial v} \right)_{N,T}$$

$$= \frac{\phi(0)}{2} + \frac{3}{2} h\nu_E + 3kT \ln \left(1 - e^{-h\nu_E/kT} \right)$$

$$+ \frac{V}{N} \left\{ -\frac{1}{2} \left[\frac{\partial \phi(0)}{\partial v} \right]_{N,T} - \frac{3}{2} h \left(\frac{\partial \nu_E}{\partial v} \right)_{N,T} \frac{(1 + e^{-h\nu_E/kT})}{(1 - e^{-h\nu_E/kT})} \right\} \qquad (4\text{-}18)$$

Exercise 4-1

Derive Eq. (4-18) from Eq. (4-13) using $v = V/N$, $(\partial v/\partial N)_{T,V} = -(V/N^2)$.

Since

$$N\mu_c = G_c = A_c + PV$$

we find the equation of state of the lattice by comparison of Eq. (4-18) and Eq. (4-13) to be

$$P = -\frac{1}{2}\left[\frac{\partial\phi(0)}{\partial v}\right]_{N,T} - \frac{3}{2}h\left(\frac{\partial v_E}{\partial v}\right)_{N,T}\frac{(1 + e^{-hv_E/kT})}{(1 - e^{-hv_E/kT})} \tag{4-19}$$

This could also have been found by direct use of $P = -(\partial A_c/\partial V)_{T,N}$. It is usually accurate to neglect the PV terms for a lattice such that

$$\mu_c \simeq \frac{A_c}{N} = \frac{\phi(0)}{2} + \frac{3}{2}hv_E + 3kT\ln(1 - e^{-hv_E/kT}) \tag{4-20}$$

In fact, solid vapor pressures are so minor that one can generally set $P = 0$ in Eq. (4-19) and use it to determine equilibrium v values of a lattice if one has a model that expresses $\phi(0)$ and v_E in terms of v.

From Eq. (4-16) it is easy to show that

$$C_{V_c} \to 3Nk \qquad \text{as } T \to \infty \tag{4-21}$$

and

$$C_{V_c} \to 3Nk\left(\frac{\theta_E}{T}\right)^2 e^{-\theta_E/T} \qquad \text{as } T \to 0 \tag{4-22}$$

Equation (4-21) is the classical limit from the equipartition principle.

Exercise 4-2

Obtain Eq. (4-21) and Eq. (4-22) by appropriate power series expansions of Eq. (4-16).

Equation (4-22) gives C_V going to zero much faster than the T^3 relation that is experimentally observed at low temperatures (for nonmetals). The crude Einstein model must be improved by obtaining a more realistic frequency distribution.

An important advance was made by Debye, in 1912, who pointed out that the low-energy (long-wavelength) modes will have wavelengths large compared to the atomic spacing of the lattice. Thus for these modes the lattice may be accurately treated as an elastic continuum. The approximation of Debye is to use the low-frequency form for $g(v)$ even for high-energy modes and to take account of the atomic nature of the lattice only by the normalization condition of Eq. (4-11). Elastic waves in an isotropic continuous solid have two possible speeds of propagation, v_l for longitudinal compressional waves (a set of vibrations along the axis of propagation) and v_t for transverse waves (two sets of vibrations perpendicular to the direction of propagation of the elastic wave) [1]. These are all called *acoustic* or simply *sound* waves. Their approximate realization for a cubic lattice with atoms is shown schematically in Figs. 4-1 and 4-2. The total number of frequencies $F(v)$ that will give rise to standing waves in a volume V of such a solid may be obtained by examining solutions of the classical wave equation

$$\nabla^2 \phi_i = \frac{1}{v_i^2} \frac{\partial^2 \phi_i}{\partial t^2} \tag{4-23}$$

where ϕ_i is the wave amplitude ($i = l$ or t). The analysis is very similar to that given in Example 1-3 for calculating the number of translational states available to a free particle considered as a wave in a box. The result [1, 2] is

$$F(v) = \int g(v)\, dv = \frac{4\pi v^3 V}{3}\left(\frac{1}{v_l^3} + \frac{2}{v_t^3}\right) \tag{4-24}$$

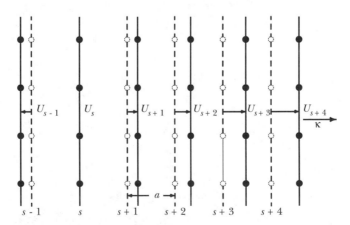

FIGURE 4-1 Dashed lines show planes of atoms when at their equilibrium sites. Solid lines show planes of atoms when displaced by a longitudinal wave in the direction $\check{\kappa}$. SOURCE: C. Kittel. *Introduction to Solid State Physics*, 3d ed. Wiley, 1966.

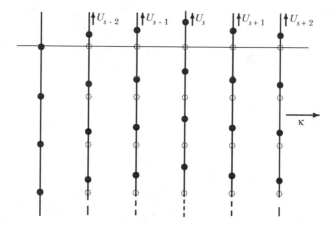

FIGURE 4-2 Planes of atoms as displaced during passage of a transverse wave in the direction $\check{\kappa}$.

SOURCE: C. Kittel. *Introduction to Solid State Physics*, 3d ed. Wiley, New York, 1966.

such that

$$g(v) = \frac{dF}{dv} = 4\pi v^2 V \left(\frac{1}{v_l^3} + \frac{2}{v_t^3} \right) \tag{4-25}$$

One usually defines a weighted speed of sound, v_s, in the solid by

$$\frac{3}{v_s^3} = \frac{1}{v_l^3} + \frac{2}{v_t^3} \tag{4-26}$$

such that incorporating a cutoff at a maximum frequency v_m

$$g(v) = \frac{12\pi V v^2}{v_s^3} \qquad v \le v_m$$

$$= 0 \qquad v > v_m \tag{4-27}$$

From Eq. (4-11) v_m is determined by

$$\int_0^{v_m} g(v)\, dv = \frac{12\pi V}{v_s^3} \int_0^{v_m} v^2\, dv = 3N$$

and

$$v_m = \left(\frac{3N}{4\pi V} \right)^{1/3} v_s \tag{4-28}$$

The Debye temperature is defined by

$$\theta_D = \frac{h v_m}{k} = \frac{h v_s}{k} \left(\frac{3N}{4\pi V} \right)^{1/3} \tag{4-29}$$

Note that this gives us the explicit dependence of θ_D on V/N if we assume v_s is independent of V/N.

Substituting Eq. (4-27) into Eq. (4-10.2), we find for the Debye model solid:

$$A_c = \frac{N\phi(0)}{2} + \frac{9}{8} Nk\theta_D + \frac{9NkT^4}{\theta_D^3} \int_0^u \ln(1 - e^{-x}) x^2\, dx \qquad (4\text{-}30)$$

in which

$$u = \theta_D/T \qquad (4\text{-}31)$$

$$x = hv/kT \qquad (4\text{-}32)$$

Exercise 4-3

Derive Eq. (4-30) in the way indicated. Note that integration over v goes from 0 to v_m only.

It is convenient to rewrite the integral occurring in Eq. (4-30) by integrating by parts using

$$\int_0^u \ln(1 - e^{-x}) x^2\, dx \equiv \int_0^u p\, dq = pq)_0^u - \int_0^u q\, dp$$

$$p = \ln(1 - e^{-x}) \qquad dp = \frac{dx}{e^x - 1}$$

$$dq = x^2\, dx \qquad q = \frac{x^3}{3}$$

such that

$$\int_0^u \ln(1 - e^{-x}) x^2\, dx = \frac{u^3}{3} \ln(1 - e^{-u}) - \int_0^u \frac{x^3}{3} \frac{dx}{e^x - 1} \qquad (4\text{-}33)$$

We also define the dimensionless Debye function $D(u)$ by

$$D(u) = \frac{3}{u^3} \int_0^u \frac{x^3\, dx}{e^x - 1} \qquad (4\text{-}34)$$

Thus we find for A_c

$$A_c = \frac{N\phi(0)}{2} + \frac{9}{8} Nk\theta_D + 3NkT \ln(1 - e^{-u}) - NkTD(u) \qquad (4\text{-}35)$$

In general $D(u)$ must be evaluated numerically and tables of it are available [3]. We examine its high and low temperature limits in the following examples.

Example 4-1

Derive equations for S_c, U_c, and C_{V_c} for the Debye model as functions of u.

Solution:

$$u = \theta_D/T \qquad \text{and} \qquad \left(\frac{\partial}{\partial T}\right)_V = \left(\frac{\partial u}{\partial T}\right)_V \frac{\partial}{\partial u} = \frac{-u}{T}\left(\frac{\partial}{\partial u}\right)_V,$$

from the calculus we have for the derivative of a definite integral

$$\frac{d}{dy}\int_A^B F(x,y)\,dx = \int_A^B \left[\frac{\partial F(x,y)}{\partial y}\right]dx + F(B,y)\frac{dB}{dy} - F(A,y)\frac{dA}{dy} \qquad (4\text{-}36)$$

Our $D(u)$ function has an integrand that depends only on the dummy variable x and only the upper limit is a function of u. Thus

$$\left(\frac{\partial}{\partial T}\right)_V D(u) = \frac{-u}{T}\left[\frac{-9}{u^4}\int_0^u \frac{x^3\,dx}{e^x - 1} + \frac{3}{u^3}\left(\frac{u^3}{e^u - 1}\right)(1)\right]$$

using Eq. (4-35)

$$S_c = -\left(\frac{\partial}{\partial T}A_c\right)_V = -3Nk\ln(1 - e^{-u}) - \frac{3NkTe^{-u}}{1 - e^{-u}}\frac{du}{dT}$$

$$+ NkD(u) + NkT\left(\frac{\partial}{\partial T}\right)_V D(u)$$

$$= 3Nk\left[\frac{4}{3}D(u) - \ln(1 - e^{-u})\right] \qquad (4\text{-}37)$$

$$U_c = A_c + TS_c = \frac{N\phi(0)}{2} + \frac{9}{8}Nk\theta_D + 3NkTD(u) \qquad (4\text{-}38)$$

$$C_{V_c} = \left(\frac{\partial U_c}{\partial T}\right)_V = 3NkD(u) + 3NkT\left(\frac{\partial}{\partial T}\right)_V D(u)$$

$$= 3Nk\left[4D(u) - \frac{3u}{e^u - 1}\right] \qquad (4\text{-}39)$$

The C_{V_c} function is tabulated in Table 4-1 on a molar basis ($N_A k = R$).

TABLE 4-1 Debye Heat Capacity Function $C_V/3R$ as a Function of θ_D/T

θ_D/T	0.0	0.1	0.2	0.3	0.4	0.5	0.6	0.7	0.8	0.9	1.0
0.0	1.0000	0.9995	0.9980	0.9955	0.9920	0.9876	0.9822	0.9759	0.9687	0.9606	0.9517
1.0	0.9517	0.9420	0.9315	0.9203	0.9085	0.8960	0.8828	0.8692	0.8550	0.8404	0.8254
2.0	0.8254	0.8100	0.7943	0.7784	0.7622	0.7459	0.7294	0.7128	0.6961	0.6794	0.6628
3.0	0.6628	0.6461	0.6296	0.6132	0.5968	0.5807	0.5647	0.5490	0.5334	0.5181	0.5031
4.0	0.5031	0.4883	0.4738	0.4595	0.4456	0.4320	0.4187	0.4057	0.3930	0.3807	0.3686
5.0	0.3686	0.3569	0.3455	0.3345	0.3237	0.3133	0.3031	0.2933	0.2838	0.2745	0.2656
6.0	0.2656	0.2569	0.2486	0.2405	0.2326	0.2251	0.2177	0.2107	0.2038	0.1972	0.1909
7.0	0.1909	0.1847	0.1788	0.1730	0.1675	0.1622	0.1570	0.1521	0.1473	0.1426	0.1382
8.0	0.1382	0.1339	0.1297	0.1257	0.1219	0.1182	0.1146	0.1111	0.1078	0.1046	0.1015
9.0	0.1015	0.09847	0.09558	0.09280	0.09011	0.08751	0.08500	0.08259	0.08025	0.07800	0.07582
10.0	0.07582	0.07372	0.07169	0.06973	0.06783	0.06600	0.06424	0.06253	0.06087	0.05928	0.05773
11.0	0.05773	0.05624	0.05479	0.05339	0.05204	0.05073	0.04946	0.04823	0.04705	0.04590	0.04478
12.0	0.04478	0.04370	0.04265	0.04164	0.04066	0.03970	0.03878	0.03788	0.03701	0.03617	0.03535
13.0	0.03535	0.03455	0.03378	0.03303	0.03230	0.03160	0.03091	0.03024	0.02959	0.02896	0.02835
14.0	0.02835	0.02776	0.02718	0.02661	0.02607	0.02553	0.02501	0.02451	0.02402	0.02354	0.02307
15.0	0.02307	0.02262	0.02218	0.02174	0.02132	0.02092	0.02052	0.02013	0.01975	0.01938	0.01902

When $\theta_D/T \geq 16$, $C_V/3R = 77.927\,(T/\theta_D)^3$.
Source: J. A. Beattie. *J. Math. Phys.* (MIT) **6**, 1 (1926).

Example 4-2

Derive high and low temperature limiting equations for $D(u)$ and obtain high and low temperature equations for U_c and C_{V_c}.

Solution:

$$D(u) = \frac{3}{u^3} \int_0^u \frac{x^3 \, dx}{e^x - 1} \qquad u = \theta_D / T$$

At high temperature u is small and since the limits on x are 0 to u, x must always be small and we may expand the denominator

$$\frac{3}{u^3} \frac{x^3}{e^x - 1} = \frac{3x^3}{u^3(x + x^2/2 + x^3/6 + \ldots)} = \frac{3x^3}{u^3 x(1 + x/2 + x^2/6 + \ldots)}$$

$$= \frac{3x^2}{u^3} \left(1 - \frac{x}{2} - \frac{x^2}{6} + \frac{x^2}{4} + \ldots \right)$$

$$= \frac{1}{u^3} \left(3x^2 - \frac{3}{2} x^3 + \frac{1}{4} x^4 + \ldots \right)$$

and integrate over x, term by term

$$D(u) = 1 - \tfrac{3}{8} u + \tfrac{1}{20} u^2 + \ldots \tag{4-40}$$

Substituting into Eq. (4-38)

$$U_c = \frac{N\phi(0)}{2} + \frac{9}{8} Nk\theta_D + 3NkT \left[1 - \frac{3}{8} \frac{\theta_D}{T} + \frac{1}{20} \left(\frac{\theta_D}{T} \right)^2 + \ldots \right]$$

$$= \frac{N\phi(0)}{2} + 3NkT + \frac{3Nk}{20} \frac{\theta_D^2}{T} + \ldots \tag{4-41}$$

The high-temperature C_{V_c} is found by differentiation of Eq. (4-41)

$$C_{V_c} = 3Nk - \frac{3}{20} Nk \left(\frac{\theta_D}{T} \right)^2 + \ldots \tag{4-42}$$

Notice how the zero point energy term, $\tfrac{9}{8} Nk\theta_D$, is canceled in the high-temperature limit for the energy and that these limits are in accord with the equipartition principle.

At *low* temperature u is large and it is necessary to write

$$D(u) = \frac{3}{u^3} \left[\int_0^\infty \frac{x^3 \, dx}{e^x - 1} - \int_u^\infty \frac{x^3 \, dx}{e^x - 1} \right] \tag{4-43}$$

From Eqs. (2-90.2) and (2-93) we note

$$\int_0^\infty \frac{x^3 \, dx}{e^x - 1} = \frac{\pi^4}{15} \tag{4-44}$$

and in the second term of Eq. (4-43) x will always be large, e^{-x} always small such that

$$\frac{x^3}{e^x - 1} = \frac{x^3 e^{-x}}{1 - e^{-x}} = x^3 e^{-x}(1 + e^{-x} + \dots)$$

$$= x^3 e^{-x} + x^3 e^{-2x} + \dots$$

and one can integrate term by term.

The leading effect is found by retaining only the first integral in Eq. (4-43)

$$D(u) = \frac{\pi^4}{5u^3} + \text{terms of order } e^{-u}$$

Thus at low temperature, from Eq. (4-38)

$$U_c = \frac{N\phi(0)}{2} + \frac{9}{8} Nk\theta_D + 3NkT\left(\frac{\pi^4}{5} \frac{T^3}{\theta_D^3} + \dots\right) \qquad (4\text{-}45)$$

$$C_{V_c} = \frac{12}{5} Nk\pi^4 \left(\frac{T}{\theta_D}\right)^3 - \dots \qquad (4\text{-}46)$$

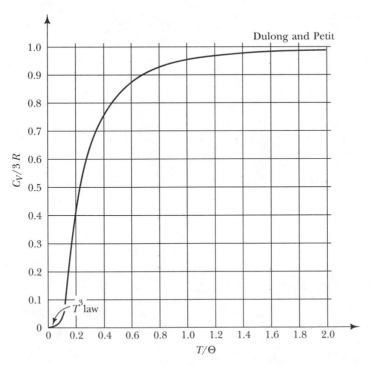

FIGURE 4-3 Debye $C_{V_m}/3R$ function for a monatomic solid lattice.
SOURCE: M. W. Zemansky and R. H. Dittman. *Heat and Thermodynamics*, 6th ed. McGraw-Hill, 1981.

Equation (4-46) is the T^3 temperature dependence law for low-temperature heat capacity, which is accurately followed by solid nonmetallic elements and most solid compounds for $T < \frac{1}{20}\theta_D$. Metals will be discussed in Sec. 4-3, compounds in Sec. 4-2. Figure 4-3 shows the Debye $(C_{V_m}/3R)$ function over a temperature range from 0 to $2\theta_D$, where it has practically reached its classical value. The Debye theory is really remarkably successful in view of the fact that its distribution law, Eq. (4-27), is very far from reality at large ν. Extensive computer calculations of the dynamics of atoms in a solid based on an assumed interatomic potential (such as

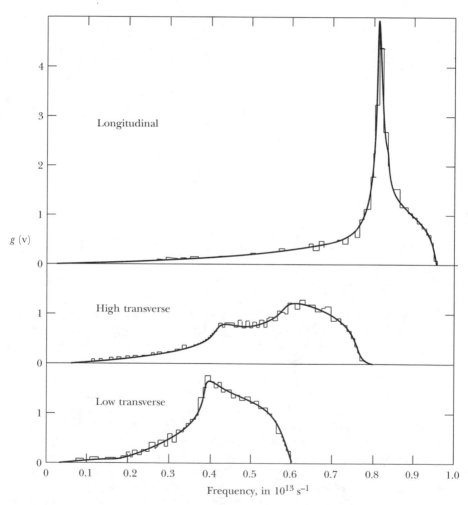

FIGURE 4-4 Frequency distributions for the three branches (one longitudinal, two transverse) totaling $\mathcal{N}_A$ frequencies ($\frac{1}{3}$ mol) for solid aluminum. The histograms were obtained from computed frequencies for 2791 wave vectors in each branch. Using $\theta_D = 394$ K the Debye theory $\nu_m = 0.82 \times 10^{13}$ s^{-1}.
SOURCE: C. B. Walker. *Phys. Rev.* **103**, 547 (1956).

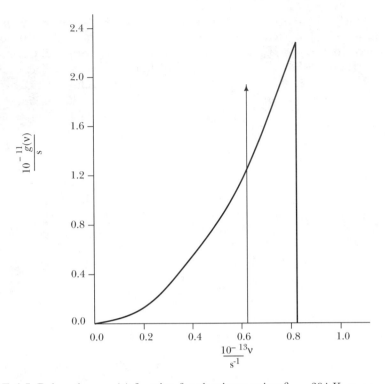

FIGURE 4-5 Debye theory $g(v)$ function for aluminum using $\theta_D = 394$ K, $v_m = 0.82 \times 10^{13}$ s^{-1}. A total of $\mathcal{N}_A$ frequencies is taken for comparison with Fig. 4-4. Equivalent Einstein theory $g(v)$ would be a delta function at $v_E = 3/4 v_m = 0.62 \times 10^{13}$ s^{-1}.

the LJ potential for molecular solids and other forms for metals) have been carried out [2, 4]. Allowed frequencies can be sampled and a smooth graph drawn through the histograms of their relative numbers in different frequency ranges. An example is shown for aluminum in Fig. 4-4 for $\mathcal{N}_A$ frequencies, that is, $\frac{1}{3}$ mole. The total $g(v)$ is the sum of the longitudinal and two transverse results shown in Fig. 4-4. This $g(v)$ is quite different from the Debye function shown in Fig. 4-5 for $\mathcal{N}_A$ frequencies of Al using $\theta_D = 394$ K, roughly characteristic of Al in the large temperature range where its θ_D is fairly constant (note Fig. 4-6). Also in Fig. 4-5 we indicate the δ function Einstein distribution, which would have $\theta_E = \frac{3}{4}\theta_D$ to match the zero point energy values for the two models [compare Eq. (4-45) with Eq. (4-15)]. We may conclude that the thermodynamic properties are not very sensitive to the details of the $g(v)$ function as long as the low-frequency behavior is correct and the normalization of Eq. (4-11) is satisfied. In all of this work, including that shown in Fig. 4-4, anharmonic contributions have been neglected. It is very complicated to rigorously include them. For discussion of anharmonic effects see references [2, 4].

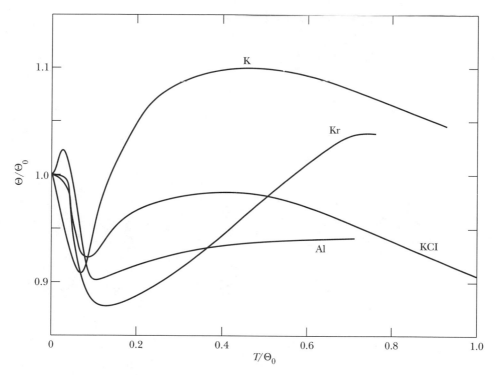

FIGURE 4-6 Effective Debye temperature θ_D for C_V (corrected to the fixed volume V_0 at 0 K and with electronic heat capacity subtracted for K and Al) as a function of T/θ_0. θ_0 is the θ_D value at 0 K. It is 71.9 K for Kr, 90.6 K for K, 235 K for KCl, and 428 K for Al.
SOURCE: D. C. Wallace. *Thermodynamics of Crystals*. Wiley, 1972.

A very sensitive way of showing the defects of the Debye theory is to make a plot of θ_D versus temperature for a substance. If the Debye theory were exact, θ_D would not vary. Examples of such curves are given in Fig. 4-6. They are obtained by taking experimental $C_{V_m}/3R$ values and interpolating in a table (such as Table 4-1) to obtain θ_D/T and thus θ_D at each temperature.

Comparing θ_D values of different atomic substances we note that θ_D usually decreases as the atomic mass increases. The elementary treatment of a simple harmonic oscillator gives frequency proportional to $\sqrt{k/m}$, with k the force constant. The speed of sound v_s in Eq. (4-29) for θ_D may be roughly taken as proportional to $m^{-1/2}$. Of course different force laws for different types of crystals (analogous to different force constants k for the simple harmonic oscillator) will also greatly affect θ_D values, and the $m^{-1/2}$ dependence for θ_D certainly fails if we compare θ_D for elements with θ_D for compounds of these elements. In Table 4-2 θ_D values appropriate to absolute zero are given for a number of substances. Figure 4-6 shows that at higher temperatures these respective θ_D's may be 10 percent or more *higher or lower* than their 0 K values.

TABLE 4-2 Debye Temperatures in Limit as $T \to 0$

Substance	θ_D/K	Substance	θ_D/K
	Nonmetals		
^{4}He	26	NaF	492
Ne	75	NaCl	321
Ar	93	NaBr	225
Kr	72	KCl	235
Xe	64	KI	132
C (graphite)	420	SiO_2	470
C (diamond)	2230	MgO	946
Si	640	TiO_2	760
	Metals		
Li	344	Zn	327
Na	158	Cd	209
K	91	Hg	72
Rb	56	Fe	420
Cs	38	Co	445
Be	1440	Ni	450
Mg	400	Al	428
Ca	230	Sn	199
Sr	147	Pb	105
Cu	343	Cr	630
Ag	225	Mo	450
Au	165	W	400

SOURCE: M. W. Zemansky and R. H. Dittman. *Heat and Thermodynamics*, 6th ed. McGraw-Hill, 1981.

4-2

Lattice Partition Function for Ideal Molecular and Ionic Crystals: Born Approximation

The Debye theory presented in Sec. 4-1 does not apply to solids containing atoms or ions of more than one type or even to those atomic solids (such as germanium) that have unit cells which contain more than one atom. However it is easy to obtain a useful approximation for the thermodynamic properties of these crystals following the method of Born [5]. If there are s atoms or ions in the unit cell and N unit cells in the crystal, it is accurate to divide the frequencies of vibration into $3N$

acoustical and $3\mathcal{N}s - 3\mathcal{N} = (3s - 3)\mathcal{N}$ optical frequencies. In the acoustical modes all the atoms in the unit cell (usually all those in one molecule or all in a neutral formula unit of ions) move in unison and the vibrations can be treated as in the Debye theory with a single θ_D corresponding to the high-frequency cutoff. The optical modes are usually of higher frequency than the acoustical cutoff and correspond to intramolecular vibrations of the atoms in the unit cell relative to one another. They may be approximated as $(3s - 3)$ different sets of $\mathcal{N}$ identical frequencies each, and may thus be fitted by $(3s - 3)$ Einstein-like terms. Hence the lattice partition function of a molecular crystal is given in this approximation by:

$$-\ln Q_c = \frac{\mathcal{N}\phi(0)}{2kT} + \frac{9}{8}\frac{\mathcal{N}\theta_D}{T} + 3\mathcal{N}\ln(1 - e^{-u}) - \mathcal{N}D(u)$$

$$+ \sum_{i=1}^{3s-3}\left[\frac{\mathcal{N}}{2}\frac{\theta_i}{T} + \mathcal{N}\ln(1 - e^{-\theta_i/T})\right] \tag{4-47}$$

in which $\phi(0)$ will here be a sum over the interaction of s reference atoms with all the other atoms in the crystal. From this we find for the heat capacity

$$C_{V_c} = 3\mathcal{N}k\left[4D(u) - \frac{3u}{e^u - 1}\right] + \sum_{i=1}^{3s-3}(\mathcal{N}k)\left(\frac{\theta_i}{T}\right)^2\frac{e^{\theta_i/T}}{(e^{\theta_i/T} - 1)^2} \tag{4-48}$$

In obtaining Eq. (4-48) we used Eq. (4-39) and modified each Einstein result, Eq. (4-16), to apply to $\mathcal{N}$ identical frequencies (*not* $3\mathcal{N}$). Since the Einstein terms at low temperature go exponentially to zero, the low-temperature C_{V_c} of compounds will be dominated by the T^3 dependence from the Debye part. This is experimentally observed. At very high temperatures the limiting $C_{V_c} = 3\mathcal{N}k + \mathcal{N}k(3s - 3) = 3\mathcal{N}ks$, being thus the sum of the classical atomic heat capacities. Because the optical frequencies are very high this limit is not often observed.

The θ_i values in Eq. (4-48) are usually approximated by the $(3s - 6)$ vibrational frequencies of the s-atomic molecule in the gas phase as long as no significant hydrogen bonding or covalent interaction between molecules occurs in the solid. This leaves three other frequencies per unit cell corresponding to torsional oscillations of low frequency (a form of rotatory motion of the atoms in the unit cell as a whole). It is often difficult to assign v_i values for these additional $3\mathcal{N}$ frequencies. They may be roughly taken account of by merely doubling the Debye contribution (which also totals $3\mathcal{N}$ frequencies) and fitting with one θ_D value using

$$C_{V_c} = 6\mathcal{N}k\left[4D(u) - \frac{3u}{e^u - 1}\right] + \sum_{i=1}^{3s-6}(\mathcal{N}k)\left(\frac{\theta_i}{T}\right)^2\frac{e^{\theta_i/T}}{(e^{\theta_i/T} - 1)^2} \tag{4-49}$$

An example of this technique is given in Problem 4-9. A thorough treatment of the vibrations of molecular crystals and their relationship to infrared and Raman spectroscopy is provided by Mitra [6].

Finally, it should be pointed out that for compounds or elements that form very anisotropic crystals with strong interatomic forces limited to planes (e.g., graphite or boron nitride) or to chains (polymers) the T^3 limiting law for C_V at low temperature is not observed (above a few degrees K). Rather a T^2 or linear dependence seems to be followed over a considerable low-temperature range.*

4-3

Thermodynamic Functions for "Free" Electrons in Metals

The high thermal and electrical conductivity of metals was interpreted long before the development of quantum mechanics as indicative of the presence of a practically "free" gas of released valence electrons. Classical theory for such a gas predicts a contribution of $\frac{3}{2}Nk$ to the constant volume heat capacity at all temperatures (for N electrons). This is *not* observed because the number density of such a gas $(1.34 \times 10^{28}$ m^{-3} in potassium at room temperature) is so high that classical statistics fails (see Problem 14 of Chap. 1). We must treat the problem as that of a dense gas of ideal fermions of spin $\frac{1}{2}$.

In Sec. 7 of Chap. 1 we have given the general equations for this problem in the grand canonical ensemble. Use Eqs. (1-156) and (1-157), each multiplied by 2 because of the spin [note Eq. (1-155)].

$$\frac{PV}{kT} = \ln \mathscr{Q} = 4\pi \left(\frac{2m}{h^2}\right)^{3/2} V \int_0^\infty \varepsilon^{1/2} \ln\left(1 + ze^{-\beta\varepsilon}\right) d\varepsilon \qquad (4\text{-}50)$$

$$\bar{N} = 4\pi \left(\frac{2m}{h^2}\right)^{3/2} V \int_0^\infty \frac{ze^{1/2}e^{-\beta\varepsilon}\, d\varepsilon}{(1 + ze^{-\beta\varepsilon})} \qquad (4\text{-}51)$$

where

$$z = e^{\beta\mu} \qquad (4\text{-}52)$$

The average occupation number of the ith energy state is given by Eq. (1-147)

$$\bar{n}_i = \frac{ze^{-\beta\varepsilon_i}}{(1 + ze^{-\beta\varepsilon_i})} \qquad (4\text{-}53)$$

At the absolute zero for ε_i up to and including a maximum value $\varepsilon_{\max}$ where

$$\varepsilon_{\max} \equiv \varepsilon_F = \mu \qquad \text{at } T = 0 \qquad (4\text{-}54)$$

$ze^{-\beta\varepsilon_i}$ is infinite (since $\beta = 1/kT$ and $\mu - \varepsilon_i$ is positive) such that all these states have $\bar{n}_i = 1$, while for $\varepsilon_i > \varepsilon_F$ the $\bar{n}_i = 0$ (since $ze^{-\beta\varepsilon_i} = 0$). This dependence of $\bar{n}_i$ on ε_i is shown in Fig. 4-7. We can determine ε_F directly from Eq. (4-51) at absolute zero, for which a factor in the integrand is $\bar{n}_i$ from Eq. (4-53) with the properties

---

* W. H. Stockmayer and C. E. Hecht. *J. Chem. Phys.* **21**, 1954 (1953).

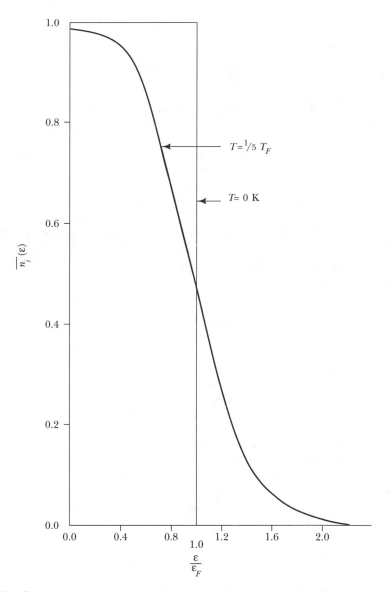

FIGURE 4-7 Average occupation number, $\bar{n}_i(\varepsilon)$, of energy states for ideal fermions as function of $\varepsilon/\varepsilon_F$ at $T = 0$ K and $T = \frac{1}{5} T_F$. Note that for $T > 0$ the chemical potential μ is no longer equal to ε_F and $\bar{n}_i(\varepsilon_F) \neq 0.50$.

just outlined (and so the integral can extend only to ε_F):

$$\bar{N} = 4\pi \left(\frac{2m}{h^2}\right)^{3/2} V \int_0^{\varepsilon_F} \varepsilon^{1/2} \, d\varepsilon = \frac{8}{3} \pi \left(\frac{2m}{h^2}\right)^{3/2} V \varepsilon_f^{3/2}$$

$$\varepsilon_F = \frac{h^2}{8m} \left(\frac{3\bar{N}}{\pi V}\right)^{2/3} \tag{4-55}$$

It is usual to define the Fermi temperature, T_F, by

$$T_F = \frac{\varepsilon_F}{k} = \frac{h^2}{8mk}\left(\frac{3}{\pi}\frac{\bar{N}}{V}\right)^{2/3} = 4.232 \times 10^{-15}\left(\frac{\bar{N}/V}{m^{-3}}\right)^{2/3} \qquad (4\text{-}56)$$

For $T \ll T_F$ the electrons will be mainly but not entirely in the lowest N possible states just as at absolute zero. Figure 4-7 also shows $\bar{n}_i$ for $T = \frac{1}{5}T_F$. For metals at room temperature T_F is greater than 2×10^4 K and so $T/T_F < 0.015$. The situation is little different from that at absolute zero! We need a low-temperature, high-density expansion of the thermodynamic properties. (In Chap. 1 we worked out a high-temperature, low-density expansion for an ordinary molecular gas of fermions.) Physically it is easy to estimate how the heat capacity of the electron gas will behave. Roughly those electrons with energies only as far as kT below ε_F are excited to energies at most kT above ε_F. Electrons deeper below the Fermi surface cannot be excited because there are no empty states within range kT to excite them into. The number N' of electrons so excited will be approximately $(kT/\varepsilon_f)N$ and their average excitation is kT. Thus

$$\Delta U \simeq N'kT = \frac{Nk^2T^2}{\varepsilon_F}$$

from which

$$C_V \simeq 2N\frac{k^2T}{\varepsilon_F} = 2Nk\frac{T}{T_F} \qquad (4\text{-}57)$$

We see that the electronic heat capacity is linear in T and should be very small even at room temperature.

To work out the low-temperature expansion we start from Eq. (4-50) using

$$x = \varepsilon/kT \qquad (4\text{-}58)$$

and

$$\gamma = \ln z = \frac{\mu}{kT} \qquad (4\text{-}59)$$

where γ will be very large in the low-temperature region:

$$\ln \mathcal{Q} = \frac{4}{\sqrt{\pi}}\frac{V}{h^3}(2\pi mkT)^{3/2}I \qquad (4\text{-}60)$$

where

$$I = \int_0^\infty x^{1/2}\ln\left(1 + e^{\gamma-x}\right)dx \qquad (4\text{-}61)$$

The expansion of I is complicated. It is worked out in Appendix 4.A, with the result being

$$I = \frac{4}{15} \gamma^{5/2} + \frac{\pi^2}{6} \gamma^{1/2} - \frac{7\pi^4}{1440} \gamma^{-3/2} + o(\gamma^{-7/2}) \tag{4-62}$$

where the last term means that terms of order $\gamma^{-7/2}$ and beyond have been neglected. From Eq. (1-141)

$$\bar{N} = z\left(\frac{\partial}{\partial z}\right)_{V,T} \ln \mathscr{Q} = \left(\frac{\partial}{\partial \gamma}\right)_{V,T} \ln \mathscr{Q}$$

$$= \frac{4}{\sqrt{\pi}} \frac{V}{h^3} (2\pi mkT)^{3/2} \left[\frac{2}{3} \gamma^{3/2} + \frac{\pi^2}{12} \gamma^{-1/2} + \frac{7}{960} \pi^4 \gamma^{-5/2} + o(\gamma^{-9/2})\right] \tag{4-63}$$

Factoring out $\frac{2}{3}\gamma^{3/2}$ in Eq. (4-63) gives us

$$\frac{h^3}{8(2mkT)^{3/2}} \left(\frac{3}{\pi} \frac{\bar{N}}{V}\right) = \gamma^{3/2}\left(1 + \frac{\pi^2}{8} \gamma^{-2} + \frac{7}{640} \pi^4 \gamma^{-4} + \cdots\right)$$

from which, on raising both sides to the $\frac{2}{3}$ power and using Eq. (4-56), we have

$$\frac{h^2}{8mkT} \left(\frac{3}{\pi} \frac{\bar{N}}{V}\right)^{2/3} = \frac{\varepsilon_F}{kT} = \frac{T_F}{T} = \gamma\left(1 + \frac{\pi^2}{8} \gamma^{-2} + \frac{7}{640} \pi^4 \gamma^{-4} + \cdots\right)^{2/3}$$

$$= \gamma\left(1 + \frac{\pi^2}{12} \gamma^{-2} + \frac{\pi^4}{180} \gamma^{-4} + \cdots\right) \tag{4-64}$$

We now invert this series to get γ as a power series in T/T_F by writing

$$\gamma = \frac{T_F}{T} + b_1 \frac{T}{T_F} + b_3 \left(\frac{T}{T_F}\right)^3 + \cdots$$

and insert this into Eq. (4-64) and equate coefficients of like powers on both sides to determine $b_1, b_3 \ldots$. The result is

$$\gamma = \frac{\mu}{kT} = \frac{T_F}{T}\left[1 - \frac{\pi^2}{12}\left(\frac{T}{T_F}\right)^2 - \frac{\pi^4}{80}\left(\frac{T}{T_F}\right)^4 + \cdots\right] \tag{4-65}$$

At $T = 0$ Eq. (4-65) reduces to Eq. (4-54).

Exercise 4-4

Use the binomial expansion to check the result given in Eq. (4-65).

From Eq. (1-139) the energy in the grand canonical ensemble is given by

$$U = -\left(\frac{\partial}{\partial \beta}\right)_{z,V} \ln \mathcal{Q} = +kT^2\left(\frac{\partial}{\partial T}\right)_{z,V} \ln \mathcal{Q} \tag{4-66}$$

which from Eq. (4-60) becomes merely:

$$U = \tfrac{3}{2}kT \ln \mathcal{Q} \tag{4-67}$$

since at constant z, $\gamma = \ln z$ is constant. Then using Eqs. (4-60) and (4-63)

$$U = \frac{3}{2}kT\bar{N}\left(\frac{\ln \mathcal{Q}}{\bar{N}}\right) = \frac{3}{2}kT\bar{N}\,\frac{[(\tfrac{4}{15})\gamma^{5/2} + (\pi^2/6)\gamma^{1/2} - \ldots]}{[(\tfrac{2}{3})\gamma^{3/2} + (\pi^2/12)\gamma^{-1/2} + \ldots]}$$

Expanding and collecting terms in γ (for γ large) and then substituting for γ using Eq. (4-65), we find

$$U = \frac{3}{5}\bar{N}\varepsilon_F\left[1 + \frac{5}{12}\pi^2\left(\frac{T}{T_F}\right)^2 - \frac{\pi^4}{16}\left(\frac{T}{T_F}\right)^4 + \ldots\right] \tag{4-68}$$

The zero point energy of the electron gas is very large because by the Pauli principle only two electrons (one with spin up, one with spin down) can have zero energy even at absolute zero. The heat capacity function is found from Eq. (4-68):

$$C_V = \left(\frac{\partial U}{\partial T}\right)_V = \frac{\pi^2}{2}\bar{N}k\left(\frac{T}{T_F}\right)\left[1 - \frac{3}{10}\pi^2\left(\frac{T}{T_F}\right)^2 + \ldots\right] \tag{4-69}$$

At room temperature only the leading term is important and we see it is very close to our estimate given in Eq. (4-57).

The pressure of the electron gas is obtained from Eq. (4-50)

$$PV = kT \ln \mathcal{Q} = \tfrac{2}{3}U \tag{4-70}$$

which is the general result for free particles (see Problems 1-20 and 1-21). Thus

$$P = \frac{2}{5}\frac{\bar{N}}{V}\varepsilon_F\left[1 + \frac{5}{12}\pi^2\left(\frac{T}{T_F}\right)^2 - \frac{\pi^4}{16}\left(\frac{T}{T_F}\right)^4 + \ldots\right] \tag{4-71}$$

The pressure of electrons at room temperature is 10^4 atm or more in most metals. This pressure is entirely canceled by the real attraction of the positive ions for the electrons. The electrons are not actually free.

Below 1 K the electronic contribution to the C_V can become dominant in metals since then the lattice C_V goes as T^3 and goes to zero faster than the linear law of Eq. (4-69). One usually plots (C_{V_m}/T) versus T^2 to obtain theoretically a straight line

$$C_{V_m}/T = \gamma' + A'T^2 \tag{4-72}$$

with

$$\gamma' = \frac{\pi^2 R}{2 T_F} = \frac{4\pi^2 Rmk}{h^2}\left(\frac{3}{\pi}\frac{\bar{N}}{V}\right)^{-2/3} \tag{4-73}$$

and from Eq. (4-46)

$$A' = \frac{12}{5}\frac{R\pi^4}{\theta_D^3} \tag{4-74}$$

Experimentally this behavior is found, but with a γ' larger than the above theoretical result. This is ascribed to the electrons having a larger effective mass than their actual mass, due to their interaction with the periodic potential of the ions and to the further interference to their motion provided by the ionic vibrations. This last is called an electron-phonon interaction and does not cease even at absolute zero because of the zero point oscillations of the ions.

Example 4-3

From the low-temperature heat capacity measurements of W. H. Lien and N. E. Phillips [*Phys. Rev.* **133**, A1370 (1964)] for potassium the following data are provided:

T/K	$10^3\,C_V/\text{J}\cdot\text{K}^{-1}\cdot\text{mol}^{-1}$	T/K	$10^3\,C_V/\text{J}\cdot\text{K}^{-1}\cdot\text{mol}^{-1}$
0.526	1.48	0.348	0.837
0.495	1.35	0.326	0.766
0.459	1.21	0.306	0.708
0.433	1.12	0.278	0.635
0.400	1.00	0.250	0.562
0.373	0.919		

Make a least-squares fit to these data to determine γ', A', and θ_D for potassium at 0 K. Compare the experimental γ' value to the theoretical one for free electrons. The molar volume of solid K at 0 K may be taken as 43.1 cm^3/mol.

Solution:

Draw up the following table.

$10^3(C_V/T)/\mathrm{J\cdot K^{-2}\cdot mol^{-1}}$	T^2/K^2
2.814	0.2766
2.727	0.2450
2.636	0.2107
2.587	0.1875
2.500	0.1600
2.463	0.1391
2.405	0.1211
2.350	0.1063
2.314	0.0936
2.284	0.0773
2.248	0.0625

A least-squares fit gives an excellent straight line of

$$\text{Slope} = 2.67 \times 10^{-3} \,\mathrm{J\cdot K^{-4}\cdot mol^{-1}} = A'$$
$$\text{Intercept} = 2.08 \times 10^{-3} \,\mathrm{J\cdot K^{-2}\cdot mol^{-1}} = \gamma'$$

From Eq. (4-74)

$$A' = \frac{1944}{\theta_D^3} \,\mathrm{J\cdot K^{-2}\cdot mol^{-1}}$$

$$\therefore \quad \theta_D^3 = 7.28 \times 10^5$$

$$\theta_D = 90.0 \ \mathrm{K}$$

At 0 K, assuming one "free" valence electron per K atom:

$$\frac{N_A}{V_m} = \frac{6.022 \times 10^{23} \,\mathrm{mol^{-1}}}{43.1 \times 10^{-6} \,\mathrm{m^3/mol}} = 1.40 \times 10^{28} \,\mathrm{m^{-3}}$$

From Eq. (4-56), which strictly depends on the V_m at 0 K

$$T_F = 2.46 \times 10^4 \ \mathrm{K}$$

the theoretical γ' from Eq. (4-73) is

$$\gamma' = \frac{\pi^2}{2}\frac{R}{T_F} = 1.67 \times 10^{-3}\,\text{J}\cdot\text{K}^{-2}\cdot\text{mol}^{-1}$$

NOTE: We may say the effective mass of the e^- in K is $(2.08/1.67)m_e = 1.25m_e$.

■

4-4
Defects in Crystals

Thus far we have treated only perfect crystals, which have a perfect periodic lattice. However, no real crystal is perfect. Any deviation from perfect periodicity is termed a defect. Defects may be classified as point, line, or plane defects. The last two are also called dislocations. Since the enthalpy associated with the formation of a dislocation is extremely large and the accompanying entropy increase is comparatively small, dislocations are thermodynamically unstable and in general crystals may be easily grown without dislocations. As we will see below, this is not the case for point defects, which always exist in large numbers in any crystal at room temperature. In this section we will discuss point defects only.

Many properties of crystals are determined as much or more by the nature of their point defects as by the nature of the host lattice. This is true of such properties as self-diffusion, color, luminescence, and electrical conductivity. It is frequently found that not only are the primary point defects important but so too is their association in pairs or larger clusters, particularly at low temperature where the role of entropy is less significant. The study of point defects in solids is currently a very active field of research.

The most common point defects (see Fig. 4-8) are vacant lattice sites with the former occupant of the site transferred to the crystal surface (Schottky defects) and vacant sites with the displaced atom or ion in an interstitial site (a position not normally occupied in the lattice). These last are called Frenkel defects. In compounds substitutional disorder occurs when one element occupies a site on the sublattice of the other element and vice versa. Foreign atoms may also be incorporated into a crystal, giving rise to nonstoichiometric compounds but with the strict condition of electrical neutrality maintained [7].

We will treat only Schottky and Frenkel defects. Simple statistical considerations will suffice to calculate defect concentrations and other properties. If an energy ε_v is required to transfer one atom from a lattice site inside the crystal to its surface, the energy cost for creating n such vacancies will be $\Delta U = n\varepsilon_v$. There will also be an entropy increase, which will be made up of a small vibrational contribution $n\Delta S_{vib}$ (where ΔS_{vib} is positive because neighboring atoms of the vacancy relax and decrease their vibrational frequency) and a configurational term ΔS_c. The latter is easily obtained by noting the number of ways that the n atoms

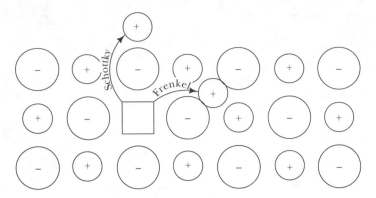

FIGURE 4-8 Schottky and Frenkel defects in an ionic crystal. The arrows indicate the displacement of the ions. In a Schottky defect the ion is moved to the crystal surface; in a Frenkel defect it goes into an interstitial position.
SOURCE: C. Kittel. *Introduction to Solid State Physics*, 3d ed. Wiley, 1966.

chosen to be removed may be picked out of N atoms originally on the perfect lattice, thus

$$\Delta S_c = k \ln \frac{N!}{(N-n)!n!} \qquad (4\text{-}75)$$

If we neglect the difference between G and A for a crystal, we find at constant pressure that the change of Gibbs free energy on going from the perfect to the disordered state is

$$\Delta G = \Delta A = \Delta U - T \Delta S$$

$$= n\varepsilon_v - nT \Delta S_{vib} - kT \ln \left(\frac{N!}{(N-n)!n!} \right) \qquad (4\text{-}76)$$

Use of the Stirling approximation gives

$$\Delta G = n\varepsilon_v - nT \Delta S_{vib} - kT[N \ln N - (N-n) \ln (N-n) - n \ln n] \quad (4\text{-}77)$$

Since for equilibrium ΔG is to reach a minimum, that is, decrease as much as possible,

$$\left(\frac{\partial \Delta G}{\partial n} \right)_{T,N} = 0 = \varepsilon_v - T \Delta S_{vib} - kT \ln \left(\frac{N-n}{n} \right)$$

For $n \ll N$ we obtain

$$\frac{n}{N} = e^{-\varepsilon_v/kT} e^{+\Delta S_{vib}/k} \qquad (4\text{-}78)$$

Since ΔS_{vib} is positive and small, we may obtain a minimum vacancy fraction by neglecting it. A typical value [7] for ε_v is 1.0 eV or 96 kJ·mol^{-1}, so this fraction is a little greater than 10^{-10} at 500 K, and a little greater than 10^{-5} at 1000 K.

The result in Eq. (4-78) applies to crystalline elements. For crystals of ionic compounds it is usually favorable energetically to form approximately equal numbers of positive and negative ion vacancies so as to keep the crystal electrostatically neutral on a local scale. The number of ways to form n separated *pairs* of positive and negative ion vacancies will then in good approximation be the square of the argument of the logarithm in Eq. (4-75) such that

$$\frac{n}{N} = e^{-\varepsilon_p/2kT}e^{+\Delta S_{vib}/k} \tag{4-79}$$

In Eq. (4-79) ε_p is the energy of formation of a *pair* and ΔS_{vib} is the average of the vibrational entropy increases for cations and anions adjacent to a vacancy.

For Frenkel (interstitial) defects the number of ways to pick n out of N lattice sites for a vacancy and simultaneously to pick n out of N' interstitial sites on which to put the displaced ion or atom leads to

$$\Delta S_c = k \ln \left\{ \left[\frac{N!}{(N - n)!n!} \right] \left[\frac{(N')!}{(N' - n)!n!} \right] \right\} \tag{4-80}$$

Then at equilibrium for n much less than both N and N'

$$n = (NN')^{1/2}e^{-\varepsilon_i/2kT}e^{\Delta S_{vib}/2k} \tag{4-81}$$

where ε_i is the energy necessary to remove a particle from a lattice site to an interstitial position. Since N' is some simple factor (not very different from unity) times N, depending on the geometry of the lattice, and since ΔS_{vib} is not generally known a priori, both Eq. (4-81) and Eq. (4-79) are of the form

$$\frac{n}{N} = (\text{const})e^{-\varepsilon/2kT} \tag{4-82}$$

To determine whether the defects in a compound are primarily of the Schottky or the Frenkel type one can heat the crystal to generate a high defect concentration and then quench it back to the original temperature with the defects frozen in. Production of Schottky type defects will lower the density of the crystal sample since removal of atoms to the surface increases the volume. Frenkel defects generally do not change the volume of the sample or its measurable density [2]. From such experiments it is found that for crystals of elements only Schottky type vacancies are important, whereas for ionic compounds one or the other type predominates depending on the radius ratio of the ions involved.

Exercise 4-5

In analogy to the derivation of Eq. (4-78), derive Eq. (4-79) and (4-81).

Of course the production of defects with increase of temperature contributes a term to the heat capacity of crystals at high temperature over and beyond that due to the lattice and possibly electronic contributions. This is discussed in Problems 21 and 22.

Appendix 4-A
Low-Temperature Expansion of Fermion Partition Function

The logarithm of the grand canonical ensemble partition function for ideal fermions is given by Eq. (4-60) and has as a factor the integral I, which we wish to expand at low temperature when $\gamma = \ln z$ is very large. From Eq. (4-61):

$$
\begin{aligned}
I &= \int_0^\infty x^{1/2} \ln \left(1 + e^{\gamma - x}\right) dx \\
&= \int_0^\gamma x^{1/2} \ln \left(1 + e^{\gamma - x}\right) dx + \int_\gamma^\infty x^{1/2} \ln \left(1 + e^{\gamma - x}\right) dx \\
&= \int_0^\gamma x^{1/2} \ln \left(1 + e^{\gamma - x}\right) dx + \int_0^\infty (y + \gamma)^{1/2} \ln \left(1 + e^{-y}\right) dy
\end{aligned}
\qquad (4.\text{A-}1)
$$

in which we have used $y = (x - \gamma)$ in the second integral. We also use the identities

$$
x^{1/2} \ln \left(1 + e^{\gamma - x}\right) = x^{1/2}(\gamma - x) + x^{1/2}[\ln \left(1 + e^{\gamma - x}\right) - (\gamma - x)]
\qquad (4.\text{A-}2)
$$

$$
\begin{aligned}
\exp \left[\ln \left(1 + e^{\gamma - x}\right) - (\gamma - x)\right] &= e^{-(\gamma - x)}(1 + e^{\gamma - x}) = e^{-(\gamma - x)} + 1 \\
&= \exp \left[\ln \left(1 + e^{x - \gamma}\right)\right]
\end{aligned}
\qquad (4.\text{A-}3)
$$

$$
\therefore \quad \ln \left(1 + e^{\gamma - x}\right) - (\gamma - x) = \ln \left(1 + e^{x - \gamma}\right)
\qquad (4.\text{A-}4)
$$

Then substituting into Eq. (4.A-1)

$$
\begin{aligned}
I &= \int_0^\gamma x^{1/2}(\gamma - x) \, dx + \int_0^\gamma x^{1/2} \ln \left(1 + e^{x - \gamma}\right) dx \\
&\quad + \int_0^\infty (y + \gamma)^{1/2} \ln \left(1 + e^{-y}\right) dy
\end{aligned}
\qquad (4.\text{A-}5)
$$

The first term may be integrated at once, and $y = (\gamma - x)$ used in the second integral to obtain

$$I = \frac{4}{15} \gamma^{5/2} + \int_0^\gamma (\gamma - y)^{1/2} \ln (1 + e^{-y}) \, dy$$

$$+ \int_0^\infty (\gamma + y)^{1/2} \ln (1 + e^{-y}) \, dy \qquad (4.A-6)$$

Since γ is very large we may extend the range of integration in the second term to ∞ as upper limit. We also use the expansion for $q \leq 1$

$$\ln (1 + q) = q - \frac{q^2}{2} + \frac{q^3}{3} - \frac{q^4}{4} + = \sum_{s=1}^\infty \frac{(-1)^{s-1}}{s} q^s \qquad (4.A-7)$$

and obtain

$$I = \frac{4}{15} \gamma^{5/2} + \sum_{s=1}^\infty \frac{(-1)^{s-1}}{s} \int_0^\infty e^{-sy} \, dy [(\gamma - y)^{1/2} + (\gamma + y)^{1/2}] \qquad (4.A-8)$$

We now expand the two square roots using $\gamma \gg y$ always, even though y goes to ∞ as upper limit, because the factor e^{-sy} damps out any contribution from regions of large y in any case. All odd powers of y/γ cancel:

$$[(\gamma - y)^{1/2} + (\gamma + y)^{1/2}] = \gamma^{1/2} \left[\left(1 - \frac{y}{\gamma} \right)^{1/2} + \left(1 + \frac{y}{\gamma} \right)^{1/2} \right]$$

$$= \gamma^{1/2} \left[2 - \frac{1}{4} \left(\frac{y}{\gamma} \right)^2 - \frac{5}{64} \left(\frac{y}{\gamma} \right)^4 + \ldots \right] \qquad (4.A-9)$$

We also use

$$\int_0^\infty y^{2n} e^{-sy} \, dy = \frac{(2n)!}{s^{2n+1}} \qquad (4.A-10)$$

We find

$$I = \frac{4}{15} \gamma^{5/2} + 2\gamma^{1/2} \sum_{s=1}^\infty \frac{(-1)^{s-1}}{s^2} - \frac{1}{2} \gamma^{-3/2} \sum_{s=1}^\infty \frac{(-1)^{s-1}}{s^4} + o(\gamma^{-7/2}) \qquad (4.A-11)$$

In general for the sums of inverse even powers $(n' = 1, 2, \ldots)$ alternating in sign

$$\sum_{s=1}^\infty \frac{(-1)^{s-1}}{s^{2n'}} = \sum_{s=0}^\infty \frac{1}{(2s+1)^{2n'}} - \sum_{s=1}^\infty \frac{1}{(2s)^{2n'}} \qquad (4.A-12)$$

where the first sum is for odd s with positive sign and the second sum takes care of even s with negative sign. Then the first sum on the right-hand side of Eq. (4.A-12) can be written for all s with positive signs less the sum of even s with positive sign. Thus

$$\sum_{s=1}^{\infty} \frac{(-1)^{s-1}}{s^{2n'}} = \sum_{s=1}^{\infty} \frac{1}{(s)^{2n'}} - \sum_{s=1}^{\infty} \frac{1}{(2s)^{2n'}} - \sum_{s=1}^{\infty} \frac{1}{(2s)^{2n'}}$$

$$= \left(1 - \frac{2}{2^{2n'}}\right) \sum_{s=1}^{\infty} \frac{1}{(s)^{2n'}} = \left(1 - \frac{2}{2^{2n'}}\right) \zeta(2n') \qquad (4.A\text{-}13)$$

where $\zeta(2n')$ are Rieman zeta functions (known sums—see Appendix II).

$$\zeta(2) = \sum_{s=1}^{\infty} \frac{1}{s^2} = \frac{\pi^2}{6}$$

$$\zeta(4) = \sum_{s=1}^{\infty} \frac{1}{s^4} = \frac{\pi^4}{90}$$

$$\zeta(6) = \sum_{s=1}^{\infty} \frac{1}{s^6} = \frac{\pi^6}{945} \qquad (4.A\text{-}14)$$

Hence the final result is

$$I = \frac{4}{15} \gamma^{5/2} + \gamma^{1/2} \zeta(2) - \frac{7}{16} \gamma^{-3/2} \zeta(4) + o(\gamma^{-7/2})$$

$$= \frac{4}{15} \gamma^{5/2} + \frac{\pi^2}{6} \gamma^{1/2} - \frac{7\pi^4}{1440} \gamma^{-3/2} + o(\gamma^{-7/2}) \qquad (4.A\text{-}15)$$

This is used in Eq. (4-62).

References and Recommended Reading

[1] A. H. Wilson. *Thermodynamics and Statistical Mechanics*. Cambridge Univ. Press, Cambridge, 1960. Chapter 6.

[2] C. Kittel. *Introduction to Solid State Physics*, 3d ed. Wiley, New York, 1966.

[3] G. N. Lewis and M. Randall. *Thermodynamics*, rev. by K. S. Pitzer and L. Brewer. McGraw-Hill, New York, 1961. Pp. 659–664.

[4] J. A. Reissland. *The Physics of Phonons*. Wiley, New York, 1973.

[5] M. Born and K. Huang. *Dynamical Theory of Crystal Lattices*. Oxford Univ. Press, Oxford, 1954.

[6] S. Mitra, in *Advances in Solid State Physics*, vol. 13, F. Seitz and D. Turnbull, eds. Academic Press, New York, 1962. P. 1.

[7] R. A. Swalin. *Thermodynamics of Solids*. Wiley, New York, 1962.

Problems

1. Derive a relationship (assuming the Debye limiting T^3 law holds for C_{V_m}) between the low-temperature molar entropy of a solid at T and its C_{V_m} at T. If C_{V_m} for iron at 32 K is 0.858 J·K^{-1}·mol^{-1}, what is its molar entropy and the limiting θ_D value for iron at 0 K?

2. The experimental C_{V_m} for diamond at 207 K is 2.68 J·K^{-1}·mol^{-1}. Determine the θ_D that will fit this datum and from it obtain θ_E and the calculated C_{V_m} from the Einstein theory.

3. Compare the high-temperature expansions of $C_{V_m}/3R$ for the Einstein and Debye theories using $\theta_E = 3/4\theta_D$ to match the zero point energy terms.

4. If an LJ potential is assumed between atoms in a crystal and only nearest neighbor interactions included, the lattice energy for the nearest neighbor distance r would be

$$\frac{N\phi(0)}{2} = \frac{Nz}{2}(4\varepsilon^*)\left[\left(\frac{\sigma}{r}\right)^{12} - \left(\frac{\sigma}{r}\right)^6\right]$$

where z is the coordination number of the lattice. Summing over all pairs (not just nearest neighbor) we would have

$$\frac{N\phi(0)}{2} = \frac{N}{2}(4\varepsilon^*)\left[\left(\frac{\sigma}{r}\right)^{12}\sum_j{}' \frac{1}{(p_{ij})^{12}} - \left(\frac{\sigma}{r}\right)^6\sum_j{}' \frac{1}{(p_{ij})^6}\right]$$

where $(p_{ij})r$ is the distance between a reference atom i and any other atom j expressed in terms of the nearest neighbor distance r. These lattice sums are rapidly convergent and for a face-centered cubic (FCC) lattice ($z = 12$) they are known to be

$$\sum_j{}' (p_{ij})^{-12} = 12.131 \qquad \sum_j{}' (p_{ij})^{-6} = 14.454$$

Clearly, effects of neighbors beyond the first are small since, including only the first nearest neighbors, both of these sums are exactly 12.

a. Using each of these forms (i.e., nearest neighbor only and full sum) predict the equilibrium $r = r_e$ for the crystal in terms of σ. The observed values of $10^{10} r_e/$m for Xe, Kr, and Ar (which form FCC crystals) are 4.34, 4.02, and 3.77, respectively. Use σ values from Table 3-5 and comment on your results.

b. Obtain an expression for $[N\phi(0)/2]_{eq}$ using the lattice sum forms in terms of (ε^*/k) and calculate this quantity for the inert gas solids.

5. For a Debye solid use equations analogous to Eqs. (4-18) and (4-19) to derive expressions for μ_c and P.

6. a. Show that the equation of state for a Debye solid obtained in Problem 5 can be written as

$$P = -\left(\frac{\partial}{\partial v}\right)_{N,T}\left[\frac{U_c(0)}{N}\right] - \frac{\theta'_D}{\theta_D}\left[\frac{U_c - U_c(0)}{N}\right]$$

where $\theta'_D = (\partial\theta_D/\partial v)_{N,T}$ and $U_c(0)$ is the temperature-independent energy of the Debye solid at 0 K.

b. Defining the Grüneisen "constant" γ_G by

$$\gamma_G = -\left(\frac{\partial \ln \theta_D}{\partial \ln V}\right)_{N,T}$$

show that the equation of state becomes

$$PV + V\left(\frac{\partial}{\partial V}\right)_{N,T} U_c(0) = \gamma_G[U_c - U_c(0)]$$

c. Assuming γ_G is independent of temperature, as follows from the Debye theory, show that

$$\gamma_G = \frac{\alpha V_m}{\beta_T C_{V_m}} = V_m\left(\frac{\partial P}{\partial U_{c_m}}\right)_V$$

where α is the coefficient of thermal expansion for the lattice

$$\alpha = \frac{1}{V}\left(\frac{\partial V}{\partial T}\right)_{P,N}$$

and β_T is the isothermal compressibility

$$\beta_T = -\frac{1}{V}\left(\frac{\partial V}{\partial P}\right)_{T,N}$$

Hint: First show $\alpha = \beta_T(\partial P/\partial T)_{V,N}$ and use the equation of state from **b**.

d. What is the formal value of γ_G in the Debye theory if the speed of sound, v_s, is assumed to be independent of $v = (V/N)$? Note: It is found that γ_G is generally constant for most solids for $T > 0.2\theta_0$, where θ_0 is the θ_D value at 0 K. Its value is generally in the range 1.5 to 3.0.

7. The directly measurable heat capacity for a solid is C_P not C_V. Using the following data for copper, obtain C_V as a function of T and by interpolation in Table 4-1 obtain the apparent θ_D at each temperature. Comment on the highest temperature results that give C_{V_m} greater than $3R$.

T/K	$C_P/\text{J}\cdot\text{K}^{-1}\cdot\text{mol}^{-1}$	$\left[-\dfrac{1}{V}\left(\dfrac{\partial V}{\partial P}\right)_T\right]10^7/\text{atm}^{-1}$	$\left[\dfrac{1}{V}\left(\dfrac{\partial V}{\partial T}\right)_P\right]10^6/\text{K}^{-1}$	$V_m/\text{cm}^3/\text{mol}$
50	6.25	7.22	11.4	7.00
100	16.1	7.30	31.5	7.01
150	20.5	7.44	40.7	7.02
200	22.8	7.59	45.3	7.03
250	24.0	7.73	48.3	7.04
300	24.5	7.88	50.4	7.06
500	25.8	8.50	54.9	7.12
800	27.7	9.35	60.0	7.26
1200	30.2	10.44	70.2	7.45

SOURCE: M. W. Zemansky and R. H. Dittman *Heat and Thermodynamics*, 6th ed. McGraw-Hill, 1981.

8. From the data in Problem 7 for Cu, calculate the Grüneisen constant γ_G for Cu for temperatures from 50 to 500 K.

$$\gamma_G = \frac{\alpha V_m}{\beta_T C_{V_m}}$$

From the experimental data, how constant is the ratio α/C_{V_m}?

9. Take θ_D of solid benzene as 150 K. Look up the values of the 30 normal modes of benzene (10 are doubly degenerate) as observed in the gas phase and use Eq. (4-49) to calculate C_{V_m} of solid benzene at 200 K. R. C. Lord et al. [*J. Chem. Phys.* **5**, 649 (1937)] use $C_{P_m} - C_{V_m} = 14.8\,\text{J}\cdot\text{K}^{-1}\cdot\text{mol}^{-1}$ for solid benzene at 200 K. Compare your result with the experimental $C_{P_m} = 83.7\,\text{J}\cdot\text{K}^{-1}\cdot\text{mol}^{-1}$ at 200 K.

10. By equating the chemical potential of the solid with that of the equilibrium gas one can obtain an equation for the sublimation pressure of a solid. Do this for a crystal of an *element* using the Einstein theory and results from Chap. 2 for a gas, in particular for a monatomic gas.

11. Apply the equation from Problem 10 to the case of neon using results obtained in Problem 4 to calculate the triple point pressure of neon at 24.6 K. (The triple point is on the sublimation curve.) The experimental result is 324 torr.

12. Derive low-temperature power series expansions for G, A, and S for a free electron gas. Note that the entropy goes to zero as $T \to 0$. From your result for S obtain C_V to check the result of Eq. (4-69).

13. From the result of Example 3 of Chap. 1 show that T_F is the temperature for which the number of particle states between lowest energy and kT_F is just equal to N, the number of particles present. Assume spin $\frac{1}{2}$ particles.

14. If a cubic piece of metal of side a has N_A free e^-, list a few sets of quantum numbers compatible with the highly degenerate highest occupied energy level at 0 K.

15. From data given in Example 4-3 for K, below what theoretical temperature will the lattice contribution to C_{V_m} become less than the electronic contribution? What is the actual temperature at which this occurs?

16. As indicated from data on Cu at 25°C in Problem 7, $\theta_D = 292$ K. Calculate the molar entropy of Cu at 25°C using the Debye theory and the Einstein theory including the small electronic contribution. The experimental value is $33.3 \text{ J} \cdot \text{K}^{-1} \cdot \text{mol}^{-1}$.

17. This is an exercise to see how close the low-density, high-temperature expansion for fermions given in Chap. 1 approaches the high-density, low-temperature expansion for fermions given in this chapter. We concentrate on the C_V function.
   a. Since the equations in Chap. 1 are for spinless fermions, we must modify them for the spin $\frac{1}{2}$ case. Start from Eq. (1-155) and use the methods of Problem 22 of Chap. 1 to show that for fermions of spin $\frac{1}{2}$

$$\frac{S}{Nk} = \left(\frac{S}{Nk}\right)_{cl} + \ln 2 + \frac{\lambda^3 \rho}{2^{9/2}}$$

   where $(S/Nk)_{cl}$ is the classical particle result.
   b. From **a** derive an expression for C_V/Nk for the low-density, high-temperature case, keeping the first correction term only.
   c. Show that $\lambda^3 \rho = (1.5045)(T_F/T)^{3/2}$.
   d. The high-temperature, low-density expansion is actually accurate at $T = T_F$ and even at $T = T_F/2$. Why?
   e. What does the low-temperature, high-density expansion of Eq. (4-69) for C_V give at $T = \frac{1}{2} T_F$? Is it accurate at so high a temperature?
   f. Plot C_V/Nk as a function of T/T_F from zero to 2 using a dotted-line estimated curve to connect the results of the two series in the region where neither is accurate. [Accurate numerical results for this region can be obtained from tables published by McDougall and Stoner, *Phil. Trans. Royal Soc. (London)*, **A237**, 67 (1938).]

18. Calculate the Fermi temperature and Fermi energy in electron volts and the molar entropy for
   a. Electrons in Ca metal at 20°C ($\rho' = 1.55 \text{ g/cm}^3$).
   b. Nucleons in a ^{235}U nucleus at 25°C (assume the nucleus is a sphere of radius 1.0×10^{-15}m).
   c. ^{3}He atoms in liquid ^{3}He at its normal boiling point of 3.22 K ($\rho' = 0.0568 \text{ g/cm}^3$). Assume in each case that the fermions are ideal and of spin $\frac{1}{2}$. For **c** refer to Problem 17.

19. From information in Problem 6 and using similar methods on the pressure equation for free electrons, show that the thermal expansion coefficient of a metal at low temperature is given by

$$\alpha = aT + bT^3$$

where the term linear in T is the electronic contribution. What is the theoretical value of a for free electrons in terms of T_F?

20. Using results in Problems 6, 7, and 8, show that at high temperatures the lattice heat capacity at constant pressure may in good approximation be written as $C_P = C_1 + C_2 T$, where C_1 and C_2 are constants. What is the value of C_1 on a molar basis?

21. Show from Eq. (4-82) that at high temperature there is an added contribution to the C_{P_m} of a crystal due to Schottky or Frenkel defects of the form

$$\Delta C_{P_m} = \frac{C\varepsilon_m^2}{2RT^2} e^{-\varepsilon_m/2RT}$$

where ε_m is the energy needed to form a mole of Schottky pair vacancies or a mole of Frenkel interstitials and C is a constant.

22. A. W. Lawson and R. W. Christy [*J. Chem. Phys.* **19**, 517 (1951)] used for the high-temperature lattice C_{P_m} of AgBr the form

$$C_{P_m}, [\text{J}\cdot\text{K}^{-1}\cdot\text{mol}^{-1}] = 6R + (0.021)\,T$$

Their high-temperature C_{P_m} values for this crystal are

T/K	$C_{P_m}/\text{J}\cdot\text{K}^{-1}\cdot\text{mol}^{-1}$	T/K	$C_{P_m}/\text{J}\cdot\text{K}^{-1}\cdot\text{mol}^{-1}$
378	57.7	575	71.1
426	59.0	599	77.4
452	60.7	622	85.8
503	63.2	646	98.7
526	65.7	670	117.0
551	68.6	687	137

AgBr melts at 694 K.

Frenkel defects, exclusively out of the lattice of smaller Ag^+ ions, give rise to the increased C_{P_m} in this case. With reference to Problem 21, use an appropriate least-squares treatment of these data to obtain ε_{i_m} and ΔS_{vib_m} for the Frenkel defects. Also calculate the fraction of interstitial Ag^+ ions at each temperature. Hint: Fit $T^2(\Delta C_{P_m})$ to a best straight line. Assume the number of interstitial sites $\mathcal{N}' = $ number of lattice sites $\mathcal{N}$.

5

Kinetic Theory and Transport Properties of Gases

Statistical mechanics provides microscopic classical or quantum mechanical models for calculating the properties of systems at equilibrium. Kinetic theory has a more ambitious goal: to provide microscopic mechanical theory to describe the approach of systems to equilibrium and to calculate some of their nonequilibrium properties such as thermal conductivity or viscosity. Even for dilute gases of noninteracting classical hard spheres, treated in this chapter, it is only practical to consider steady state transport for which the velocity distribution function is very nearly that for equilibrium (i.e., Maxwellian) as given by Eq. (1-192). A steady state flow of energy or particles in a system because of a constant gradient of temperature or concentration is by definition a situation with no time-dependent parameters. However it is *not* an equilibrium situation because such a system is not isolated and must be connected to energy or particle reservoirs. These reservoirs will themselves be either slowly changing or they must be continually worked upon and thus are *not* in equilibrium.

To derive exact results even for hard sphere particles is very complex and will not be attempted here. However we find it valuable to use a simple mean free path approach, which gives good physical insight into molecular behavior and which can be extended in Chap. 6 to more complicated cases where more rigorous methods are quite unmanageable. Indeed it is found that approximate calculations of this sort give the correct dependence of viscosity, thermal conductivity, and other characteristics on temperature and pressure for hard spheres and with numerical coefficients that deviate sometimes less than 50 percent from the rigorous results in those cases for which they are known.

5-1

Ideal Gas Pressure and Molecular Effusion

We consider in Fig. 5-1 a unit area of wall, $\int dS = 1$, in contact with a volume of gas of uniform number density, $\rho = \mathcal{N}/V$. We show the volume from which molecules moving in a specified θ, ϕ direction with speed c will strike the surface in unit time. From Eq. (1-192) the differential number of molecules that will strike the wall in unit time from this direction is just the number per unit volume in that direction times the volume of the figure shown of base unity and slant height $(c)(1s)$ [i.e., of volume $(1)c(1) \cos \theta$]. Thus

$$dN = \rho(c \cos \theta) \left(\frac{m}{2\pi k T} \right)^{3/2} e^{-mc^2/2kT} (c^2 \, dc \, \sin \theta \, d\theta \, d\phi) \tag{5-1}$$

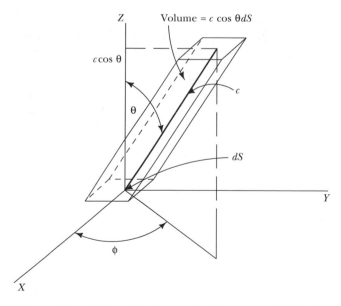

FIGURE 5-1 Volume from which molecules coming from a specified θ, ϕ direction (relative to the z axis, which is perpendicular to the unit surface area) with a speed c strike the surface in unit time.

In order to obtain the total number striking from all directions and traveling with all possible speeds c we must integrate Eq. (5-1) over all c and over the *hemisphere* of directions (ϕ from 0 to 2π, but θ only from 0 to $\pi/2$ since for θ in the range $\pi/2$ to π the molecules are moving away from the wall surface area). Using Eqs. (1-195), (1-196), and (1-197) and noting that

$$\int_0^{\pi/2} \cos \theta \sin \theta \, d\theta = \frac{1}{2}$$

we have

$$Z_{wall} = \frac{\rho \bar{c}}{4} = \frac{\rho}{4} \left(\frac{8kT}{\pi m} \right)^{1/2} \tag{5-2}$$

where $\bar{c}$ is the average speed of the molecules. This formula for the total number of surface collisions per unit area per unit time has many important applications. We have already used it in discussing blackbody radiation in Sec. 2-7.

A slight modification of our treatment will give us the ideal gas pressure from kinetic theory. At each wall collision the component of momentum normal to the wall, namely $mc \cos \theta$, is reversed such that the change of momentum per unit time on unit area, which is just the pressure, is

$$dp = \rho(c \cos \theta)(2mc \cos \theta) \left(\frac{m}{2\pi kT} \right)^{3/2} e^{-mc^2/2kT}(c^2 \, dc \, \sin \theta \, d\theta \, d\phi) \tag{5-3}$$

Integrating again over the hemisphere of directions and over c, and using

$$\int_0^{\pi/2} \cos^2 \theta \sin \theta \, d\theta = \frac{1}{3}$$

we have from Eq. (1-198)

$$P = \tfrac{1}{3}\rho m\bar{c^2} = \rho kT \tag{5-4}$$

which is the ideal gas law.

Exercise 5-1

Use the kinetic theory result in Eq. (5-4) to obtain Avogadro's principle: that equal volumes of any two gases in the ideal gas state at the same pressure and temperature contain equal numbers of molecules.

We have entirely neglected collisions between molecules in obtaining Eqs. (5-2) and (5-4). This is justified because we assumed equilibrium, and at equilibrium the Maxwell distribution is exact for classical particles and changes caused by individual collisions cancel each other out on the average. For every molecule that is knocked out of movement in any particular θ, ϕ direction with speed c there must be on average another one that is knocked into this range by another collision. This is one example of the principle of *detailed balance*. It is derivable from the fact that the equations of both classical and quantum mechanics are unchanged when time is reversed and so at equilibrium, when behavior is unchanged under time reversal, it is equally likely that a molecule be knocked out of or into a particular θ, ϕ direction. An extended discussion of this principle is given by Tolman [1].

If a sufficiently small hole or slit is made in the wall of a container of gas the equilibrium inside will be only negligibly disturbed. Then the rate of molecular escape through the hole will be given by $(Z_{wall})a$, where a is the area of the hole. This process is called molecular effusion. By sufficiently small is meant that the hole diameter or slit width d must be less than the mean free path λ in the contained gas (see Sec. 5-2). If this condition is not met and $d > \lambda$, the molecules left behind will experience a net force driving them out through the hole because they cannot collide with the molecules that have already left (when $d < \lambda$ such collisions would be negligibly few) and the resultant flow is collective, that is, "hydrodynamic," not effusive. (See the discussion in Appendix 5.B.) Assuming that the effusive condition holds for a hole of area a between two containers at pressures P_1 and P_2, respectively, and common temperature T, the net rate of particle loss from container 1 will be given by

$$-\frac{dN_1}{dt} = (Z_{wall_1} - Z_{wall_2})a = \frac{(P_1 - P_2)a}{(2\pi mkT)^{1/2}} \tag{5-5}$$

in which we have used ρ from the ideal gas law in Eq. (5-2). This relation is the basis for Knudsen's effusion method for the experimental determination of very low vapor pressures, such as those of solids for which ordinary techniques fail. It is necessary to know the effective molecular weight of the vapor species to use this method.

Example 5-1

L. Brewer and coworkers in using the Knudsen method for graphite found that at a temperature of 2603 K, 0.648 mg of carbon vapor passed through a hole of 3.25 mm² area in 3.50 hours into a second chamber kept at high vacuum by pumping. Assuming the vapor phase to be monatomic (which may not be correct and would have to be checked by mass spectrometry), calculate the vapor pressure of graphite at 2603 K.

Solution:

We rewrite Eq. (5-5) on a molar basis by dividing by N_A, using n_1 = moles of species:

$$-\frac{dn_1}{dt} = \frac{(P_1 - P_2)a}{(2\pi MRT)^{1/2}}$$

Here P_1 is the vapor pressure of graphite, which does not change as molecules effuse since more graphite will evaporate to maintain equilibrium inside, and $P_2 \simeq 0$ by vacuum pumping. Thus $-dn_1/dt$ is constant, and assuming the vapor is monatomic it is equal to $(6.48 \times 10^{-4})/(12.0)(3.50)(3600) = 4.29 \times 10^{-9}$ mol·s^{-1}. With $a = 3.25 \times 10^{-6}$ m^2

$$P_1 = \frac{4.29 \times 10^{-9}}{3.25 \times 10^{-6}} \left[2\pi \left(\frac{12.0}{1000} \right) (8.314)(2603) \right]^{1/2}$$

$$= 5.33 \times 10^{-2} \text{ N/m}^2 = 5.26 \times 10^{-7} \text{ atm} = 4.00 \times 10^{-4} \text{ torr}$$

■

If a source of effusing molecules is maintained in a vacuum and the effusing molecules are collimated further by slits as shown in Fig. 5-2, one obtains a well-defined molecular beam. Such beams have been extensively studied [2] since they are a source of practically collision-free molecules with negligible molecule-molecule interaction. In fact one may pick out from such a beam only those molecules with a particular fixed velocity component along the beam direction by use of a velocity selector. In its simplest form such a selector is a pair of disks mounted on a common axis that can be rotated at a known rate; narrow slots cut into these disks are displaced by $\theta°$ from each other. Then only those molecules can get through the pair whose time of flight l/c between the disks separated by distance l along the axis is equal to the rotation time $\theta/360v$ where v is the rotation rate of the axis in units of time^{-1}. By using different v values, different velocity components c (i.e., different one-dimensional speeds along the beam axis) may be selected. See Problem 2.

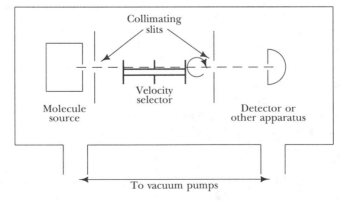

FIGURE 5-2 General experimental arrangement for generation and study of a molecular beam formed by effusing molecules.

Example 5-2

Discuss the distribution of speeds in an effusive molecular beam and show that if one normalizes this distribution to unity, at the most probable speed it has the form

$$\left(\frac{c}{c_m}\right)^3 e^{3/2} e^{-(3/2)(c/c_m)^2}$$

where c_m, the most probable speed is given by $c_m = (3kT/m)^{1/2}$

Solution:

The speed distribution in the beam is the speed-weighted Maxwell distribution given in Eq. (5-1) that is characteristic of the temperature of the source reservoir. It is not the same as the speed distribution inside the source because it depends on a factor c^3 rather than the factor c^2 of the Maxwell distribution. Hence we may write it with a constant of proportionality K as

$$K e^{-mc^2/2kT} c^3$$

This has a maximum for a c_m such that

$$\frac{d}{dc}\left(Kc^3 e^{-mc^2/2kT}\right) = e^{-mc_m^2/2kT}\left(3c_m^2 - \frac{m}{kT}c_m^3\right) = 0$$

$$c_m = \left(\frac{3kT}{m}\right)^{1/2}$$

If this form is to be unity for $c = c_m$

$$K = c_m^{-3} e^{mc_m^2/2kT} = c_m^{-3} e^{+3/2}$$

and the distribution becomes

$$\left(\frac{c}{c_m}\right)^3 e^{3/2} e^{-(3/2)(c/c_m)^2}$$

This curve is shown in Fig. 5-12 for helium at 300 K, for which $c_m = 1.37$ km/s.

5-2
Molecular Collisions and Mean Free Path

We now calculate the frequency of intermolecular collisions, starting with a single-component gas considered as hard spheres of diameter σ. The cylindrical volume swept out by a single such molecule moving with average speed $\bar{c}$ in unit time is

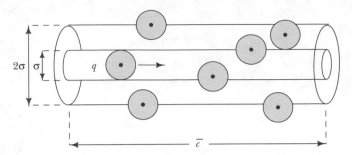

FIGURE 5-3 The collision cylinder, $\pi\sigma^2\bar{c}$, swept out by molecule q traveling a distance $\bar{c}$ in unit time.

$\pi(\sigma/2)^2\bar{c}$. A collision occurs whenever the distance between molecular centers becomes as small as σ. To start we treat all the other molecules as frozen in place. Then the one moving molecule will hit all those whose centers lie in a volume of a cylinder with diameter 2σ, as shown in Fig. 5-3. In other words we may further imagine the stationary molecules to be points and let the moving molecule have radius σ. Thus the number of collisions per unit time of the one moving molecule is equal to the number of stationary molecules in its collision cylinder, $\pi\sigma^2\bar{c}\rho$. In actuality, since all molecules are moving the average of the relative speed, $\bar{c}_r$, should be used for this calculation. In Fig. 5-4 some examples of the relative speed of two

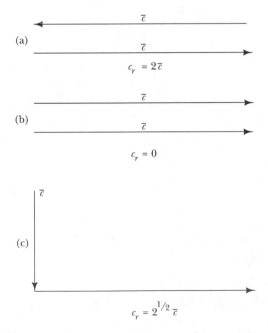

FIGURE 5-4 Examples of the relative speed of two molecules. (a) Same average speed, at angle of 180°. (b) Same average speed, at angle of 0°. (c) Same average speed, at angle of 90°.

molecules are shown. If an exact average [3] of all relative speeds is computed it turns out to be that characteristic of the 90° angle:

$$\bar{c}_r = \sqrt{2}\bar{c} \qquad (5-6)$$

Thus the exact number of collisions of one hard sphere molecule in unit time,* z_{11}, is

$$z_{11} = \sqrt{2}\pi\sigma^2\rho\bar{c} \qquad (5-7)$$

Furthermore, the total number of collisions for all the like molecules in unit volume in unit time, Z_{11}, is

$$Z_{11} = \frac{z_{11}\rho}{2} = \frac{\sqrt{2}}{2}\pi\sigma^2\rho^2\bar{c} = 2\rho^2\sigma^2\left(\frac{\pi kT}{m}\right)^{1/2} \qquad (5-8)$$

Division by 2 is needed so that collisions of A with B and then of B with A are not counted as two collisions.

The mean free path, λ, of a molecule is defined as the average distance a molecule travels between two successive collisions. Since $(1/z_{11})$ is the time of free flight between two collisions for a molecule traveling with speed $\bar{c}$

$$\lambda = \frac{\bar{c}}{z_{11}} = \frac{1}{\sqrt{2}\pi\sigma^2\rho} = \frac{kT}{\sqrt{2}\pi\sigma^2 P} \qquad (5-9)$$

If the number density is constant, the mean free path is independent of temperature. However at constant temperature the mean free path is inversely proportional to the pressure. Ideal hard sphere gas kinetic theory will only be accurate if $\lambda \gg \sigma$. Several tabulations of σ values for molecules are given in Chap. 3. In Sec. 5-4 we will show how σ values are obtained from gas viscosity measurements.

Exercise 5-2

Show that at 1 atm and 0°C λ is of order 10^{-7} m for a typical σ value (300 pm).

The free path lengths of a molecule certainly vary randomly between successive collisions. Let us calculate the probability that a single molecule travels a distance

---

* We use a double subscript to fit the notation used in what follows for collisions between unlike molecules.

l without a collision. This is identical to the fraction N/N_0 of an original beam of N_0 molecules shot into the gas at some $t = 0$ in a given direction that are still moving in that direction at time t, when they have gone the distance $\bar{c}t = l$. In the time interval dt the number colliding and dropping out of this beam is

$$dN = -Nz_{11}\, dt$$

such that

$$\int_{N_0}^{N} \frac{dN}{N} = -z_{11} \int dt$$

and

$$\frac{N}{N_0} = e^{-z_{11}t} = e^{-z_{11}l/\bar{c}} \tag{5-10}$$

Thus P_l, the probability for a molecule to go as far as l without a collision, is

$$P_l = \frac{N}{N_0} = e^{-l/\lambda} \tag{5-11}$$

in which we have used Eq. (5-9). The distance l at which a given molecule in the beam makes its collision is its free path. Hence the mean free path is found by multiplying l by the number colliding between l and $l + dl$, integrating over all l, and dividing by the total number of molecules in the beam

$$\bar{l} = \frac{1}{N_0} \int_0^\infty l \left(-\frac{dN}{dl} \right) dl = \frac{1}{N_0} \int_0^\infty l \left(\frac{N_0 e^{-l/\lambda}}{\lambda} \right) dl$$

$$= \int_0^\infty \frac{l e^{-l/\lambda}}{\lambda}\, dl = \lambda \tag{5-12}$$

This result shows the consistency of Eq. (5-11). We notice that the fraction of the molecules that go as far as λ is only $e^{-1} = 0.368$, such that more than half have collided before getting as far as one mean free path. Nevertheless, λ *is* the mean free path because, of those that reach λ some go very far: 0.368 of them will get to 2λ, 0.135 to 3λ, and so on.

When we deal with a gas mixture of species 1 and 2 it may be shown by exact integration over relative speeds [3] that the total number of collisions between these species in unit volume in unit time, Z_{12}, is given by

$$Z_{12} = \rho_1 \rho_2 \sigma_{12}^2 \left(\frac{8\pi kT}{\mu} \right)^{1/2} \tag{5-13}$$

where μ is the reduced mass of the pair

$$\mu = \frac{m_1 m_2}{m_1 + m_2} \tag{5-14}$$

and

$$\sigma_{12} = \tfrac{1}{2}(\sigma_1 + \sigma_2) \tag{5-15}$$

Notice that if 1 and 2 are identical the $\mathcal{Z}_{12}$ formula will count all collisions twice and must be divided by 2 in order to regain Eq. (5-8) for $\mathcal{Z}_{11}$.

The number of collisions per second between a given type 1 molecule with type 2 molecules is found by dividing Eq. (5-13) by ρ_1:

$$z_{12} = \rho_2 \sigma_{12}^2 \left(\frac{8\pi k T}{\mu} \right)^{1/2} \tag{5-16}$$

Similarly, the number of collisions per second between a given type 2 molecule with those of type 1 is

$$z_{21} = \rho_1 \sigma_{12}^2 \left(\frac{8\pi k T}{\mu} \right)^{1/2} \tag{5-17}$$

The first or left subscript refers to the given molecule involved. Both of these equations reduce to z_{11} given by Eq. (5-7) when 1 and 2 are identical.

The total number of collisions between a given type 1 molecule and all other kinds of molecules in a mixture of $i \geq 2$ species is given by

$$z_1 = \sum_i z_{1i} = \sum_i \rho_i \sigma_{1i}^2 \left(\frac{8\pi k T}{\mu_{1i}} \right)^{1/2} \tag{5-18}$$

Thus in a gaseous mixture the mean free path of a molecule of type 1 is

$$\lambda_1 = \frac{\bar{c}_1}{z_1} \tag{5-19}$$

where $\bar{c}_1$ is the average speed of molecule 1.

Example 5-3

For σ in pm, pressure in atm, and M in g·mol^{-1} derive expressions with appropriate numerical coefficients for λ in m, z_{11} in s^{-1}, and $\mathcal{Z}_{11}$ in s^{-1}·m^{-3}. For $T = 300$ K, $M = 28.0$ g/mol (N$_2$), and $\sigma = 380$ pm, evaluate these expressions for $P/\text{atm} = 100$, 10, 1, 10^{-4}, 10^{-6}, 10^{-10} and comment on the pressure range in this case, for which simple kinetic theory for λ should be accurate.

Solution:

From Eq. (5-9)

$$\lambda = \frac{(1.38 \times 10^{-23}\ \mathrm{N \cdot m \cdot K^{-1}})(T/\mathrm{K})}{\sqrt{2}\pi(\sigma/\mathrm{pm} \times 10^{-12}\ \mathrm{m/pm})^2(P/\mathrm{atm} \times 1.013 \times 10^5\ \mathrm{N \cdot m^{-2}/atm})}$$

$$\frac{\lambda}{\mathrm{m}} = \frac{(3.068 \times 10^{-5})(T/\mathrm{K})}{(\sigma/\mathrm{pm})^2(P/\mathrm{atm})}$$

From Eq. (5-7) using $\rho = P/kT$

$$z_{11} = \frac{\sqrt{2}\pi(\sigma/\mathrm{pm} \times 10^{-12}\ \mathrm{m/pm})^2(P/\mathrm{atm} \times 1.013 \times 10^5\ \mathrm{N \cdot m^{-2}/atm})}{(1.381 \times 10^{-23}\ \mathrm{N \cdot m \cdot K^{-1}})(T/\mathrm{K})}$$

$$\times\ (8RT/(\pi M/1000))^{1/2}$$

$$\frac{z_{11}}{\mathrm{s}^{-1}} = \frac{(4.742 \times 10^6)(P/\mathrm{atm})(\sigma/\mathrm{pm})^2}{(T/\mathrm{K})^{1/2}(M/\mathrm{g \cdot mol^{-1}})^{1/2}}$$

From Eq. (5-8)

$$Z_{11} = \left(\frac{P/\mathrm{atm} \times 1.013 \times 10^5\ \mathrm{N \cdot m^{-2}/atm}}{(1.38 \times 10^{-23}\ \mathrm{N \cdot m \cdot K^{-1}})(T/\mathrm{K})}\right)^2$$

$$\times\ (\sigma/\mathrm{pm} \times 10^{-12}\ \mathrm{m/pm})^2\left(\frac{\pi RT}{(M/1000)}\right)^{1/2}$$

$$\frac{Z_{11}}{\mathrm{m}^{-3} \cdot \mathrm{s}^{-1}} = \frac{(1.739 \times 10^{34})(P/\mathrm{atm})^2(\sigma/\mathrm{pm})^2}{(T/\mathrm{K})^{3/2}(M/\mathrm{g \cdot mol^{-1}})^{1/2}}$$

For $T = 300$ K, $M = 28.0$ gmol^{-1}, $\sigma = 380$ pm these become

$$\frac{\lambda}{\mathrm{m}} = \frac{6.37 \times 10^{-8}}{(P/\mathrm{atm})} \qquad \frac{Z_{11}}{\mathrm{m}^{-3} \cdot \mathrm{s}^{-1}} = 9.13 \times 10^{34}\ (P/\mathrm{atm})^2$$

$$\frac{z_{11}}{\mathrm{s}^{-1}} = (7.47 \times 10^9)(P/\mathrm{atm})$$

For the pressure values given

P/atm	λ/m	Z_{11}/m$^{-3} \cdot$s^{-1}	z_{11}/s^{-1}
100	6.37×10^{-10}	9.13×10^{38}	7.47×10^{11}
10	6.37×10^{-9}	9.13×10^{36}	7.47×10^{10}
1	6.37×10^{-8}	9.13×10^{34}	7.47×10^{9}
10^{-4}	6.37×10^{-4}	9.13×10^{26}	7.47×10^{5}
10^{-6}	6.37×10^{-2}	9.13×10^{22}	7.47×10^{3}
10^{-10}	6.37×10^{2}	9.13×10^{14}	7.47×10^{-1}

For kinetic theory to be accurate in calculating λ and quantities dependent on λ, the condition is $\sigma \ll \lambda \ll L$, where L is some minimum linear dimension of the container of the gas. At sufficiently low pressure the gas molecules will effectively only collide with the walls of the container and the actual λ will approach the value L. Assuming the gas container has no minimum linear dimension much larger than 10 cm, Eq. (5-9) for λ should be accurate for pressures from 10 to 10^{-5} atm. In the very high pressure (100 atm) case the real λ must be much less than the 637 pm given by Eq. (5-9).

Exercise 5-3

Calculate from the ideal gas law the average separation between gas molecules at 300 K and 100 atm, which is given by $(V/N)^{1/3}$. What is the relevance of this result to the λ value obtained in Example 5-3 for $P = 100$ atm?

In what follows it will be useful to know the average perpendicular distance between any arbitrarily chosen plane in a gas and the point of last collision before crossing that plane for an average molecule. The average distance a molecule has traveled since its last collision is its mean free path, λ. However, only for those molecules moving in a direction perpendicular to the plane will the perpendicular distance be λ. For other directions the perpendicular distance is $\lambda \cos \theta$, as shown in Fig. 5-5. To obtain the average perpendicular distance advanced, $\bar{l}_{\perp}$ we multiply $\lambda \cos \theta$ by the differential fraction of the molecules coming from the θ, ϕ direction (in unit time for unit area on the plane), which is given by Eq. (5-1) divided by Eq. (5-2), and integrate over all speeds and directions that lead toward the plane, that is, over a hemisphere:

$$\bar{l}_{\perp} = \frac{1}{(\rho \bar{c}/4)} \int_0^{\infty} c^2 \, dc \int_0^{\pi/2} \sin \theta \, d\theta$$

$$\times \int_0^{2\pi} d\phi \, \rho(c \cos \theta) \left(\frac{m}{2\pi k T}\right)^{3/2} e^{-mc^2/2kT} (\lambda \cos \theta)$$

$$\bar{l}_{\perp} = \frac{\lambda}{\pi} \int_0^{\pi/2} d\theta \cos^2 \theta \sin \theta \int_0^{2\pi} d\phi = \frac{2}{3} \lambda \qquad (5\text{-}20)$$

The usefulness of this result is exemplified by the discussion associated with Fig. 5-6.

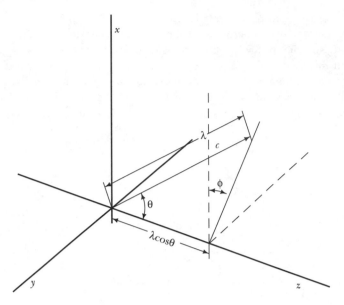

FIGURE 5-5 A molecule that has moved a distance λ in reaching an arbitrary plane (outlined with dashes) in a gas from the θ, ϕ direction has moved $\lambda \cos \theta$ in the direction perpendicular to the plane.

5-3

Transport Properties of Gases

We are now ready to discuss steady state transport of a general property, $\rho_z Q_z$, which is proportional to the number density at location z. We assume that ρQ depends only on the z coordinate and has a time-independent gradient $(d/dz)(\rho Q_z)$. We also assume that the gradient is small enough so that $\rho_z Q_z$ does not change much in a distance of the order of a mean free path. This will be correct as long as the density is not so low that the mean free path becomes equal to some minimum dimension of the apparatus. Our final assumption is a crude heuristic one of local equilibrium in steady state flow. That is, we assume that every molecule forgets its past history whenever it undergoes a collision, and that it acquires on the average the value of Q characteristic of the z value of the plane perpendicular to the z axis at which it makes this collision. Hence as a molecule moves along its next free path it carries a sample of the property Q of the gas from the neighborhood of its last collision. If there is a gradient of Q when the molecule makes its next collision, it will gain or lose in Q from or to the molecules in the neighborhood of the new collision. (These collisions will eventually eliminate any gradient in Q unless it is continually maintained by external processes.)

By flux we mean the magnitude of the property Q passing through unit area of a plane in unit time. Strictly, flux is a vector quantity, and for steady state flows we take the flux as positive if there is a net flow of Q in the positive z direction and negative if there is a net flow in the negative z direction. The net flux of Q

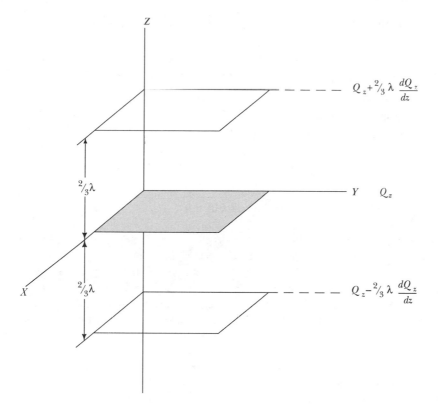

FIGURE 5-6 In a gas in which a gradient in the property Q exists along the z axis, there will be a net flux of Q through any plane perpendicular to the z axis. Molecules crossing any particular plane, such as the shaded one, will on the average have undergone their last collision at a z value equal to that of the plane $\pm \frac{2}{3}\lambda$, depending on whether they are moving downward or upward, respectively.

through any plane in the gas perpendicular to the z axis, $\mathcal{J}_z(Q)$, may be formulated with reference to Fig. 5-6. As shown by the result of Eq. (5-20), molecules coming up from below toward the shaded plane will on the average have made their last collision a perpendicular distance $\frac{2}{3}\lambda$ below the shaded plane where the value of Q_z will have been $Q_z - \frac{2}{3}\lambda(dQ_z/dz)$. This is the value of Q_z they will carry across the shaded plane. By Eq. (5-2) the number of molecules per unit area per unit time coming up from below is $\frac{1}{4}\bar{c}\rho$, where $\rho = \rho_z - \frac{2}{3}\lambda(d\rho_z/dz)$. Subtracting the downward flux of Q_z carried by particles coming from a last collision on average in a plane $\frac{2}{3}\lambda$ above the shaded plane, where $\rho = \rho_z + \frac{2}{3}\lambda(d\rho_z/dz)$, we find for the net flux of Q through the shaded plane in the positive direction

$$
\mathcal{J}_z(Q) = \frac{1}{4}\bar{c}\left[\left(\rho_z - \frac{2}{3}\lambda\frac{d\rho_z}{dz}\right)\left(Q_z - \frac{2}{3}\lambda\frac{dQ_z}{dz}\right)\right.
$$
$$
\left. - \left(\rho_z + \frac{2}{3}\lambda\frac{d\rho_z}{dz}\right)\left(Q_z + \frac{2}{3}\lambda\frac{dQ_z}{dz}\right)\right]
$$

Neglecting the terms with the product of both gradients, this becomes

$$\mathfrak{J}_z(Q) = -\frac{1}{3}\bar{c}\lambda\frac{d}{dz}(\rho_z Q_z)$$ (5-21)

This flux will be positive when the gradient is negative, that is, when $\rho_z Q_z$ decreases as z increases. In the following sections we will examine the physical meaning and details of the transport for different choices of Q.

5-4
Viscosity of Gases

Consider the idealized picture shown in Fig. 5-7 in which a thin plate of area a is pulled through a gas by a constant external force F_y parallel to the y axis with a constant velocity v_{y_0}. We assume v_{y_0} is small enough (much less than $\bar{c}$ in the gas) that after a short time a steady state or laminar flow will be imposed on the gas molecules, such that: those in the layer adjoining the plate will have $\bar{v}_y = v_{y_0}$; those in higher layers (at increased z values) will have smaller $\bar{v}_y$ values; and finally those in the gas layer adjoining the stationary top containing wall of the gas will have $\bar{v}_y = 0$. The velocity gradient $(d\bar{v}_y/dz)$ will be assumed constant and equal to v_{y_0}/L. There will be a net flux of momentum upward through the gas (in the direction of positive z) even for the case of constant ρ and no net flux of particles, because through any plane perpendicular to the z axis between the moving plane and the upper stationary wall the molecules coming up from below will carry larger values of $m\bar{v}_y$ than will the molecules coming down from above. The net momentum flux will give rise to a force per unit area of the plate due to the gas that is equal and opposite to the constant force F_y in the steady state when all layers of the gas and also the thin plate move without acceleration (each layer with a different constant $\bar{v}_y$) and thus under zero *net* force. It is found experimentally for small v_{y_0} that the necessary external force per unit area of the plate for laminar flow is proportional to the velocity gradient. The function of proportionality *defines* the coefficient of

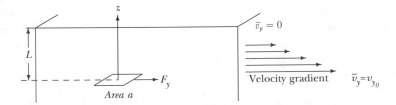

FIGURE 5-7 A thin plate of area a is pulled through a gas by a constant external force F_y, which causes a constant velocity gradient to develop in the gas along the z axis equal to v_{y_0}/L.

viscosity, η, (usually merely called the viscosity)

$$\frac{F_y}{a} = \eta \left(\frac{d\bar{v}_y}{dz} \right) \tag{5-22}$$

Thus the viscosity is a measure of internal friction in a fluid, friction that opposes the motion of objects in the fluid. Then for steady state flow with constant $\rho_z = \rho$, we have from Eq. (5-21) with $Q = m\bar{v}_y$:

$$-\frac{F_y}{a} = -\eta \frac{d\bar{v}_y}{dz} = \mathcal{J}_z(m\bar{v}_y) = -\frac{1}{3}\bar{c}\lambda\rho \frac{d}{dz}(m\bar{v}_y)$$

Consequently, η is given by

$$\eta = \tfrac{1}{3}\bar{c}\lambda\rho m \tag{5-23}$$

This result was first derived by Maxwell in 1860, and as he pointed out it shows that η is independent of gas density or pressure (use Eq. 5-9). This might appear counterintuitive, but as the density goes up and more particles are available to transport momentum they become less effective in doing so in the same proportion because of the decrease of their mean free path. This result cannot hold over an arbitrarily large density range. At very low pressures λ will go to a constant characteristic of the size of the container and then of course η will go to zero as $\rho \to 0$. At very high pressures Eq. (5-9) will no longer hold and η will increase as ρ increases [3]. Substituting from Eq. (5-9) and using

$$\bar{c} = \left(\frac{8kT}{\pi m} \right)^{1/2} \tag{5-24}$$

we convert Eq. (5-23) into

$$\eta = \frac{1}{3}\frac{2}{\pi^{3/2}}\frac{(mkT)^{1/2}}{\sigma^2} \tag{5-25}$$

which predicts that η increases with increase of temperature as $T^{0.5}$. For very dense gases and always for liquids, η will *decrease* as temperature increases. This is due to the fact that momentum transport in a dense phase can occur across a plane by direct molecular interaction via attractive potentials even without physical molecular flow across the plane. This will be discussed further in Chap. 6. Figure 5-8 shows some experimental results for gaseous argon that roughly confirm the behavior predicted from Eq. (5-25). Generally, for real gases η increases faster than $T^{0.5}$, often more like $T^{0.7}$. This comes about because σ^2, which is proportional to the effective cross-sectional area of a molecule, depends on T, becoming smaller as temperature increases. At low temperature a molecule appears larger and has a

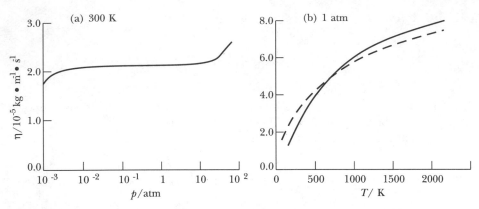

FIGURE 5-8 The viscosity of argon gas as a function of (a) pressure and (b) temperature. The broken line in (b) is calculated from Eq. (5-25) using $\sigma = 260$ pm.
SOURCE: P. W. Atkins. *Physical Chemistry*, 2d ed. W. H. Freeman and Company, San Francisco, 1982.

greater effect on other molecules because the attractive long-range forces are more effective in changing the motion of molecules with low average speeds. An approximate way to take this into account is to fit viscosity data to an equation using

$$\sigma^2 = \sigma_0^2 \left(1 + \frac{S}{T} \right) \tag{5-26}$$

where S is called the Sutherland constant of the molecule.

Rigorous theory for hard sphere (HS) molecules* [4, 5] gives an equation for η of the same form as Eq. (5-23) but with a different numerical coefficient:

$$\eta_{rHS} = \left(\frac{5\pi}{32} \right) \bar{c}\lambda\rho m \tag{5-27}$$

This coefficient is only 1.473 times the value (1/3) obtained from our simple theory. Of course real molecules are *not* hard spheres and Eq. (5-27) will not give the correct temperature dependence of η for real gases. To remedy this we must explicitly [4, 5] express σ^2 as a function of T and of the intermolecular potential, often taken to be of the Lennard-Jones form (see Chap. 3). Thus, from the temperature dependence of viscosity we can obtain information about the intermolecular potential energy [6].

Directly from the definition of η in Eq. (5-22), the viscosity is seen to have dimensions of momentum/area or mass per length per time. The cgs unit, $1 \text{ g·cm}^{-1}\text{·s}^{-1}$, is called the poise. The SI unit, $1 \text{ kg·m}^{-1}\text{·s}^{-1}$, has no special name.

---

* We use a subscript *rHS* for expressions obtained for this model in good first approximation. (The second approximation in this case is numerically 0.498.)

Exercise 5-4

Show that the SI unit of viscosity is equivalent to 1 Pa·s, where Pa is the pressure unit, Pascal. Also show 1 Pa·s = 10 poise.

Exercise 5-5

Use parameters for N_2 from Example 5-3 to calculate η at 273 K (and 1 atm) from Eq. (5-23) and from Eq. (5-27). Compare with the experimental result in Table 5-1.

Gaseous viscosities are usually measured by applying Poiseuille's formula* for the flow of a fluid (liquid or gas) through a capillary tube of radius r and length l:

$$\frac{dV}{dt} = \frac{(P_1^2 - P_2^2)\pi r^4}{16l\eta P_0} \tag{5-28}$$

where V is the volume of the fluid flowing, P_1 and P_2 are the pressures at each end of the tube, and P_0 is the pressure at which the gaseous volume that flows through the tube in a known time t is eventually measured. All pressures in Eq. (5-28) are at the temperature of the capillary tube.

Example 5-4

The following table gives the data[†] of a viscosity measurement for gaseous n-octane.

Temperature of thermostat with capillary tube	150.6°C
Length of tube	0.522 m
Radius of tube	1.45×10^{-4} m
Pressure before tube	30.4 torr
Pressure beyond tube	8.0 torr
Rate of flow	0.0400 g·h^{-1}

---

* For a derivation see a physical chemistry laboratory text, e.g., [7].
[†] V. Fried, H. Hameka, and U. Blukis. *Physical Chemistry*. Macmillan, New York, 1977.

From these data calculate the viscosity and the collision diameter of *n*-octane molecules using the rigorous hard sphere theory.

Solution:

Under steady state conditions the rate of volume flow is constant. Expressing it in terms of mass flow, P_0 will cancel:

$$\frac{dV}{dt} = \frac{RT}{MP_0}\frac{dg}{dt} = \frac{(P_1^2 - P_2^2)\pi r^4}{16l\eta P_0}$$

$$\frac{dg}{dt} = \left(\frac{0.0400}{1000}\right)\frac{1}{(3600)} = 1.111 \times 10^{-8} \text{ kg/s}$$

$$P_1 = \left(\frac{30.4}{760}\right)(1.013 \times 10^5) = 4.052 \times 10^3 \text{ N/m}^2$$

$$P_2 = \left(\frac{8.0}{760}\right)(1.013 \times 10^5) = 1.07 \times 10^3 \text{ N/m}^2$$

$$M = 114 \text{ g/mol}$$

$$\eta = \frac{(P_1^2 - P_2^2)\pi r^4\ (M/RT)}{16l(dg/dt)} = \frac{(1.53 \times 10^7)\pi(4.421 \times 10^{-16})(0.114)}{16(0.522)(1.11 \times 10^{-8})(8.314)(424)}$$

$$= 7.41 \times 10^{-6} \text{ kg·m}^{-1}\text{·s}^{-1}$$

From rigorous theory

$$(\eta)_{rHS} = \frac{5}{16\sigma^2}\left(\frac{mkT}{\pi}\right)^{1/2}$$

$$\sigma^2 = \frac{5[(0.114)(8.314)(424/\pi)]^{1/2}}{16N_A(7.41 \times 10^{-6})} = 7.92 \times 10^{-19} \text{ m}^2$$

$$\sigma = 890 \text{ pm}$$

5-5
Thermal Conduction in Gases

If there is a temperature gradient in the z direction of a gas there will be a flux of heat, that is, of energy from the higher temperature to the lower temperature region in accordance with the Second Law of thermodynamics. Phenomenologically, this flux of energy, $\mathcal{J}_z$ (energy), is very accurately found to be proportional to the temperature gradient, with the function of proportionality defining the coefficient of

FIGURE 5-9 A substance in thermal contact with two heat reservoirs at constant temperatures T_1 and T_2. For $T_2 > T_1$ heat flows in the $-z$ direction and there is a net negative flux of energy $J_z(\bar{\varepsilon})$.

thermal conductivity, κ

$$J_z(\text{energy}) = -\kappa \frac{dT}{dz} \qquad (5\text{-}29)$$

Equation (5-29) is Fourier's law of heat conduction and is accurately obeyed not only by gases but also by liquids and isotropic solids. Note that if the system is a fluid, temperature differences can give rise to density differences and to possible macroscopic flow called convection. We are explicitly going to treat here steady state energy flow in which convection is absent, the density is constant, and energy conduction occurs because of random molecular motion only. In practice this is achieved by using layers of gas that are of small size in the direction of gravitational force and/or by keeping the cooler layers below the warmer.

For the situation shown in Fig. 5-9, with T increasing as z increases there will be a net flux of energy in the negative z direction across any plane between and parallel to the reservoir walls. On a molecular basis this occurs because molecules coming from above carry a higher average energy per molecule, $\bar{\varepsilon}$, across such a plane than do those coming up from below. This energy flux is given by Eq. (5-21) with $Q_z = \bar{\varepsilon}(z)$

$$J_z(\text{energy}) = J_z[\bar{\varepsilon}(z)] = -\frac{1}{3} \bar{c}\lambda\rho \frac{d}{dz}\bar{\varepsilon}(z)$$

$$= -\frac{1}{3} \bar{c}\lambda\rho \left[\frac{d\bar{\varepsilon}(z)}{dT}\right] \frac{dT}{dz} \qquad (5\text{-}30)$$

Comparison of Eq. (5-29) with Eq. (5-30) enables us to identify κ as

$$\kappa = \frac{1}{3} \bar{c}\lambda\rho \frac{d\bar{\varepsilon}(z)}{dT} \qquad (5\text{-}31)$$

For ideal gases the average energy is only a function of temperature and from thermodynamics (see Appendix I) we have

$$d\bar{\varepsilon} = \frac{1}{N_A} dE_m = \left(\frac{C_{V_m}}{N_A}\right) dT \qquad (5\text{-}32)$$

Table 5-1 Transport Properties of Gases at 0°C, 1 atm

Gas	$10^6 \eta/\text{kg} \cdot \text{m}^{-1} \cdot \text{s}^{-1}$	$10^2 \kappa/\text{J} \cdot \text{m}^{-1} \cdot \text{K}^{-1} \cdot \text{s}^{-1}$	$10^5 D/\text{m}^2 \cdot \text{s}^{-1}$	$M\kappa/C_{V_m}\eta$	$\left(1 + \dfrac{9}{4}\dfrac{R}{C_{V_m}}\right)$	$\dfrac{\rho m D}{\eta}$
He	19.20	14.25	—	2.37	2.50	—
Ne	29.67	4.64	4.52	2.53	2.50	1.38
Ar	20.99	1.65	1.56	2.51	2.50	1.33
Kr	23.27	0.870	0.81	2.50	2.50	1.30
Xe	21.07	0.515	0.48	2.57	2.50	1.33
H_2	8.53	16.60	12.9	1.90	1.91	1.36
N_2	16.63	2.30	1.85	1.89	1.92	1.39
O_2	19.18	2.44	1.87	1.98	1.91	1.39
CO_2	13.66	1.46	0.97	1.66	1.66	1.40
H_2O	8.61	1.79	—	1.51	1.76	—
CH_4	10.30	3.02	2.06	1.87	1.75	1.43
C_2H_6	8.51	1.82	—	1.45	1.43	—

SOURCE: W. Kauzmann. *Kinetic Theory of Gases.* Benjamin, New York, 1966.

where C_{V_m} is the molar heat capacity at constant volume. Thus

$$\kappa = \frac{1}{3} \bar{c}\lambda\rho \left(\frac{C_{V_m}}{N_A}\right) \tag{5-33}$$

For the same reasons as for viscosity, this relation predicts κ to be independent of density (or pressure). Using Eqs. (5-9) and (5-24) we find

$$\kappa = \frac{2}{3\pi^{3/2}\sigma^2} \left(\frac{kT}{m}\right)^{1/2} \left(\frac{C_{V_m}}{N_A}\right) \tag{5-34}$$

The rigorous hard sphere result* for κ [4, 5] is

$$\kappa_{rHS} = \frac{25\pi}{64} \bar{c}\lambda\rho \left(\frac{C_{V_m}}{N_A}\right) \tag{5-35}$$

which has a coefficient 3.682 times that obtained from simple kinetic theory. From the simple theory the dimensionless ratio,

$$\frac{M\kappa}{C_{V_m}\eta} = 1.00 \tag{5-36}$$

where M is the molecular weight. The rigorous results give

$$\left(\frac{M\kappa}{C_{V_m}\eta}\right)_{rHS} = \frac{5}{2} \tag{5-37}$$

Experimentally (see Table 5-1) this ratio lies in the range between 1.3 and 2.5, with monatomic molecules having values close to 2.5 since they are close to being hard spheres and carry no internal energy in rotational and vibrational states. In our crude theory we have not properly taken account of the fact that faster molecules with larger c values cross a plane more frequently than slower ones. These faster ones carry more kinetic energy but they do not (in a gradient of imposed collective flow) carry any greater y component of momentum (see Fig. 5-7). Hence for hard spheres the ratio κ/η should be larger than the result in Eq. (5-36). This idea is the basis of Eucken's[†] correction of Eq. (5-36) to make it apply rather accurately to real molecules including polyatomic ones. First write the heat capacity as a sum of translational and internal contributions with $(C_{V_m})_{tr} = \frac{3}{2}R$

$$C_{V_m} = (C_{V_m})_{tr} + (C_{V_m})_i \tag{5-38}$$

Assuming that the factor 2.5 only applies to the transport of translational energy, and noting that $(5\pi/32)(2.5) = 25\pi/64$, to fit Eq. (5-35) for monatomic species we

---

* Again in good first appproximation, with the second approximation being numerically 1.255.
[†] A. Eucken. *Physik. Zeitschrift* **14**, 324 (1913).

use

$$\kappa = \left(\frac{5\pi}{32}\right)\bar{c}\lambda\rho\left(\frac{1}{N_A}\right)[2.5(C_{V_m})_{tr} + (C_{V_m})_i] \qquad (5\text{-}39)$$

and from Eq. (5-27)

$$\eta = \left(\frac{5\pi}{32}\right)\bar{c}\lambda\rho\left(\frac{M}{N_A}\right) \qquad (5\text{-}40)$$

Then assuming ideal gas behavior

$$C_{P_m} - C_{V_m} = R \qquad (5\text{-}41)$$

and the definition

$$\gamma = \frac{C_{P_m}}{C_{V_m}} \qquad (5\text{-}42)$$

we obtain the generalization of Eq. (5-37) given by Eucken:

$$\frac{M\kappa}{C_{V_m}\eta} = \frac{1}{4}(9\gamma - 5) = 1 + \frac{9R}{4C_{V_m}} \qquad (5\text{-}43)$$

For monatomic molecules with only translational contribution to $C_{V_m}, \gamma = \frac{5}{3}$ and Eq. (5-43) is identical to Eq. (5-37). Equation (5-43) shows that κ is proportional to η.

Exercise 5-6

Use Eq. (5-39) through Eq. (5-42) and necessary algebra to derive Eq. (5-43). Start with $R = C_{P_m} - C_{V_m} = (\gamma - 1)C_{V_m}$.

Equation (5-39) is a good approximation (for moderate pressures) for κ for polyatomic molecules that transport energy in internal modes as well as in translational form. In practical terms, with κ in units of $J\cdot K^{-1}\cdot m^{-1}\cdot s^{-1}$, it becomes

$$\frac{\kappa}{J\cdot K^{-1}\cdot m^{-1}\cdot s^{-1}} = \frac{(26.70)}{(\sigma/pm)^2}\left(\frac{T/K}{M/g\cdot mol^{-1}}\right)^{1/2}$$

$$\times \left(\frac{C_{V_m}}{J\cdot K^{-1}\cdot mol^{-1}}\right)\left(1 + \frac{9}{4}\frac{R}{C_{V_m}}\right) \qquad (5\text{-}44)$$

■

Exercise 5-7

Verify the result in Eq. (5-44) and use it to calculate κ for CH_4 gas at 273 K (and 1 atm). Use $\sigma = 380$ pm and $C_{V_m} = 3R$. Compare with the experimental result in Table 5-1.

■

■

Example 5-5

Consider air as a single substance of $M = 28.8$ g/mol with $\sigma = 380$ pm. A double window has panes of glass separated by 3.5 cm of air. Estimate the rate of heat loss by conduction from a warm room (25°C) to a cold exterior (-25°C) through such a window of area 2.0 m².

Solution:

From Eq. (5-44), using $C_{V_m} = \frac{5}{2}R$ for a diatomic gas and $T = 298$, we have

$$\kappa = \frac{26.70}{(380)^2} \left(\frac{298}{28.8}\right)^{1/2} \left(\frac{5}{2}\right) (8.314)(1.90) = 2.35 \times 10^{-2}\, \text{J·K}^{-1}\text{·m}^{-1}\text{·s}^{-1}$$

$$\frac{dT}{dz} = \frac{-50\ \text{K}}{0.035\ \text{m}} = -1.43 \times 10^3\ \text{K/m}$$

From Eq. (5-29) the energy loss is

$$a\mathcal{J}_z(\bar{\varepsilon}) = -a\kappa \frac{dT}{dz}$$

$$= -(2.0\ \text{m}^2)(2.35 \times 10^{-2})\ \text{J·K}^{-1}\text{·m}^{-1}\text{·s}^{-1}(-1.43 \times 10^3\ \text{K/m})$$

$$= 67\ \text{J·s}^{-1}$$

■

■

Example 5-6

Estimate to what pressure the air between the panes of glass in Example 5-5 must be reduced in order to reduce the rate of heat loss by a factor of 10.

Solution:

κ itself must be reduced by a factor of 10. This can happen when λ becomes of the order of the distance between the panes and becomes essentially fixed and ρ is reduced further by a factor of 10, since κ is proportional to $\lambda\rho$ and becomes proportional to ρ when λ is fixed. At higher pressures κ is independent of pressure as explained above.

From the data in Example 5-5 and the equation for λ in Example 5-3, P for $\lambda = 0.035$ meter is

$$\frac{P}{atm} = \frac{(3.068 \times 10^{-5})(298)}{(380)^2(0.035)} = 1.81 \times 10^{-6}\ atm$$

Thus we estimate $P = 1.81 \times 10^{-7}$ atm for the heat loss to be reduced by a factor of 10. This is the reason why the volume between the walls of a vacuum bottle is evacuated.

5-6
Diffusion in Gases

Consider a gas of practically identical molecules but assume that a subset of these are labeled in some way—perhaps by a variant radioactive isotope.* Let ρ_1 be the number density of labeled molecules and assume that a gradient exists in ρ_1 along the z axis. However, the mean number density ρ throughout the gas will be assumed constant so that there will be no collective mass motion in the system. Then experimentally Fick's first law of diffusion, which defines the coefficient of self-diffusion, D, is accurately obeyed for the number flux of labeled particles along the z axis:

$$\mathcal{J}_z(1) = -D_1 \frac{d\rho_1}{dz} \tag{5-45}$$

If ρ_1 decreases as z increases (see Fig. 5-10), the derivative in Eq. (5-45) is negative and the net particle flux $\mathcal{J}_z(1)$ is positive. The minus sign was introduced in Eq. (5-45) so that D will be positive. From Eq. (5-21) with Q formally set equal to unity, since in this case the particles only transport themselves, we have for this flux

$$\mathcal{\tilde{J}}_z(1) = -\frac{1}{3}\bar{c}\lambda \frac{d\rho_1}{dz} \tag{5-46}$$

Comparing Eq. (5-45) with Eq. (5-46), we identify the self-diffusion coefficient of labeled particles (1) as

$$D_1 = \frac{1}{3}\bar{c}\lambda \tag{5-47}$$

---

* This means the molecules are not all exactly identical and will differ at least slightly in mass.

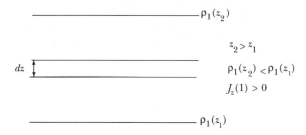

FIGURE 5-10 Net flux of particles in the positive z direction, in particular through a slab of gas of thickness dz at an arbitrary z position.

The self-diffusion coefficient depends inversely on the density (or pressure) and on temperature as $T^{3/2}$. From Eqs. (5-24) and (5-9) we have

$$D_1 = \frac{2}{3\pi^{3/2}} \left(\frac{kT}{m}\right)^{1/2} \frac{1}{\sigma^2 \rho} = \frac{2}{3\sqrt{m}} \left(\frac{kT}{\pi}\right)^{3/2} \frac{1}{\sigma^2 P} \tag{5-48}$$

The rigorous hard sphere result* [4, 5] is

$$(D_1)_{rHS} = \frac{3\pi}{16} \bar{c}\lambda \tag{5-49}$$

which has a coefficient $9\pi/16 = 1.767$ times that obtained from the simple kinetic theory. From Eqs. (5-23) and (5-47) we note

$$\frac{\rho m D}{\eta} = 1.00 \tag{5-50}$$

independent of temperature or pressure, whereas the rigorous theory of Eqs. (5-27) and (5-48) gives

$$\left(\frac{\rho m D}{\eta}\right)_{rHS} = \frac{6}{5} = 1.20 \tag{5-51}$$

Real gases do show this ratio to be independent of temperature and pressure, but to differ for different gases and to take values in the range 1.3 to 1.4 (see Table 5-1).

It is useful to exhibit the partial differential equation that ρ_1, considered as a function of z and of time t, satisfies for more general cases than steady state flux. We limit ourselves to one spatial dimension (along z) and apply the conservation

---

* The second approximation is numerically 0.599.

of mass (i.e., of particles) principle. In Fig. 5-10 the change of particles in unit time in any slab of gas of area a and of thickness dz at z, that is,

$$\frac{\partial}{\partial t}(\rho_1 a\, dz)$$

must be due to any difference of the net flux at $(z + dz)$ compared to the net flux at z:

$$\frac{\partial}{\partial t}(\rho_1 a\, dz) = a \mathcal{J}_z(1) - a \mathcal{J}_{z+dz}(1)$$

$$\frac{\partial \rho_1}{\partial t}\, dz = \mathcal{J}_z(1) - \left[\mathcal{J}_z(1) + \frac{\partial \mathcal{J}_z(1)}{\partial z}\, dz \right]$$

$$\frac{\partial \rho_1}{\partial t} = -\frac{\partial \mathcal{J}_z(1)}{\partial z} \tag{5-52}$$

Using Eq. (5-45), now with partial derivative notation since ρ_1 depends on t as well as z, we have

$$\frac{\partial \rho_1}{\partial t} = D_1 \frac{\partial^2 \rho_1}{\partial z^2} \tag{5-53}$$

This result is the diffusion equation. It is also referred to as Fick's second law of diffusion. We will examine its implications in Chap. 6. For steady state flow, of course, with $(\partial \rho_1/\partial z)$ a constant we have $(\partial \rho_1/\partial t) = 0$. Other examples of using conservation principles in describing fluid flow will be given in Appendices 5.A and 5.B.

For binary diffusion in a two-component gaseous mixture at constant temperature and pressure, starting with a difference in number density along the z axis*, there will be only *one* diffusion coefficient to consider. It is called the mutual or binary diffusion coefficient. This is due to the pressure constancy, which leads to the constraint at any z position

$$\rho_1(z) + \rho_2(z) = \text{const}$$

or

$$\frac{\partial}{\partial z}(\rho_1(z) + \rho_2(z)) = \frac{\partial \rho_1(z)}{\partial z} + \frac{\partial \rho_2(z)}{\partial z} = 0$$

---

* That region of the z axis with an excess of molecules of heavier mass will be the region of higher number density at fixed T and P.

which yields

$$\frac{\partial \rho_1(z)}{\partial z} = -\frac{\partial \rho_2(z)}{\partial z} \tag{5-54}$$

Then from Eq. (5-46) we have

$$\mathcal{J}_z(1) + \mathcal{J}_z(2) = 0 \tag{5-55}$$

which states that the flux of 1 into the region rich in 2 (symbol D_{12}) must just cancel the flux of 2 into the region rich in 1 (symbol D_{21}). Thus from Eq. (5-45)

$$-D_{12}\frac{\partial \rho_1(z)}{\partial z} - D_{21}\frac{\partial \rho_2(z)}{\partial z} = 0$$

and using Eq. (5-54) we find, as stated above,

$$D_{12} = D_{21} \tag{5-56}$$

In binary diffusion the gas with the higher self-diffusion coefficient diffuses rapidly into the gas having the lower self-diffusion coefficient. There will then result a mass flow of predominantly the other gas in the opposite direction to maintain constant pressure. Taking this into account by simple kinetic theory [7], one obtains* for the binary diffusion coefficient:

$$D_{12} = \frac{1}{6}(X_1\lambda'_2\bar{c}_2 + X_2\lambda'_1\bar{c}_1) \tag{5-57}$$

In this result X_1 is the local mole fraction of 1 at a plane in the mixture perpendicular to the z axis and λ'_2 is the modified local mean free path of species 2, taking into account only collisions with the different species 1 in the vicinity of the same plane. That is, λ'_2 is given by

$$\lambda'_2 = \frac{\bar{c}_2}{z_{21}} \tag{5-58}$$

where z_{21}, from Eq. (5-17), is with use of $\rho_1 = X_1\rho$:

$$z_{21} = X_1\rho\sigma_{12}^2\left(\frac{8\pi k T}{\mu}\right)^{1/2} \tag{5-59}$$

---

* This is *not* a misprint, the local mole fraction of 1 is multiplied by the product $\lambda'\bar{c}$ referring to 2, and vice versa.

and ρ is the total (fixed) number density in the system. Notice that the local mole fractions will cancel out in each term and that D_{12} will be independent of the local (and final) mole fractions of the two species in the mixture. Similarly we use

$$\lambda_1' = \frac{\bar{c}_1}{z_{12}} \tag{5-60}$$

$$z_{12} = X_2 \rho \sigma_{12}^2 \left(\frac{8\pi k T}{\mu}\right)^{1/2} \tag{5-61}$$

We omit collisions of like with like in using λ_2' and λ_1' because such collisions cannot affect the total momentum possessed by all the molecules of one kind and thus do not affect the mean mass velocity of that species in its diffusion. Hence the total number of molecules of that species crossing a reference plane in unit time is not changed by such collisions and is the same as if such collisions did not occur. Notice also that we have a factor $\frac{1}{6}$ in Eq. (5-57) in place of the $\frac{1}{3}$ in Eq. (5-47). This is necessary to avoid counting molecule flux twice over in using the primed mean free path expressions, and is necessary for D_{12} given by Eq. (5-57) to become equal to D_1 when molecules 1 and 2 are identical. Using Eq. (5-58) through Eq. (5-61) in Eq. (5-57), we find after some algebra:

$$D_{12} = \frac{1}{6\pi\rho\sigma_{12}^2} \left(\frac{8kT}{\pi}\right)^{1/2} \sqrt{\frac{1}{\mu}} \tag{5-62}$$

Exercise 5-8

Derive Eq. (5-62) and show that it is identical to Eq. (5-48) for self-diffusion when molecules 1 and 2 are identical.

Finally, to obtain the rigorous hard sphere result we must multiple Eq. (5-62) by the same factor, $9\pi/16$, found in going from Eq. (5-47) to Eq. (5-49):

$$(D_{12})_{rHS} = \frac{3}{8\sqrt{2\pi}\rho\sigma_{12}^2} (kT)^{1/2} \sqrt{\frac{1}{\mu}} \tag{5-63}$$

If we use $\rho = P/kT$ and $N_A\mu = M_1 M_2/(M_1 + M_2)$ we obtain, with M_i in $g\cdot mol^{-1}$, the following practical expression for D_{12} in units of $m^2\cdot s^{-1}$:

$$\frac{(D_{12})_{rHS}}{m^2\cdot s^{-1}} = \frac{1.859 \times 10^{-3}}{(P/atm)(\sigma_{12}/pm)^2} \left(\frac{T}{K}\right)^{3/2} \sqrt{\frac{M_1 + M_2}{M_1 M_2}} \tag{5-64}$$

Real gases show a temperature dependence for D_{12} of T^n, with $1.5 < n < 2.0$, for reasons given in the discussion of Eq. (5-26). Further discussion of diffusion including diffusion in condensed phases is given in Chap. 6.

Appendix 5.A
The Speed of Sound in a Gas–Relation to Thermodynamic Properties

The passage of a sound wave through a gas is a nonequilibrium flow process of great interest. We will treat the simplest case of plane waves, that is, one-dimensional waves in which the pressure and other fluid properties vary only along the direction of propagation, which we take to be the z axis. Any periodic mechanical vibration imparted to the gas, say by a tuning fork, gives rise to local density increases and decreases in the vicinity of the source. These can only propagate throughout the gas because of collisions between the molecules. Hence the sound velocity v_s should be rather similar in magnitude to the average speed $\bar{c}$ of molecular motion in the gas. The effect of the sound wave is to give the gas molecules a net oscillatory motion back and forth along the z axis over and beyond their average speed $\bar{c}$; this local average velocity will be denoted $u(z, t)$. The magnitude of $u(z, t)$ will be very small and should not be confused with v_s, which is the speed at which a density peak (or trough) propagates.

We first write an equation for the change of mass density, $\rho' = m\rho$, with time at a fixed z position by invoking conservation of mass in precisely the same way we derived Eq. (5-52)—except that now there is also an imposed velocity $u(z)$ because of the sound wave. Since the rate of mass flow across unit area of a plane perpendicular to the z axis is $\rho'u$, we have for the change of mass in unit time in any slab of gas of area a and thickness dz at z:

$$\left(\frac{\partial}{\partial t}\right)_z (\rho' a\, dz) = a[\rho'(z, t)u(z, t) - \rho'(z + dz, t)u(z + dz, t)]$$

$$= -a\left(\frac{\partial}{\partial z}(\rho'u)\right)_t dz$$

or

$$\left(\frac{\partial \rho'}{\partial t}\right)_z + \left(\frac{\partial}{\partial z}(\rho'u)\right)_t = 0 \tag{5.A-1}$$

Equation (5.A-1) is usually called the equation of continuity. It may be rewritten as

$$\left(\frac{\partial \rho'}{\partial t}\right)_z + \rho'\left(\frac{\partial u}{\partial z}\right)_t + u\left(\frac{\partial \rho'}{\partial z}\right)_t = 0 \tag{5.A-2}$$

We will neglect the last term, since u is small and the density changes caused by the sound wave are also small,* to obtain the linearized equation of continuity:

$$\left(\frac{\partial \rho'}{\partial t}\right)_z + \rho'\left(\frac{\partial u}{\partial z}\right)_t = 0 \tag{5.A-3}$$

Next we consider conservation of momentum along the z axis, that is, Newton's second law of motion for the moving mass of gas in the slab of gas between z and $z + dz$. The net force in the positive z direction between the surfaces at z and $(z + dz)$ is

$$a[P(z, t) - P(z + dz, t)] = -a\left(\frac{\partial P(z, t)}{\partial z}\right)_t dz \tag{5.A-4}$$

since pressure is force per unit area. We will neglect all dissipative terms due to overcoming the actual viscosity of the gas and to effects of the thermal conductivity siphoning off energy into random local heating. This means we are assuming that compression and expansion in the sound wave are adiabatic and reversible. It also means the sound wave suffers no absorption or dimunition in intensity as it passes through the gas. This of course is an idealization and will permit us to relate the speed of sound, only with zero dissipation (v_{so}), to thermodynamic quantities. Assuming then that all the net force of Eq. (5.A-4) produces a change in velocity, that is, an acceleration, du/dt, in the mass, $\rho' a\, dz$, of the slab we find

$$(\rho' a\, dz)\frac{du}{dt} = -a\left(\frac{\partial P}{\partial z}\right)_t dz$$

or

$$\rho'\frac{du}{dt} + \left(\frac{\partial P}{\partial z}\right)_t = 0 \tag{5.A-5}$$

The total derivative of u as a function of t and z can be written in terms of the appropriate partial derivatives using

$$du = \left(\frac{\partial u}{\partial t}\right)_z dt + \left(\frac{\partial u}{\partial z}\right)_t dz$$

$$\frac{du}{dt} = \left(\frac{\partial u}{\partial t}\right)_z + u\left(\frac{\partial u}{\partial z}\right)_t \tag{5.A-6}$$

---

* $(\partial u/\partial z)_t$ on the other hand is not necessarily small, since at fixed t even the sign of u changes periodically with change in z.

in which we have used $u = dz/dt$. Inserting this into Eq. (5.A-5), the equation of motion becomes

$$\rho'\left(\frac{\partial u}{\partial t}\right)_z + \rho'u\left(\frac{\partial u}{\partial z}\right)_t + \left(\frac{\partial P}{\partial z}\right)_t = 0 \qquad (5.A-7)$$

Compared to the other terms in Eq. (5.A-7), the middle term with factor u is negligible and we omit it to obtain the linearized equation of motion:

$$\rho'\left(\frac{\partial u}{\partial t}\right)_z + \left(\frac{\partial P}{\partial z}\right)_t = 0 \qquad (5.A-8)$$

Taking the pressure to be a function of entropy (S) and ρ' we use

$$dP = \left(\frac{\partial P}{\partial \rho'}\right)_S d\rho' + \left(\frac{\partial P}{\partial S}\right)_{\rho'} dS$$

$$\left(\frac{\partial P}{\partial z}\right)_t = \left(\frac{\partial P}{\partial \rho'}\right)_S\left(\frac{\partial \rho'}{\partial z}\right)_t + \left(\frac{\partial P}{\partial S}\right)_{\rho'}\left(\frac{\partial S}{\partial z}\right)_t \qquad (5.A-9)$$

For the case of no dissipation, that is, reversible flow as described above, $(\partial S/\partial z)_t = 0$, and writing

$$\left(\frac{\partial P}{\partial \rho'}\right)_S \equiv v_{so}^2 \qquad (5.A-10)$$

we find

$$\left(\frac{\partial P}{\partial z}\right)_t = v_{so}^2\left(\frac{\partial \rho'}{\partial z}\right)_t \qquad (5.A-11)$$

Thus the equation of motion, Eq. (5.A-8), becomes

$$\left(\frac{\partial u}{\partial t}\right)_z = -\frac{v_{so}^2}{\rho'}\left(\frac{\partial \rho'}{\partial z}\right)_t \qquad (5.A-12)$$

which is to be substituted into the result of differentiating the linearized equation of continuity, Eq. (5.A-3), with respect to time at constant z

$$\left(\frac{\partial^2 \rho'}{\partial t^2}\right)_z + \rho'\left(\frac{\partial}{\partial t}\right)_z\left(\frac{\partial}{\partial z}u\right)_t = 0$$

$$\left(\frac{\partial^2 \rho'}{\partial t^2}\right)_z + \rho'\left(\frac{\partial}{\partial z}\right)_t\left(\frac{\partial u}{\partial t}\right)_z = 0$$

$$\left(\frac{\partial^2 \rho'}{\partial t^2}\right)_z = -\rho'\left(\frac{\partial}{\partial z}\right)_t\left[\left(\frac{-v_{so}^2}{\rho'}\right)\left(\frac{\partial \rho'}{\partial z}\right)_t\right] = v_{so}^2\left(\frac{\partial^2 \rho'}{\partial z^2}\right)_t \qquad (5.A-13)$$

Equation (5.A-13) is the standard form of a wave equation for a periodic disturbance in mass density, ρ', as a function of time at fixed z and as a function of z at
fixed time, with v_{so} having the meaning of the velocity of disturbance. Hence the
velocity of sound under the assumption of zero dissipation is given by the thermodynamic relation, Eq. (5.A-10). In fact, this relation is correct for the velocity of
sound in any fluid medium, including liquids, as long as the assumption of no dissipation holds. By straightforward thermodynamic transformations (see Problem 24)

$$\left(\frac{\partial P}{\partial \rho'}\right)_S = \gamma \left(\frac{\partial P}{\partial \rho'}\right)_T \tag{5.A-14}$$

with

$$\gamma = \frac{C_P}{C_V} \tag{5.A-15}$$

being the ratio of the heat capacity at constant pressure to that at constant volume.
For ideal gases $(\partial P/\partial \rho')_T = kT/m$ and the speed of sound is given by

$$v_{so} = \left(\frac{\gamma k T}{m}\right)^{1/2} \tag{5.A-16}$$

This result is close to but always smaller than $\bar{c}$ in the gas since γ is less than or
equal to $5/3$, which is less than $8/\pi$ occurring in the $\bar{c}$ formula. For nonideal gases
v_{so} will depend on the pressure as well as the temperature because $(\partial P/\partial \rho')_T$, and
γ will be at least slightly pressure dependent.

Appendix 5.B
Supersonic Molecular Beams

Flow of a gas through a convergent-divergent nozzle from a high to a low pressure
can provide extremely rapid cooling and produce a supersonic molecular beam.
The molecules acquire a net velocity along the beam direction greater than the
speed of sound at the low translational temperature reached. They emerge at
temperatures below 0.10 K in some cases and at such low density that intermolecular interactions are insignificant. In particular, because the cooling occurs so
quickly in a way that minimizes three-body collisions, there is little or no condensation to liquid or solid. For these reasons such beams have been termed "almost
ideal spectroscopic samples" and are the subject of much current research [8].

The diameter of the nozzle opening even at its most constricted cross section
is far larger than the mean free path in the high-pressure gas, so that hydrodynamic
and not effusive flow occurs. We will describe steady state flow along the beam
direction (z axis) with u, the acquired net speed along the beam, a function of z

only. Thus the equation of motion, Eq. (5.A-7), becomes

$$\rho'u\frac{du}{dz} = -\frac{dP}{dz} \qquad (5.B\text{-}1)$$

or

$$u\,du + \frac{dP}{\rho'} = 0 \qquad (5.B\text{-}2)$$

Since the cross-sectional area, a, will vary through the nozzle the equation of continuity for mass flow is obtained from the constraint

$$\rho'ua = \text{const} \qquad (5.B\text{-}3)$$

Taking the logarithm of Eq. (5.B-3) and differentiating, we have

$$\frac{d\rho'}{\rho'} + \frac{du}{u} + \frac{da}{a} = 0 \qquad (5.B\text{-}4)$$

We will assume that except for a thin boundary layer close to the walls of the nozzle the flow is so rapid as to permit all effects of viscosity to be neglected. We thus assume adiabatic reversible conditions.*

We introduce the Mach number, $\mathcal{M}$,

$$\mathcal{M} = \frac{u}{v_{so}} \qquad (5.B\text{-}5)$$

where v_{so} is the speed of sound given by Eq. (5.A-10):

$$v_{so}^2 = \left(\frac{\partial P}{\partial \rho'}\right)_S \qquad (5.B\text{-}6)$$

For adiabatic reversible flow we may transform dP/ρ' in Eq. (5.B-2) using Eq. (5.B-6)

$$\frac{dP}{\rho'} = \frac{dP}{d\rho'}\frac{d\rho'}{\rho'} = v_{so}^2\frac{d\rho'}{\rho'} = \frac{u^2}{\mathcal{M}^2}\frac{d\rho'}{\rho'} \qquad (5.B\text{-}7)$$

Then Eq. (5.B-2) becomes

$$\frac{d\rho'}{\rho'} = -\mathcal{M}^2\frac{du}{u} \qquad (5.B\text{-}8)$$

---

* This is *not* a bad assumption. Viscosity-controlled flow in a tube, which leads to Poiseuille's formula, Eq. (5-28), only occurs at low (below 1000) Reynolds number, R, where $R = ru\rho'/\eta$. See reference [9] for further discussion.

which is inserted into Eq. (5.B-4) to obtain the area-velocity relation for compressible flow:

$$\frac{du}{u} = -\frac{da/a}{1 - \mathcal{M}^2} \tag{5.B-9}$$

The comparable expression for incompressible flow (for a liquid) is obtained from Eq. (5.B-4) by setting $d\rho'/\rho' = 0$ and reads

$$\frac{du}{u} = -\frac{da}{a} \tag{5.B-10}$$

which means that as cross-sectional area decreases u increases and vice versa, as shown in curve (1) of Fig. 5-11. The same is true from Eq. (5.B-9) for compressible (gas) flow as long as $\mathcal{M} < 1$ (subsonic speed), as shown in curve (2) of the figure. These curves have a maximum in u at the throat (minimum cross-sectional area) of the nozzle since du and da go to zero together, but even this maximum is sub-

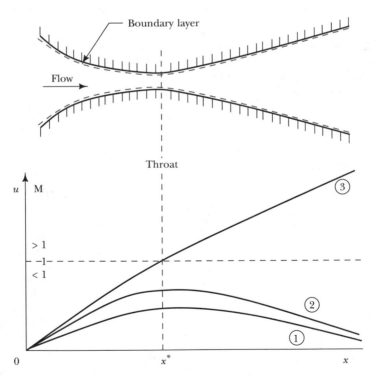

FIGURE 5-11 Flow in a converging-diverging nozzle.
(1) An incompressible fluid.
(2) A compressible gas of a speed that never exceeds Mach 1.
(3) A compressible gas of a speed that reaches Mach 1 at the throat and thus continues to increase in speed as the nozzle widens on the right.
Source: Reference [9].

sonic. However, when $\mathcal{M} > 1$ (supersonic speed) the denominator in Eq. (5.B-9) is negative, and with increase of cross-sectional area u will increase further, as shown in curve (3) of Fig. 5-11. Note also from Eq. (5.B-9) for $\mathcal{M} = 1$ and du/u remaining finite (i.e., positive), we must have $da/a = 0$, meaning we are at the minimum in cross-sectional area. Thus, when there is a sufficiently large pressure difference [note Eq. (5.B-18) below] between the gas reservoir and outside the nozzle, the gas must reach Mach 1 at the throat and will then further increase in speed beyond the throat.

Consideration of energy conservation will now give us the values of the thermodynamic functions in the supersonic beam at any point 1 compared to those in the reservoir (label 0) in steady state flow. Under adiabatic flow (not necessarily reversible flow) from the reservoir to point 1, the work done on the gas per mole (just as in the Joule-Thomson effect) is

$$W = P_0 V_{m_0} - P_1 V_{m_1} \tag{5.B-11}$$

This must be set equal to the change in energy of the gas, which is the change in internal energy U_m plus the change in kinetic energy of directed mass flow. Hence we find, per mole,

$$P_0 V_{m_0} - P_1 V_{m_1} = U_{m_1} - U_{m_0} + \tfrac{1}{2} N_A m u_1^2 - \tfrac{1}{2} N_A m u_0^2 \tag{5.B-12}$$

Introducing the enthalpy

$$H_m = U_m + P V_m \tag{5.B-13}$$

this becomes

$$H_{m_0} + \tfrac{1}{2} N_A m u_0^2 = H_{m_1} + \tfrac{1}{2} N_A m u_1^2 \tag{5.B-14}$$

If nonequilibrium effects or condensation occur between the reservoir and point 1 they can be incorporated into the enthalpy difference $H_{m_1} - H_{m_0}$. We will assume, however, adiabatic *reversible* flow of an ideal gas with no condensation such that

$$H_{m_1} - H_{m_0} = C_{P_m}(T_1 - T_0) \tag{5.B-15}$$

In addition, in the reservoir there is no directed mass flow and $u_0 = 0$. Hence from Eqs. (5.A-16) and (5.B-5) we find

$$C_{P_m}(T_0 - T_1) = \frac{1}{2} N_A m u_1^2 = \frac{1}{2} N_A m \left(\frac{\gamma k T_1 \mathcal{M}^2}{m} \right) = \frac{1}{2} \gamma R T_1 \mathcal{M}^2 \tag{5.B-16}$$

Then, using $C_{P_m} = C_{V_m} + R$ such that $R/C_{P_m} = (\gamma - 1)/\gamma$, we obtain

$$\frac{T_1}{T_0} = \frac{1}{1 + (\gamma - 1)(\mathcal{M}^2/2)} \tag{5.B-17}$$

The other thermodynamic properties may be obtained from the equations that hold for an adiabatic reversible process with ideal gases (see any physical chemistry

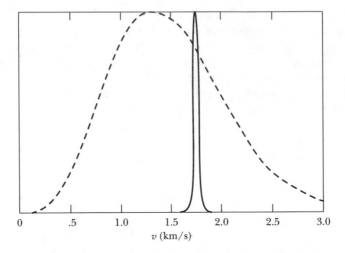

FIGURE 5-12 Speed distribution in an effusive molecular beam (dashed curve) and in a supersonic molecular beam (solid curve). Both curves are normalized to unity at the most probable speed and are for He at a reservoir temperature of 300 K. The curve for the supersonic beam assumes the gas has been expanded to Mach 30.
SOURCE: Reference [8].

text):

$$\frac{T_1}{T_0} = \left(\frac{P_1}{P_0}\right)^{(\gamma - 1)/\gamma} = \left(\frac{\rho_1'}{\rho_0'}\right)^{\gamma - 1} \tag{5.B-18}$$

Figure 5-12 shows the speed distribution in a supersonic beam of helium starting from a reservoir temperature of 300 K and assuming expansion to $\mathcal{M} = 30$. For this case $\gamma = 5/3$, $T_1 = (3.32 \times 10^{-3})(300) = 1.00$ K, and $u_1 = \sqrt{5/3 RT_1 \mathcal{M}^2/M} = 1.77$ km·s^{-1}, which is the maximum of the distribution. Also shown in Fig. 5-12 for comparison is the speed distribution in an effusive molecular beam of helium from a reservoir at 300 K, which was discussed in Example 5-2. The u_1 value is only 1.29 times that of the most probable speed in the effusive beam because a large Mach value does not mean an extremely high speed when the temperature and the speed of sound both are low. The isentropic expansion converts random thermal motion into directed mass flow. Since there is only a finite amount of random energy to convert, the final mass flow speed cannot increase without limit even as $\mathcal{M} \to \infty$ as $T_1 \to 0$.

References and Recommended Reading

[1] R. C. Tolman. *Statistical Mechanics*. Oxford Univ. Press, Oxford, 1938.
[2] N. F. Ramsey. *Molecular Beams*. Oxford Univ. Press, Oxford, 1956.
[3] W. Kauzmann. *Kinetic Theory of Gases*. Benjamin, New York, 1966.
[4] J. O. Hirschfelder, C. F. Curtiss, and R. B. Bird. *Molecular Theory of Gases and Liquids*. Wiley, New York, 1954.

[5] D. A. McQuarrie. *Statistical Mechanics*. Harper and Row, New York, 1976.
[6] J. Olbregts and J. P. Walgraeve. "Determining the Intermolecular Potential Energy in a Gas." *J. Chem. Ed.* **53**, 602 (1976).
[7] D. P. Shoemaker, C. W. Garland, J. I. Steinfeld, and J. W. Nibler. *Experiments in Physical Chemistry*, 4th ed. McGraw-Hill, New York, 1981.
[8] R. E. Smalley, L. Wharton, and D. H. Levy. *Accounts of Chem. Research* **10**, 139 (1977).
[9] P. P. Wegener. *Nonequilibrium Flows*. Dekker, New York, 1969. Vol. 1, Part 1, Chap. 4.

Problems

1. For a two-dimensional ideal gas, integration of Eq. (1-192) gives

$$\frac{dN}{N} = \left(\frac{m}{2\pi k T}\right) \exp\left[-m(v_x^2 + v_y^2)/2kT\right] dv_x\, dv_y$$

whence, changing to polar coordinates in a plane in speed space, we have

$$\frac{dN}{N} = \left(\frac{m}{2\pi k T}\right) e^{-mc^2/2kT} c\, dc\, d\phi$$

a. Go through the derivations of Z_{edge} for the number of molecules that strike unit length per second of an edge of a two-dimensional gas to show

$$Z_{edge} = \frac{\rho}{\pi} \bar{c}$$

where ρ is the density in molecules per unit area and $\bar{c}$ is the average molecular speed in two dimensions.

b. Also derive an expression for the pressure of this gas noting that in two dimensions pressure is force/length.

2. In a test of the Maxwell-Boltzmann one-dimensional speed distribution (see Problem 27, Chap. 1) a molecular beam of krypton atoms was studied using a velocity selector. This consisted of five coaxial disks of common 1.0 cm separation with slots in their rims displaced 2° between nearest neighbor disks. As reported in Atkins, *Physical Chemistry* (W. H. Freeman and Co., 2nd ed. 1982, p. 885) the relative intensities I of the detected krypton atom beam for two different temperatures and at a series of rotation rates were as follows:

v/s^{-1}	20	40	80	100	120
I (40.0 K)	0.846	0.513	0.069	0.015	0.002
I (100.0 K)	0.592	0.485	0.217	0.119	0.057

By taking the intensities as proportional to the one-dimensional speed distribution, find the constant of proportionality by fitting to the data points for $v = 20$ s^{-1}. Then calculate all the other intensities from theory and note how well they match the above data.

3. From the density of Pt $(21$ g/cm$^3)$ one may estimate (how?) that there are about 1.6×10^{15} Pt atoms per cm^2 of Pt surface. Calculate the number of collisions per cm^2 of Pt surface made by O_2 gas at a pressure of (a) 1 atm, (b) 10^{-6} atm, (c) 10^{-10} atm and a temperature of 300 K. Estimate the number of collisions on any particular Pt atom per second in these three cases.

4. Some metals have vapor pressures too small even to measure by the Knudsen technique. These can be heated to a given temperature in vacuum and their rate of evaporation (as mass loss) measured. If one then assumes that in a dynamic equilibrium of the metal with its vapor every time a molecule from the vapor strikes the surface it sticks there, the metal's vapor pressure can be estimated. Tungsten held at 2000 K in vacuum is found to lose 1.14×10^{-12} kg per second per meter^2 of exposed surface. Estimate the vapor pressure of tungsten at 2000 K.

5. For water vapor at 25°C and a partial pressure of 24.0 mm Hg (100 percent humidity) calculate Z_{wall} in cm$^{-2} \cdot$s^{-1}. Using this result and assuming that at equilibrium between liquid water and water in the vapor all molecules striking the surface condense, estimate the mean lifetime of a water molecule on the surface of liquid water at 25°C. Take the surface area of a water molecule as 10^{-19} m^2.

6. A box of volume V containing an ideal gas of molecular weight M at temperature T is divided into two equal halves by a partition. At the start the pressure on the left side is $P_1(0)$ and that on the right is $P_2(0)$. A small hole of area a is introduced in the partition so that molecules can effuse in both directions. Obtain an expression for $P_1(t)$—that is, of the pressure on the left side as a function of time. Assume the partition is very thin.

   Hint: You will need to solve the differential Eq. (5-5) by appropriate integration. Note that $n_1(t) + n_2(t) = n_1(0) + n_2(0)$, where $n_i(t)$ is the moles in part i as a function of time.

7. For the situation given in Problem 6, suppose $P_2(0) = 0$ and calculate the time necessary for P_1 to reach $\frac{3}{4}P_1(0)$ if the gas is helium, $V = 2.00$ m^3, $T = 300$ K, and $a = 0.0100$ cm^2.

8. Derive a general equation for the time needed for the pressure of a contained gas in a volume V at temperature T with a hole of area a in its wall open to a surrounding vacuum to fall from P_0 to P. This is a simpler integration than that of Problem 6. Assume conditions of effusion.

9. A container has for one wall a membrane containing many small holes and is surrounded by a vacuum. If it is filled with gas at a moderate pressure, gas will escape by effusion into the vacuum. It is found that if such a system is filled with helium at some temperature T_0 and pressure P_0 the pressure will

fall to $\frac{1}{2}P_0$ in 1.0 hour. If the same system at T_0 is then filled with an equimolar mixture of Xe and He to a total pressure of P_0, what will be the ratio of the pressure of Xe to that of He after 1.0 hour? What will be the mole fraction of Xe in the container after 1.0 hour? Use the result of Problem 8.

10. Preferential effusion of gaseous $^{235}UF_6$ in a mixture with gaseous $^{238}UF_6$ is used to obtain uranium enriched in ^{235}U for fission bombs and for breeding plutonium in reactors. What is the enrichment factor of a single stage of such a process? Assuming you start with natural uranium, which is 0.992830 ^{238}U and 0.007110 ^{235}U, how many stages would be needed to bring the ratio of the two isotopes to unity? Assume that the reservoir for stage 1 is kept at the natural ratio by continuously flushing fresh natural UF_6 into it. There is only one natural isotope of F with mass $18.99840u$. The precise other masses are $235.0439u$ and $238.0508u$.

11. a. Using the distribution function for molecular speeds in a gas, calculate the average value of the quantity of translational kinetic energy carried off by effusing molecules per unit time per unit hole area when a low-pressure gas effuses out of a container through a small hole into a vacuum. Note that the rate at which molecules of speed c strike unit area of wall is $\rho c/4$.

   b. By dividing your result in **a** by the total number of molecules per unit time per unit area that effuse, obtain the average kinetic energy per molecule in the effusing beam.

   c. How does the average kinetic energy per molecule in the beam compare to the average kinetic energy per molecule in the original reservoir? Give a physical reason for this result.

12. For a gaseous mixture of Ne(1) and Xe(2) of mole fraction 0.25 in Ne at 0°C and 1.00 atm, calculate all the relevant collision numbers (z and Z values) and the two mean free paths. Use $\sigma_1 = 280$ pm, $\sigma_2 = 410$ pm.

13. When a glass tube filled with air at room temperature and 1 atm is fitted with electrodes to which a high DC voltage is applied, cathode rays will travel between them, giving some energy to the gas molecules. However no radiation is observed. If the tube is evacuated with a simple vacuum pump to about 0.01 atm the residual gas will emit pink-purple radiation. Make an order of magnitude estimate of the lifetime of the excited state in the N_2 molecule that is responsible for the radiation for the low-pressure situation. Explain your reasoning. Use data in Example 5-3.

14. Use Eq. (5-26) for a temperature-dependent effective hard sphere diameter and show that $T^{3/2}/\eta$ is a linear function of the temperature

$$\frac{T^{3/2}}{\eta} = c_1 T + c_2$$

in which c_1 and c_2 are constants. Determine c_1 and c_2 in terms of the constants appearing in Eq. (5-26) and other constants of the gas. Use the rigorous hard sphere formula.

15. The data of C. P. Ellis and C. J. Raw [*J. Chem. Phys.* **30**, 574 (1959)] for the temperature dependence of η of N_2 is as follows

T/K	972.7	1021	1069	1120	1166	1273
$(10^7)\eta/\text{kg}\cdot\text{m}^{-1}\cdot\text{s}^{-1}$	391.6	401.7	411.9	421.6	437.4	458.2

Perform a least-squares fit of this data to the equation obtained in Problem 14 and find σ_0 and S for N_2. What is the effective hard sphere diameter of N_2 at 273 K?

16. Consider the experimental setup shown in Fig. 5-13 for measuring the vapor pressure of liquid ^{4}He below its critical temperature of 5.20 K. What is the condition for the pressure readings at room temperature to accurately be equal to the He vapor pressure at the low temperature of the bath? Take σ for He as 260 pm. Comment on the application of this to ^{4}He using the following accurate vapor pressures.

T/K	P/torr
0.50	1.63×10^{-5}
0.80	0.0114
1.00	0.1200
1.50	3.60
2.00	23.77
2.50	77.49
3.00	182.1
4.00	616.5
4.21	760.0

17. What is the ratio of the thermal conductivity of gaseous CH_3Cl at 298 K to that at 50 K? Use data from Example 2-2.

18. What percentage of the energy transport in CH_4 gas at 273 K is carried by the internal degrees of freedom of CH_4? Note Exercise 5-7.

19. In a Pirani pressure gauge a very thin wire is held at a fixed temperature T_w well above the ambient gas temperature T_g by passing a current through it. (The temperature is kept fixed by keeping constant the resistance of the wire as measured in a Wheatstone bridge circuit.) Thus the power input for a steady state condition can be determined. Assume the gas is at such a low pressure

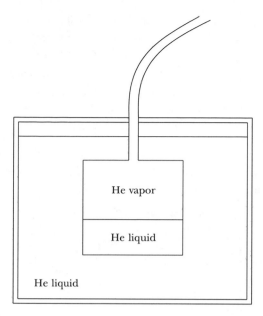

FIGURE 5-13 A liquid He bath at constant low temperature (which may be varied) surrounds a two-phase He vapor-liquid system. A flexible metal tube of diameter ~1 cm leads to a mercury manometer or other pressure measuring device at 300 K.

that the mean free path in the gas is larger than the distance from the wire to a cylindrical wall concentric with the wire. Also assume that every molecule that strikes the wire and brings on the average a kinetic energy of $\frac{3}{2}kT_g$ from its previous collisions with the concentric wall will acquire and remove from the wire on leaving it a kinetic energy of $\frac{3}{2}kT_w$ on the average. Obtain an expression under these assumptions for the ambient gas pressure in terms of the wire dimensions (treated as a cylinder) and steady state power input for a known $(T_w - T_g)$. (In actual experimental work a Pirani gauge must be calibrated for each gas used.)

20. If a power input of 0.726 mW is necessary to keep a wire of diameter 2.0×10^{-4} m and length 2.5×10^{-2} m at a temperature 175 K above the ambient temperature of 298 K in argon gas, calculate the gas pressure under the assumptions given in Problem 19.

21. For argon gas ($\sigma = 340$ pm) at 25°C calculate the self-diffusion coefficient for $P = 1$ atm and $P = 10$ atm. If a pressure gradient in a labeled species of argon of 0.050 atm·cm^{-1} is established, what will be the flux of these species in each case?

22. a. At a fixed pressure and temperature, how does the binary diffusion coefficient vary with the mass of the second species for fixed mass of the first? Use Eq. (5-64).

b. Using σ values from Table 3-5, calculate D_{12} at 273 K and 1 atm for O_2-N_2 and for O_2-CCl_4. Compare with the experimental values, which are 1.81×10^{-5} and 0.64×10^{-5} $m^2 \cdot s^{-1}$, respectively.

23. The diffusion of NH_3 and HCl gases toward each other from opposite ends of a tube filled with air such that a ring of NH_4Cl smoke forms where they meet is often erroneously described in textbooks. The distance from the NH_3 end to the ring divided by the distance from the HCl end to the ring is wrongly stated to be the ratio of the square roots of the molecular weights of HCl to NH_3, that is, 1.46. Anyone doing this experiment will find this is incorrect. What should be the ratio? Use $\sigma_{air\text{-}NH_3} = 340$ pm; $\sigma_{air\text{-}HCl} = 350$ pm. See E. Mason and B. Kronstadt. *J. Chem. Ed.* **44**, 740 (1967).

24. From thermodynamic relations, prove Eq. (5.A-14)

$$\left(\frac{\partial P}{\partial \rho'}\right)_S = \gamma \left(\frac{\partial P}{\partial \rho'}\right)_T$$

25. A standing sound wave in iodine vapor at 400 K produces nodes that are 6.77 cm apart when the frequency of the wave is 1000 s^{-1}. What is the molecular formula of the iodine vapor? Recall that the distance between nodes of a standing wave is one-half wavelength.

26. If the speed of sound in a mixture of He and Ne at 300 K is found to be 758 $m \cdot s^{-1}$, what is the mole fraction of He present?

27. For a monatomic gas in a supersonic beam calculate T_1/T_0 as a function of $\mathcal{M}$ (the Mach number). Also calculate u_1/c_m as a function of $\mathcal{M}$, where u_1 is the acquired net speed of mass flow reached at temperature T_1 (and $\mathcal{M}$), and c_m is the most probable speed for molecules in an effusive beam from a reservoir at T_0. (See Example 5-2.) Take $\mathcal{M}$ in steps of 0.2 between 0 and 1, of 0.5 between 1 and 5, and of 5 between 5 and 30. What are the limiting values of these quantities as $\mathcal{M} \to \infty$?

28. Interpret the meaning of obtaining the expression for incompressible fluid flow, Eq. (5.B-10), by setting $\mathcal{M} = 0$ in Eq. (5.B-9).

29. What must the pressure on the far side of a nozzle be reduced to if $T_0 = 300$ K, $P_0 = 100$ atm, and $T_1 = 0.030$ K for He?

30. For helium leaving a reservoir at $P_0 = 10$ atm, $T_0 = 300$ K in a supersonic beam, show that for resonable nozzle radii r (greater than 1.0 cm) the Reynolds number of the flow, $R = ru\rho'/\eta$, will be much larger than 1000. Calculate explicitly for $\mathcal{M} = 0.20$ and $\mathcal{M} = 30$.

6

Kinetic Theory of Dense Phases

In this chapter we treat some general aspects of transport processes in dense phases. The random walk method, which also applies to gases, is introduced. The classical results of Einstein and Langevin on the motion of a (large) Brownian particle in a fluid are derived. Changes necessary when one tries to extend these results to the motion of an ordinary fluid molecule moving among others of its own kind are described with reference to computer simulation studies and the velocity autocorrelation function for molecules. In the last sections the viscosity of liquids and several methods of determining the molecular weights of macromolecules in solution are studied.

6-1

Random Walk

In a dense medium like a liquid, a particle will be closely surrounded by a set of nearest neighbors and will undergo a series of repeated collisions with them without significantly changing the location of its center of mass. However there will be fluctuations in the identities and number of these nearest neighbors such that a given particle can after a time τ make a net change in its position in a random way. On a time scale large compared to τ, the motion of a given particle will appear to be a random walk in space. For time periods short compared to τ the particle motion will depend on the details of the molecular interactions and be hard to describe. However for the more important longer time periods a statistical treatment will be quite adequate. Hence we are led to the study of random walk processes.

We begin with the simplest case of one-dimensional motion. Suppose we have a particle that starts from $x = 0$ and takes, at regular time intervals, steps of fixed length l with equal probability (of $\frac{1}{2}$) to move in the positive or negative direction. We want to find the probability $P(m, N)$ that after taking N steps the particle reaches the position ml, where m is a positive or negative integer or zero with absolute value $|m| \leq N$. We assume the steps are all independent and so the probability of any definite sequence of N steps is $(1/2)^N$. To obtain $P(m, N)$ we must multiply $(1/2)^N$ by the number of distinct sequences that will reach ml after N steps of length l, which is

$$\frac{N!}{[1/2(N + m)]![1/2(N - m)]!}$$

This combinatorial factor is obtained by noting that if N_p is the number of steps in the positive direction and N_n is the number in the negative direction we have

$$N = N_p + N_n \tag{6-1}$$

and

$$N_p l - N_n l = ml \tag{6-2}$$

by our definition of m, whence

$$N_p - N_n = m \tag{6-3}$$

$$m = 2N_p - N \tag{6-4}$$

$$N_p = \left(\frac{N + m}{2}\right) \tag{6-5}$$

$$N_n = \left(\frac{N - m}{2}\right) \tag{6-6}$$

From Eq. (6-4) we note that N and m are either both odd or both even. Therefore

$$P(m, N) = \frac{N!}{[1/2(N + m)]![1/2(N - m)]!} \left(\frac{1}{2}\right)^N \tag{6-7}$$

Our procedure here is a special case of a Bernoulli trial treated in Problem 15 of Chap. 1. We know from that case that the discrete sum of $P(m, N)$ for N_p ranging from O to N (thus for m ranging from $-N$ to $+N$ or $x = ml$ ranging from $-Nl$ to $+Nl$) is normalized to unity because of the binomial expansion formula. We want to go over to a continuous distribution, replacing the discrete sum on N_p with an integration over the differential from Eq. (6-5)

$$dN_p = \frac{dm}{2} = \frac{dx}{2l} \tag{6-8}$$

We also want to evaluate $P(m, N)$ in terms of simpler functions than the clumsy factorials for the most important physical situation of large N and $m \ll N$. To this end we use a more accurate form than we have heretofore of the Stirling approximation (see Appendix II) for large N

$$\ln (N!) = (N + \tfrac{1}{2}) \ln N - N + \tfrac{1}{2} \ln (2\pi) \tag{6-9}$$

Taking the logarithm of Eq. (6-7) and using Eq. (6-9) we have

$$\ln P(m, N) - \left(N + \frac{1}{2}\right) \ln N - \frac{1}{2} (N + m + 1) \ln \left[\frac{N}{2}\left(1 + \frac{m}{N}\right)\right]$$
$$- \frac{1}{2} (N - m + 1) \ln \left[\frac{N}{2}\left(1 - \frac{m}{N}\right)\right]$$
$$- \frac{1}{2} \ln 2\pi - N \ln 2 \tag{6-10}$$

For $m < N$ we expand

$$\ln \left(1 \pm \frac{m}{N}\right) = \pm \frac{m}{N} - \frac{m^2}{2N^2} \pm \cdots \tag{6-11}$$

and retain powers only through $(m/N)^2$ such that Eq. (6-10) becomes

$$\ln P(m, N) \cong -\frac{1}{2} \ln N + \ln 2 - \frac{1}{2} \ln 2\pi - \frac{m^2}{2N}$$

or

$$P(m, N) \cong \left(\frac{2}{\pi N}\right)^{1/2} \exp\left(\frac{-m^2}{2N}\right) \tag{6-12}$$

This is a Gaussian distribution very sharply peaked at $m = 0$. We may check that it is correctly normalized by extending the range of m from $-\infty$ to $+\infty$ since only for $m \ll N$ will the integrand be appreciably nonzero:

$$\sum_{N_p=0}^{N} P(m, N) \cong \int P(m, N) \, dN_p \cong \int_{-\infty}^{+\infty} P(m, N) \frac{dm}{2}$$

$$= \frac{1}{\sqrt{2\pi N}} \int_{-\infty}^{+\infty} e^{-m^2/2N} \, dm = 1 \qquad (6\text{-}13)$$

The distribution function for m is

$$f(m) = \frac{1}{\sqrt{2\pi N}} e^{-m^2/2N} \qquad (6\text{-}14)$$

Clearly, the average value of m is zero since displacements are equally likely in the positive and negative sense, but $\overline{m^2}$ is not zero:

$$\overline{m} = \int_{-\infty}^{+\infty} m f(m) \, dm = \frac{1}{\sqrt{2\pi N}} \int_{-\infty}^{+\infty} m e^{-m^2/2N} \, dm = 0 \qquad (6\text{-}15)$$

$$\overline{m^2} = \int_{-\infty}^{+\infty} m^2 f(m) \, dm = \frac{1}{\sqrt{2\pi N}} \int_{-\infty}^{+\infty} m^2 e^{-m^2/2N} \, dm = N \qquad (6\text{-}16)$$

In terms of displacement along the x axis the above means that after N steps of a random walk with constant step length l in one dimension

$$\overline{x_N^2} = \overline{(ml)^2} = \overline{m^2} l^2 = N l^2 \qquad (6\text{-}17)$$

and the root mean square (*rms*) displacement is

$$x_{rms} = \sqrt{\overline{x_N^2}} = \sqrt{N} l \qquad (6\text{-}18)$$

This result shows the essential nature of a random walk. If the process were coherently in one direction, N steps would result in a displacement proportional to N whereas in a random walk the x_{rms} is proportional to $\sqrt{N}$.

It is interesting to derive Eq. (6-17) using an argument by Feynman [1]. If after $(N-1)$ steps we are at x_{N-1}, then after N steps

$$x_N = x_{N-1} + l \qquad \text{with } x_N^2 = x_{N-1}^2 + 2l x_{N-1} + l^2$$

or

$$x_N = x_{N-1} - l \qquad \text{with } x_N^2 = x_{N-1}^2 - 2l x_{N-1} + l^2$$

In a number of independent sequences we expect to obtain each value one-half of the time such that the average of x_N^2 should be the average of the above two results:

$$\overline{x_N^2} = \overline{x_{N-1}^2} + l^2$$

Since obviously $x_1 = \pm l$ and $\overline{x_1^2} = l^2$, we see by induction that

$$\overline{x_N^2} = Nl^2$$

We obtain a distribution function in terms of displacements, $x = ml$, using Eq. (6-8) to introduce the factor l^{-1} into Eq. (6-14), that is, $f(m)\, dm = f(x)\, dx$, such that

$$f(x) = \frac{\exp\left(-x^2/2Nl^2\right)}{\sqrt{2\pi Nl^2}} = \frac{\exp\left[-1/2(x/x_{rms})^2\right]}{\sqrt{2\pi}\, x_{rms}} \tag{6-19}$$

In Eq. (6-19) $f(x)$ is the probability density (per unit length in this case) for a single particle to get precisely to $+x$ (or $-x$) in N steps of a random walk. From an ensemble point of view (see Chap. 1) $f(x)$ also represents the fraction per unit length of a set of N_0 labeled (sub 1) particles initially introduced at $x = 0$ that will get precisely to $+x$ (or $-x$) after all of them have taken N random steps. We introduce the time t for this to occur [and eliminate N from Eq. (6-19)] by using $N\tau = t$, with τ the time for a single step as described at the beginning of this section. Then if N_0 is the initial number at $x = 0$ and $t = 0$, the number density, $\rho_1 \equiv N_0 f(x)$ as a function of t and x is

$$\rho_1(x, t) = \frac{N_0 \exp\left[-x^2/2(t/\tau)l^2\right]}{\sqrt{2\pi(t/\tau)l^2}} \tag{6-20}$$

Notice that at $t = 0$ the function ρ_1 is zero for all x except at $x = 0$ where it is infinite because a finite number of particles (N_0) are supposed to be in an infinitesimal width of line at $x = 0$. Mathematically we have

$$\rho_1(x, 0) = N_0 \delta(x) \tag{6-21}$$

where $\delta(x)$ is the delta function (see Appendix II). Of course at all times

$$\int_{-\infty}^{+\infty} \rho_1(x, t)\, dx = N_0 \qquad \text{for all } t \tag{6-22}$$

since no particles are lost. They simply *diffuse* outward as time passes by the random walk process. Figure 6-1 shows $\rho_1(x, t)$ as a function of x for various times t. Clearly we are dealing here with a diffusion process and if we take the one-dimensional diffusion equation [see Eq. (5-53)], which is

$$\frac{\partial \rho_1(x, t)}{\partial t} = D_1 \frac{\partial^2 \rho_1(x, t)}{\partial x^2} \tag{6-23}$$

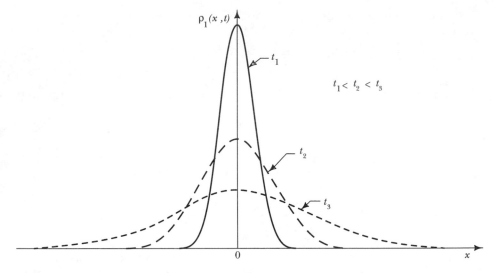

FIGURE 6-1 The number density $\rho_1(x, t)$ as a function of x at various times t after molecules are introduced at time $t = 0$ at the plane $x = 0$. The areas under all the curves are equal to $\mathcal{N}_0$, the number of labeled molecules introduced at time zero.
SOURCE: F. Reif. *Fundamentals of Statistical and Thermal Physics.* McGraw-Hill, 1965. p. 487.

a solution that fits the initial boundary condition is

$$\rho_1(x, t) = \frac{\mathcal{N}_0 e^{-x^2/4D_1 t}}{\sqrt{4\pi D_1 t}} \tag{6-24}$$

as can be easily checked by differentiation.

Exercise 6-1

Verify that Eq. (6-24) is a solution of Eq. (6-23).

By comparing Eq. (6-20) with Eq. (6-24) they become identical if we identify

$$4D_1 t = \frac{2t}{\tau} l^2 = 2\mathcal{N}l^2 = 2\overline{x_N^2}$$

and find

$$D_1 = \frac{\overline{x_N^2}}{2t} \tag{6-25}$$

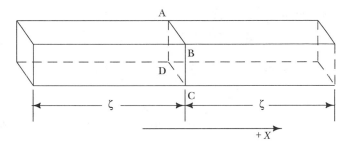

FIGURE 6-2 Two compartments of unit cross section and length ζ on either side of an arbitrary plane ABCD in a fluid perpendicular to the x axis. A number density gradient, $\partial\rho_1/\partial x$, of negative sign exists along the x axis.

This tells us that the diffusion coefficient in one dimension is obtained by dividing the average of the squared displacement in a time t by twice that time period. This result, first obtained by Einstein in 1905, is so important that it is useful to present another derivation of it. In Fig. 6-2 we illustrate a density gradient along the x axis with ρ_1 decreasing as x increases and consider two compartments of unit cross section and length ζ on either side of a central plane ABCD. Let t be the mean time such that molecules within a distance ζ to the left moving in the positive x direction will cross the central plane and molecules within a distance ζ to the right moving in the negative x direction also cross the central plane. Only *half* of the molecules on either side will have x velocity components in the respectively correct directions. If the average number density in the compartment on the left is ρ_1, then the average density in the compartment on the right will be $\rho_1 + \zeta(\partial\rho_1/\partial x)$ since geometrically similar points in the two compartments are just ζ distant from each other. The *net number* of molecules diffusing in the positive x direction in time t is simply one-half the difference in numbers of molecules in the two compartments. This net number, by Eq. (5-45), is the number flux times the time, t:

$$\frac{1}{2}\left[\zeta(1)\rho_1 - \zeta(1)\left(\rho_1 + \zeta\frac{\partial\rho_1}{\partial x}\right)\right] = tJ_x(1) \equiv -tD_1\frac{\partial\rho_1}{\partial x}$$

$$-\frac{1}{2}\zeta^2\frac{\partial\rho_1}{\partial x} = -tD_1\frac{\partial\rho_1}{\partial x} \tag{6-26}$$

such that
$$D_1 = \frac{1}{2}\frac{\zeta^2}{t} \tag{6-27}$$

As we have defined it above, ζ^2 is the mean of the squared distance moved along the x axis in the time t and can be identified with $\overline{x_N^2}$, and so Eq. (6-27) is equivalent to Eq. (6-25).

In a three-dimensional random walk the square of a radial displacement R_N from some starting point after N steps, on taking averages is

$$\overline{R_N^2} = \overline{x_N^2} + \overline{y_N^2} + \overline{z_N^2} \tag{6-28}$$

Since in an isotropic fluid

$$\overline{x_N^2} = \overline{y_N^2} = \overline{z_N^2} \tag{6-29}$$

and each must be separately given by Eq. (6-25), we have

$$\overline{R_N^2} = 6D_1 t \tag{6-30}$$

It is important to stress that these relations hold for dense phases (liquids) as well as for gases as long as the time t exceeds the very short time τ, for which strongly correlated molecular interactions occur.

Example 6-1

Calculate $(\overline{R_N^2})^{1/2}$ for N_2 molecules at $0°C$ and 1 atm for $t/s = 1, 10^2, 10^4$.

Solution:

From Table 5-1, $D_1 = 1.85 \times 10^{-5} \ m^2/s$

$$\frac{(\overline{R_N^2})^{1/2}}{m} = \sqrt{6D_1 t} = 1.05 \times 10^{-2} \ (t/s)^{1/2}$$

t/s	$(\overline{R_N^2})^{1/2}/m$
1	1.05×10^{-2}
10^2	1.05×10^{-1}
10^4	1.05

We note that the rms distance is proportional to the square root of the elapsed time

---

Example 6-2

What fraction of a set of molecules diffusing along the x axis starting from $x = 0$ lie beyond $+x_{rms}$ at any time?

Solution:

To answer this question we calculate first the fraction of the diffusing particles that are between $-x_{rms}$ and $+x_{rms}$. Then one-half of unity less this fraction will provide the fraction called for.

From Eq. (6-19) the fraction between $-x_{rms}$ and $+x_{rms}$ is

$$\left(\frac{\mathcal{N}}{\mathcal{N}_0}\right)_{rms} = \int_{-x_{rms}}^{+x_{rms}} \frac{1}{\sqrt{2\pi}\,x_{rms}} \exp\left(-\frac{1}{2}\frac{x^2}{x_{rms}^2}\right) dx$$

$$= \frac{2}{\sqrt{2\pi}\,x_{rms}} \int_0^{x_{rms}} \exp\left[-\frac{1}{2}\left(\frac{x}{x_{rms}}\right)^2\right] dx$$

Use new variable $u = (1/\sqrt{2})\,(x/x_{rms})$ and recall the definition of the function

$$\text{erf}\,(s) = \frac{2}{\sqrt{\pi}} \int_0^s e^{-u^2}\,du$$

from Example 1-4. Thus

$$\left(\frac{\mathcal{N}}{\mathcal{N}_0}\right)_{rms} = \frac{2}{\sqrt{\pi}} \int_0^{1/\sqrt{2}} e^{-u^2}\,du = \text{erf}\,(0.7071) = 0.6826$$

from numerical integration or use of published tables. Hence the fraction of particles at any time that are beyond $+x_{rms}$ is

$$\frac{1 - 0.6826}{2} = 0.1587$$

Comment: It is these molecules, some of which are very far beyond $+x_{rms}$, that will reach one's nose first, and very quickly compared to the time necessary for x_{rms} to equal the distance between your nose and say an unstoppered bottle of NH_3 in a room.

To conclude this section it is instructive to rederive Eq. (5-47) for D_1 of gases using the random walk model in three dimensions. Each step l will no longer be parallel to the x axis nor will l be constant. We have rather that

$$\overline{R_N^2} = N\overline{l^2} \tag{6-31}$$

where we must take an average of the square of the length of the random step. It is meaningful to introduce the mean free path λ for gases, but $\overline{l^2} \neq \lambda^2$. From Eq. (5-11) for the distribution of path lengths in a gas we obtain

$$\overline{l^2} = \int_0^\infty l^2 \left(\frac{e^{-l/\lambda}}{\lambda}\right) dl = 2\lambda^2 \tag{6-32}$$

and so

$$\overline{R_N^2} = 2N\lambda^2 \tag{6-33}$$

We eliminate $\mathcal{N}$ by using

$$\mathcal{N}\lambda = \bar{c}t \tag{6-34}$$

whence, from Eq. (6-30),

$$\overline{R_N^2} = 6D_1 t = 2\bar{c}\lambda t \tag{6-35}$$

and

$$D_1 = \frac{1}{3}\bar{c}\lambda \tag{6-36}$$

which is identical to Eq. (5-47).

6-2

Brownian Motion: Einstein Relation for Mobility

The random walk of molecules cannot be directly observed. However the incessant random motion of macroscopic yet tiny particles of radius 10^{-6}–10^{-7} meters suspended in a fluid under the buffeting of the fluid molecules can easily be seen, as in Fig. 6-3. This is called Brownian motion after Robert Brown, a botanist who first observed it in 1827 when using a microscope to study pollen suspensions. The phenomenon was first correctly treated by Einstein, who was fundamentally interested in finding observable consequences of atomic-molecular theory. In addition to deriving Eq. (6-25), he obtained a general relation between the diffusion coefficient and the mobility (or response to a force) for a set of special particles in a fluid. These special particles are acted upon by a force F, which might be gravitational or electrical or otherwise, because of their especially large mass or size or because of a net electrical charge. By definition, the mobility, μ, is the drift velocity, v_d, acquired under unit force F such that

$$v_d = \mu F \tag{6-37}$$

The flux of special particles $\mathcal{J}_d$ (number passing unit area in unit time) in response to the force F is

$$\mathcal{J}_d = \rho_1 v_d = \rho_1 \mu F \tag{6-38}$$

To be specific, let us take F to be negative (in the direction of $-x$) in Fig. 6-2 such that the associated potential energy U is

$$U = -Fx \tag{6-39}$$

and becomes more positive as x increases such that the force tends to drive particles to the left. This is opposed by the net diffusion, which is to the right in Fig. 6-2.

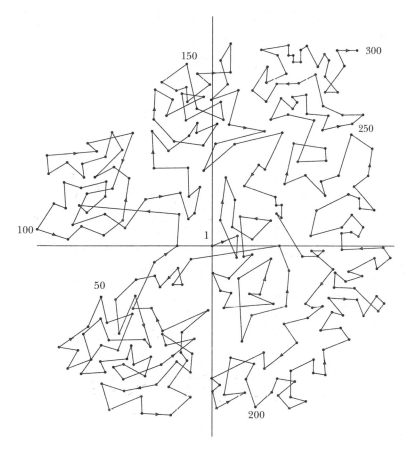

FIGURE 6-3 Brownian motion of a tiny yet macroscopic particle. Positions of a dust particle suspended in water at 25°C are noted at the end of successive 10-second intervals using a projection microscope of magnification 100X.
Source: E. S. R. Gopal. *Statistical Mechanics and Properties of Matter*. Halstead, New York, 1974.

At equilibrium the total flux is zero:

$$\mathcal{J}_d + \mathcal{J}_x(1) = 0$$

or

$$\rho_1 \mu F - D_1 \frac{\partial \rho_1}{\partial x} = 0 \qquad (6\text{-}40)$$

such that

$$\frac{\partial \rho_1}{\partial x} = \frac{\rho_1 \mu F}{D_1} \qquad (6\text{-}41)$$

In addition, at equilibrium in a force field the Boltzmann distribution gives us the particle density

$$\rho_1 = (\text{const})e^{-U/kT}$$

which upon differentiation and use of Eq. (6-39) becomes

$$\frac{\partial \rho_1}{\partial x} = (\text{const})e^{-U/kT}\left(-\frac{1}{kT}\right)\frac{\partial U}{\partial X} = \frac{\rho_1 F}{kT} \tag{6-42}$$

Comparing Eq. (6-41) with Eq. (6-42) we can relate D_1 to μ:

$$D_1 = \mu kT \tag{6-43}$$

Although derived with a one-dimensional example, Eq. (6-43) holds in three dimensions as well.

We can relate μ to the viscosity η of a fluid medium by using the Stokes law, which states that the frictional force (now not an external force) on a large particle assumed spherical with radius r and moving with velocity v is given by

$$F = 6\pi\eta rv \tag{6-44}$$

A rigorous derivation of this law from macroscopic hydrodynamics is quite complex [2, 3]. Comparison of Eq. (6-44) with the definition of mobility in Eq. (6-37) gives

$$\mu = \frac{1}{6\pi\eta r} \tag{6-45}$$

and so we obtain the Stokes-Einstein diffusion law for a large spherical particle:

$$D_1 = \frac{kt}{6\pi\eta r} \tag{6-46}$$

Historically, use of this relation based on observations of Brownian motion was one of the first ways of experimentally determining Avogadro's number.

Example 6-3

Lee, Sears, and Turcotte [4] report the following results. One spherical particle (of radius $r = 4.0 \times 10^{-7}$ m) was viewed in water ($\eta = 0.90 \times 10^{-3}$ kg·m^{-1}·s^{-1}) at 298 K. It was found that a set of 403 values of the net displacements Δx, parallel to the x axis, observed after successive intervals of 2.0 seconds were distributed as follows:

$10^6(\Delta x)/\text{m}$	Frequency (ν) of Occurrence
Less than -5.5	0
Between	
-5.5 and -4.5	1
-4.5 and -3.5	2
-3.5 and -2.5	15
-2.5 and -1.5	32
-1.5 and -0.5	95
-0.5 and $+0.5$	111
0.5 and 1.5	87
1.5 and 2.5	47
2.5 and 3.5	8
3.5 and 4.5	5
4.5 and 5.5	0
Greater than 5.5	0

Determine $\overline{x^2}$ for a 2.0 s time interval and plot on the same graph: (a) a theoretical frequency of occurrence function, and (b) the above observed step function, drawing the steps at the x value for the midpoint of each of the ranges. Calculate the diffusion coefficient for the Brownian sphere and obtain a value of Boltzmann's constant using Stokes' law. Assuming R is known from macroscopic measurements, calculate N_A from this data.

Solution:

Taking the Δx to have the value at the midpoint of each displacement range

$$(10^{12})\,\overline{\Delta x^2} = \frac{(1)(25) + 7(16) + 23(9) + 79(4) + 182(1) + 111(0)}{403}$$

$$= \frac{842}{403} = 2.09 \text{ m}^2$$

$$x_{rms} = \sqrt{\overline{\Delta x^2}} = 1.45 \times 10^{-6} \text{ m}$$

From Eq. (6-19) the smoothed theoretical frequency of observation curve is

$$v = 403 f(x) = \frac{403}{\sqrt{2\pi} x_{rms}} \exp - \frac{1}{2} \left(\frac{x}{x_{rms}} \right)^2$$

such that

$$\frac{10^{-6} v}{m^{-1}} = 111 \exp \left[-0.238 (x')^2 \right] \qquad \text{where } x' = 10^6 x/m$$

This curve is shown in Fig. 6-4 along with the observed step function. The data follow the curve quite well. A perfect fit would show the steps with mirror image symmetry about $x = 0$ and with the curve passing through the midpoint of each step.

From Eq. (6-25)

$$D_1 = \frac{\overline{\Delta x^2}}{2t} = \frac{2.09 \times 10^{-12} \ m^2}{4.0 \ s} = 5.22 \times 10^{-13} \ m^2/s$$

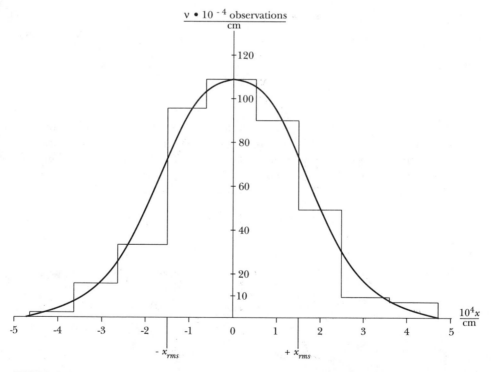

FIGURE 6-4 The theoretical statistical distribution of occurrences of 403 displacements x in 2.0 s time intervals for a Brownian particle in water as compared with the observed step function.

Using the Stokes law, Eq. (6-46),

$$k = \frac{6\pi\eta r D_1}{T} = \frac{(6\pi)(0.90 \times 10^{-3})(4.0 \times 10^{-7})(5.22 \times 10^{-13})}{298}$$

$$= 1.2 \times 10^{-23} \, \text{J·K}^{-1}$$

$$\mathcal{N}_A = \frac{R}{k} = \frac{8.31}{k} = 6.9 \times 10^{23} \, \text{mol}^{-1}$$

This value for $\mathcal{N}_A$ is only 13 percent too high.

■

6-3
Brownian Motion: Langevin Theory

From the definition of mobility in Eq. (6-37) and taking absolute values, μ is equal to $|v_d|/|F|$, which is dimensionally equivalent to time divided by mass. In fact we may set

$$\mu = \frac{\tau}{M} \tag{6-47}$$

where M is the mass of the Brownian particle and τ is a relaxation time that will be defined below where we follow the development given by Pathria [5].

If we imagine that a Brownian particle in a fluid acquires a velocity* v (we drop the subscript d for drift since we assume no external force is present) at some initial time that is different from its equilibrium mean value of $\bar{v} = 0$, we can ask how v will return to its equilibrium value. The equation of motion of the particle will be

$$M \frac{dv}{dt} = \text{force} \equiv f \tag{6-48}$$

Langevin suggested that f is made up of a net force $-v/\mu$ acting in a direction opposite to the velocity (since a moving particle will always suffer more collisions from the front by virtue of its own motion) and a random fluctuating force $\mathcal{F}(t)$ that is due to the random motions of the molecules in the medium. This second contribution will always average to zero at any fixed time over an ensemble of

---

* This might have been the result of a spontaneous fluctuation or it might have been produced by applying an external force for a short period and then switching the force off.

replicas of our moving Brownian particle in a fluid. We write this as

$$\overline{\mathcal{F}(t)} = 0 \qquad \text{for any } t \tag{6-49}$$

where the overbar denotes ensemble averaging. Hence the Langevin equation of motion for the Brownian particle is

$$M\frac{dv}{dt} = -\frac{v}{\mu} + \mathcal{F}(t) \tag{6-50}$$

Taking the ensemble average of Eq. (6-50)

$$M\frac{d\bar{v}}{dt} = -\frac{1}{\mu}\bar{v} \tag{6-51}$$

and integrating with use of Eq. (6-47) we find

$$\bar{v} = v(0)e^{-t/\tau} \tag{6-52}$$

This means the ensemble average of v reverts to zero as $t \to \infty$, with a relaxation time τ given by Eq. (6-47).

We next derive an equation for the ensemble average of the rate of change of r^2, that is, for $(d/dt)r^2$ for the Brownian particle. We divide Eq. (6-50) by M and use

$$A(t) = \frac{\mathcal{F}(t)}{M} \tag{6-53}$$

for the random acceleration. Thus

$$\frac{dv}{dt} = -\frac{v}{\tau} + A(t) \tag{6-54}$$

or

$$\frac{d^2r}{dt^2} = -\frac{1}{\tau}\frac{dr}{dt} + A(t) \tag{6-55}$$

using

$$v = \frac{dr}{dt} \tag{6-56}$$

Since $(d/dt)r^2 = 2r(dr/dt)$ we multiply Eq. (6-55) by $2r\tau$ and take the ensemble average:

$$\frac{d}{dt}\overline{r^2} = 2\overline{r\frac{dr}{dt}} = -2\tau\overline{r\frac{d^2r}{dt^2}} + 2\tau\overline{rA(t)} \qquad (6\text{-}57)$$

The last term in Eq. (6-57) vanishes since there can be no correlation between the position $r(t)$ of a Brownian particle and the acceleration $A(t)$ exerted on it by the molecules of the fluid and

$$\overline{rA(t)} = (\bar{r})\overline{A(t)} = 0$$

by virtue of Eq. (6-49). Next we use the identity

$$r\frac{d^2r}{dt^2} = \frac{d}{dt}\left(r\frac{dr}{dt}\right) - \left(\frac{dr}{dt}\right)^2$$

in Eq. (6-57) to obtain

$$\frac{d}{dt}\overline{r^2} = -2\tau\overline{\frac{d}{dt}\left(r\frac{dr}{dt}\right)} + 2\tau\overline{\left(\frac{dr}{dt}\right)^2} \qquad (6\text{-}58)$$

Finally we use $r(dr/dt) = \frac{1}{2}(d/dt)(r^2)$ and Eq. (6-56) to obtain

$$\frac{d}{dt}\overline{r^2} = -\tau\frac{d^2}{dt^2}\overline{r^2} + 2\tau\overline{v^2} \qquad (6\text{-}59)$$

Although it will take the Brownian particle a time $\sim\tau$ to reach thermal equilibrium with the molecules of the medium, it will be accurate enough* to replace $\overline{v^2}$ in Eq. (6-59) with the thermal equilibrium result from the equipartition principle (see Sec. 1-8)

$$\overline{v^2} = \frac{3kT}{M} \qquad (6\text{-}60)$$

Hence we have a simple second-order differential equation to solve for $\overline{r^2}$:

$$\frac{d^2}{dt^2}\overline{r^2} + \frac{1}{\tau}\frac{d}{dt}\overline{r^2} = \frac{6kT}{M} \qquad (6\text{-}61)$$

---

* See reference [5], pp. 458–461.

Assuming without loss of generality that $\overline{r^2} = 0$ at $t = 0$ and that $(d/dt)(\overline{r^2}) = 0$ at $t = 0$, the solution is

$$\overline{r^2} = \frac{6kT\tau^2}{M}\left[\frac{t}{\tau} - (1 - e^{-t/\tau})\right] \tag{6-62}$$

Exercise 6-2

Verify that Eq. (6-62) is the solution of Eq. (6-61) satisfying the stated boundary conditions.

For $t \ll \tau$ one can expand the exponential in Eq. (6-62) to find

$$\overline{r^2} = \frac{6kT\tau^2}{M}\left(\frac{t}{\tau} - 1 + 1 - \frac{t}{\tau} + \frac{t^2}{2\tau^2} - \cdots\right)$$

$$= \frac{3kTt^2}{M} = \overline{v^2}t^2 \tag{6-63}$$

that is, coherent behavior at very short times with root mean square displacement directly proportional to the time:

$$r_{rms} = v_{rms}t \tag{6-64}$$

However for $t \gg \tau$ Eq. (6-62) becomes

$$\overline{r^2} = 6kT\left(\frac{\tau}{M}\right)t = 6kT\mu t \tag{6-65}$$

which is the random walk behavior shown when the time is long enough for the Brownian particle to have made a huge number of collisions. Equation (6-65) is identical to Eq. (6-30), noting Eq. (6-43). A typical Brownian particle of diameter 10^{-6} m and density slightly above that of water has $M \sim 10^{-15}$ kg. In water at 298 K $(\eta = 0.90 \times 10^{-3}$ kg$\cdot$m$^{-1}\cdot$s$^{-1})$, such a particle from Eq. (6-45) has $\tau \sim 10^{-7}$ s. An observing instrument cannot resolve the Brownian v_{rms} value of $(3kT/M)^{1/2} \sim 4 \times 10^{-3}$ m/s. One observes only the net distance advanced in one second to be about $r_{rms} \sim \sqrt{6kT\mu(1)} \sim 2 \times 10^{-6}$ m. In the interval between successive observations the velocity of such a Brownian particle has grown and decayed as given by Eq. (6-52) many times.

6-4

Diffusion as Related to Velocity Fluctuations at Equilibrium

Recent theories of dense phase kinetic processes [6–8] make use of autocorrelation functions, in particular the velocity autocorrelation function (VAF). We can define this function and see its relevance most easily by starting from a somewhat more formal statement of Eq. (6-25) for the diffusion coefficient of one-dimensional motion:

$$D = \lim_{t \to \infty} \frac{1}{2t} \overline{(x(t) - x(0))^2} \tag{6-66}$$

The overbar in Eq. (6-66) indicates an ensemble average over an equilibrium ensemble stationary in time. Since

$$x(t) - x(0) = \int_0^t v(t_1)\, dt_1 \tag{6-67}$$

the square of this term may be represented by a double integration over two dummy variables

$$D = \lim_{t \to \infty} \frac{1}{2t} \int_0^t dt_1 \int_0^t dt_2\, \overline{v(t_1)v(t_2)} \tag{6-68}$$

The quantity

$$\kappa(t_1 t_2) = \overline{v(t_1)v(t_2)} \tag{6-69}$$

is the VAF since it is a measure of the statistical correlation between the value of the fluctuating velocity at time t_1 and its value at time t_2. The following general properties of the VAF are important:

1. For an equilibrium ensemble stationary in time $\kappa(t_1 t_2)$ can only depend on $s = t_2 - t_1$

$$\kappa(s) = \overline{v(t_1)v(t_1 + s)}$$

independent of the value, t_1.

2. $\kappa(0) = \overline{v(t_1)^2}$ is a constant independent of t_1.

3. $\kappa(s)$ is symmetric about $s = 0$, that is,

$$\kappa(-s) = \kappa(s) = \kappa(|s|) \tag{6-70}$$

This follows because a shift by time s in both instants of measurement cannot change the average:

$$\kappa(s) = \overline{v(t_1)v(t_1 + s)} = \overline{v(t_1 - s)v(t_1)} = \kappa(-s)$$

4. As s becomes large in comparison with some relaxation time, τ^*, the values become uncorrelated ("memory" is lost) and $\kappa(s)$ goes to zero:

$$\kappa(s) = \overline{v(t_1)v(t_1 + s)} \xrightarrow[s > \tau^*]{} [\overline{v(t_1)}][\overline{v(t_1 + s)}] = 0 \qquad (6\text{-}71)$$

For a Brownian particle $\tau^* \sim \tau$, which in Sec. 6-3 was estimated to be about 10^{-7} s. For a fluid molecule itself τ^* will be much smaller, most probably in the range of 10^{-12} to 10^{-13} s.

To evaluate D itself from Eq. (6-68) we introduce the variables

$$S = \tfrac{1}{2}(t_1 + t_2) \qquad (6\text{-}72)$$

$$s = t_2 - t_1 \qquad (6\text{-}73)$$

where it can be shown that

$$dt_1\, dt_2 = ds\, dS \qquad (6\text{-}74)$$

and it is useful[†] to break up the integration over S, which varies from 0 to t, into two parts: from 0 to $t/2$, for which s varies from $-2S$ to $+2S$, and from $t/2$ to t, for which s varies from $-2(t - S)$ to $+2(t - S)$. Hence the integral I that occurs in Eq. (6-68) is

$$I = \int_0^t dt_1 \int_0^t dt_2\, \kappa(s).$$

$$= \int_0^{t/2} dS \int_{-2S}^{+2S} \kappa(s)\, ds + \int_{t/2}^t dS \int_{-2(t-S)}^{+2(t-S)} \kappa(s)\, ds$$

$$= 2 \int_0^{t/2} dS \int_0^{2S} \kappa(s)\, ds + 2 \int_{t/2}^t dS \int_0^{2(t-S)} \kappa(s)\, ds \qquad (6\text{-}75)$$

where the last line follows from Eq. (6-70). Since by Eq. (6-71) $\kappa(s)$ goes rapidly to zero for s greater than a very short time τ^*, we can replace the limits of inte-

---

[†] For details on the integration limits see Pathria [5].

gration on s to be 0 to ∞, whence

$$I = 2t \int_0^\infty \kappa(s)\, ds \tag{6-76}$$

and

$$D = \lim_{t \to \infty} \frac{1}{2t} I = \int_0^\infty \kappa(s)\, ds = \int_0^\infty \overline{v(0)v(|s|)}\, ds \tag{6-77}$$

In Eq. (6-77) we have set the arbitrary time $t_1 = 0$ since any t_1 will do. This result is an example of a more general fluctuation-dissipation theorem [5, 9]. It is so called because it is a relation between a nonequilibrium transport property D, which is inherently related to the dissipation of inhomogeneities or gradients, and the fluctuations of the velocity in an *equilibrium* ensemble.

For a Brownian particle the fluctuating force term in the Langevin equation of motion, Eq. (6-50), may be neglected since the τ of such a particle is much greater than the time scale ($\sim 10^{-13}$ s) of the fluctuating molecular force. Hence the solution of Eq. (6-50) is

$$v(|s|) = v(0)e^{-s/\tau}$$

and the VAF of a Brownian particle has a simple exponential decay

$$\kappa(s) = \overline{v^2(0)}e^{-s/\tau} \tag{6-78}$$

from which Eq. (6-43) can be derived anew (see Problem 9).

We would like to extend these results to treat an ordinary fluid molecule diffusing among others of its own kind, but the $\mathscr{F}(t)$ in Eq. (6-50) cannot be neglected now since τ^* for a molecule is about the same magnitude as that of the time scale of the fluctuating molecular force. No simple analytical formula such as Eq. (6-78) can be derived, but it was widely believed that $\kappa(s)$ for molecules would approach zero after a few molecular collision times since such collisions were expected to randomize the velocity of a particle and cause it to "forget" its initial value. However research in this area was revolutionized by the first molecular dynamics computer simulation results for $\kappa(s)$, obtained by Alder and Wainwright.[†] These showed (Fig. 6-5) an exponential decrease of $\kappa(s)/\overline{v^2(0)}$ for $s^* < 10$ (where $s^* = s/t^*$, where t^* is the mean free time between collisions) but also exhibited long time tails ($s^* > 10$) of nonexponential form going as $(t/t^*)^{-d/2}$, where d is the dimensionality (2 or 3) of the molecular system (disks or spheres). These curves are now qualitatively well understood. The negative values of $\kappa(s)$ in high-density systems represent a velocity reversal of a particle as it bounces off its nearly complete cage of nearest

[†] B. J. Alder and T. E. Wainwright. *Phys. Rev. Lett.* **18**, 988 (1967); *Phys. Rev.* **A1**, 18 (1970).

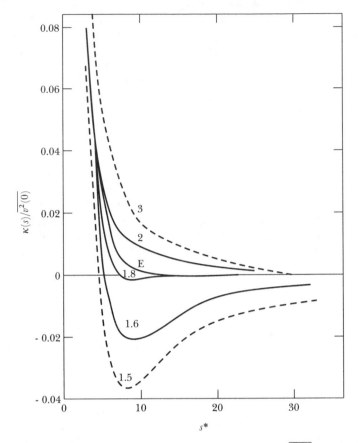

FIGURE 6-5 The reduced velocity autocorrelation function $\kappa(s)/\overline{v^2(0)}$ for hard sphere molecules as a function of time s^*, measured in units of the mean free time between collisions at a series of densities V/V_0 that label each curve. V_0 is the hard sphere close packed volume. Solid curves are for systems of 108 spheres, dashed curves for 500 sphere systems. SOURCE: B. J. Alder, D. M. Gass, and T. E. Wainwright. *J. Chem. Phys.* **53**, 3813 (1970).

neighbor molecules. The computer simulations show that two sets of correlated motions give rise to the long time tails. One is a series of collisions whereby some of the momentum transferred by a given molecule to those in front of it is eventually returned to it by collisions from behind, that is, a vortex pattern. The other set consists of repeated collisions of the given molecule with another with which it has collided before, as illustrated in Fig. 6-6. Both of these series of events give rise to correlations that extend the "memory" of the given molecule for its past. In fact for the situation shown in Fig. 6-6 the probability that two given particles will collide at time t if they are known to have collided at time $t = 0$ is easily estimated. We treat it as two independent random walks that lead the two particles to the same position after time t, ignoring the effect of the finite particle size. By a generalization of Eq. (6-24) to the d-dimensional diffusion equation, the probability density

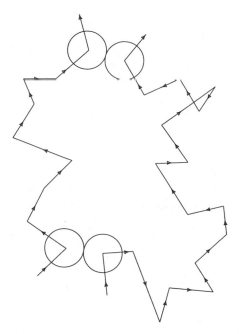

FIGURE 6-6 A recollision event with several intermediate collisions. Two particles collide with each other and then each undergoes a random walk produced by intermediate collisions before they collide again.
SOURCE: Reference [6].

$P(r, t)$ that a particle starting from the origin at $t = 0$ will be at r at time t is

$$P(r, t) = (4\pi Dt)^{-d/2} e^{-r^2/4Dt} \tag{6-79}$$

Thus the probability of recollision is $P^2(r, t)$ integrated over all r in d-dimensional space and goes as

$$\int d\bar{r} P^2(r, t) = \frac{1}{(4\pi Dt)^d} \int d\bar{r} e^{-r^2/2Dt} \propto (Dt)^{-d/2} \tag{6-80}$$

This is believed [6] to be the origin of the $(t)^{-d/2}$ dependence of the long time tails of the VAF for molecules.

Exercise 6-3

Show explicitly for $d = 1, 2, 3$ the $(t)^{-d/2}$ dependence of the integral in Eq. (6-80). Use $d\bar{r} \propto r^{d-1} dr$.

The discovery of the long time tails in the VAF has led to profound changes in thinking about transport properties in dense phases. The most important is that a molecule's motion is strongly correlated with its previous history over many tens or even hundreds of collisions. This in turn has made obsolete much of the theoretical work in dense phase kinetic theory prior to 1970. In particular, transport properties cannot be expressed as power series in the density of the phase. New approaches [6–8] are being developed.

6-5

Viscosity and Its Use in Determining Molecular Weights of Polymers

For gases, from Eq. (5-51) the viscosity η is directly proportional to the diffusion coefficient and increases slowly with increase of temperature, whereas from Eq. (6-46) η for liquids is inversely proportional to the diffusion coefficient and decreases very rapidly with increase of temperature. A qualitative way to understand this is to consider the liquid diffusion process as requiring a time τ for a molecule to succeed in moving out of its initial cage of nearest neighbors and thus to move radially a net distance σ equal roughly to the average intermolecular separation in the fluid, $(V/\mathcal{N})^{1/3}$. Then by Eq. (6-30)

$$D \cong \frac{\sigma^2}{6\tau} \tag{6-81}$$

and the time τ may be estimated by

$$\tau = \tau_0 e^{W/kT} \tag{6-82}$$

where τ_0 is $\sim \sigma/\bar{c}$ the time for a single direct step of length σ as the molecule oscillates against its cage of neighbors and $\bar{c}$ is given by Eq. (1-197). W is an activation energy for diffusion in dense systems where other molecules must be displaced for a given molecule to advance. Finally, by assuming that the Stokes law can apply to a single molecule itself with radius r, Eqs. (6-46), (6-81), and (6-82) give

$$\eta = \frac{kT\tau_0}{\pi r \sigma^2} e^{W/kT} = \frac{kT}{\pi r \bar{c} \sigma} e^{W/kT} \tag{6-83}$$

Experimentally (see Table 6-1) the exponential decrease of η with temperature is found to be obeyed, $\eta \sim A e^{W/kT}$, but the preexponential term given in Eq. (6-83) is only good to within a factor of 5 for nonhydrogen bonded liquids. For water this factor (see Problem 11) gives much too large a value. This is probably related to the fact that liquid water is made up of clusters of water molecules held together by hydrogen bonds. W_m is usually about one-third the value of the molar energy of vaporization of the liquid except for liquid metals, which have very small W_m values compared to their energies of vaporization and compared to W_m values of

TABLE 6-1 Liquid Viscosity Values, $(10^3)\eta/\text{kg}\cdot\text{m}^{-1}\cdot\text{s}^{-1}$

Liquid	$t/°C$				
	0	20	30	40	60
H_2O	1.787	1.002	0.7975	0.6529	0.4665
CCl_4	1.329	0.969	0.843	0.739	0.585
C_2H_5OH	1.773	1.200	1.003	0.834	0.592
Glycerol $(CH_2OHCHOHCH_2OH)$	12.11×10^3	1.49×10^3	629	—	—
Ethylene glycol (CH_2OHCH_2OH)	—	19.9		9.13	4.95
C_6H_6	0.912	0.652	0.564	0.503	0.392
Hg	1.685	1.554	1.499	1.450	1.367

SOURCE: R. C. Weast, ed. *Handbook of Chemistry and Physics*, 56th ed. CRC Press, Cleveland, OH, 1975.

most liquids generally. This is because the moving particles in liquid metals that transport momentum (being the ions that have released electrons into the electron cloud) are all positively charged and need less activation energy to displace each other.

Liquid viscosities can be measured by observing the time for a known volume of liquid to flow through a capillary tube of known length and internal cross section under a known pressure difference. Such absolute measurements are rarely done in practice since it can be shown [10] that if equal volumes of two liquids are timed in flow through the same capillary the ratio of their viscosities is given by

$$\frac{\eta_1}{\eta_2} = \frac{\rho_1' t_1}{\rho_2' t_2} \tag{6-84}$$

where ρ_i' is the mass density of liquid i.

Solution viscosities are useful in determining the molecular weights of dissolved macromolecules. Einstein showed in 1906 that in dilute solutions there is an additive term due to the solute that increases the viscosity beyond that of the pure solvent. The relation is conventionally written

$$\eta = \eta_0 + \eta_0[\eta]c + \text{(terms in higher powers of } c) \tag{6-85}$$

where η_0 is the viscosity of the pure solvent, c is the concentration of solute in mass per volume units, and $[\eta]$ is called the intrinsic viscosity of the solute. (The intrinsic viscosity has units of volume per mass.) From Eq. (6-85)

$$[\eta] = \lim_{c \to 0} \left(\frac{\eta - \eta_0}{\eta_0 c} \right) \tag{6-86}$$

and shows that the intrinsic viscosity is found by extrapolation of solution viscosity values to zero concentration. Since Einstein showed that $[\eta]c$ is proportional to the effective volume swept out by tumbling solute molecules (when they are very large compared to the solvent molecule) there is a proportionality

$$[\eta] \propto \frac{N_A v_h}{M} \tag{6-87}$$

where M is the molecular weight of the macromolecule and v_h is the hydrodynamic volume of a dissolved macromolecule. If the macromolecule is a rigid sphere its v_h is proportional to M and $[\eta]$ is independent of the molecular weight of the dissolved species. However for a flexible random coil molecule v_h is proportional to M^b, where b is $\frac{3}{2}$ or a little larger.* If the molecule is a rigid rod of length d (proportional to M) its tumbling volume will be nearly proportional to d^3 or M^3 and thus its intrinsic viscosity will go as M^2. In practice we write

$$[\eta] = \kappa' \left(\frac{M}{1 \text{ g} \cdot \text{mol}^{-1}} \right)^a \tag{6-88}$$

where the constants κ' and a depend on the nature of the macromolecule and on the solvent and on the temperature. Notice that M is to be expressed in g/mol so that κ' has the same units as $[\eta]$.

For synthetic polymers of the same type but with varying molecular weights (whose κ' and a in a given solvent are known) use of Eq. (6-88) is a good way to estimate the molecular weight of a particular polymer preparation.

Example 6-4

For polyisobutylene in cyclohexane at 30°C $\kappa' = 2.6 \times 10^{-2}$ cm^3/g and $a = 0.70$. The viscosity ratios η/η_0 for a series of such solutions with a particular polyisobutylene preparation are

$10^3 \, c/\text{g} \cdot \text{cm}^{-3}$	0.500	1.72	2.71	4.41
η/η_0	1.290	2.207	3.312	6.579

Find the intrinsic viscosity and the molecular weight of the polyisobutylene.

---

* This is because the volume of a random coil molecule is approximately proportional to the cube of the root mean square separation of the ends of the coil. This separation in turn from the random walk model (Sec. 6-1) is proportional to $\sqrt{N}$, where N is the number of monomer units in the macromolecule. Finally, N is proportional to M.

Solution:

We use Eq. (6-86) and calculate $1/c \; (\eta/\eta_0 - 1)$

$10^3 \; c/\text{g} \cdot \text{cm}^{-3}$	0.500	1.72	2.71	4.41
$\dfrac{1}{c} \; (\eta/\eta_0 - 1)/\text{cm}^3 \cdot \text{g}^{-1}$	580	702	853	1265

The points are plotted in Fig. 6-7. The extrapolated intercept at $c = 0$ yields $[\eta] = 472 \text{ cm}^3/\text{g}$.

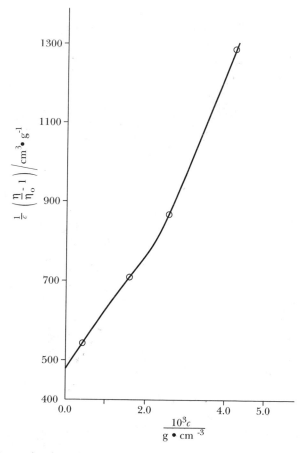

FIGURE 6-7 Determination of the intrinsic viscosity of a polyisobutylene preparation dissolved in cyclohexane.

From Eq. (6-88)

$$M = \left(\frac{[\eta]}{\kappa'}\right)^{1/a} = (1.82 \times 10^4)^{1/0.70} = 1.2 \times 10^6$$

∎

If the molecular weights are known, as for a series of proteins, Eq. (6-88) may be used to determine their a values in different solvent media. This assumes they have a common a value in a given medium, which is often the case for the value of a as shown above is indicative of general chain conformation. If a is much less than 0.5, compact, nearly spherical conformation is indicated, whereas if a is in the range 0.5 to 0.9 random coil behavior is dominant, and if a is greater than 1.1 rodlike behavior is indicated. Two examples of such characterization are shown in Fig. 6-8, where log $[\eta]$ is plotted against log $(M/1 \ g \cdot mol^{-1})$ for various proteins denatured in Guanidine HCl (slope is $a = 0.66$ for line 1) behaving as random coils and for DNA samples from different sources (slope is $a = 1.13$ for line 2) behaving as long rods.

The denaturation (unfolding) of proteins is often studied by intrinsic viscosity measurements. When a globular (spherelike) protein unfolds to a random coil its a value will increase and so will its intrinsic viscosity. However when a long rodlike protein denatures, its a value will decrease and so will its intrinsic viscosity.

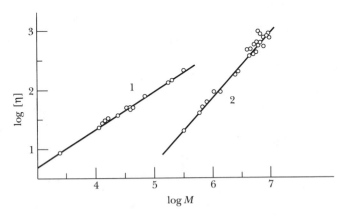

FIGURE 6-8 Plots of log $[\eta]$ versus log $(M/1 \ g \cdot mol^{-1})$. Line 1, various proteins denatured with 6M Guanidine HCl. Line 2, native DNA from a variety of sources also in 6M Guanidine HCl. The slope of each line is an index of molecular shape and is equal to the parameter a of Eq. (6-88).
SOURCE: A. G. Marshall. *Biophysical Chemistry*. Wiley, 1978. Chapter 7.

6-6
Sedimentation of Macromolecules

The velocity of sedimentation of a tiny (yet macroscopic) particle of mass m, density ρ', and thus volume m/ρ' in a liquid or solution of density ρ_0' is easily measurable. The net force of sedimentation, f, is that of gravity less the buoyant force of displaced fluid; thus

$$f = mg - \left(\frac{m}{\rho'}\right)\rho_0' g \qquad (6\text{-}89)$$

where g is the acceleration due to gravity. The velocity of fall will very quickly reach a constant value (meaning zero net force on the particle) when the net force of sedimentation is exactly balanced by the frictional force opposing its fall, which we express in terms of a general mobility μ by means of Eq. (6-37)

$$\frac{v}{\mu} = mg\left(1 - \frac{\rho_0'}{\rho'}\right)$$

$$v = mg\left(1 - \frac{\rho_0'}{\rho'}\right)\mu \qquad (6\text{-}90)$$

If the particle is spherical with radius r and the Stokes law [Eq. (6-45)] holds, this becomes

$$v = \frac{2}{9}\frac{r^2}{\eta}(\rho' - \rho_0')g \qquad (6\text{-}91)$$

Exercise 6-4

Derive Eq. (6-91).

If the particle is a large solute molecule such as a protein in solution, its v will be so small that its sedimentation is essentially canceled by back diffusion in a gravity field. Nevertheless, sedimentation of large molecules can be observed in ultracentrifuges where physically the acceleration g is replaced by $\omega^2 x$, with $\omega = 2\pi\nu$ (and ν the number of revolutions per second being of order 10^3 s^{-1}), and x is the radial displacement of the particle from the axis of rotation.

Exercise 6-5

Calculate the radial acceleration as a factor times g at a position 5.0 cm from the axis of rotation in an ultracentrifuge operating at 75,000 rpm.

Using $v = dx/dt$ we find

$$\frac{dx}{dt} = m\omega^2 x \left(1 - \frac{\rho_0'}{\rho'}\right)\mu \tag{6-92}$$

Defining the sedimentation coefficient s by

$$s = \frac{(dx/dt)}{\omega^2 x} \tag{6-93}$$

we have

$$s = m \left(1 - \frac{\rho_0'}{\rho'}\right)\mu \tag{6-94}$$

a function of the given solute-solution pair and independent of x and t. Hence Eq. (6-93) can be integrated if observations are made on x as a function of t:

$$s = \frac{\ln (x_2/x_1)}{\omega^2 (t_2 - t_1)} \tag{6-95}$$

As the high molecular weight molecules are thrown further and further out along the rotor, a moving boundary eventually forms on one side of which there is practically only solvent. This position can be detected by optical measurements sensitive to the rate of change of the refractive index along the rotor, which rate will be a maximum at this boundary; see Fig. 6-9. The moving boundary positions are the x values, which are observed at different times and inserted into Eq. (6-95). This is thus a way to determine molecular weights, $M = m\mathcal{N}_A$, of proteins and other macromolecules. It is not necessary to assume the particles are spherical if their diffusion coefficient D is measured separately, since use of Eq. (6-43) in Eq. (6-94) gives

$$M = \frac{RTs}{D(1 - \rho_0'/\rho')} \tag{6-96}$$

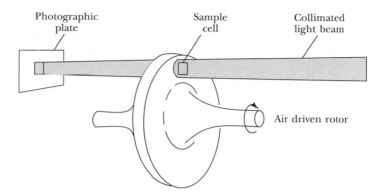

Photographic plate   Sample cell   Collimated light beam

Air driven rotor

FIGURE 6-9 Diagram of an ultracentrifuge rotor, cell, and optical measurement system. SOURCE: T. Svedberg. *Endeavour* **6**, 89 (1947).

If the centrifugation is continued long enough, diffusive flow along the rotor toward the axis of rotation will finally result in a sedimentation equilibrium with no further change in concentration of the macromolecules at any x value. This may require continuous centrifugation for several days but obviates the need to measure D. Since the potential energy U of a solute molecule in a centrifugal force field as function of x is given by

$$U(x) = -\frac{1}{2} m \left(1 - \frac{\rho'_0}{\rho'} \right) x^2 \omega^2 \tag{6-97}$$

the Boltzmann distribution for the ratio of solute concentrations at equilibrium is

$$\frac{c(x_2)}{c(x_1)} = \exp - \{ [U(x_2) - U(x_1)]/kT \}$$

$$= \exp \left[\frac{M}{2RT} \omega^2 \left(1 - \frac{\rho'_0}{\rho'} \right) \left(x_2^2 - x_1^2 \right) \right] \tag{6-98}$$

whence

$$M = \frac{2RT \ln [c(x_2)/c(x_1)]}{\omega^2 (1 - \rho'_0/\rho')(x_2^2 - x_1^2)} \tag{6-99}$$

Optical methods are used to determine the ratio of $c(x_2)$ to $c(x_1)$. For further details on these types of experiments specialized texts [11–12] should be consulted.

The most common method of measuring the D value of a solute (whether a macromolecule or not) is one in which a sharp boundary is formed between a

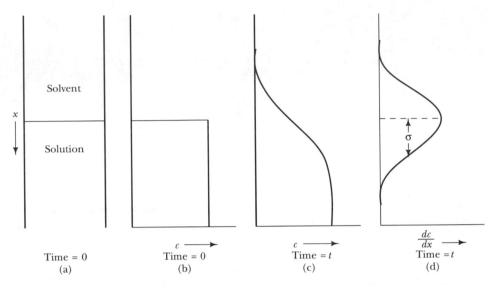

FIGURE 6-10 Diffusion of an initial sharp boundary in a cell of uniform cross section as a means of determining the diffusion coefficient D of a solute. The quantity σ in (d) is directly obtained from experimental results on the rate of change of refractive index with position x. It is the half-width of the curve between the two points at time t, which are a factor $e^{-1/2} = 0.607$ as high as the maximum of the curve at $x = 0$ and time t. $\sigma^2/2t = D$. SOURCE: F. Daniels and R. A. Alberty. *Physical Chemistry*, 4th ed. Wiley, New York, 1975.

solution and its solvent, as shown in Fig. 6-10(a,b). The boundary then becomes diffuse as time increases, Fig. 6-10(c), and the rate of change of concentration with vertical position x will have the form shown in Fig. 6-10(d). The rate of change of concentration will be proportional to the rate of change of the refractive index with x, which is the quantity usually measured. The concentration as function of x and t is given by the solution to the one-dimensional diffusion equation

$$\frac{\partial c}{\partial t} = D \frac{\partial^2 c}{\partial x^2} \tag{6-100}$$

subject to the boundary conditions that at $t = 0$, $c = c_0$ for all $x > 0$ (lower region of the figures), and $c = 0$ for all $x < 0$; and for $t > 0$, $c \to c_0$ as $x \to \infty$ and $c \to 0$ as $x \to -\infty$. The solution is

$$c = \frac{c_0}{2}\left(1 + \frac{2}{\sqrt{\pi}} \int_0^{x/\sqrt{4Dt}} e^{-y^2}\, dy\right) \tag{6-101}$$

which is easily seen to obey all the above conditions and of course Eq. (6-100). We are interested in the experimental curve z, which will be proportional with constant k' to the derivative $\partial c/\partial x$

$$z \equiv k' \frac{\partial c}{\partial x} = \frac{c_0 k'}{\sqrt{4\pi Dt}} e^{-x^2/4Dt} \qquad (6\text{-}102)$$

This curve has a maximum at $x = 0$ where

$$z(x = 0) = \frac{c_0 k'}{\sqrt{4\pi Dt}}$$

The two points of inflection of this curve where $d^2z/dx^2 = 0$ occur at

$$x = \pm\sigma \qquad (6\text{-}103)$$

with

$$\sigma^2 = 2Dt \qquad (6\text{-}104)$$

and at which

$$z(x = \pm\sigma) = \frac{c_0 k'}{\sqrt{4\pi Dt}} e^{-1/2} \qquad (6\text{-}105)$$

such that

$$\frac{z(x = \pm\sigma)}{z(x = 0)} = e^{-1/2} = 0.607 \qquad (6\text{-}106)$$

Hence the experimental D is found from Eq. (6-104), where σ is half the width at time t of the experimental curve, as measured between the two x values that are 0.607 as high as the experimental maximum in the rate of change of refractive index. From practically instantaneously obtained z curves at different times t, the σ^2 may be calculated and plotted versus t and the best straight line determined, whose slope from Eq. (6-104) will be $2D$.

References and Recommended Reading

[1] R. P. Feynman, *Lectures on Physics*, vol I. Addison Wesley, Reading, MA, 1963. Chapters 41, 42, 43.

[2] G. Joos. *Theoretical Physics*, 3d ed. Hafner, New York, 1958.

[3] M. A. Lauffer. "Motion in Viscous Liquids." *J. Chem. Ed.* **58**, 250 (1981).

[4] J. F. Lee, F. W. Sears, and D. L. Turcotte. *Statistical Thermodynamics*. Addison Wesley, Reading, MA, 1963.

[5] R. K. Pathria. *Statistical Mechanics*. Pergamon, New York, 1972. Chapter 13.

[6] J. R. Dorfman in *Perspectives in Statistical Physics*, H. J. Raveché, ed. North Holland, Amsterdam, 1981. Chapter 2.

[7] B. J. Alder and W. E. Alley. *Physics Today* **37**, 56 (Jan. 1984).

[8] E. G. D. Cohen. *Physics Today* **37**, 64 (Jan. 1984).

[9] F. Reif. *Fundamentals of Statistical and Thermal Physics*. McGraw-Hill, New York, 1965. Chapter 15.

[10] D. P. Shoemaker, C. W. Garland, J. I. Steinfeld, and J. W. Nibler. *Experiments in Physical Chemistry*, 4th ed. McGraw-Hill, New York, 1981.

[11] D. Freifelder. *Physical Biochemistry*. W. H. Freeman and Company, San Francisco, 1976.

[12] C. Tanford. *Physical Chemistry of Macromolecules*. Wiley, New York, 1961.

Problems

1. Calculate the exact and the approximate (Gaussian) probability of being five steps from the origin in a one-dimensional random walk for $N = 3, 5, 15, 25, 51$.

2. Consider a child meandering home in one dimension with per step probabilities no longer equal. On the average he takes two forward steps for every one backward. He also takes 20 steps per minute. Calculate the probabilities that in 15 minutes the child will have moved **(a)** 300 steps forward, **(b)** made no net progress at all.

3. Two particles start from the origin and independently take a one-dimensional random walk with equal probabilities of moving to the right or the left. What is the probability that they will meet again after each has taken N steps? Use the Gaussian approximation of Eq. (6-12) and evaluate your answer for $N = 1, 5, 10, 20$. What is the exact answer for $N = 1$?

4. Estimate the diffusion coefficient of a dust particle in air at $0°C$ and 1 atm if its diameter is 5×10^{-7} m. Use data from Table 5-1. What is the rms distance moved by the particle in one minute?

5. To a sample of $^{12}CO_2$ gas at $25°C$ and 1 atm in a spherical container of diameter 80 cm, a small amount of $^{14}CO_2$ is added through a valve at the bottom. Use data from Table 5-1 to estimate the time necessary for uniform mixing of the $^{12}CO_2$ and $^{14}CO_2$ molecules to be achieved throughout the container.

6. If the diffusion constant of helium gas in a certain type of rock is 1.0×10^{-12} m^2/s, what is the rms distance moved by this gas in 10^6 years?

7. If one recklessly assumes that hydrodynamics in the form of the Stokes law can be applied to a single water molecule moving in liquid water, calculate the relaxation time τ that appears in Eq. (6-52) for such a molecule. Actually it is now [7] believed that hydrodynamics does apply accurately down to distances of a few molecular diameters and your estimate will be only about one factor of ten too small.

8. What are the implications for the diffusion coefficient in two dimensions of the long time tails found in the VAF of a molecule?

9. Derive Eq. (6-43) for a Brownian particle using Eq. (6-77) and the result in Eq. (6-78) for the VAF of such a particle.

10. The following data for the viscosity of water and mercury are given.

Water*		Hg†	
$t/°C$	$(10^3)\eta/kg \cdot m^{-1} \cdot s^{-1}$	$t/°C$	$(10^3)\eta/kg \cdot m^{-1} \cdot s^{-1}$
0	1.787	0	1.661
10	1.307	17	1.572
20	1.002	20	1.547
30	0.7975	35	1.476
40	0.6529	63	1.360
50	0.5468	80	1.299
60	0.4665	98	1.263
70	0.4042	138	1.168
		179	1.106

* *Handbook of Chemistry and Physics*, 56th ed. CRC Press, Cleveland, OH, 1975.
† *Landolt-Börnstein Tables*, 6th ed. Springer, Berlin, 1969. Vol. II, Part 5a.

Determine the activation energy, W_m, for diffusion for each of these substances.

11. Using the W_m values found in Problem 10 for water and mercury and estimated preexponential factors of the form given in Eq. (6-83), calculate the respective viscosities at 20°C and compare with the experimental results.

12. The relative viscosity η/η_0 of a solution of 5.0 grams of polymer in 1.0 liter of solvent is 1.80. A solution twice as concentrated has a relative viscosity of 2.80.
   a. Calculate the intrinsic viscosity of the polymer in the solvent.
   b. If the parameters of Eq. (6-88) for this system are $\kappa' = 0.050 \text{ cm}^3 \cdot g^{-1}$ and $a = 0.60$, calculate the molecular weight of the polymer.

13. P. Doty et al. [*J. Am. Chem. Soc.* **78**, 947 (1956)] have measured the intrinsic viscosity of poly γ-benzyl-L-glutamate fractions of known molecular weight in dichloroacetic acid (1) and in dimethylformamide (2).
   Some of their results in (1) were:

$10^{-3} M/g \cdot mol^{-1}$	22	180	300	450
$[\eta]/cm^3 \cdot g^{-1}$	16.7	104	162	230

Some of their results in (2) were

$10^{-3} M/\text{g} \cdot \text{mol}^{-1}$	40	80	150	280	450
$[\eta]/\text{cm}^3 \cdot \text{g}^{-1}$	15.8	53.3	160	477	1090

Obtain best fit log-log plots of Eq. (6-88) and determine κ' and a in the two solvents. Characterize the shape of the polymer molecules in the two solvents.

14. A macroscopic steel ball ($\rho' = 7.86$ g/cm^3) 0.20 cm in diameter takes 10.4 s to fall 10.0 cm through a viscous liquid ($\rho' = 1.26$ g/cm^3). Calculate the viscosity of the liquid at the temperature of the experiment.

15. How long will it take a spherical air bubble 1.0 mm in diameter to rise 20 cm through water at 20°C? Will a larger diameter bubble rise faster or slower than this?

16. Suppose sedimentation results are not available for a particular macromolecule in a solvent but only its diffusion coefficient D in the solvent at a certain temperature, for which the solvent viscosity is also known. Assume the Stokes law and express the molecular weight in terms of these quantities and of ρ', which is the reciprocal of the volume per gram of the dissolved protein as found from volumetric measurements of the protein in the given solvent.

17. A certain protein in water at 20°C has a sedimentation constant $s = 7.75 \times 10^{-13}$ s, a diffusion coefficient $D = 4.80 \times 10^{-11}$ m^2/s, and $\rho' = 1.35$ g/cm^3. Use $\rho'_0 = 0.998$ g/cm^3.
   a. Calculate the molecular weight of the protein.
   b. Calculate the radius of this protein assuming it to be spherical.
   c. Is this protein spherical? Explain.

18. Freifelder (reference [11], p. 335) reports for a form of DNA from E-coli the following data on the position of the solvent boundary as function of time (in minutes) in an ultracentrifuge operating at 33,000 rpm.

t/min	2	10	14	18	22	26
x/cm	6.225	6.310	6.358	6.400	6.445	6.491

Calculate (preferably using a least-squares program) the sedimentation coefficient of the DNA.

19. It is not possible to measure the diffusion coefficient of large nonspherical molecules such as various DNA types because it is so small. Use the result in Problem 18 to estimate D, given from other experiments that the molecular weight is $\sim 10^7$ g/mol and $\rho' = 1.8$ g/cm^3. What would be the time required for σ of Eq. (6-104) to become 1 mm with such a D? Comment on the significance of this length of time for experimental work.

20. In an equilibrium sedimentation experiment it is desired that the concentration ratio between positions $x = 4.0$ cm and $x = 6.0$ cm should be at least 10.0 for macromolecules that have an M value about 5.0×10^5 g/mol. Assuming $(\rho_0'/\rho') = 0.75$, calculate the minimum speed of operation (in rpm) of the ultracentrifuge necessary to achieve this at 25°C.

21. Use Eq. (6-101) and tabulated values of the error function to calculate c/c_0 at x values from 2.5 to -2.5 cm in steps of 0.5 cm at such a time that the half-width of the diffusing boundary is $\sigma = 0.50$ cm. Calculate the times involved for the following three systems

H$_2$ in CCl$_4$ (liquid)	$D = 9.75 \times 10^{-9}$ m^2/s
Sucrose in water	$D = 0.52 \times 10^{-9}$ m^2/s
Ovalbumin in water	$D = 7.8 \times 10^{-11}$ m^2/s

22. Generalize Eq. (6-101) to obtain the solution to Eq. (6-100) for the diffusion situation in which the initially sharp boundary is formed between two solutions of the same solute in a common solvent with concentrations c_0'' below and c_0' above. In practice $c_0'' \geq c_0'$. Show that all the width and inflection properties of the $(\partial c/\partial x)$ function remain the same as in the case with solute initially present only below the boundary (except of course for $c_0'' = c_0'$). Repeat the calculation of Problem 21 for the case $c_0''/c_0' = 4.00$, $c_0' = c_0$.

7

New Concepts for Treating Equilibrium and Nonequilibrium Statistical Processes

7-1
Introduction

These concepts rely on the extensive use of computers to do "experiments" or to generate graphics from which fruitful heuristics can be deduced. This methodology, which was also discussed in Secs. 3-5 and 6-4, will become vital for progress in almost all areas of scientific research. The actual topics have already led to new insights with which students should be acquainted. This chapter and the literature cited can serve to bring the reader to the threshold of research in several exciting fields.

Fractal geometry, treated in Secs. 7-2 and 7-3, has been characterized* as "a return of the eye to science." Along with this comes the use of geometric intuition

---

* B. Mandelbrot, lecture at the New York Academy of Sciences, November 1983.

in entirely new domains with consequent discoveries of deep common relations underlying previously unconnected subjects.

Certain aspects of order hidden in seemingly chaotic processes are the subject of Sec. 7-4. The generality of the methods presented there and their successful application to many diverse problems suggest the beginnings of a new discipline. It will be a discipline for dynamical systems analogous to what thermodynamics is for systems at equilibrium.

Section 7-5 introduces cellular automata, which are particularly simple models for the study of dynamical systems in which space and time are explicitly taken to be discrete. The time evolution of cellular automata from simple initial configurations can give rise to patterns of fractal dimension.

7-2
Fractal Geometry

Many patterns in nature are so irregular and/or fragmented that they cannot be described by Euclidean geometry. They are now called *fractals* after Benoit Mandelbrot, who coined the term in 1975 and whose treatise on the subject [1] is the most important reference for fractals and their relevance to almost all fields of science. Most natural fractal patterns have the property of scale invariance, which means they appear the same (at least in a statistical sense) at any level of magnification or resolution between lower and upper cutoffs.

Consider the coastline of an island of finite area and of approximate dimensions 100 km by 100 km. On a linear scale of resolution, ε, below 100 km—say from 20 km down to a lower cutoff of about 1 m—the coastline will appear self-similar. (We omit effects of human building and modification, which cannot affect features on a scale large compared to human size.) Below 1 m individual boulders and even pebbles or individual molecules will cause the self-similarity to end. What is the length of such a coastline? It will depend on the precision with which it is measured, for if our scale ε is made smaller the length will increase. Consider the increase of length attendant on measuring the ins and outs of all little bays, not to mention rocks and other elements. Indeed, for a mathematical coastline the length will go to infinity as $\varepsilon \to 0$ because the fractal dimension, D, of such a coastline is greater than unity.

Mandelbrot has drawn on a large body of mathematical concepts of the period 1875–1925 and shown their utility for science. The Hausdorff dimension, D, of a set (1919) is one of these. Suppose a set whose dimension we seek is a subset of a p-dimensional other space (where p must be equal to or greater than D; in practice it is often easiest to think about $p = D$). Let $\mathcal{N}(\varepsilon)$ be the number of p-dimensional cubes of side ε needed to cover the set. Then the Hausdorff dimension D is given by:

$$D = \lim_{\varepsilon \to 0} \frac{\ln \mathcal{N}(\varepsilon)}{\ln (1/\varepsilon)} \qquad (7\text{-}1)$$

where $(1/\varepsilon)$ is the *number* of lengths ε that fit into the unit length. This implies for small ε that:

$$\mathcal{N}(\varepsilon) = F\varepsilon^{-D} \qquad (7\text{-}2)$$

where F is a constant with units of $(\text{length})^D$, since then:

$$\ln \mathcal{N}(\varepsilon) = \ln F - D \ln \varepsilon \cong -D \ln \varepsilon \qquad \text{for } \varepsilon \to 0$$

The topological dimension, D_T, of a set is the intuitive or Euclidean dimension based on the idea of dividing or cutting. A point cannot be cut and has $D_T = 0$. If a point cut will divide a set into two parts, such a set has $D_T = 1$. If sets of $D_T = 1$ are needed to divide or cut another set then that set has $D_T = 2$, and so on. The technical definition of a fractal is a set with $D > D_T$, that is, one whose Hausdorff dimension is greater than its topological dimension. Any set with D non-integral is a fractal. An exceptional fractal such as the Peano curve [1] that can pass through every point in a plane has $D = 2$ but $D_T = 1$. Of course for Euclidean patterns (sets) the Hausdorff D coincides with D_T. If the set is a single point, $\mathcal{N}(\varepsilon) = 1$ and $D = 0$. If the set is a straight line, in unit length of it one can fit $\mathcal{N}(\varepsilon) = 1/\varepsilon$ one-dimensional cubes of side ε and $D = 1$. If the set is a section of the xy plane given by $0 < x < 1$ and $0 < y < 1$, this region can be covered by $(1/\varepsilon)(1/\varepsilon) = \varepsilon^{-2}$ cubes of side ε, and so $\mathcal{N}(\varepsilon) = \varepsilon^{-2}$ and $D = 2$, and so on. In these standard cases no limit $\varepsilon \to 0$ need be taken.

■

Example 7-1

Deduce the Hausdorff dimensionality of the original Cantor set, which is constructed as shown in Fig. 7-1 from a unit length by removing the middle third, then removing from the two remaining pieces their middle thirds, and so on ad infinitum. This is the set used by Cantor (1884) to prove that any two line segments regardless of their length contain an equal number of points and that a line segment has as many points as a two-dimensional surface or a three-dimensional volume.

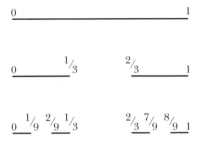

FIGURE 7-1 The zeroth and first two steps (s) in the construction of the Cantor set of $D = 0.6309$.

Solution:

As shown in Fig. 7-1 for the zeroth step of the process $(s = 0)$, one cube of side $\varepsilon = 1$ will cover the unit line itself. For $s = 1$, $\varepsilon = \frac{1}{3}$ and $N = 2$ cubes of side ε are needed to cover the set. At $s = 2$, $\varepsilon = \frac{1}{9}$ and four cubes of side ε are needed. In general, at step s

$$\varepsilon = (1/3)^s, \qquad N = 2^s$$

The limit $s \to \infty$ in the process will give $\varepsilon \to 0$ and enables us to visualize the end result. This is an infinite set of points very nonuniformly distributed with a Euclidean length of zero and a topological dimension of zero since no cuts can be made in a set of points. The Hausdorff dimension is:

$$D = \lim_{(\therefore s \to \infty)} \varepsilon \to 0 \; \frac{\ln (2^s)}{\ln (3^s)} = \frac{\ln 2}{\ln 3} = 0.6309$$

Mandelbrot calls any fractal with $0 < D < 1$ a *dust*. Notice the self-similarity or *scaling* nature of the final result. The full Cantor set between 0 and 1 looks the same as the part between 0 and $\frac{1}{3}$ if this be magnified by a factor of 3 or as the part between 0 and $\frac{1}{9}$ if this be magnified by a factor of 9, and so on.

■

Since the number of $p \geq D$ dimensional cubes of side ε is given by Eq. (7-2) and their individual measure in dimension D is ε^D, then their total measure, m, in dimension D is:

$$m(D) = (F\varepsilon^{-D})(\varepsilon^D) = F \tag{7-3}$$

which is a constant independent of ε, as it must be if D is the correct dimension of the set. We see that the Hausdorff dimension retains the Euclidean property of being an exponent in defining a measure in space. If we evaluate the measure in a dimension $D'' > D$ we find:

$$m(D'') = (F\varepsilon^{-D})(\varepsilon^{D''}) = F\varepsilon^{D''-D} \to 0 \qquad \text{as } \varepsilon \to 0 \tag{7-4}$$

just as the ordinary area or volume of a Euclidean line is zero. However if we evaluate the measure of a set in a dimension $D' < D$ we obtain:

$$m(D') = (F\varepsilon^{-D})(\varepsilon^{D'}) = F\varepsilon^{-(D-D')} \to \infty \qquad \text{as } \varepsilon \to 0 \tag{7-5}$$

just as the length of a square is infinite and just as the length of a coastline becomes infinite when we try to evaluate its length in $D' = 1$. Empirically (see Problem 7-1), coastlines have D values close to 1.2. The Hausdorff dimensionality is a transition point for the measure function since for any set of dimensionality D the $m(D' < D)$

is infinite, the $m(D'' > D)$ is zero, and only the measure in dimension D is finite and nonzero. The volume of human flesh (i.e., nonvascular tissue) is certainly finite and nonzero, hence, its dimension is $D = 3$. However, as Mandelbrot [1] points out its topological dimension is $D_T = 2$, since flesh forms the common boundary between arteries and veins, which are themselves topologically three-dimensional. Thus, flesh can be considered a fractal surface with positive volume!

An interesting curve that is continuous but nowhere differentiable, that has an infinite length but bounds a finite area, was introduced by Koch (1904). It can serve as a model for coastlines or snowflake boundaries. The construction of this curve proceeds as shown in Fig. 7-2, starting with an equilateral triangle. At each step every side is divided into thirds and a smaller equilateral triangle is erected on the middle third, and one proceeds ad infinitum. At each step, when ε is divided by 3 the length of the curve is multiplied by $4/3$. Thus, the *scaling relation* for the length of this fractal is:

$$L\left(\frac{\varepsilon}{3}\right) = \frac{4}{3} L(\varepsilon) \tag{7-6}$$

The general rule for the measure, length, is Eq. (7-5) with $D' = 1$:

$$m(D' = 1) \equiv L(\varepsilon) = (F\varepsilon^{-D})(\varepsilon) = F\varepsilon^{1-D} \tag{7-7}$$

Using Eq. (7-6) in Eq. (7-7) we can calculate the D of the Koch curve:

$$F\left(\frac{\varepsilon}{3}\right)^{1-D} = \frac{4}{3}(\varepsilon)^{1-D}F$$

$$3^{D-1} = \frac{4}{3}$$

$$D = \frac{\ln 4}{\ln 3} = 1.2619$$

FIGURE 7-2 The zeroth and first two steps in the construction of the Koch curve of $D = 1.2619$.
SOURCE: Reference [1], p. 42.

FIGURE 7-3 The zeroth and first three steps in the construction of the Sierpinski gasket of $D = 1.5850$.
SOURCE: Reference [1], p. 142.

A Sierpinski gasket is generated as shown in Fig. 7-3 by starting with a solid (black) triangle and cutting out triangular gaps as shown ad infinitum. Every solid triangle in going one step has its base divided by 2, but their number is times 3 such that the length of the perimeter of the black triangles scales as:

$$L\left(\frac{\varepsilon}{2}\right) = \frac{3}{2} L(\varepsilon)$$

which used in Eq. (7-7) gives:

$$D = \frac{\ln 3}{\ln 2} = 1.5850$$

as the dimension of the gasket.

Exercise 7-1

Write down the area scaling relation for the black triangles in the Sierpinski gasket and the analog of Eq. (7-7) for area, that is, for the measure $m(D = 2)$. Calculate the dimensionality of the gasket using these equations. What is the area of the gasket as $\varepsilon \to 0$?

Objects of uniform density in Euclidean dimensions have a mass as function of radius R behaving as:

$$M(R) \propto R^E \qquad \text{where } E = 1, 2, \text{ or } 3 \tag{7-8}$$

When fractals are self-similar this rule applies to fractal sets in the form:

$$M(R) \propto R^D \tag{7-9}$$

We illustrate this for the Cantor line of Fig. 7-1. Suppose a unit mass was originally distributed along the line from $R = 0$ to $R = 1$. (It may help to imagine the line thickened into a bar and originally of low density.) Then as the process of cutting proceeds imagine the mass always curdling out of the middle third into the end thirds until you are left with an infinite number of points (or an infinite number of infinitely thin wedges of the imagined bar) of infinitely high density. The mass distribution $M(R)$ along the Cantor line is called the Cantor function and is shown in Fig. 7-4. It is also called the "Devil's staircase" because of its strange properties.* $M(R)$ represents the mass between 0 and R; it will double each time R is multiplied by 3 and remains below 1 because of Eq. (7-9) and the result from Example 7-1 that $3^D = 2$ for the Cantor set. This function increases in infinitesimal rises at the infinitely many points of infinite density. The sum of these rises is unity and the sum of the horizontal steps is also unity. Thus, the length of the staircase is the finite number 2. Hence the staircase is not a fractal since its $D = D_T = 1$. One can generalize the $M(R)$ function by extrapolation by factors of 3 such that the dust is first extended to be between $R = 0$ and $R = 3$, with the single largest gap between $R = 1$ and $R = 2$ and with the regions 0 to 1 and 2 to 3 simply replicas of the original 0–1 dust. One can then go on to the region 0–9, and so on. Mandelbrot found that if $R = $ time and $M(R)$ counts errors between 0 and R in transmitting symbols in computer data transmission circuits, the extrapolated Devil's staircase is a reasonable first model of actual error occurrence. However, any Cantor dust is too regular to accurately model real irregular behavior. One needs to introduce a random method of cutting the gaps out of the solid line pieces, a method that will give a staircase that is at least statistically invariant to translation along R. This has been successfully accomplished.

If one considers the distribution of mass in the universe as a whole, with $M(R)$ the mass inside a sphere of radius R, the mass density $\rho' = M(R)/\frac{4}{3}\pi R^3$ will vary as R^{D-3}. For small R, if we take our origin as the center of the earth, $D = 3$ and ρ' is approximately constant. As R increases far beyond the earth's radius D will go to zero and we are at the scale of matter as isolated "points," that is, the interstellar scale. For much greater R, of the size of a galaxy or a cluster of galaxies (and perhaps for R as large as the radius of the universe itself), indirect astronomical observation [2] indicates $D \sim 1.23$. This can be termed the fractal dimension of galaxy distribution. It indicates that mass is very far from being homogeneously distributed in the universe and is the result of the clustering of stars and of galaxies. A projection onto a two-dimensional surface such as the "sky," that is, the surface of the celestial sphere as seen from the earth, will have the same dimensionality as that of the object projected as long as its dimensionality is less than 2. Hence all of the stars of the galaxies will give a projection of $D \sim 1.23$ and thus of *zero* area on the celestial sphere of the earth. This provides a geometrical resolution of the so-called blazing sky effect, which states that if $D = 3$ for galaxy distribution, any

---

* Its continuous variation occurs at the points of the Cantor set whose length add to zero. At all other points its derivative vanishes, i.e., $M(R)$ remains constant along intervals that add up to the entire length of the line.

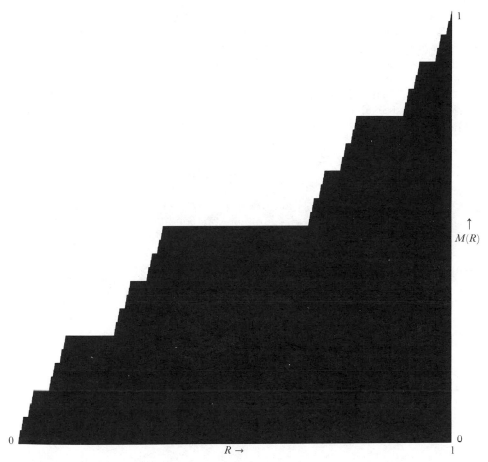

FIGURE 7-4 The Devil's staircase: $M(R)$, the mass between 0 and R along the Cantor line with unit mass between 0 and 1.
SOURCE: Reference [1], p. 83.

direction taken from the earth would sooner or later encounter a star, with the result that the sky always would appear (day or night) uniformly bright or blazing. Physical arguments based on relativity theory and the finite lifetime of galaxies* can also explain away the effect even if $D = 3$, but they now are superfluous since $D < 2$. Books on cosmology (e.g., [3]) should be consulted for more detail on this question.

Fractal geometry, once our intuition has been exposed to it, would seem to be very relevant to the description of turbulence in fluids, and so it is. In turbulent

---

* P. Wesson et. al., *Astrophysical J.* **317**, 601 (1987).

flow, eddies of a huge range of sizes occur, suggestive of the self-similar nature of fractals with scale change. Turbulence ends in heat evolution, the energy of the fluid motion having been dissipated because of fluid viscosity. However, the dissipation is not homogeneously distributed throughout the volume of a fluid—some regions have much more heat evolution than others. We know after all that wind comes in gusts. Thus the spatial set of points on which dissipation is concentrated must have a fractal dimensionality $D < 3$. It must also have $D > 2$ since, as Mandelbrot points out, if D were less than 2 experimental apparatus inserted in turbulent regions would generally fail to measure turbulence for the same reason the night sky is not blazing! Since such apparatus does measure turbulence, we may conclude that $D > 2$. Further technical arguments [1] indicate the D for turbulence is about $\frac{8}{3}$.

It has long been known [4] that the curve $(D_T = 1)$ of motion traced out by a particle in quantum mechanics is continuous but nondifferentiable. It has recently been shown to be a fractal with $D = 2$ [5]. This is analogous to the result [1] for the fractal nature of the curve (also with $D = 2$) traced out by the trail extrapolated to infinite time for a particle undergoing Brownian motion. See Fig. 6-3 for a sample of such a trail (for a finite time, of course). Two other fractal aspects of Brownian motion are presented in Fig. 7-5, which shows $X(t)$, the position of the X coordinate of a particle (started from $X = 0$ at time zero), as a function of time. This curve also has the meaning of accumulated wins (say for X positive) or losses in an honest

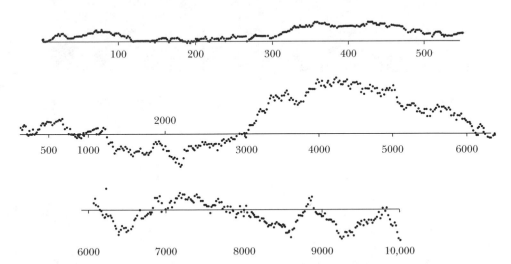

FIGURE 7-5 The X position of a particle undergoing Brownian motion as a function of time. Alternatively, the graph may represent the cumulative wins (say heads) in a coin-tossing game as function of the number of tosses. The curve itself has $D = 1.50$. The set of points corresponding to the time at which $X(t) = 0$ has $D = 0.50$. These points constitute a random Cantor dust.
SOURCE: Reference [1], p. 241.

heads or tails coin tossing game. We know from Sec. 6-1 that during a time Δt the difference of [max $X(t)$ − min $X(t)$] is of the order $\sqrt{\Delta t}$. Hence to cover this portion of the graph of $X(t)$ by squares of side Δt requires about $1/\sqrt{\Delta t}$ such squares. To cover a unit time interval being thus $1/\Delta t$ sets of Δt in time would require $(1/\Delta t)\, 1/\sqrt{\Delta t} \sim (\Delta t)^{-3/2}$ such squares. Hence by Eq. (7-1), with $\varepsilon \equiv \Delta t$ the Hausdorff dimensionality of the fractal $X(t)$ graph is $\frac{3}{2}$. Notice that because of the random nature of Brownian motion the graph $X(t)$ is *not* self-similar. As in Euclidean geometry, except for special relative orientations the dimensionality of the intersection $D_{1 \times 2}$ of two sets of dimensions D_1 and D_2 embedded in a common Euclidean space of dimension E is given by

$$D_{1 \times 2} = D_1 + D_2 - E \qquad \text{for } D_1 + D_2 \geq E \qquad (7\text{-}10.1)$$
$$= 0 \qquad \text{for } D_1 + D_2 < E \qquad (7\text{-}10.2)$$

If we use Eq. (7-10.1) to calculate the dimension of the intersection of the line $X(t) = 0$ (of $D = D_T = 1$) with $X(t)$ (of $D = \frac{3}{2}$) in the common space of $E = 2$ (for time and position X), we find $D_{1 \times 2} = 0.50$. Thus the set of points in time for which $X(t)$ returns to zero is a (random) Cantor dust.

Exercise 7-2

Satisfy yourself that Eq. (7-10) gives standard results for intersections of standard Euclidean sets in $E = 2$ and $E = 3$.

7-3

Applications of Fractal Geometrical Thinking: Generalized Brownian Motion, Scaling Properties of Polymers

Many important physical applications of fractals stem from a generalization of Brownian motion [1, 6, 7] for which $\overline{X(t)}$ is zero as usual but

$$\overline{X^2}(t) = t^{2H} \qquad (7\text{-}11)$$

with H *not* equal to $\frac{1}{2}$. Then

$$t = \{[\overline{X^2}(t)]^{1/2}\}^{1/H} \qquad (7\text{-}12)$$

with the quantity in curly braces a scale of length and t analogous to a mass scale (proportional to the *number* of steps in the random walk) such that in analogy to Eq. (7-9) the fractal dimensionality of the generalized Brownian trail (whence the subscript B) is

$$D_B = \frac{1}{H} \tag{7-13}$$

For $H > \frac{1}{2}$ the situation is termed one of persistence since successive increments of position instead of being fully independent tend to go in the same direction. The Brownian trail is now less irregular at all scales than for an ordinary random walk. A self-avoiding random walk in two- or three-dimensional Euclidean space ($E = 2$ or 3) is effectively one with $H > \frac{1}{2}$ and is a useful model for describing conformations of polymer chains [8]. In a Euclidean dimensionality of 4 or greater an ordinary random walk is automatically self-avoiding. This can be proved by applying Eq. (7-10) to the intersection of two halves of an infinitely long random walk, with each half of dimension $D_B = 2$.

The case of $H < \frac{1}{2}$, called antipersistence, gives rise to a Brownian trail with far more self-intersections than ordinary. Such random walks can be used to model diffusion and reaction rates of species on a fractal structure [9]. H values in the range $0 < H < \frac{1}{2}$ interpolate between exponent zero, meaning no increase of $X^2(t)$ with time because of the trapping of a species on a finite cluster of sites, and exponent $\frac{1}{2}$, as in normal (Euclidean) diffusion. From Sec. 6-1 the diffusion coefficient $\mathscr{D}$ (here written as a script quantity to avoid confusion with fractal dimensionality) is proportional to $\overline{X^2}(t)/t$, whence

$$\mathscr{D} \propto \frac{t^{2H}}{t} \propto t^{2H-1} \tag{7-14}$$

Furthermore, if we define the root mean square value of X by

$$X_{rms} = [\overline{X^2}(t)]^{1/2}$$

then

$$\mathscr{D} \propto X_{rms}^{-\theta} \tag{7-15}$$

with

$$\theta = \frac{1 - 2H}{H} \tag{7-16}$$

Thus for anomalous diffusion ($H \neq \frac{1}{2}$) the diffusion coefficient depends explicitly on time or alternatively on X_{rms}.

The mean number of *distinct* sites $\bar{s}(t)$ visited by a random walker as $t \to \infty$ on a fractal of dimensionality D will be proportional to X_{rms} raised to the power D.

Hence

$$\bar{s}(t) \propto t^{HD} \tag{7-17}$$

For Euclidean diffusion in any $E \geq 2$ we have $HD = E/2$. It may be shown [6] that $(1 - HD)$ is the dimensionality* of the fractal set on the time axis for return of a random walker to the origin (its starting position). Furthermore $2HD$ has been called the spectral dimensionality, D_s, by Orbach [10, 11] in connection with a study of phononlike (i.e., vibrational) excitations of a fractal. It turns out that $2HD$ (which would be precisely E for a Euclidean structure) is the formal dimensionality of the wave number space of a fractal structure.

When we turn to properties of polymers we find a particularly striking example of the bringing together of previously unconnected areas of science by the use of fractal geometric concepts, in this instance polymer science and critical state theory [8]. Polymers are macromolecules formed by the chemical bonding of many monomers. If the monomers can bond at only two sites, only linear or chain polymers can form. If at least some of the monomers can bind at three or more sites, branched polymers can form. If the reactions of branching can proceed far enough, gelation will occur at a sharp gel point. This is the point at which a polymer solution suddenly changes from a viscous liquid (or sol) into an elastic connected polymer network (the gel) that extends across the entire sample. A lattice model of gelation is provided by percolation theory [12]. In a site percolation model, lattice sites are occupied randomly (with probability p) by objects which in this instance represent monomers with functionality (reaction site number) equal to the number of first nearest neighbors, z, on the lattice. A linkage or bonding is automatically assumed to occur between any two nearest neighbor occupied sites, giving rise to a cross-linked cluster or polymer. Because of this automatic linkage of nearest neighbor occupied sites there is effectively no interaction between the monomers. This means there is no need to assure that the sites are occupied according to a Boltzmann distribution (see Sec. 3-5). Singularities occur in the properties of the clusters of nearest neighbor occupied sites as a function of p and not in thermodynamic properties as a function of temperature, as in thermal critical phenomena (see Sec. 3-8). In particular, at a critical value p_c there will exist a largest cluster that just spans the entire lattice, meaning it provides a tortuous (fractal!) connected path of nearest neighbors that reaches from one side of the lattice to the opposite side. By definition, the sudden existence of this connected path is the onset of percolation (which models the onset of gelation for real network polymers). This is a purely geometrical phase transition associated with the change from local connectivity to global connectivity.

The order parameter [see Eq. (3-128)] of the percolation problem is P, the probability of a site chosen at random to belong to the single largest cluster that percolates. This parameter vanishes for $p < p_c$ and is taken to obey the power law:

$$P \propto (p - p_c)^\beta \tag{7-18}$$

---

* If HD is greater than unity the actual fractal dimensionality for return is zero, meaning no return to the origin is possible.

The order parameter for the Ising model or for the gas-liquid transition can assume two different values, corresponding to up or down spin clusters or to gaseous or liquid density droplets, respectively; in contrast to this, P has merely one value for $p > p_c$. This means the percolation critical state problem is not in the same universality class as the Ising model/gas-liquid critical state class and its critical state exponents differ from those of the latter. See Table 7-1.

The generalized susceptibility, χ, in the percolation model [analogous to the compressibility of the gas-liquid case as given in Eq. (3-129)] is the average number of sites in a cluster to which a randomly selected site belongs. The total number of clusters containing s sites each, divided by the total number of sites on the lattice, is denoted n_s. Thus sn_s is the probability that a randomly selected site is in a cluster of size s. Hence χ is given by

$$\chi = \sum_s s^2 n_s \qquad (7\text{-}19)$$

The sum in Eq. (7-19) omits the single largest cluster. For $p \gg p_c$ most sites will belong to this cluster (of infinite size in the limit of an infinitely large lattice) and χ will be small. For $p \ll p_c$ most sites will be empty or belong to small clusters, so again χ will be small. As $p \to p_c$ from above or below, χ will diverge (in the limit of an infinite lattice) according to the power law:

$$\chi \propto |p - p_c|^{-\gamma} \qquad (7\text{-}20)$$

The correlation length, ζ, for the percolation problem is effectively the radius of the dominant clusters that cause χ to diverge. Its power law is taken to be

$$\zeta \propto |p - p_c|^{-\nu} \qquad (7\text{-}21)$$

A scaling law analysis [12] similar to that which leads to Eq. (3-143) and Eq. (3-146) shows that on a lattice of Euclidean dimension E,

$$\gamma + 2\beta = \nu E \qquad (7\text{-}22)$$

TABLE 7-1 Percolation Model (Network Polymer) Exponents

Exponent	$E = 2$	$E = 3$	Mean Field Limit
β	$5/36$	0.44	1
γ	$43/18$	1.76	1
ν	$4/3$	0.88	$1/2$
D	$91/48$	2.5	4

Exponents are given in various Euclidean dimensions E and in the mean field or Cayley tree limit for which the generalized Flory result for the exponent, $\nu_F \equiv 1/D = 2/(E + 2)$ is exact using $E = E^* = 6$. Symbol D denotes the fractal dimensionality of the incipient infinite cluster at the onset of percolation at $p = p_c$. Data from reference [12].

Furthermore, the fractal dimension D of the incipient infinite cluster at $p = p_c$ is given by

$$D = E - (\beta/\nu) \qquad (7\text{-}23)$$

The classical or mean field limit of gelation was first worked out long ago* using different terminology. The lattice used was a Cayley tree starting from a central point from which z bonds emerge. Each bond ends at another site from which again z bonds arise, $z - 1$ of which lead to new sites (one bond connects to the origin). The branching is then imagined as being endlessly repeated. There are no closed cycles on the tree and it is assumed that there are no steric hindrances! This is a very artificial (nonphysical) assumption that could only make sense if the dimensionality of the tree is very high. Just how high is now known from modern theory [13, 14]. The exponents in the mean field limit are listed in column four of Table 7-1. The single dimensionality, $E = E^*$, at which Eq. (7-22) is satisfied by the mean field limit exponents is called the upper critical dimension. For the percolation model of network polymers, $E^* = 6$. At $E = 6$ and above the excluded volume effect is negligible and the exponents remain fixed at their mean field limits. Below $E = 6$ the exponents will vary as shown in Table 7-1. At $E = E^* = 6$, Eq. (7-23) gives $D = 4$ as the limiting fractal dimension of a network polymer. As asserted in Sec. 3-8, the dimension above which thermal correlations in a critical system become unimportant is $E^* = 4$. Evidently, geometrical correlations remain nontrivial between $E = 4$ and $E = 6$ for network polymers in the vicinity of the gel point.

There are two fundamental length scales for polymers. One is the scale of the monomer size. Here we have not been concerned with this scale (we have taken the monomers to be pointlike in the lattice model). The other scale is much longer and serves to characterize some longest linear dimension R of the polymer, for example, its radius of gyration or its mean end-to-end length. On the scale of R, properties of a polymer become independent of the detailed properties of its monomers. Then, in analogy with thermal critical phenomena, there should exist some universal scaling relations. We have discussed some of these for network polymers above. Long ago Flory[†] postulated that as N, the number of monomers, goes to infinity R should vary as

$$R = aN^{\nu_F} \qquad (7\text{-}24)$$

where a is a constant of the order of the length of a monomer. Hence in the terminology of fractal geometry, in analogy to Eq. (7-9) and noting that Eq. (7-24) has $N \propto R^{1/\nu_F}$, we can identify the fractal dimension D of a polymer as

$$D \equiv \frac{1}{\nu_F} \qquad (7\text{-}25)$$

* W. H. Stockmayer. *J. Chem. Phys.* **11**, 45 (1943); **12**, 125 (1944).
† P. J. Flory. *J. Am. Chem. Soc.* **63**, 3083, 3091, 3096 (1941). See also a comprehensive review of this work in reference [15].

(The exponent v_F is not the same as the correlation length exponent, v.) If excluded volume effects did not exist, a linear polymer could be correctly modeled by an ordinary N-step random walk with $v_F = \frac{1}{2}$, $D = 2$ in any dimension $E \geq 2$. Thus, with

$$R = aN^{1/2} \tag{7-26}$$

Similarly, as first shown by Zimm and Stockmayer,* a branched polymer modeled by independent random walks on each branch and with neglect of excluded volume effects would have

$$R = aN^{1/4} \tag{7-27}$$

in any dimensionality $E \geq 2$ (thus with $v_F = \frac{1}{4}$).

To obtain better v_F values for real polymers many authors have generalized Flory's original approach. We follow deGennes [8] and Family [13] in what follows. Flory supposed that at any temperature the Helmholtz free energy divided by kT, $A' \equiv A/kT$, of a polymer in solution could be expressed by the sum of two terms. The first is an elastic term, A'_{el}, with a Hooke's law form,

$$A'_{el} = \frac{R^2}{R_0^2} \tag{7-28}$$

R_0 is a measure of polymer extension in the idealized condition in which there are no repulsive interactions between monomers, and R_0 is taken to be given by Eq. (7-26) or Eq. (7-27) for linear or branched polymers, respectively. This term, A'_{el}, favors small values of R (when we minimize A')—thus highly tangled arrangements of high entropy. The second term, A'_r, is a contribution from repulsive interactions. These interactions per unit volume of polymer should be proportional (with constant c) to the mean of the square of the actual monomer density, N/R^E. Then in a mean field approximation we neglect correlations between monomers and replace the mean of the square by the square of its average:

$$\frac{A'_r}{R^E} = c \overline{\left(\frac{N}{R^E}\right)^2} \cong c \frac{N^2}{(R^E)^2}$$

or

$$A'_r \cong c \frac{N^2}{R^E} \tag{7-29}$$

Thus we have

$$A' = A'_{el} + A'_r = \frac{R^2}{R_0^2} + \frac{cN^2}{R^E} \tag{7-30}$$

---

* B. H. Zimm and W. H. Stockmayer. *J. Chem. Phys.* **17**, 1301 (1949).

TABLE 7-2   Dimensional Dependence of Exponent v_F

Dimension (E)	Linear Polymer	Branched Polymer	Network Polymer at Gelation Point
2	$\frac{3}{4}$	0.64	$^{48}/_{91}$
3	0.59	$\frac{1}{2}$	0.40
4	$\frac{1}{2}$	0.45	0.31
5	—	0.40	0.27
6	—	0.32	$\frac{1}{4}$
7	—	0.28	—
8	—	$\frac{1}{4}$	—

Values believed to be exact are given as rational fractions. Data from reference [14].

Minimization of this free energy with respect to R, using R_0 for linear polymers from Eq. (7-26), gives

$$\frac{dA'}{dR} = \left(\frac{2R}{a^2 N} - \frac{E c N^2}{R^{E+1}} \right) = 0$$

leading to

$$N \propto R^D \qquad \text{with } D = \frac{E + 2}{3} \tag{7-31}$$

or

$$v_F = \frac{3}{E + 2} \tag{7-32}$$

The interpretation of these results is that for an upper critical dimension of $E = E^* = 4$ the exponent v_F will attain its mean field value of $\frac{1}{2}$ (and the fractal dimension of the linear polymer will reach the value 2). In dimensions less than 4 v_F will be given approximately by Eq. (7-32). In fact, Eq. (7-32) seems to be nearly exact when compared to simulation studies* of self-avoiding random walks in various dimensionalities. See Table 7-2.

A similar minimization of Eq. (7-30) using R_0 for branched polymers from Eq. (7-27) leads to

$$N \propto R^D \qquad \text{with } D = \frac{2}{5}(E + 2) \tag{7-33}$$

---

* C. Domb. *Advances Chem. Phys.* **15**, 229 (1969).

or

$$v_F = \frac{5}{2(E + 2)} \tag{7-34}$$

The interpretation is that for an upper critical dimension of $E = E^* = 8$ the exponent v_F will attain its mean field value of $\frac{1}{4}$ (and the fractal dimension of the branched polymer will reach the value 4).

Until now we have considered v_F values for single polymer molecules in isolation. To treat the gel point case we must model the effect on the incipient largest cluster of the competition of other clusters for volume. To do this deGennes* showed that one needs to reduce or "screen" the repulsive free energy term by dividing it by χ [see Eq. (7-20)], which is the average number of monomers in clusters other than the largest. Except near the critical point for gelation, χ is of order unity and D is unchanged from the value given by Eq. (7-33). A simple scaling argument shows that at the critical point for gelation

$$\chi \propto \mathcal{N}^{\gamma/Dv} \qquad \text{(critical region)} \tag{7-35}$$

Using the mean field values of the exponents from Table 7-1 in Eq. (7-35) shows $\chi \propto \mathcal{N}^{1/2}$ at the critical point. Then the A' for the incipient infinite cluster at the critical point is

$$A' = \frac{R^2}{R_0^2} + \frac{c\mathcal{N}^2}{\chi R^E} = \frac{R^2}{a^2 \mathcal{N}^{1/2}} + \frac{c'\mathcal{N}^{3/2}}{R^E} \tag{7-36}$$

Minimization of this leads to

$$\mathcal{N} \propto R^D \qquad \text{with } D = \frac{E + 2}{2} \tag{7-37}$$

or

$$v_F = \frac{2}{E + 2} \tag{7-38}$$

Equation (7-38) leads to our previous result that $E^* = 6$ (for v_F to attain its mean field value of $\frac{1}{4}$) for a network polymer at the gelation critical point.

Exercise 7-3

Provide the scaling argument that leads to Eq. (7-35) by noting that $\zeta \propto R$ and then using Eq. (7-21) and Eq. (7-20).

---

* P. G. deGennes. *C. R. Acad. Sci.* (Paris) **291**, 17 (1980).

FIGURE 7-6 Computer-generated aggregate of 3000 particles on a square lattice. The calculation starts with a one-particle frozen "cluster" at the center. Then each further stage sends a particle in a random walk starting from a circle circumscribing the existing cluster. When this walker reaches any lattice site adjacent to an already caught particle it is frozen in place. The walkers tend to be captured by the ends, which grow more rapidly than the interior, as seen here with the first 1500 to attach shown as open circles and the second 1500 as dots.
SOURCE: T. A. Witten and L. M. Sander. *Phys. Rev.* **B 27**, 5686 (1983).

Table 7-2 collects known v_F values for polymers from exact enumeration or computer simulation for comparison with the generalized Flory results given by Eqs. (7-32), (7-34), and (7-38).

An avalanche of papers on fractals has been roaring down the mountainous heights of the scientific literature for some years.* Many common physical and chemical systems besides polymers have been found to exhibit fractal geometry [16]. These include protein surfaces [17] and all types of aggregates and dendritic structures built up by diffusion [7, 18]. Figure 7-6 shows a computer-generated

---

* The term "fractal" was only first indexed in *Chemical Abstracts* in January, 1985.

dendritic fractal (of D about 1.7) on a square lattice. The corresponding dendritic structure in three-dimensional space has a D of about 2.4, which is the same as the fractal dimensionality of dielectric breakdown (the complex path of lightning bolts) [19]. We shall encounter fractals again in the next sections of this chapter.

7-4
Universal Properties of Nonlinear Systems

Many systems of interest to physical scientists and most studied by nonphysical scientists (including economists and politicians) are inherently nonlinear. The behavior of such systems is complex and often bizarre. It is important to enrich one's intuition by learning some of the strange properties that even simple nonlinear equations can exhibit. Recent research [20–25] has revealed hitherto unsuspected universal behavior in nonlinear systems. This behavior is most easily studied in one-variable, one-parameter cases that are now known to show behavior qualitatively similar to more complex systems. Even the onset of chaos (turbulence) in fluid flow is mainly controlled by a single parameter–the Reynolds number (see Appendix 5.B).

A useful definition of chaos is that situation in which a dynamical variable x_n (at time n) is extremely sensitive to the precise value of x_0 (at time zero) such that after a short period of time only a statistical analysis is possible for the variable even if the nonlinear dynamical process is completely deterministic. This *common* occurrence is now called deterministic chaos to distinguish it from cases in which random external effects cause chaotic behavior.

As a concrete example of the time histories of a dynamical variable, let us consider the population of insects in a localized region as time varies from one summer to the next [26–27]. The insects are born, lay eggs, and die in one summer and the next summer the process repeats. We use λ to denote a constant or static birth rate observable in the dilute population case when there is little population pressure on the environment. Then the population in the $(n + 1)$ summer is related to that in the nth summer by:

$$p_{n+1} = \lambda p_n \qquad (7\text{-}39)$$

Exercise 7-4

Show that the solution of the *difference* equation, Eq. (7-39), is

$$p_n = \lambda^n p_0$$

which represents exponential population growth if $\lambda > 1$.

As the population grows and competition for nutrients increases, Eq. (7-39) will fail and one must introduce a dynamic birth rate $\lambda_{eff}(p)$, which is a function of p. Then

$$p_{n+1} = \lambda_{eff}(p_n)p_n \qquad (7\text{-}40)$$

with

$$\lambda_{eff} < \lambda \qquad (7\text{-}41.1)$$

$$\lim_{p \to 0} \lambda_{eff}(p) = \lambda \qquad (7\text{-}41.2)$$

and

A simple form satisfying the conditions of Eq. (7-41) is

$$\lambda_{eff}(p) = \lambda - ap \qquad (7\text{-}42)$$

such that Eq. (7-40) becomes

$$p_{n+1} = (\lambda - ap_n)p_n = \lambda p_n \left(1 - \frac{a}{\lambda} p_n \right)$$

$$= \lambda p_n (1 - x_n) \qquad (7\text{-}43)$$

by defining

$$x_n = \frac{a}{\lambda} p_n \qquad (7\text{-}44)$$

Eliminating the population variables from Eq. (7-43), we have the standard non-linear form

$$x_{n+1} = \lambda x_n (1 - x_n) \qquad (7\text{-}45)$$

where

$$(1 - x_n) = \frac{p_{n+1}}{\lambda p_n}$$

is the ratio of the population in the $(n + 1)$ summer to what it would have been if the static model held true. Thus x_n and $(1 - x_n)$ can only vary from 0 to 1 and λ must be in the range of 0 to 4 to keep x_n in this range on iteration. If λ is even slightly above 4 the x_n will go to $-\infty$ on repeated iteration of Eq. (7-45).*

---

* Except for a Cantor-like *dust* (i.e., a fractal set of dimension between 0 and 1) in the range (0, 1) constituted by those points, which after one or more iterations will reach the number unity (exactly) and thus iterate to zero. See Problem 7-9.

We are interested in what happens at a fixed λ when we iterate choosing any starting x_0. Equation (7-45) is a special example of the general one-parameter equation.

$$x_{n+1} = f(\lambda, x_n) \tag{7-46}$$

The only essential condition for interesting results is that $f(\lambda, x_n)$ not be monotonic; that is, it should have a maximum in the range of values open to x_n. For Eq. (7-45) the maximum occurs at $\bar{x} = 1/2$, independent of λ. In this section we use a bar on x to indicate the x value for which $f(\lambda, x)$ is a maximum, a prime to indicate differentiation ($f' = df/dx$), and the symbol $f^{(i)}$ to indicate the ith iterate of f:

$$f^{(1)} = f(x)$$
$$f^{(2)} = f^{(1)}[f(x)]$$
$$f^{(3)} = f^{(1)}\{f^{(1)}[f(x)]\} \tag{7-47}$$

and so on.

An asterisk will denote a fixed point, that is, a point for which

$$x^* = f(\lambda, x^*) \tag{7-48}$$

meaning the function f gives back the same value as that originally chosen. For Eq. (7-45) the x^* values are found from

$$x^* = \lambda x^*(1 - x^*)$$

with solutions

$$x^* = 0 \tag{7-49}$$

and

$$x^* = \frac{\lambda - 1}{\lambda} > 0 \qquad \text{thus for } \lambda > 1 \tag{7-50}$$

However, we need to find the condition for a fixed point to be stable, by which is meant the condition for points near x^* to move closer to x^* with each iteration. We have

$$f(x^* + \delta_n) = x^* + \delta_{n+1}$$
$$= f(x^*) + f'(x^*)\delta_n + \cdots$$
$$= x^* + f'(x^*)\delta_n$$
$$\therefore \quad \delta_{n+1} = f'(x^*)\delta_n \tag{7-51}$$

and the condition for δ_{n+1} to be less than δ_n and thus for x^* to be stable is

$$|f'(x^*)| < 1 \qquad (7\text{-}52)$$

For Eq. (7-45)

$$f'(x) = \lambda(1 - 2x) \qquad (7\text{-}53)$$

such that

$$f'(0) = \lambda \qquad f'\left(\frac{\lambda - 1}{\lambda}\right) = 2 - \lambda$$

Hence for $\lambda \le 1$, $x^* = 0$ is stable and interation will lead only to the one value of zero population; that is, the insects die out if the birth rate is so low. For $1 \le \lambda \le 3$, interation from any starting population (that is neither 0 nor 1) will lead to the one value given by Eq. (7-50). This last result is shown graphically in Fig. 7-7 for

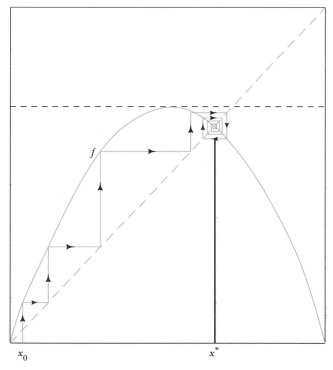

FIGURE 7-7 Iterates of x_0 starting near the unstable fixed point ($x^* = 0$) eventually move to $x^* = 0.64$ for $\lambda = 2.8$.
Source: D. R. Hofstadter. *Scientific American* **245**, 22 (Nov., 1981).

$\lambda = 2.8$. On such a graph we move vertically from the starting x_0 to the curve f to find $f(x_0)$. Then x_1, which equals $f(x_0)$, is found by moving horizontally to the equality line where $f = x$ (the straight line of slope unity). Then repeat. The vertical moves from an x_n to its $f(x_n)$ may eventually be downward, but finally we get to the x^* value, which of course is a point where the $f(x)$ curve crosses the equality line. A more global view is shown in Fig. 7-8.

For $\lambda > 3$, iteration to a single x^* value is no longer possible and we must consider next the stable fixed points of the $f^{(2)}(x)$ function, which will give rise to a stable cycle of two points. In terms of our insect populations, after a time the populations will assume only two values alternately. If we denote the two stable fixed points of $f^{(2)}$ as α and β (the fixed points of $f^{(1)}$ are also fixed points of $f^{(2)}$ but they will

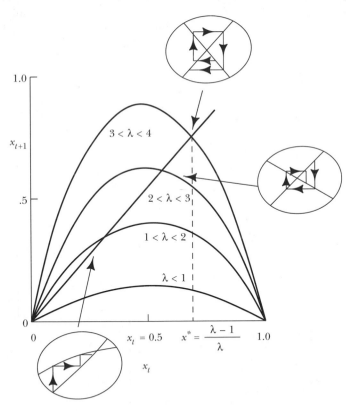

FIGURE 7-8 As λ is increased above 1, a second fixed point appears at $x^* = (\lambda - 1)/\lambda$. For values of $\lambda < 2$ the approach to this point is monotone. For $2 \leq \lambda \leq 3$ the fixed point is approached in oscillatory fashion as in Fig. 7-7. For $\lambda > 3$ the second fixed point is no longer stable since $f'(x^*)$ is then more negative than -1.
SOURCE: J. Guckenheimer, G. Oster, and A. Ipaktchi. *J. Math. Biology* **4**, 101 (1977).

not be stable for $\lambda > 3$) we have

$$\alpha = f^{(2)}(\alpha) \qquad (7\text{-}54.1)$$

and

$$\beta = f^{(2)}(\beta) \qquad (7\text{-}54.2)$$

with

$$\alpha = f^{(1)}(\beta) \qquad (7\text{-}55.1)$$

and

$$\beta = f^{(1)}(\alpha) \qquad (7\text{-}55.2)$$

necessary for consistency with the definitions in Eq. (7-47). Then

$$\begin{aligned}
f^{(2)}(\alpha + \delta_n) &\equiv \alpha + \delta_{n+2} \\
&= f^{(2)}(\alpha) + f^{(2)\prime}(\alpha)\delta_n + \dots \\
&= \alpha + f^{(2)\prime}(\alpha)\delta_n
\end{aligned}$$

but also with use of Eq. (7-55) and another Taylor series expansion

$$\begin{aligned}
f^{(2)}(\alpha + \delta_n) &= f^{(1)}[f^{(1)}(\alpha + \delta_n)] \\
&= f^{(1)}[\beta + f^{(1)\prime}(\alpha)\delta_n] \\
&= \alpha + f^{(1)\prime}(\beta)f^{(1)\prime}(\alpha)\delta_n
\end{aligned}$$

Thus

$$\delta_{n+2} = f^{(2)\prime}(\alpha)\delta_n = f^{(1)\prime}(\alpha)f^{(1)\prime}(\beta)\delta_n \qquad (7\text{-}56)$$

and the fixed point is stable if

$$\left| f^{(2)\prime}(\alpha) \right| < 1$$

The same pair of expansions applied to a point $\beta + \delta_n$ leads to

$$f^{(2)\prime}(\alpha) = f^{(2)\prime}(\beta) = f^{(1)\prime}(\alpha)f^{(1)\prime}(\beta) \qquad (7\text{-}57)$$

Equation (7-57) is the chain rule, which is easily generalized to the result that the derivatives of $f^{(i)}$ are equal at *all* the stable fixed points of the i cycle and they

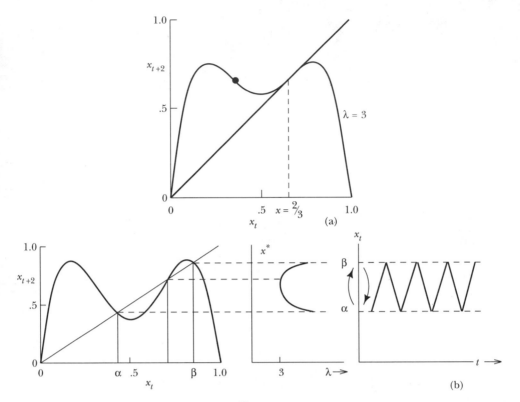

FIGURE 7-9 a. At $\lambda = 3$ the graph of $f^{(2)}$ becomes tangent to the equality line $x_{t+2} = x_t$ at the limiting x^* value $(\frac{2}{3})$ for the one cycle.
b. For $\lambda > 3$ the $f^{(2)}$ curve intersects the equality line at two new stable fixed points, α and β.
SOURCE: J. Guckenheimer, G. Oster, and A. Ipaktchi. *J. Math. Biology* **4**, 101 (1977).

are equal to the product of the derivatives of the *original function* at the x values of the cycle points of the i cycle. So if one of the cycle points is $\bar{x}$ (where the derivative of the original function is zero), all the derivatives of $f^{(i)}$ are zero at the cycle points.

As shown in Fig. 7-9(a) the $f^{(2)} \equiv x_{t+2}$ curve is precisely tangent to the equality line $x_{t+2} = x_t$ at the limiting x^* value $(\frac{2}{3})$ of the 1 cycle for $\lambda = 3$. For $\lambda > 3$, as shown in Fig. 7-9(b), the $f^{(2)}$ function intersects the equality line in two new stable fixed points (α, β) with a common slope at these points. Further detail is shown in Fig. 7-10. The new stable fixed points are "born" with common slope $f^{(2)\prime} = +1$ and when these slopes get to -1 because of further increase in λ, the period 2 case ceases to be stable. There follows another pitchfork bifurcation (so called because of its appearance, see Fig. 7-12) to two new stable cycle points arising from *each* of the others at the limit of their stability, thus to a 4 cycle representing the now stable fixed points of $f^{(4)}$ (see Fig. 7-11). The pitchfork period-doubling bifurcations follow one after another (with smaller and smaller windows of stability as regards

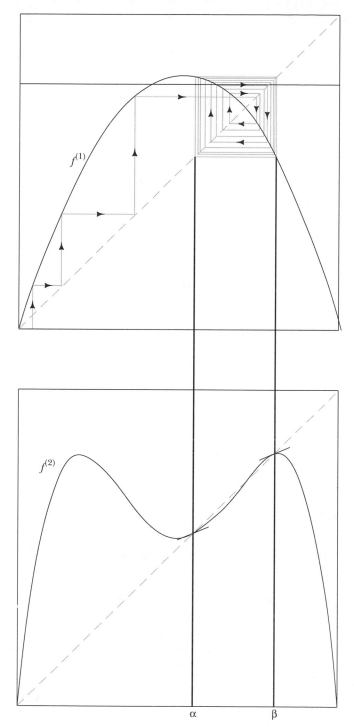

α

β

FIGURE 7-10 At the top a spiral into the stable cycle of two is shown which is correlated at the bottom with the behavior of the $f^{(2)}$ function. Note the equal slopes of the $f^{(2)}$ function at the stable cycle points.

SOURCE: D. R. Hofstadter. *Scientific American* **245**, 22 (Nov. 1981).

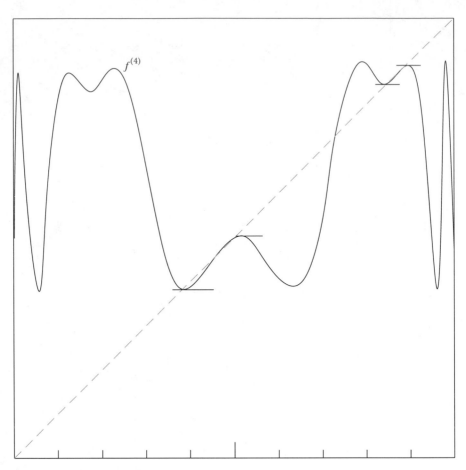

FIGURE 7-11 The $f^{(4)} \equiv x_{t+4}$ curve as a function of x_t for a λ value (3.48) such that one of the stable four cycle points is almost at $\bar{x} = 0.50$, so that the common slopes are all practically zero at the cycle points.
SOURCE: D. R. Hofstadter. *Scientific American* **245**, 22 (Nov. 1981).

range of λ) until they end at a $\lambda_\infty \equiv \lambda_c = 3.5699$ with 2^∞ cycle points. This is shown in Fig. 7-12, with numerical data summarized in Table 7-3.

Notice that when $\lambda = \lambda_c$ no odd cycles have yet occurred nor have all the even cycles been encountered, since we have had cycles of 2, 4, 8, 16 . . . but not cycles of 6, 10, 12 . . . . Odd stable cycles occur in tiny widths of λ for $\lambda > \lambda_c$ by means of a *tangent bifurcation*, as shown in Fig. 7-13 for the $f^{(3)}$ function. The new fixed points are "born" in pairs at $\lambda = 3.8284$ with $f^{(3)\prime} = +1$ for both members of the pair. However, one slope immediately goes above $+1$ so one pair member is never part of a stable cycle of 3, while the slope at the other new crossing varies from $+1$ to -1 as λ goes from 3.8284 to 3.8415. When the common slope gets to -1, period doubling to cycles of $2 \times 3 = 6, 2^2 \times 3 = 12$ . . . occurs, which cycles reach another λ_∞ point at $\lambda = 3.8495$!

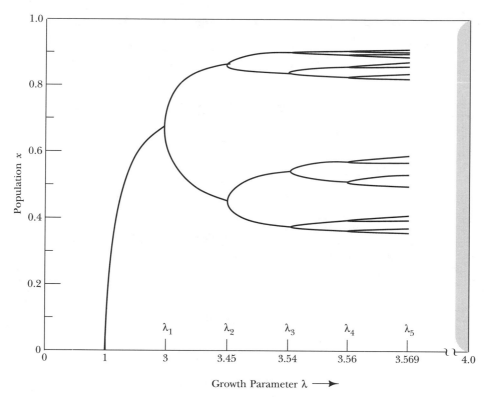

Growth Parameter $\lambda \longrightarrow$

FIGURE 7-12 Stable values for the population as a function of the parameter λ. The solid lines at any fixed λ show the values of the populations x_i that are possible for the i cycle after long enough iteration. For λ below 1 only a population of zero is possible. For $1 < \lambda \leq 3$ only a single stable nonzero population results. For $3 < \lambda \leq 3.45$ the population oscillates between the two values on the upper and lower branches of the curve. Above 3.45 the population oscillates among the four branches, and so on. Note the very nonlinear scale and the break at $\lambda_c = 3.5699$. The notation is defined in Table 7-3. On the right only the continuum of x_i values is shown as the population approaches fully chaotic behavior at $\lambda = 4$.
SOURCE: Reference [27].

Exercise 7-5

For $\lambda = 3.8400$ choose any starting x_0 value in the range of $0 < x_0 < 1$ and iterate Eq. (7-45) sufficiently to determine the three stable cycle 3 x values to four significant figures each. It will probably require many fewer than 100 iterations on a hand-held calculator to do this. Also calculate the common slope of the $f^{(3)}$ function at the cycle points.

TABLE 7-3 Numerical Data for Figures 7-12, 7-14

$$x_{n+1} = \lambda x_n (1 - x_n)$$

There is a cycle of 2^n beginning at λ_n and extending beyond. There are 2^n bands starting at λ'_n and extending below between which x jumps in a regular way but takes on random values in each band. Λ_n is the λ value for which $x = \bar{x}$ (the x value for $f^{(1)}$ to be a maximum) is one of the fixed points of $f^{(2^n)}$.

n	2^n	λ_n	Λ_n	λ'_n
0	1	1.00	—	4.00
0	—	—	2.00	—
1	2	3.00	—	3.6785735*
1	—	—	3.2360680	—
2	4	3.4494897	—	3.5925722
2	—	—	3.4985614	—
3	8	3.5440903	—	3.5748049
3	—	—	3.5546408	—
4	16	3.5644073	—	3.5709859
4	—	—	3.5666674	—
5	32	3.5687594	—	3.5701685
5	—	—	3.5692435	—
6	64	3.5696916		3.5699935
⋮	⋮			
∞	∞		3.5699457	

* Also value above which first odd cycle enters.
SOURCE: Reference [25].

It is truly amazing that the simple Eq. (7-45) gives rise to such a complex structure of stable cycles as the control parameter λ is varied. Moreover, there is a *structural universality* for *all* equations of type (7-46) as long as they have a single maximum in the range of possible x_n values. Two examples of other equations of type (7-46) that have been extensively studied [23, 26] are

$$x_{n+1} = \lambda \sin (\pi x_n) \tag{7-58}$$

and

$$x_{n+1} = x_n \exp [\lambda(1 - x_n)] \tag{7-59}$$

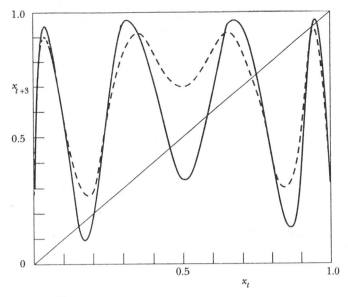

FIGURE 7-13 The $f^{(3)} \equiv x_{t+3}$ curve as a function of x_t for $\lambda = 3.70$ (dashed curve) and for $\lambda = 3.90$ (solid curve), which has six new intersections with the equality line but all are unstable; that is, $|f^{(3)\prime}| > 1$ at all six crossings. However at $\lambda = 3.8284$ the two left minima and the rightmost maximum of the $\lambda = 3.70$ curve move to be simultaneously tangent to the equality line of slope unity, giving "birth" to three stable solutions, which at $\lambda = 3.8415$ reach the end of their range of stability for the period 3 case.
SOURCE: E. Ott. *Rev. Mod. Phys.* **53**, 655 (1981).

The structural universality consists of an order relation for the *first* appearance of a periodic cycle of length i as λ is increased [23, 28]. If we let

$$A = \text{Set } \{2^n l; \quad n \geq 0, l \geq 3 \text{ for } l \text{ odd}\}$$

$$B = \text{Set } \{2^m; \quad m \geq 0\}$$

then for the first appearance A comes after all B, with the higher n values in A appearing *before* the lower n values such that the order of first occurrence is

$$3 \leftarrow 5 \leftarrow 7 \leftarrow 9 \ldots (2)(3) \leftarrow (2)(5) \leftarrow (2)(7) \ldots (4)(3) \leftarrow$$
$$(4)(5) \leftarrow (4)(7) \ldots (8)(3) \leftarrow (8)(5) \leftarrow (8)(7) \ldots \leftarrow 2^\infty \ldots \leftarrow$$
$$16 \leftarrow 8 \leftarrow 4 \leftarrow 2 \leftarrow 1 \tag{7-60}$$

The periodic cycle of length 3 is the *last* to first enter as λ is increased. We note that for Eq. (7-45) all the above periodic first entries past the fundamental set of period doublings occur between $\lambda = 3.5699$ and $\lambda = 3.8284$. Once a period type occurs there will generally be more of the same period found stable at larger λ values in tiny ranges. They differ in the order of visitation about the position of the maximum ($\bar{x}$) in the iteration pattern of the period. There are two different stable cycles of 4, three different stable cycles of 5, and so on. Each stable iteration pattern occurs only once and for a given value of λ *at most* only one periodic cycle of a particular iteration pattern can be stable. We say at most because for $\lambda > \lambda_c$ of the fundamental set B of period doublings, much of the range of λ only exhibits chaotic behavior in a sense to be defined below.

We continue our treatment of Eq. (7-45) with a description of the fully chaotic behavior at $\lambda = 4$. Make a change of variable from x_n to θ_n

$$x_n = \left[\frac{1 - \cos{(2\pi\theta_n)}}{2} \right] \tag{7-61}$$

so that Eq. (7-45) for $\lambda = 4$ becomes

$$\left[\frac{1 - \cos{(2\pi\theta_{n+1})}}{2} \right] = 4 \left(\frac{1}{2} \right) [1 - \cos{(2\pi\theta_n)}] \left(\frac{1}{2} \right) [1 + \cos{(2\pi\theta_n)}]$$

$$= \frac{1}{2} \{ 2[1 - \cos^2{(2\pi\theta_n)}] \}$$

$$= \frac{1}{2} \{ 2 - [1 + \cos{(4\pi\theta_n)}] \}$$

$$= \frac{1 - \cos{(4\pi\theta_n)}}{2} \tag{7-62}$$

In the third line of Eq. (7-62) we used the identity

$$2 \cos^2 \left(\frac{a}{2} \right) = 1 + \cos a$$

Thus by comparing the left and right sides of Eq. (7-62) the solution of the transformed Eq. (7-45) is explicitly found (for $\lambda = 4$ *only*) to be

$$\theta_{n+1} = 2\theta_n \tag{7-63}$$

or

$$\theta_n = 2^n \theta_0 \tag{7-64}$$

Since x_n depends on $\cos{(2\pi\theta_n)}$, with $2\pi\theta_n$ in radians and θ_n dimensionless, adding an integer to θ_n or changing its sign will *not* change the value of x_n. We need only consider the decimal part of θ_n. We use modulo 1 arithmetic and can throw away any integer part of θ_n. Better yet, if we work in base 2 rather than base 10 such that, for example,

$$\text{If } \theta_0 = \tfrac{1}{2} + \tfrac{1}{8} + \tfrac{1}{16} + \tfrac{1}{64} + \tfrac{1}{256} \cdots$$

$$\text{then } \theta_0 = 0.71 \ldots \cdot_{10}$$

$$= 0.10110101 \ldots _2$$

the multiplication by 2 required in the solution of Eq. (7-63) merely gives a shift of the "binary point" one position to the right. After discarding the integer part we would find, starting with the above θ_0,

$$\theta_1 = 0.0110101 \ldots$$

$$\theta_2 = 0.110101 \ldots$$

$$\theta_3 = 0.10101 \ldots$$

such that θ_n will depend on the nth and higher bits in θ_0. Hence, we can see the chaos in the solution of Eq. (7-63) very directly. Unless we have infinite precision in the specification of our initial state (θ_0), all initial state data are rapidly *lost*. The important philosophical implications of this type of result for science is discussed in reference [29]. Hence Eq. (7-45) with $\lambda = 4.0$, on iteration from any $0 < x_0 < 1$ will, after a short initial period, generate random numbers in the full range between 0 and 1 and was suggested for use in this way for the first generation of computers.*

Exercise 7-6

Show that if two initial θ_0 values differ by the small positive quantity ε, then upon separate iteration with Eq. (7-63) after about $n = \ln (1/\varepsilon)/\ln 2$ steps of iteration (using modulo 1 arithmetic) all closeness in the two results must be lost.

Exercise 7-7

The result in Exercise 7-6 means that after the same number of iterations of Eq. (7-45) with $\lambda = 4.0$, all closeness in starting x_0 values must be lost. What is the value of n if $x_0 = 0.62000$ and $x_0' = 0.62010$? Calculate the ε from the difference in θ_0 values using Eq. (7-61).

---

* S. M. Ulam and J. von Neumann. *Bull. Am. Math. Soc.* **53**, 1120 (1947).

Exercise 7-8

Illustrate your result in Exercise 7-7 by using a calculator to iterate Eq. (7-45) with $\lambda = 4.0$, starting with $x_0 = 0.62000$ and then with $x_0' = 0.62010$.

We see from Eq. (7-61) that θ_n need only vary from 0 to $\frac{1}{2}$ since $(1 - \theta_n)$ gives the same x_n as does θ_n and, for example, $\theta_n = 0.4$ gives the same result as $\theta_n = 0.6$. Further, since $2\theta_n$ and $(1 - 2\theta_n)$ are also equivalent it is useful to write the solution given by Eq. (7-63) in the form:

$$\theta_{n+1} = \begin{array}{ll} 2\theta_n & \text{for } 0 \leq \theta_n \leq \frac{1}{4} \\ (1 - 2\theta_n) & \text{for } \frac{1}{4} \leq \theta_n \leq \frac{1}{2} \end{array} \qquad (7\text{-}65)$$

This enables us to see that for θ_n the precise ratio of any two integers (i.e., any rational number) a particular cycle in θ_n and thus also in x_n exists. We can also use Eq. (7-65) to calculate the values giving rise to a particular cycle. *Unstable* periodic solutions of all possible periods exist at $\lambda = 4$.* If $y < \frac{1}{4}$ is the lowest θ_n value of a cycle then, for example, the two sets of period 3, using Eq. (7-65), are

(1)	$y, 2y, 4y$	assuming $2y < \frac{1}{4}$
with	$1 - 2(4y) = y$	as condition for a cycle of 3
such that	$1 - 8y = y$	
	$y = \frac{1}{9}$	

and the θ values are $\frac{1}{9}, \frac{2}{9}, \frac{4}{9}$, corresponding to x values of 0.11698, 0.41318, and 0.96985; and

(2)	$y, 2y, 1 - 2(2y)$	assuming $2y > 1/4$
with	$1 - 2(1 - 4y) = y$	such that $y = 1/7$

and the θ values are $\frac{1}{7}, \frac{2}{7}, \frac{3}{7}$, corresponding to x values of 0.18826, 0.61126, and 0.95048.

---

* By the ordering relation of Eq. (7-60) unstable periodic solutions of all possible periods exist for Eq. (7-45) for all $\lambda > 3.8415$ but they are most numerous in the sense of the large number of different sets with the same period at $\lambda = 4$.

Exercise 7-9

Show that the two 3-cycle results given above for $\lambda = 4.0$ both pertain to unstable cycles. Use the generalization of Eq. (7-57) for the $f^{(3)}$ function.

More complex behavior occurs with Eq. (7-45) between $\lambda = 4$ and $\lambda = \lambda_c$. As λ decreases from 4 the range of generally random values of x can no longer extend from 0 to 1. Since the maximum is always at $\bar{x} = \frac{1}{2}$, the maximum x obtainable by iteration is $\lambda/4$ and the minimum is $[1 - (\lambda/4)]\lambda^2/4$. This contraction of range is shown in Fig. 7-14 along with the reverse bifurcation into bands at $\lambda'_1, \lambda'_2 \ldots$ dis-

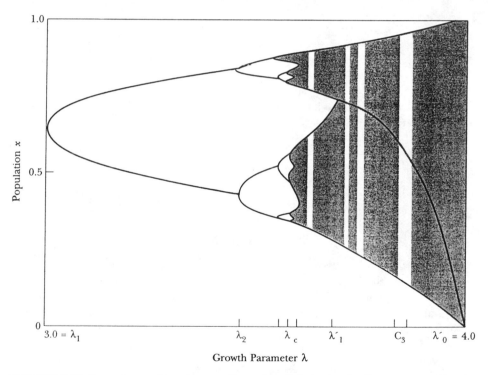

FIGURE 7-14 Smooth transition from cyclic to chaotic behavior for Eq. (7-45). At $\lambda = \lambda_c$ the 2^∞ cycle becomes an infinity of bands with the population oscillating in a regular way from band to band but taking on random values within each band. The bands merge such that for $\lambda > \lambda'_1$ there is only a single band of values that the population can randomly take on. The thin white stripes are particularly wide windows of λ in which a stable period dominates the behavior of the iterates, which are then only intermittently chaotic. The widest of these is C_3, corresponding to the stable cycle of three. The λ scale is here linear, in contrast to Figure 7-12.
SOURCE: Reference [27].

covered by Lorenz [25]. Below $\lambda = \lambda_1' = 3.6786$ the population is randomly some-where in the lower band on even steps and randomly somewhere in the higher band on odd steps of iteration. Also, a widening central region of x values is totally excluded. Below $\lambda_2' = 3.5926$ there are four allowed bands in which the x's may be randomly found. Finally, in a perfect fit to the infinite number of lines, as λ rises to λ_c one has an infinite number of infinitesimally thin bands as λ reaches λ_c from above. The actual numerical values for Eq. (7-45) are collected in Table 7-3.

Exercise 7-10

$\lambda = 3.585$ is in the four-band region. Iterate Eq. (7-45) starting with $x_0 = 0.6200$ and determine the ranges of the four bands and their order.

Exercise 7-11

$\lambda = 3.628$ is in the two-band region but also near the middle of the window for a stable period of 6. Iterate Eq. (7-45) starting with $x_0 = 0.6200$ to determine the ranges of the two bands and the six stable cycle 6 x values to five significant figures each.

Example 7-2

A few (only) of the numerical results in Table 7-3 may be obtained analytically. Write down the equation for the second iterate of Eq. (7-45) and solve for its new stable (cycle 2) fixed points. Use Eq. (7-57) to express $f^{(2)\prime}$ at the cycle points in terms of λ only and solve for λ_2 where cycle 4 first becomes stable and cycle 2 ceases to be stable. Systematic notation is defined in Table 7-3.

Solution:

From Eq. (7-47)

$$f^{(1)}(x) = \lambda x(1 - x)$$
$$f^{(2)}(x) = f^{(1)}[\lambda x(1 - x)] = \lambda^2 x(1 - x)[1 - \lambda x(1 - x)]$$

In general, $f^{(n)}(x)$ is a polynomial in x of order 2^n, and analytically to obtain the roots of $f^{(n)}(x^*) = x^*$, and so the fixed points, is not possible. To simplify our notation, use $y = x^*$. We may divide out the factors for the roots corresponding to the fixed points of $f^{(1)}$ itself that *are* roots of $f^{(2)}(y) = y$. These factors are y and $(1 - \lambda + \lambda y)$, corresponding to $y = 0$ and $y = (\lambda - 1)/\lambda$ from Eqs. (7-49) and (7-50). Our equation is

$$y = \lambda^2 y (1 - y) [1 - \lambda y (1 - y)]$$

which after division by y becomes

$$-1 + \lambda^2 (1 - y) - \lambda^3 (1 - y)^2 y = 0$$

which after division by $(1 - \lambda + \lambda y)$ becomes

$$-1 - \lambda + \lambda y + \lambda^2 y - \lambda^2 y^2 = 0$$

The solution of this quadratic is:

$$y = \frac{(1 + \lambda) \pm \sqrt{(\lambda - 3)(\lambda + 1)}}{2\lambda}$$

for the new cycle 2 fixed points. For these to be real, $\lambda \geq 3$. At $\lambda = 3$ there is only the one solution, $y = \frac{2}{3}$, corresponding to the end of the period 1 case.

From Eq. (7-57), denoting the two solutions of the above quadratic as y_α and y_β, we have

$$f^{(2)\prime}(y_\alpha) = f^{(2)\prime}(y_\beta) = f^{(1)\prime}(y_\alpha) f^{(1)\prime}(y_\beta)$$
$$= \lambda^2 (1 - 2y_\alpha)(1 - 2y_\beta)$$
$$= \lambda^2 [1 - 2(y_\alpha + y_\beta) + 4 y_\alpha y_\beta]$$

The sum and product of our two solutions are functions only of λ and ones in which all square roots cancel:

$$f^{(2)\prime} = \lambda^2 \left\{ 1 - \frac{2(1 + \lambda)}{2} + 4 \left[\frac{(1 + \lambda)^2}{4\lambda^2} - \frac{(\lambda - 3)(\lambda + 1)}{4\lambda^2} \right] \right\}$$
$$= \lambda^2 + (1 + \lambda)(4 - 2\lambda)$$

We can check that period 2 is "born" at $\lambda = 3$ with $f^{(2)\prime} = +1$. It ends when $f^{(2)\prime} = -1$. Thus to find λ_2 we solve

$$-1 = \lambda_2^2 + (1 + \lambda_2)(4 - 2\lambda_2)$$

to find

$$\lambda_2 = 1 + \sqrt{6} = 3.4494897$$

which checks the value given in Table 7-3.

■

Example 7-3

It is difficult to numerically solve for the λ_n with $n > 2$ since the simultaneous unknowns are the x values of all the cycle points and the λ_n. We work from the generalized form of Eq. (7-57) plus the necessary conditions:

$$\lambda x_1^*(1 - x_1^*) = x_2^*$$
$$\lambda x_2^*(1 - x_2^*) = x_3^*$$
$$\vdots \qquad \vdots$$

However to solve for the Λ_n of Table 7-3 is straightforward. Λ_n is the λ value for which one of the fixed points is $x = \bar{x}$. It is the λ of the superstable cycle. For Eq. (7-45) find Λ_0 and Λ_1 analytically and devise a simple numerical scheme to calculate $\Lambda_2, \Lambda_3 \ldots$.

Solution:

For Eq. (7-45) $\bar{x} = \frac{1}{2}$ irrespective of λ.

For period $2^0 = 1$ to have $\frac{1}{2}$ as the fixed point:

$$\frac{1}{2} = \Lambda_0(\tfrac{1}{4})$$

$$\Lambda_0 = 2$$

This is also the result of solving Eq. (7-50) with $x^* = \frac{1}{2}$.

For period $2^1 = 2$ to have $\frac{1}{2}$ as fixed point:

$$x_1 = \Lambda_1\left(\frac{1}{4}\right)$$

$$x_2 = \frac{1}{2} = \Lambda_1 x_1(1 - x_1) = \frac{\Lambda_1^2}{4}\left(1 - \frac{\Lambda_1}{4}\right)$$

The solution of this quadratic is:

$$\Lambda_1 = 1 + \sqrt{5} = 3.2360680$$

For period $2^2 = 4$, guess a Λ_2 and start with $x_0 = \frac{1}{2}$ and calculate x_4. Slightly change Λ_2 and find a new x_4 to determine $dx_4/d\Lambda_2$, which is used to correct the guess until x_4 is precisely $\frac{1}{2}$.

We start by guessing $\Lambda_2 = 3.4500$ and find $x_4 = 0.477577$, then guess $\Lambda_2 = 3.4510$ and find $x_4 = 0.477894$. Thus:

$$\frac{dx_4}{d\Lambda_2} \cong \frac{0.000317}{0.0010} = 0.317$$

We want dx_4 to be $+0.022106$ (to reach $\frac{1}{2}$) from our 3.4510 guess

$$\therefore \quad d\Lambda_2 = \cong \frac{0.022106}{0.317} = 0.0697$$

Thus the next guess for $\Lambda_2 = 3.4510 + 0.0697 = 3.5207$, which yields $x_4 = 0.5154178$. Also guess $\Lambda_2 = 3.5208$, which yields $x_4 = 0.5154954$. We modify our

$$\frac{dx_4}{d\Lambda_2} = \frac{0.0000776}{0.0001} = 0.776$$

Now we want $dx_4 = -0.0154178$ when using $\Lambda_2 = 3.5207$, thus

$$d\Lambda_2 = \frac{-0.0154178}{0.776} = -0.01987$$

and the next

$$\Lambda_2 = 3.5207 - 0.01987 = 3.50083$$

Continuing in this way we obtain finally

$$\Lambda_2 = 3.4985614$$

For period $2^3 = 8$ we must find the Λ_3 that forces x_8 to be precisely $\frac{1}{2}$ when $x_0 = \frac{1}{2}$.

■

Even more significant than the structural universality mentioned above is the *metrical universality* for one-dimensional (one-variable) nonlinear systems discovered by Feigenbaum [20–22] in 1975. This depends only on the nature of the function f near its one maximum. In detail it depends on the leading power in the expansion

$$f(x) - f(\bar{x}) \sim |x - \bar{x}|^q \tag{7-66}$$

In the case for which $f'(x) = 0$ and $f''(x) < 0$, that is, the usual condition for a maximum, $q = 2$.* The rate of bifurcation is found to be given by a constant δ, where

$$\delta = \lim_{n \to \infty} \frac{\lambda_n - \lambda_{n-1}}{\lambda_{n+1} - \lambda_n} \tag{7-67}$$

and for $q = 2$ the value of δ has been found to be

$$\delta = 4.6692016 \ldots \qquad (q = 2) \tag{7-68}$$

This also means we can write

$$\lambda_c - \lambda_n = B(\delta)^{-n} \qquad \text{for large } n \tag{7-69}$$

---

* However if $f''(x)$ also vanishes then a maximum can still occur with $q = 4, 6 \ldots$ in special cases.

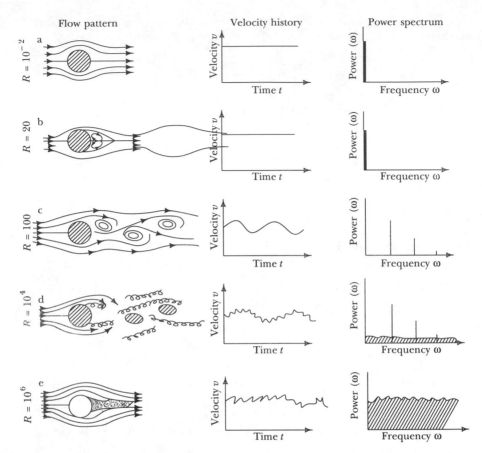

FIGURE 7-15 Patterns of fluid flow past a cylinder for different Reynolds numbers R. Fluid velocity at a point downstream is measured as a function of time from which the velocity power spectrum can be calculated. For small R the flow has a velocity constant in time (a); as R is increased the velocity takes on two values periodically (c); and finally becomes chaotic or turbulent (e).
Source. Reference [27].

where B is another constant but one that depends on f itself and not only on q. Equations (7-67), (7-68), and (7-69) also hold with λ_n replaced by Λ_n or by λ'_n. There also exists a universal scaling parameter α (only dependent on q) in the following sense. For $\lambda = \Lambda_n$ the 2^n iterate of $\bar{x}$ will give $\bar{x}$ exactly and the 2^{n-1} iterate of $\bar{x}$ will be the member of the cycle closest to $\bar{x}$ because these two points were coincident before the nth period doubling began to separate them. Define

$$d_n = \left| f^{(2^{n-1})}(\Lambda_n, \bar{x}) - \bar{x} \right| \tag{7-70}$$

which is the absolute value of the difference in x positions of the maximum and its closest fellow cycle member in the superstable cycle of 2^n members. The value of

d_n decreases as n increases and the definition of α is:

$$\alpha = \lim_{n \to \infty} \left(\frac{d_n}{d_{n+1}} \right) \qquad (7\text{-}71)$$

The numerical value of α has been found to be:

$$\alpha = 2.5029079 \ldots \qquad (q = 2) \qquad (7\text{-}72)$$

α shows how the separation of iterates in the vicinity of the x value of the maximum for f scales down as the periodicity is doubled. The separation of all close pairs does not scale uniformly. For close pairs furthest (in the sense of their midpoint) from $\bar{x}$ the separation decreases as α^q, thus by 6.2546 for $q = 2$. Intermediately separated (from $\bar{x}$) pairs will have their separations decrease by factors between α and α^q as $n \to \infty$.

The universal features discussed above have been observed in experiments on real physical and chemical reaction systems approaching chaos as a control variable (analogous to λ) is varied [30–31]. Figure 7-15 gives a schematic view of a study of fluid flow past a cylinder as a function of the Reynolds number of the flow. In this case the dynamical variable measured is fluid velocity at a point downstream from the cylinder. Another well-studied flow system is a thin layer of fluid heated from below (Rayleigh-Bénard flow) and controlled by the Rayleigh number, which is proportional to the temperature difference between the plates. There is no net mass flow (convection) in the fluid until the temperature difference is sufficiently large. The first time-independent convection pattern consists of stable convection rolls with two possible rotational directions, alternately clockwise and counterclockwise, as shown in Fig. 7-16. At still larger temperature differences, if

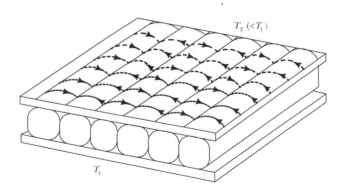

FIGURE 7-16 Reyleigh-Bénard cell with the first convective stable cycle at sufficiently large $(T_1 - T_2)$ such that the hotter, less dense fluid succeeds in flowing upward while the cooler, more dense fluid flows downward in stable convective rolls with two possible rotational directions.
SOURCE: E. Cohen and R. Schmitz in *Fundamental Problems in Statistical Mechanics*, vol. VI, E. Cohen, ed. North Holland, Amsterdam, 1985. p. 230.

one measures temperature at a point in the cell, period doubling in the temperature power spectrum (see below and Fig. 7-17) is observed. In the famous Belousov-Zhabotinski reaction [31]—the cerium-catalyzed oxidation and bromination of malonic acid by a sulfuric acid solution of bromate ion—the control variable is the common rate of flow of initially fixed concentration reagents through a well-stirred reaction vessel. Time histories of the bromide ion concentration in the reactor are collected with a specific ion probe. In all these examples the time dependence of a dynamical variable $V(t)$ is measured at sequential time intervals:

$$t_n = n(\Delta t) \qquad \text{where } n = 1, 2, \ldots, N$$

This time series $V(t_n)$ is recorded in a computer and its power spectral density $P(\omega)$ defined by

$$P(\omega) = |V(\omega)|^2 \tag{7-73}$$

where

$$V(\omega) = \frac{1}{\sqrt{N(\Delta t)}} \int_0^{N(\Delta t)} dt \, e^{2\pi i \omega t} V(t) \tag{7-74}$$

is calculated. As can be shown mathematically from the above equations

For $V(t)$ time independent, $P(\omega) = 0$ except for a peak at $\omega = 0$

For $V(t)$ fully chaotic, $P(\omega)$ is essentially continuous—no peaks

For $V(t)$ in a cycle of two, $P(\omega)$ has a main peak at ω_0 equal to the inverse of the time to go one step and a much less intense peak at $\frac{1}{2}\omega_0$, which is the inverse of the time to go through the two steps of the cycle.

And so on.

In Fig. 7-15 the schematic power spectrum of velocity is shown. In Fig. 7-17 the logarithm of an experimental* power spectrum of temperature in a Rayleigh-Bénard cell is compared with the theoretical postions and their heights. Normalizing in a certain sense from the $n = 2$ peaks, the theoretically predicted averages of the $n = 3$ and $n = 4$ peaks are drawn as horizontal lines. Each of these lines should be down in magnitude by 8.18 decibels from the respective higher (by 1) n value if α is given by Eq. (7-72) [32]. The experimental $n = 3$ component peaks are down an average of 8.4 $\pm$0.5 from the $n = 2$ while the $n = 4$ peaks are down from the $n = 3$ peaks by 8.3 $\pm$0.4. Within the experimental uncertainties, in this experiment theory is confirmed to almost two significant figures. The experiment used a temperature recording device in a surface near the upper sections of the convective rolls (shown schematically in Fig. 7-16) in convecting liquid ^{4}He. Regular oscillations

---

* J. Maurer and A. Libchaber, *J. Phys. Lett.* (Paris) **40**, 419 (1979).

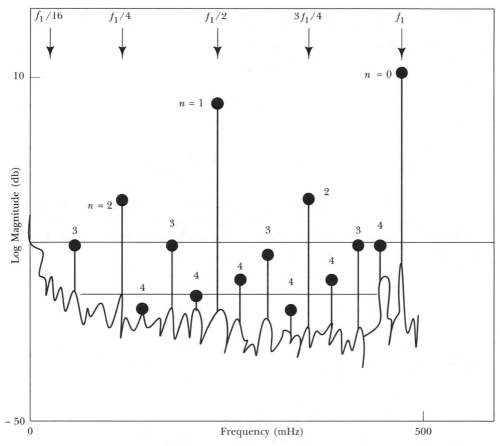

FIGURE 7-17 Experimental observation of period-doubling behavior in Rayleigh-Bénard flow. Ten times the logarithm (decibels) of the power spectrum of temperature versus frequency (in millihertz). See text for further details where $f \equiv \omega$.
SOURCE: *Physics Today* (Mar. 1981). p.17.

of temperature were observed and their time periodicity τ' was determined. As the Rayleigh number of the flow was increased to the bifurcation point the measured temperature no longer precisely repeated its value after time τ' but did so after time $2\tau'$. In the power spectrum a small amount of "noise" of low intensity appeared to enter at half the fundamental frequency (which was about 0.45 Hz for $n = 0$ in Fig. 7-17, corresponding to a time period $\tau' = 2.2$ s before any period doubling).

The following qualitative ideas are the most important of this section. Although not all real systems approach chaotic behavior via period doubling [30] and experiment is necessary to determine under what conditions systems will show such doubling, once the period doubling regime is reached the metrical universality discussed above will hold irrespective of the specific details, such as whether we have a fluid flow or an oscillating chemical reaction system, and so on. There are many analogies between these universal dynamical properties and critical point phenomena (see reference [28] and Sec. 3-8). The control parameter λ is analogous to temperature, λ_c is analogous to the critical temperature, and it is the cycle length, l, as function of λ_n that diverges at λ_c. In fact, one can show

$$l(\lambda_n) \propto (\lambda_c - \lambda_n)^{-\tau} \tag{7-75}$$

where

$$\tau = \frac{\ln 2}{\ln \delta} = 0.4498 \tag{7-76}$$

Thus δ is analogous to a critical point exponent.

A quantitative measure of chaos is provided by the Lyapunov exponent, which we now introduce as the final topic of this section. If two initial values for iteration in Eq. (7-46) differ by the small positive quantity $\varepsilon_0 = x_0 - x_0'$, then after one iteration they will differ by the positive quantity:

$$\varepsilon_1 = \varepsilon_0 |f'(x_0)| \tag{7-77}$$

since from Eq. (7-46)

$$x_{n+1} = f(x_n)$$

and in particular

$$\begin{aligned} f(x_0) - f(x_0') &= f(x_0) - [f(x_0) + (x_0' - x_0)f'(x_0) + \ldots] \\ &= \varepsilon_0 f'(x_0) \end{aligned}$$

Similarly, after two iterations they will differ by the positive quantity

$$\varepsilon_2 = \varepsilon_1 |f'[f(x_0)]| = \varepsilon_1 |f'(x_1)| = \varepsilon_0 |f'(x_0)| |f'(x_1)| \tag{7-78}$$

and after n iterations the difference is

$$\varepsilon_n = \varepsilon_0 \prod_{i=0}^{n-1} |f'(x_i)| = \varepsilon_0 \exp \sum_{i=0}^{n-1} \ln |f'(x_i)| \tag{7-79.1}$$

$$\equiv \varepsilon_0 e^{n\mathscr{L}} \tag{7-79.2}$$

in which the Lyapunov exponent, $\mathscr{L}$, is defined by

$$\mathscr{L} = \lim_{n \to \infty} \frac{1}{n} \sum_{i=0}^{n-1} \ln |f'(x_i)| \tag{7-80}$$

If $\mathscr{L}$ is less than zero the separation of nearby points on iteration decreases exponentially and nearby orbits (of iteration) converge to the same stable cycle. However if $\mathscr{L} > 0$ the separation grows exponentially, which is analogous to the way we have defined chaos earlier in this section. $\mathscr{L}$ is a function of λ and can only rise above zero for $\lambda > \lambda_c$. Its behavior as numerically determined for Eq. (7-45) is

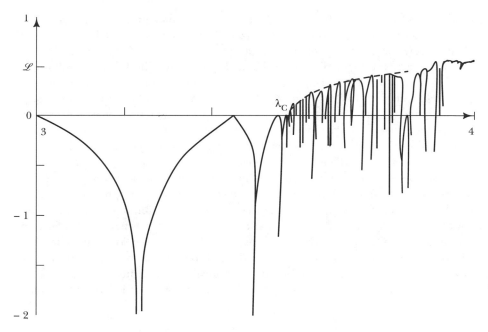

FIGURE 7-18 Lyapunov exponent $\mathscr{L}$ for the population model of Eq. (7-45). Generally chaotic behavior occurs for $\mathscr{L} > 0$.
SOURCE: Reference [28].

shown in Fig. 7-18. $\mathscr{L}$ diverges to $-\infty$ for λ equal to any Λ_n, which means one recurring value of x is bound to be $\bar{x}$ and thus bound to have an $f' = 0$. $\mathscr{L}$ is zero at the end of every stable cycle because the sets of x_i come in groups whose sum of $\ln |f'(x_i)|$ values must add to zero by Eq. (7-57). There are still dips in $\mathscr{L}$ to negative values for $\lambda > \lambda_c$ corresponding to the stable cycles in small ranges of λ past λ_c, but the dotted line "envelope" for $\mathscr{L}$ in Fig. 7-18 rises steadily as

$$\mathscr{L} = A(\lambda - \lambda_c)^\tau \qquad \text{for } \lambda \geq \lambda_c \tag{7-81}$$

where A is a constant of order unity and τ is given by Eq. (7-76).

7-5

Cellular Automata as Exactly Computable Models of Self-Organizing Behavior

What are now called cellular automata were first introduced by John von Neumann,* the mathematician and father of the modern computer, in collaboration

---

* J. von Neumann. *Collected Works*, vol. 5, A. H. Taub, ed. Pergamon, Oxford, 1963. p. 288.

with S. Ulam in 1948–1950. He sought to show that biological processes such as the reproduction and evolution of organized forms could be modeled by simple cells following local rules as to the change of a cell variable with time. He was successful. Revised versions of his ideas have been used by John Conway and others to design a very interesting computer program (which is more than a game) appropriately called "Life," which is a cellular automaton in two dimensions [33].

In general, a cellular automaton consists of a regular lattice of sites in one or more dimensions with a variable at each site (or cell) that can assume k possible discrete values. Each variable at every site is simultaneously changed at discrete time steps according to a deterministic rule that specifies what new value the variable must take based on the current variable values at the site and in its immediate neighborhood. A cellular automaton can serve as a discrete approximation to a partial differential equation. If we take the atomistic tradition of science seriously, we may say with Toffoli [34] that cellular automata are an alternative (rather than an approximation) to differential equations in modeling science. We may go further and, quoting Vichniac [35], say, "This old tradition has taken on a modern specialization that, stated grossly, sees nature as locally and digitally computing its immediate future." Also relevant to this is Feynman's comment [36] that Vichniac cites:

It always bothers me that according to the laws as we understand them today, it takes a computing machine an infinite number of logical operations to figure out what goes on in no matter how tiny a region of space, and no matter how tiny a region of time. How can all that be going on in that tiny space . . . ? So I have often made the hypothesis that ultimately physics will not require a mathematical statement, that in the end the machinery will be revealed, and the laws will turn out to be simple, like the checker board with all its apparent complexities.

Incidentally, checkers and chess are both examples of two-dimensional cellular automata with more than nearest neighbors playing a role.

In this section we will deal with elementary cellular automata that are one-dimensional, with only two possible values at each site and with a relevant neighborhood consisting of the site itself and its two nearest neighbor sites, one on the left and one on the right. This case has been thoroughly treated by Wolfram in a significant tutorial article [37] on which we base our discussion. Wolfram has also developed the computer software [38, 39] to follow the evolution of these automata under the local rules, which for the elementary case are functions of site value triples. The two possible variable values will be taken as 1 and 0 and in the figures they will be represented by dots and blanks, respectively. The eight possible states of three adjacent sites that can assume two values at each site will be represented by the numbers 7 through 0 in base 2:

111   110   101   100   011   010   001   000

0 1 0 0 0 1 1 1 0 1 0 1 0 1 1 0 1 1 0 1 0

 0 1 0 1 1 0 1 0 0 0 0 0 1 1 0 1 1 0 0

  0 0 1 1 0 0 1 0 0 0 1 1 1 0 1 1 1

FIGURE 7-19 Evolution of a given configuration of an elementary cellular automaton under rule 90 for two time steps. At each step the value of the two end sites cannot be given since information on their prior triples is not complete.

Then any rule of transformation may be written as an eight-digit number in base 2 such that the first digit gives the value to be taken on the next time step by the central site when one has a triple 111, the second gives the next value for the central site when one has a triple 110, and so on. These rules are referred to by their numerical values in base 10. Thus rule 90 is:

$$01011010 = 1 \times 2^6 + 1 \times 2^4 + 1 \times 2^3 + 1 \times 2^1 = 90$$

An example of its use acting on a given starting configuration is shown for two time steps in Fig. 7-19. It is useful to give a functional equivalent to the abstract rules for purposes of programming them on standard serial (in distinction to parallel) processing computers. Thus rule 90 is equivalent to taking the sum modulo 2 of the values of the two adjacent sites to get the new value for the central site on the next step.

All told there are $2^8 = 256$ different elementary cellular automaton rules. However, two minor restrictions are usually applied to reduce the rules to 32 "legal" ones. We wish a starting configuration consisting only of zeroes to remain unchanged. Hence the 000 triplet should give 0 and rules whose final binary digit is 1 are not allowed. We require also that the rules be reflection-symmetric so that 100 and 001 (and separately 110 and 011) should give identical values. This means the fourth and seventh digits are to be identical and the second and fifth are to be identical. Thus the legal rules are of the form

$$\alpha_1\alpha_2\alpha_3\alpha_4\alpha_2\alpha_5\alpha_4 0 \tag{7-82}$$

Each α_i can be 1 or 0 and add up to $2^5 = 32$.

Exercise 7-12

Use Eq. (7-82) to enumerate all the legal rules for elementary cellular automata and name each one with its numerical value in base 10. Check your results with the tabulation in Fig. 7-20.

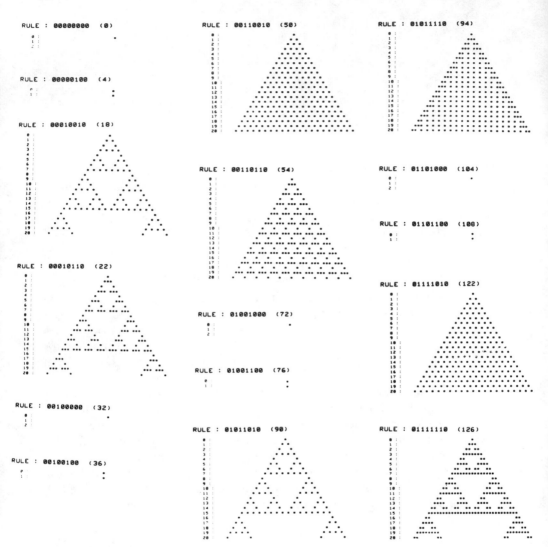

FIGURE 7-20 Evolution of elementary cellular automata according to the 32 legal rules each starting from a configuration with a single site with value 1. Sites with value 1 are represented by dots and those with value 0 by blanks. Successive lines show the configurations at successive time steps. The time evolution is carried to the time at which a specific configuration occurs for a second time or for at most 20 steps.
SOURCE: Reference [37].

Figure 7-20 shows the evolution from a single site with value 1 (analogous to growth from a seed) of elementary cellular automata under all 32 legal rules. There are two types of simple rules, which either take the initial 1 and immediately erase it (rules 0, 32, 72, 104, 128, 160, 200, and 232) or merely copy the single 1 forever

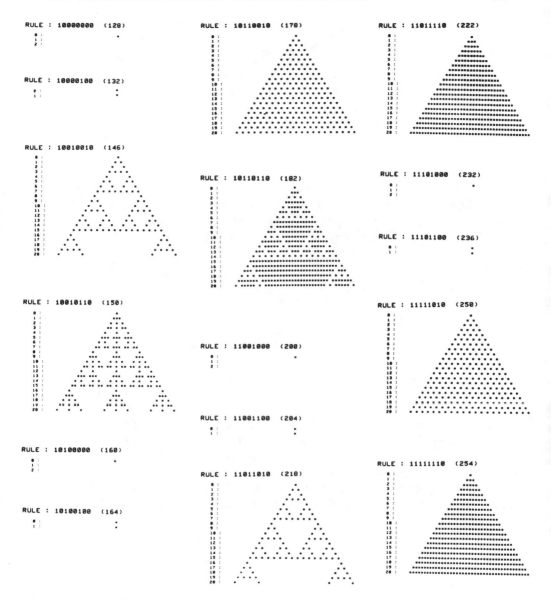

(rules 4, 36, 76, 108, 132, 164, 204, and 236). A third type of simple rule extends the set of 1's by one site in each direction at each time step and yields a completely uniform type of row (rules 50, 122, 178, 222, 250, and 254) or uniform pairs of rows (rules 54 and 94). This leaves eight complex rules constituting a fourth type (rules 18, 22, 90, 126, 146, 150, 182, and 218), which develop nontrivial patterns.

■
_____

Exercise 7-13

Explain in detail the criteria for the first two types of simple rules mentioned above with reference to the enumeration done in Exercise 7-12. (Specify the necessary local rules for certain triplets such as 100, etc.) Pick out the rules that will fit. What if anything can you say about local rules for the third and fourth types of legal rules?

_____
                                                                                                    ■

The nontrivial patterns at infinite time become self-similar fractals. By reference to Fig. 7-3 and Sec. 7-2 we recognize that all the complex rules except 150 generate a Sierpinski gasket of $D = 1.5850$. The long time limit of the pattern generated by rule 150 can be shown to be [37] a fractal with

$$D = \frac{\ln (1 + \sqrt{5})}{\ln 2} = 1.6942$$

This is a very important result. In Sec. 7-2 we described some fractals without suggesting any theories of their origin. Wolfram and others conjecture that natural fractals arise from simple (single?) initial configurations via local processes that follow the rules of a sufficiently generalized cellular automaton!

From Eq. (7-10.1) the dimension of the intersection of the long time fractal pattern of dimension D with a line perpendicular to the time axis—that is, the dimension of the set of dots at a fixed time $t = \tau$—will be D-1. This will hold as $\tau \to \infty$ except for the exceptional time steps of the form $\tau = 2^j$, which can be shown to have only two or four 1's (dots in Fig. 7-20) no matter how large j becomes.

We now turn to the self-organizing behavior of cellular automata. Figure 7-21 shows the evolution from the same specific initial disordered* configuration, 60 sites in width, under each of the 32 legal rules. The striking result is that for the complex rules (and several others as well) the independence of the initial sites is quickly lost and structure in the form of triangles of blanks or of stars emerges. One does not have self-similarity in the fractal structure arising from a random starting configuration. Whereas in the case arising from a single "seed" the number of triangles $T(n)$ of base width n for a complex rule goes as

$$T(n) \sim n^{-D} \qquad \text{with } D = 1.59 \text{ or } 1.69 \qquad (7\text{-}83)$$

when one starts from a random configuration

$$T(n) \sim C^{-n} \qquad (7\text{-}84)$$

_____

* A random initial configuration with no starting correlations between variable values at different sites, which in this example were taken to be 0 or 1, each with probability 0.50.

with C a constant.* Note also Fig. 7-22, which shows 300 steps in the evolution of a random starting configuration under rule 126. If the width of the starting configuration goes to infinity, triangles of all sizes would be generated, following Eq. (7-84).

The global properties of the elementary cellular automata may be analyzed under any rule for any finite number of sites N (thus of 2^N possible configurations). Every possible configuration is specified by a length N, base 2 integer, and the given rule defines a transformation from one such integer to another with periodic boundary conditions (as if the sites lie on a circle with the first and last sites next to each other). Assuming each configuration exists in a starting ensemble of configurations with equal probability of occurrence†, time evolution merges many configurations into a few and soon trajectories starting from almost all initial configurations are concentrated onto short cycles, called attractors, containing only a few configurations. For example, with the trivial case of rule 0, all binary numbers are in one step mapped to zero. The deterministic local cellular automaton rules are irreversible in the sense of associating a given configuration with a unique following configuration but not generally permitting a unique predecessor configuration to be inferred. Rule 204 is the exception since it is the identity and is trivially reversible. In a situation with reversible local rules the total number of possible configurations will remain constant with time. In the irreversible case here the number of possible configurations will decrease with time. This may be quantitatively followed by defining an information content entropy, S_I, which decreases with time increase starting from the equiprobable ensemble. The definition is:

$$S_I(\tau) = -\sum_i p_i(\tau) \log_2 p_i(\tau) \tag{7-85}$$

where $p_i(\tau)$ is the probability of configuration i at time step τ and the logarithm is taken to base 2. $\tau = 0$ corresponds to the equiprobable ensemble with all $p_i(0) = 2^{-N}$. This entropy is the average number of binary digits necessary to specify one configuration in an ensemble of possible configurations of N sites. Its decrease with increase of τ is shown for the case of $N = 10$ under rule 126 in Fig. 7-23.

■ ───

Exercise 7-14

Show that $S_I(0) = N$. Recall that if the logarithm to base i of a quantity A is called q, that is, $q = \log_i A$, then $i^q = A$ such that

$$q = \frac{\ln A}{\ln i} = \frac{\log_{10} A}{\log_{10} i}$$

─── ■

───────────────

* For rules 90 and 150, $C = 2$ whereas for the other complex rules $C = \frac{4}{3}$, as determined by computer simulation. Equation (7-84) holds for n greater than 5 or 6.
† This assumes we take 1's and 0's to occur with equal probability.

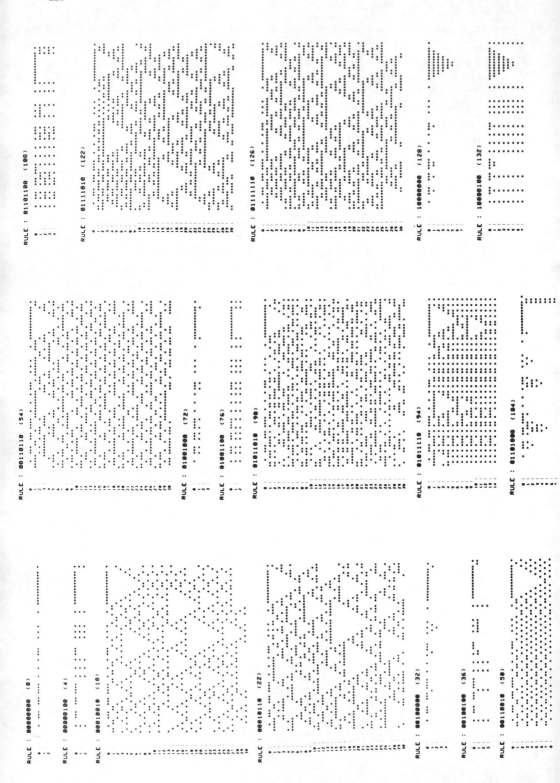

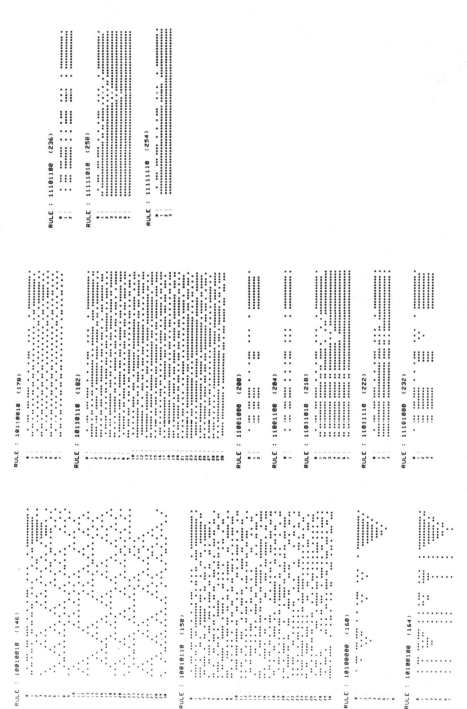

FIGURE 7-21 Evolution of a common specific starting random configuration under the 32 legal rules for elementary cellular automata. All the cellular automata are taken to satisfy periodic boundary conditions, meaning that the 60th site (the rightmost) is assumed to stand to the immediate left of the first site. Evolution is shown until a specific configuration occurs for a second time or for at most 30 steps. A rudimentary form of self-organization in the form of triangular structures develops in nearly half of the cases. Source: Reference [37].

FIGURE 7-22 Evolution of a random starting configuration under rule 126 for elementary cellular automata carried forward for 300 steps. Periodic boundary conditions are assumed. SOURCE: Reference [37].

The self-organizing behavior shown in Fig. 7-21 arises because of the decrease of entropy discussed above. The fraction of the 2^N configurations of a length N cellular automaton (with periodic boundary conditions) *not* reached by evolution from an arbitrary starting configuration will increase toward unity as N increases. This is shown in Fig. 7-24 for rule 126.

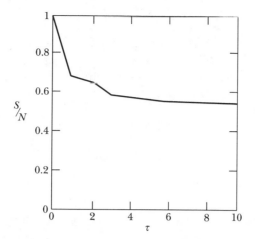

FIGURE 7-23 Decrease of the average information content entropy per site with increase of time τ for an ensemble of elementary cellular automata ($\mathcal{N} = 10$) evolving according to rule 126 from a starting equiprobable ensemble.
SOURCE: Reference [37].

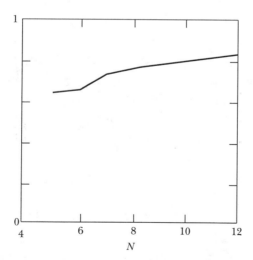

FIGURE 7-24 The fraction of the 2^N possible configurations of a length $\mathcal{N}$ elementary cellular automaton (with periodic boundary conditions) *not* reached by evolution (no matter how many time steps τ are taken) from an arbitrary initial configuration under rule 126 as a function of $\mathcal{N}$.
SOURCE: Reference [37].

■

Example 7-4

Analyze the evolution according to rule 126 of the configuration probabilities, $p_i(\tau)$, and thus of the information content entropy per site for an ensemble of elementary cellular automata with $N = 5$ evolving from an initial equiprobable ensemble. Show that $S_I(\tau)/N$ will no longer change for $\tau > 2$.

Calculate the fraction of the 32 configurations not reached for $\tau \geq 2$.

Solution:

Rule 126 causes the central site value for the triplets to become

111	110	101	100	011	010	001	000
0	1	1	1	1	1	1	0

respectively at each succeeding time step. The 32 possible starting configurations will be written as the numbers 0 to 31 in base 2 notation and referred to by their decimal names. Thus configuration $(6) = 00110$, and under one application of rule 126 with periodic boundary conditions $[(0)00110(0)]$ becomes $01111 = (15)$. Proceeding in this way the single time transformation results are

$(0) \rightarrow (0)$	$(11) \rightarrow (31)$	$(21) \rightarrow (31)$
$(1) \rightarrow (19)$	$(12) \rightarrow (30)$	$(22) \rightarrow (31)$
$(2) \rightarrow (7)$	$(13) \rightarrow (31)$	$(23) \rightarrow (28)$
$(3) \rightarrow (23)$	$(14) \rightarrow (27)$	$(24) \rightarrow (29)$
$(4) \rightarrow (14)$	$(15) \rightarrow (25)$	$(25) \rightarrow (15)$
$(5) \rightarrow (31)$	$(16) \rightarrow (25)$	$(26) \rightarrow (31)$
$(6) \rightarrow (15)$	$(17) \rightarrow (27)$	$(27) \rightarrow (14)$
$(7) \rightarrow (29)$	$(18) \rightarrow (31)$	$(28) \rightarrow (23)$
$(8) \rightarrow (28)$	$(19) \rightarrow (30)$	$(29) \rightarrow (7)$
$(9) \rightarrow (31)$	$(20) \rightarrow (31)$	$(30) \rightarrow (19)$
$(10) \rightarrow (31)$		$(31) \rightarrow (0)$

In the first time step from the 32 starting configurations we obtain each of 11 configurations—(0), (7), (14), (15), (19), (23), (25), (27), (28), (29), and (30)—twice, and configuration (31) ten times.

At $\tau = 0$, all 32 $p_i(0) = \frac{1}{32}$ and $S_I(0)/5 = 1$.

At $\tau = 1$, $p_{31}(1) = \frac{10}{32} = 0.3125$ and the 11 configurations that occur twice each have $p_i(1) = \frac{2}{32} = 0.0625$, while the other 20 configurations have $p_i(1) = 0$.

Using Eq. (7-85) and the relation given in Exercise 7-14 we have

$$S_I(1) = -(11)(0.0625) \frac{\ln 0.0625}{\ln 2} - (0.3125) \frac{\ln 0.3125}{\ln 2}$$

$$= 3.2744$$

$$\frac{S_I(1)}{5} = 0.6549$$

At the second step we see from our transformation tabulation that the ten configurations (31) all go to (0) and the two (0) configurations remain (0) such that $p_0(2) = {}^{12}\!/_{32} = 0.375$. The other ten configurations simply interchange among themselves as five cycles of two since, for example, the two configurations (19) go to two of (30) while the two of (30) go to two of (19). Hence the ten all have $p_i(2) = {}^2\!/_{32} = 0.625$ and

$$S_I(2) = -10(0.0625) \frac{\ln 0.0625}{\ln 2} - 0.375 \frac{\ln 0.375}{\ln 2}$$

$$= 3.0306$$

$$\frac{S_I(2)}{5} = 0.6061$$

At larger time steps no further change can occur since configuration (0) always remains (0) and the other ten interchange among themselves as at step $\tau = 2$. Since configuration (31) is eliminated at the second step, only 11 configurations remain at $\tau \geq 2$ and the fraction of configurations *not* reached is ${}^{21}\!/_{32} = 0.656$.

■

Cellular automata are certain to be of great importance in simulating many fundamental problems of science and in extending the theory of abstract computation itself [34, 35, 37, 40].

References and Recommended Reading

[1] B. B. Mandelbrot. *The Fractal Geometry of Nature*. W. H. Freeman and Company, New York, 1982.

[2] G. DeVaucouleurs. *Science* **167**, 1203 (1970).

[3] E. R. Harrison. *Cosmology*. Cambridge Univ. Press, 1981.

[4] R. P. Feynman and A. R. Hibbs. *Quantum Mechanics and Path Integrals*. McGraw-Hill, New York, 1965.

[5] L. F. Abbott and M. B. Wise. *Am. J. Phys.* **49**, 37 (1981).

[6] B. B. Mandelbrot. *J. Statistical Physics* **34**, 895 (1984).

[7] H. E. Stanley in *On Growth and Form: Fractal and Non-Fractal Patterns in Physics*, H. E. Stanley and N. Ostrowsky, eds. M. Nijhoff, Boston, 1986.

[8] P. G. deGennes. *Scaling Concepts in Polymer Physics*. Cornell Univ. Press, Ithaca, NY, 1979.

[9] A. Blumen, J. Klafter, and G. Zumofen. *Phys. Rev.* **B 28**, 6112 (1983).
J. Klafter and A. Blumen. *J. Chem. Phys.* **80**, 875 (1984).
L. W. Anacker, R. Kopelman, and J. S. Newhouse. *J. Statistical Physics* **36**, 591 (1984).

[10] S. Alexander and R. Orbach. *J. Physique Lett.* (Paris) **43**, L625 (1982).

[11] R. Orbach. *J. Statistical Physics* **36**, 735 (1984).

[12] D. Stauffer. *Introduction to Percolation Theory*. Taylor and Francis, London, 1985.

[13] F. Family. *J. Statistical Physics* **36**, 881 (1984).

[14] F. Family in *Random Walks and Their Applications in the Physical and Biological Sciences*, M. Shlesinger and B. West, eds. Am. Inst. Physics, New York 1984. p. 33.

[15] P. J. Flory. *Principles of Polymer Chemistry*. Cornell Univ. Press, Ithaca, NY, 1971. Chapter 12.

[16] R. Orbach in *Organic Molecular Aggregates*, P. Reineker, H. Haken, and H. C. Wolf, eds. Springer, Berlin, 1983.

[17] P. Pfeifer, U. Welz, and H. Wippermann. *Chem. Phys. Lett.* **113**, 535 (1985).

[18] *Proceedings of the International Conference on the Kinetics of Aggregation and Gelation*, F. Family and D. P. Landau, eds. North Holland, Amsterdam, (1984).

[19] L. Niemeyer, L. Pietronero, and H. Wiesmann. *Phys. Rev. Lett.* **52**, 1033 (1984).

[20] M. J. Feigenbaum. *J. Statistical Physics* **19**, 25 (1978).

[21] M. J. Feigenbaum. *J. Statistical Physics* **21**, 699 (1979).

[22] M. J. Feigenbaum. *Physica* **7D**, 16 (1983).

[23] N. Metropolis, M. L. Stein, and P. R. Stein. *J. Combinatorial Theory* **15(A)**, 25 (1973).

[24] B. Derrida, A. Gervois, and Y. Pomeau. *J. Phys.* **A 12**, 269 (1979).

[25] E. N. Lorenz. *Annals NY Acad. Sciences* **357**, 282 (1980).

[26] R. B. May. *Nature* **261**, 459 (1976).

[27] L. P. Kadanoff. *Physics Today* (Dec. 1983). p. 46.

[28] B. Hu. *Physics Reports* **91**, 233 (1982).

[29] J. Ford. *Physics Today* (Apr. 1983). p. 40.

[30] H. L. Swinney. *Physica* **7D**, 3 (1983).

[31] I. R. Epstein, K. L. Kustin, P. de Kepper, and M. Orbán. *Scientific American* **248**, 112 (Mar. 1983).

[32] M. J. Feigenbaum. *Physics Letters* **74A**, 375 (1979).

[33] M. Gardner. *Scientific American* **223**, 120 (Oct. 1970); *op. cit.* **224**, 117 (Feb. 1971).
E. R. Berlekamp, J. H. Conway, and R. K. Guy. *Winning Ways for Your Mathematical Plays*. Academic Press, New York, 1982. Chapter 25.

[34] T. Toffoli. *Physica* **10D**, 117 (1984).

[35] G. Y. Vichniac. *Physica* **10D**, 96 (1984).

[36] R. P. Feynman. *The Character of Physical Law*. MIT Press, Cambridge, MA, 1967. p. 57.

[37] S. Wolfram. *Rev. Mod. Phys.* **55**, 601 (1983).

[38] *Physics Today* (May 1983). p. 66.

[39] S. Wolfram. *Scientific American* **251**, 188 (Sept. 1984).

[40] T. Toffoli and N. Margolus. *Cellular Automata Machines*. MIT Press, Cambridge, MA, 1987.

Problems

1. Consult an atlas that shows the border of Spain with France and Andorra in several different scales of km to 1 mm on the maps. Mark off the length of the border on each map by positioning the edge of a paper repeatedly along it. Use Eq. (7-7) to estimate the dimensionality of the border.

2. Show that as the length of the Koch curve constructed as shown in Figure 7-2 goes to infinity the area it bounds goes to $\frac{8}{5}$ the area of the starting equilateral triangle. Start with a triangle of unit length for a side. Use the geometric series.

3. Mandelbrot [1] defines a similarity dimension that can be shown to be identical to the Hausdorff dimension for self-similar fractals in terms of quantities N and r used in constructing such fractals. Since:

- A straight line of unit length may be divided into $N = b$ subintervals of length $r = 1/b$ with $Nr = 1$ and
- A unit square can be divided into $N = b^2$ squares of side $r = 1/b$ with $Nr^2 = 1$, and
- A unit cube can be divided into $N = b^3$ cubes of side $r = 1/b$ with $Nr^3 = 1$

the generalization or symmetry dimension is

$$Nr^D = 1$$

or

$$D = \frac{\ln N}{\ln (1/r)} = \frac{\ln N}{\ln b}$$

Then in constructing a fractal you choose an *initiator* shape—a line or polygon or regular solid of common unit length for all of its sides—and a *generator* shape, which is usually a broken line of N oriented equal sides of length r. In each stage of construction, starting from the initiator shape, each straight line portion of the previous stage is replaced by a reduced (in linear measure by factor $r = 1/b$) copy of the generator displaced so as to have the same end points as those of the interval being replaced.

For the Koch curve of Eq. (7-6) the initiator is an equilateral triangle of unit side. The generator is ⌐∧⌐ with $N = 4$ and $r = \frac{1}{3}$ (since the starting distance between the large solid dots is unity) such that $D = \ln 4/\ln 3$, checking our previous result.

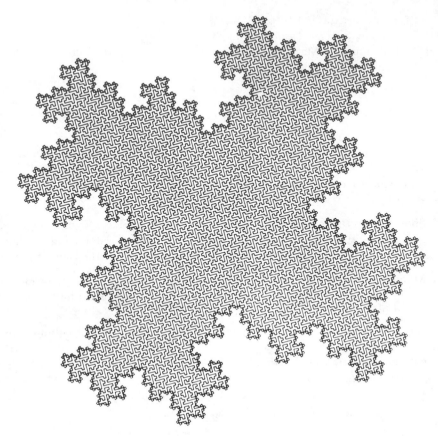

FIGURE 7-25 An advanced stage in the construction called for in Problem 3. Ignore the interior tiling shown as it is not relevant to the problem.
SOURCE: Reference [1], p. 49.

For the Cantor dust of Example (7-1) the initiator is the unit straight line and the generator is •— —• composed of $N = 2$ lengths of $r = \frac{1}{3}$. Thus $D = \ln 2/\ln 3$ as before.

Take a square of unit side as the initiator and as generator use

Show that $r = 1/\sqrt{5}$. Construct at least the first two stages past the original square on the way to the fractal. Remember that the fractal is the complex "coastline" at the edge of your figure as you go forward in construction ad infinitum. An advanced stage of this construction is shown in Fig. 7-25.

4. Use the method of Problem 3 with a square as initiator and $\bigvee$ as generator applied interiorly to the unit square to start with. Notice that on first application you get eight sides as four sets of coincident pairs and then go off both ways for each pair to get 16 sides in the second stage, and so on. Describe how this procedure leads to a fractal of dimensionality 2.

5. For a Sierpinski carpet the initator is a solid square and the generator (here more than just an oriented set of lines) replaces a solid square each time with eight solid squares of side $r = \frac{1}{3}$ by cutting out a central square, as shown in Fig. 7-26. What is the dimensionality of the Sierpinski carpet?

FIGURE 7-26 The generator and next two stages on the way to a Sierpinski carpet. SOURCE: Reference [1], p. 144.

6. Use the method of Exercise 7-1 to calculate the dimensionality of the Sierpinski carpet.

7. For a Menger sponge the initiator is a solid cube of unit side. The generator is obtained by cutting out an interior cross of five cubes of side $\frac{1}{3}$ the original cube plus two additional cubes at the center front and back of the cross, thus leaving as generator 20 solid cubes of side $\frac{1}{3}$ the original. What is the dimensionality of the Menger sponge?

8. Use the method of Exercise 7-1 in terms of *volume* measure to calculate the dimensionality of the Menger sponge described in Problem 7.

9. It is mentioned in a footnote in the text that for $\lambda > 4$ in Eq. (7-45) iteration of almost any x value will lead to unbounded decrease (to negative infinity). For $\lambda = 4.5$ illustrate this for a few values of x in the range $0-1$. However there are exceptional points of x in the range $0-1$ that will iterate precisely to unity and so iterate finally to zero. They constitute a Cantor dust of fractal dimension $0 < D < 1$. By considering the inverse mapping of

$$x' \rightarrow \lambda x - \lambda x^2$$

namely

$$\lambda x - \lambda x^2 \rightarrow x'$$

with solutions

$$x = \frac{1}{2} \pm \sqrt{0.25 - \frac{x'}{\lambda}}$$

as the two x values that will iterate to a given x' determine the first six points of this dust when $\lambda = 4.5$. At the limits $\lambda = 4$ and $\lambda = \infty$ the bounded points no longer constitute a fractal. Explain.

10. Show that for Eq. (7-45) if $\lambda > 1$ and x is even slightly less than zero, iteration must lead to negative infinity. Show that the same occurs if x is even slightly greater than unity.

11. From the data on Λ_n, with $n = 0$ to 5 in Table 7-3 for Eq. (7-45), calculate the successive approximations to Feigenbaum's constants δ and α. Use Eq. (7-70) for the d_n values with $\bar{x} = \frac{1}{2}$.

12. For the superstable cycle of 3 for Eq. (7-45) solve numerically for the λ value and the two unknown cycle points (the third is $\bar{x} = \frac{1}{2}$).

13. Use Eq. (7-65) to obtain θ values and then the x values of the cycle of 2 for Eq. (7-45) at $\lambda = 4.0$. Show that these values are the same as those obtainable from the results of Example 7-2. Also use Eq. (7-65) to generate θ values for at least two different cycles of 4 at $\lambda = 4.0$.

14. Equation (7-59) is used to model population dynamics [26] and will be discussed in this and the following two problems. Using some numerical but mainly analytical methods, work out the cycle 1 fixed point behavior, λ values for the start and end of the cycle 2 period, and equations and numerical results for the fixed points of cycle 2, denoted y_1 and y_2. Note that by the cycle property $y_1 e^{\lambda(1-y_1)} = y_2$ and $y_2 e^{\lambda(1-y_2)} = y_1$. Also prove $y_1 + y_2 = 2$.

15. Calculate the Λ_n values for Eq. (7-59) for $n = 0$ to $n = 4$ that involve the superstable cycles for which the x of the maximum, $\bar{x} = 1/\Lambda_n$, is one of the fixed points of $f^{(2^n)}$. In the numerical method of Example 7-3 you must now start with a guess (g) for $1/\Lambda_n$ and find the deviation (Δ) from your guess after iteration, then find $dg/d\Delta$ and proceed. Note that for Eq. (7-59) $\lambda_c = 2.6924$. As the Λ_n are obtained, also calculate the d_n of Eq. (7-70).

16. From your results in Problem 15 calculate the successive approximations to Feigenbaum's constants δ and α arising from iteration of Eq. (7-59). Also for each Λ_n calculate the full set of cycle points to get an idea of the allowed x values in the different periods.

17. Determine the Lyapunov exponents at $\lambda = 3.80$ and 3.90 from Fig. 7-18 and use them to calculate the constant A in Eq. (7-81).

18. An idea from critical point theory that has been applied to the λ_c condition in nonlinear dynamical problems is the following. For large n such that $\lambda_n \simeq \lambda_{n+1} \simeq \lambda_c$, the properties of $f^{(2^n)}$ should become the same as those of $f^{(2^{n+1})}$.

In particular, the 2^n cycle common slope at its cycle points should equal the common slope at the cycle points of the 2^{n+1} cycle. This provides a means of calculating λ_c. If in crude approximation we equate the $f^{(1)\prime}$ and $f^{(2)\prime}$ common cycle slopes, what are the calculated λ_c values for Eq. (7-45) and for Eq. (7-59)? In the latter case a numerical calculation is necessary. Compare your results with the precise λ_c values of these equations. This crude approximation will be found to be quite good!

19. Show that the functional equivalent to rule 150 for elementary cellular automata is the sum modulo 2 of the variable values at the adjacent sites and at the site itself at the previous time step. Also show that rule 204 is the identity function.

20. If we extend our treatment even slightly to nonelementary cellular automata, the number of rules ("legal" or not) increases tremendously. Treat the case of one-dimensional cellular automata with only the site and its two adjacent neighbors affecting the rules but now with each site permitted to take on three values (2, 1, 0). How many total rules and how many "legal" rules are now possible? Work in base 3.

21. The sequence of configurations reached by evolution from any initial configuration of a finite cellular automaton of length N must become periodic after at *most* how many steps? In practice much shorter cycles than the maximum are found. As examples find the cycles of the following initial configurations of an $N = 10$ cellular automaton according to rule 126. Use periodic boundary conditions.

$$
\begin{array}{llllllllll}
\text{(a)} & 0 & 0 & 0 & 0 & 0 & 1 & 0 & 0 & 1 & 1 \\
\text{(b)} & 0 & 0 & 1 & 1 & 0 & 0 & 1 & 0 & 1 & 1 \\
\text{(c)} & 1 & 1 & 1 & 0 & 1 & 0 & 1 & 1 & 0 & 1
\end{array}
$$

22. For an ensemble of elementary cellular automata with $N = 5$ evolving according to rule 254, carry out the analysis of Example (7-4).

23. "Life" is a cellular automaton on a two-dimensional square lattice with all eight first and second nearest neighbors (four adjacent orthogonally and four adjacent diagonally) treated equally and two variable values per cell. The values are occupied ("alive") or empty ("dead"). The *one* rule of the process has three parts for what occurs at the next time step (or generation):

Survivals: Every live cell with two or three neighboring live cells survives.
Deaths: Every live cell not with two or three neighboring live cells dies (becomes empty).
Births: Every empty cell adjacent to exactly three neighboring live cells becomes occupied.

This automaton may be followed on a large checkerboard with flat counters of two colors, as recommended by Conway [33], as follows: Start with any configuration of black counters. Locate all those that will die and place another black counter on top of each. Put white counters in each empty cell that will have a birth. After checking to make sure no errors have been made in following the rule remove all piles of two and replace all newborn whites with black counters. Then go on.

Clearly, starting configurations of less than three live cells cannot evolve to anything of interest. Work out the life histories of all the possible starting triplets and all the possible starting sets of four.

24. The starting configuration

$$
\begin{array}{ccc}
\text{X} & & \\
 & \text{X} & \\
\text{X} & \text{X} & \text{X}
\end{array}
$$

in "Life" is called a glider. Follow its evolution and explain why it is so named.

25. Show by exact summation that the sum of the gaps cut out in constructing the Cantor dust of Example 7-1 add to the complete length of the unit line. This checks the general principal that the measure of a set in a dimension higher than its fractal dimension is zero. However, it is possible to construct different Cantor dusts that no longer have the self-similarity property but that have a finite (nonzero) length. They have $D_T = 0$ but $D = 1$. One such construction involves cutting out the central $\frac{1}{3}$, then the central $\frac{1}{9}$, then the central $\frac{1}{3^3}$, and so on ad infinitum. Show in this case that the sum of cut-out gaps is accurately 0.4399 of the unit length such that the length of the resulting dust is 0.5601. For more detail on topological dimension, consult R. Courant and H. Robbins, *What Is Mathematics?* Oxford Univ. Press, 1941, 1978. pp. 248–251. For physical applications of dusts of nonzero length and area, see D. Umberger and J. D. Farmer, *Phys. Rev. Lett.* **55**, 661 (1985).

26. Explain why the order in which $f^{(n)}(\bar{x})$ reaches the value $\bar{x}$ as the value of λ increases must give the order of occurrence of stable cycles of length n. [The order of *first* occurrence will follow Eq. (7-60), but you are not being asked to prove that.] For Eq. (7-45), $\bar{x} = \frac{1}{2}$ and $f^{(1)}(\frac{1}{2}) \equiv \lambda/4$ is a straight line giving the maximum value of x obtainable by iteration as λ increases. The minimum value of x as λ increases is given by another simple curve, $f^{(2)}(\frac{1}{2}) \equiv f^{(1)}(\lambda/4) = (\lambda^2/4)(1 - \lambda/4)$, which is monotonically decreasing in the interesting range of $\lambda > 3$. Most of the other $f^{(n)}(\frac{1}{2})$ curves ($n > 2$) have several minima and maxima as λ increases from 3 to 4 and all must pass through the common value

$$
f^{(3)}\left(\frac{1}{2}\right) \equiv f^{(2)}\left(\frac{\lambda}{4}\right) = x^* = \frac{\lambda - 1}{\lambda} \tag{A}
$$

where x^* is the unstable fixed point of the cycle of one given by Eq. (7-50). Why is this so? Determine numerically from Eq. (A) the value of λ at which all the $f^{(n)}(\frac{1}{2})$ curves for $n > 2$ cross. By examining the curves of $f^{(n)}(\frac{1}{2})$ given by R. Fox et al. [*Phys. Rev.* **A 33**, 2809 (1986)], explain why Eq. (A) is also the condition on λ for the first odd cycle (an infinitely large odd number) to become stable.

27. The sequence of bifurcations of Eq. (7-45) ends in a set of infinitely many points at $\lambda_c = 3.5699457$. This set of points is a Cantor (fractal) dust called the Feigenbaum attractor. To calculate its D proceed as follows. Equation (7-3) expresses the measure of a fractal in dimension D as the sum of the measures in dimension D of cubes of side l_i covering the parts of the fractal. If we adapt this equation to the bifurcation of a line of length L_n into two parts of lengths $l_1 = L_n/s_1$ and $l_2 = L_n/s_2$, at the $(n + 1)$ bifurcation we have

$$(1)\left(\frac{L_n}{s_1}\right)^D + (1)\left(\frac{L_n}{s_2}\right)^D = (L_n)^D \tag{A}$$

which dividing by L_n^D becomes

$$s_1^{-D} + s_2^{-D} = 1 \tag{B}$$

The s_i are the factors by which L_n is divided to obtain the lengths after the $(n + 1)$ bifurcation. Every l_i at each stage of construction of the set is the distance between a point and the iterate closest to it. Figure 7-27 shows the lengths between closest iterates of Eq. (7-45) starting from $x_0 = \bar{x} = \frac{1}{2}$ at $\lambda = \lambda_c$. If this set were a uniform Cantor set like the one in Example 7-1, where any length is always divided into two of length $L_n/3$, the calculation would be immediate:

$$2(3)^{-D} = 1 \qquad D = \frac{\ln 2}{\ln 3}$$

FIGURE 7-27 Construction of the Feigenbaum period-doubling fractal dust. The numbers refer to the number of the iterate of $x_0 = \bar{x} = 0.50$ at λ_c for Eq. (7-45).

However here the s_i are not both the same and they must be determined by examining in high orders of iteration their limiting values. Do this for the lines that straddle the $x = \bar{x} = 0.5$ value and successively divide. Numerically solve Eq. (B) above for D. (Carry through the 32nd iteration.)

28. Circle maps are different from iterative equations of the type of Eq. (7-46). A standard example is

$$\theta_{n+1} = f(\Omega, K, \theta_n) = \theta_n + \Omega - \frac{K}{2\pi} \sin (2\pi\theta_n)$$

with Ω and K constants. A phase shift $\theta_n \to \theta_n + 1$ represents a full rotation. These maps can serve as models for the behavior of the phase difference, θ_n, between coupled oscillators or between an oscillator and its coupled periodic external driving force. Subscript n can be thought of as counting phase observation times in units of the reciprocal of the frequency $(1/v_2)$ of one of the oscillating components of the system.

Show that the condition that a maximum (or minimum) occurs in $f(\theta_n)$ is that $K > 1$. Why is there no maximum for $K = 1$? (The map with negative K is equivalent to one with positive K under the substitution $\theta \to \theta + \frac{1}{2}$, so we may take $K > 0$.)

29. The circle map in Problem 28 mimics many real systems of coupled oscillators that exhibit resonance (called mode locking) whenever the frequency of a harmonic Pv_2 of one oscillator approaches some harmonic Qv_1 of another and the frequencies adjust to lock exactly into the rational ratio P/Q. The constant Ω is the ratio v_1/v_2, and for a *range* of Ω for $K \leq 1$ the long-term iteration of the map will lead to periodic behavior,

$$\theta_{n+Q} = \theta_n + P$$

with P, Q relatively prime integers such that Q steps of iteration advances θ_n by P units and the winding number

$$W = \lim_{n \to \infty} \left[\frac{f^{(n)}(\theta) - \theta}{n} \right]$$

becomes equal to P/Q. It is also possible for certain *point* values of Ω to lead to irrational values of W and quasi-periodic behavior.

W is the mean number of rotations per iteration and would be precisely equal to Ω were it not for the nonlinear term, $(K/2\pi) \sin (2\pi\theta_n)$, in the circle map equation. It can be shown* that at $K = 1$ the widths of the Ω values that lead to all possible rational P/Q ratios exactly add up to the length unity (Ω

---

* M. H. Jensen, P. Bak, and T. Bohr, *Phys. Rev.* **A 30**, 1960 (1984).

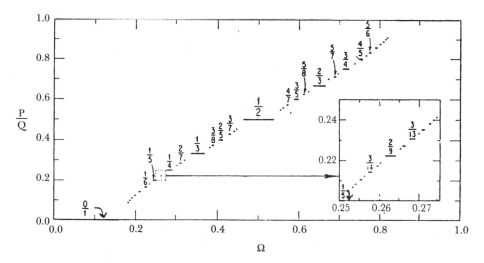

FIGURE 7-28 The Devil's staircase for the circle map with $K = 1$. Winding number versus $\Omega = v_1/v_2$ with steps wider than 0.0015 explicitly shown. Note the self-similar nature of the staircase under magnification in the inset, which shows steps wider than 0.00015.
SOURCE: T. C. Halsey, M. H. Jensen, L. P. Kadanoff, I. Procaccia, and B. Shraiman. *Phys. Rev.* **A 33**, 1141 (1986).

varies from zero to unity only) and that these widths constitute a Devil's staircase (see Fig. 7-28) whose risers (points of Ω where the steps rise up) constitute a Cantor dust of dimension $D = 0.87$ from which the irrational W values can be reached. For $K < 1$ the dimensionality of the dust leading to irrational W values will be one (i.e., its points will add to a nonzero length). For $K > 1$ the widths for rational W values must overlap and so the long time orbits will jump between resonances in an unpredictable way and chaos will result!

   Play with the circle map equation with $K = 1$ using a calculator. Show that the width for the first $(0/1)$ and last $(1/1)$ steps are equal and determine its value. Explicitly iterate in the range of the first step for $\Omega = 0.10$ starting from $\theta_0 = 0.40$, and show that W will go to zero. Determine the P and Q values for $\Omega = 0.1600$ (slightly past the end of the width of the first step). Calculate and learn more about the properties of this map as the spirit moves you.

Thermodynamic Formulae

In this appendix we present with brief comment the formulas of thermodynamics necessary for following the various derivations in the text. A list of references for the science of thermodynamics itself is also provided.

Temperature is a concept that does not appear in mechanics or electromagnetism. It is introduced by the *Zeroth Law of thermodynamics*, which asserts that any two systems that are in equilibrium with a third across conducting (nonadiabatic) walls are in equilibrium with each other. Hence they must all have a common scalar numerical value, which is their common empirical temperature. When we match this empirical scale to the limiting isothermal behavior of any gas as its pressure goes to zero, we obtain the usual absolute temperature scale, T.

A boundary that isolates a system from interactions with its surroundings is called *adiabatic*. Such a boundary may be movable or deformable and may be pierced by electrical or other leads that will permit the performance of work on the system within it. It is found that the change of state of a system enclosed within adiabatic walls as measured by its temperature change depends only on the amount of work, W, performed upon it and not on the means by which the work is put into the system nor on the intermediate stages through which the system may be taken. Hence a function of state called the energy, U, must exist such that in the adiabatic case its increase is given by

$$\Delta U = W_a \qquad (\text{I-1})$$

where subscript a denotes the adiabatic nature of the change. W_a will be negative if the system performs work on the surroundings. If the system is not enclosed adiabatically the same particular energy change, $\Delta U'$ (as observed by the *same* temperature change in the system), can be obtained by performing a different amount of work such that

$$\Delta U' = W'_a = Q' + W' \qquad (\text{I-2})$$

This defines the heat input, Q', to the system as

$$Q' = W'_a - W' \qquad (\text{I-3})$$

which is the difference between the work done in the adiabatic process and in the actual process. Heat is not something stored in a system but is energy in transit (of negative sign if it is emitted) which becomes converted into disorganized or random work.

Equation (I-2) in differential form is the *First Law of thermodynamics*:

$$dU = dQ + dW \qquad (\text{I-4})$$

in which dU is an exact differential (its integral dependent only on the initial and final states) and dQ and dW are inexact differentials (dependent on the actual path).

For the case of a reversible path (subscript *rev*), which is an idealized change proceeding infinitely slowly and passing through a continuous sequence of equilibrium states, the inexact differential of the work may be expressed in terms of generalized coordinates y_α and their conjugate forces Y_α:

$$dW_{rev} = \sum_\alpha Y_\alpha dy_\alpha \qquad (\text{I-5})$$

The dy_α are exact differentials of (usually) extensive thermodynamic properties of the system itself, and the Y_α are (usually) intensive thermodynamic properties of the system undergoing the work input. Mechanics or electromagnetic theory provides us with the forms of Y_α and y_α. For brevity here we will retain only the reversible work term for volume change, $Y_1 = -P, y_1 = V$, introducing the variables pressure and volume, respectively. Then Eq. (I-4) becomes

$$dU = dQ_{rev} - P \, dV \qquad (\text{I-6})$$

which is not useful as yet because of the inexact differential, dQ_{rev}. The *Second Law of thermodynamics* gives us the means to advance by defining a new state function, the entropy, S, in terms of a reversible process

$$dS \equiv \frac{dQ_{rev}}{T} \tag{I-7}$$

and by providing a fundamental inequality for an irreversible process:

$$dS > \frac{dQ}{T} \tag{I-8}$$

Thus Eq. (I-6) becomes

$$dU = T\,dS - P\,dV \tag{I-9}$$

and holds for any type of process because all the variables are functions of state. However, only for a reversible process will $T\,dS$ be equal to dQ. Equation (I-9) incorporates the two major laws of thermodynamics.

We next extend Eq. (I-9) to the case of possible variation of the mole numbers, n_i, of c independent chemical components. Equation (I-9) suggests that a natural set of variables for U is $S, V, n_1 \ldots n_c$. Thus we may in complete generality write

$$dU = \left(\frac{\partial U}{\partial S}\right)_{V,n_i} dS + \left(\frac{\partial U}{\partial V}\right)_{S,n_i} dV + \sum_{i=1}^{c} \left(\frac{\partial U}{\partial n_i}\right)_{S,V,n_{j \neq i}} dn_i \tag{I-10}$$

Comparison with Eq. (I-9) shows:

$$\left(\frac{\partial U}{\partial S}\right)_{V,n_i} = T \tag{I-11.1}$$

$$\left(\frac{\partial U}{\partial V}\right)_{S,n_i} = -P \tag{I-11.2}$$

Following Gibbs, we may define a new intensive variable, μ_i, the chemical potential of component i, by

$$\mu_i \equiv \left(\frac{\partial U}{\partial n_i}\right)_{S,V,n_{j \neq i}} \tag{I-11.3}$$

Hence

$$dU = T\,dS - P\,dV + \sum_{i=1}^{c} \mu_i\,dn_i \tag{I-12}$$

The intensive variables of Eqs. (I-11) play analogous roles in multiphase system equilibria, as can be derived from the fundamental inequality, Eq. (I-8). If two such phases are at different temperatures and thermal contact is possible, energy will flow from the phase at higher temperature to the one of lower temperature until the temperatures are equal and thermal equilibrium is attained. If two phases have different pressures and a wall between them can move, the phase of higher pressure will gain in volume at the expense of the other until the pressures are equal and mechanical equilibrium is reached. Finally, if two phases at the same T, P which can exchange particles of a component i have different values of μ_i net i particle flow will occur from the phase of higher chemical potential until the μ_i values become identical in the two phases and matter flow equilibrium is reached.

Extensive thermodynamic variables at constant values of any intensive independent variables are homogeneous functions of degree unity in the mole numbers and any other extensive variables among the independent ones. Hence Euler's theorem of homogeneous functions (see Appendix II) applied to the energy reads:

$$(1)\, U = S\left(\frac{\partial U}{\partial S}\right)_{V,n_i} + V\left(\frac{\partial U}{\partial V}\right)_{S,n_i} + \sum_{i=1}^{c} n_i \left(\frac{\partial U}{\partial n_i}\right)_{S,V,n_{j\neq i}} \tag{I-13}$$

Use of Eqs. (I-11) gives us the integrated form for the energy:

$$U = TS - PV + \sum_{i=1} n_i \mu_i \tag{I-14}$$

Differentiating Eq. (I-14) totally and comparing with Eq. (I-12), we obtain the Gibbs-Duhem equation:

$$S\,dT - V\,dP + \sum_{i=1}^{c} n_i\,d\mu_i = 0 \tag{I-15}$$

The *Third Law of thermodynamics* is needed to tell us the behavior of the entropy as $T \to 0$, since the defining Eq. (I-7) is singular at the absolute zero of temperature. There are several alternative statements of it:

1. As $T \to 0$ the magnitude of the entropy change of any isothermal reversible process tends to zero.

2. It is impossible to precisely attain the condition of $T = 0$ K by any finite series of steps.

Statement (1) means that all equilibrium states at 0 K lie on a surface of constant entropy, S_0. If this constant S_0 is set equal to zero, by convention the entropy of all systems internally at equilibrium will approach zero as $T \to 0$ K. This is known as the Nernst Heat Theorem and is often considered another statement of the Third Law. Apparent (but not real) exceptions to the Third Law due to the freezing-in of nonequilibrium configurations as the temperature is lowered are discussed in any standard reference.

The inequality in Eq. (I-8) enables us to predict the direction of spontaneous (i.e., nonreversible) net processes that bring a system to equilibrium. The entropy of an isolated system can only increase. If a system is not isolated the total entropy change of the system and its surroundings must increase, although the sign of the entropy change of the system itself can be either positive or negative. It is always of more practical use to specify a property change of a system itself rather than of both system and surroundings. We can show from Eq. (I-12) that for systems at constant S and V the function U of the system must decrease to a minimum in all changes leading to equilibrium. This is still not a practical criterion since S cannot in any simple experimental way be directly held constant. Thus we define new functions:

Helmholtz free energy, A

$$A = U - TS \tag{I-16}$$

Enthalpy, H

$$H = U + PV \tag{I-17}$$

Gibbs free energy, G

$$G = H - TS = U + PV - TS = A + PV \tag{I-18}$$

Taking total derivatives of these and using Eq. (I-12), we find

$$dA = -S\,dT - P\,dV + \sum_i \mu_i\,dn_i \tag{I-19}$$

$$dH = T\,dS + V\,dP + \sum_i \mu_i\,dn_i \tag{I-20}$$

and

$$dG = -S\,dT + V\,dP + \sum_i \mu_i\,dn_i \tag{I-21}$$

From Eq. (I-19) and Eq. (I-8) it can be shown that at constant T and V the function A of the system will decrease to a minimum in all changes leading to equilibrium. Moreover, for a reversible change to equilibrium the *magnitude* of the decrease of A will be the total net work (work other than that due to volume change) that the system is capable of doing on the surroundings. If there is no net work possible the decrease of A in a reversible change will be zero. The net work is a maximum for the reversible change and will be less for any irreversible approach to equilibrium. If only the temperature is held constant the decrease of A in a reversible change will be the maximum *total* work that the system can do on the surroundings.

Similarly, it can be shown that at constant T and P the function G of the system will decrease to a minimum in all changes leading to equilibrium. The magnitudes of the decrease of G in the reversible change will be the maximum net work that the system can do on the surroundings at constant T and P. If there is no net work possible in the reversible change, ΔG will be zero. (The analogous but not practically useful results for the H function need not be elaborated.)

The differentials of U, A, H, and G are exact and we have the obvious results:

$$T = \left(\frac{\partial U}{\partial S}\right)_{V,n_i} = \left(\frac{\partial H}{\partial S}\right)_{P,n_i} \tag{I-22}$$

$$P = -\left(\frac{\partial U}{\partial V}\right)_{S,n_i} = -\left(\frac{\partial A}{\partial V}\right)_{T,n_i} \tag{I-23}$$

$$V = \left(\frac{\partial H}{\partial P}\right)_{S,n_i} = \left(\frac{\partial G}{\partial P}\right)_{T,n_i} \tag{I-24}$$

$$S = -\left(\frac{\partial A}{\partial T}\right)_{V,n_i} = -\left(\frac{\partial G}{\partial T}\right)_{P,n_i} \tag{I-25}$$

Notice how μ_i, in addition to its definition given by Eq. (I-11.3), can also be written as:

$$\mu_i = \left(\frac{\partial G}{\partial n_i}\right)_{T,P,n_j \neq i} = \left(\frac{\partial A}{\partial n_i}\right)_{T,V,n_j \neq i} = \left(\frac{\partial H}{\partial n_i}\right)_{S,P,n_j \neq i} \tag{I-26}$$

The first equality in Eq. (I-26) identifies μ_i as a partial molar Gibbs free energy, since both T and P are held constant in the variation involved. Application of Euler's theorem to G as a function of T, P, $n_i \ldots n_c$ gives at constant T, P:

$$(1)G = \sum_i n_i \left(\frac{\partial G}{\partial n_i}\right)_{T,P,n_j \neq i} = \sum_i n_i \mu_i \tag{I-27}$$

For a pure substance μ_i is simply the molar Gibbs free energy of i.

Taking the mixed second derivatives of each type of energy function, we can derive the Maxwell equations. Working with A we have

$$\left(\frac{\partial P}{\partial T}\right)_V = -\frac{\partial^2 A}{\partial T \, \partial V} \equiv -\frac{\partial^2 A}{\partial V \, \partial T} = \left(\frac{\partial S}{\partial V}\right)_T \tag{I-28}$$

The other important result is from G:

$$\left(\frac{\partial S}{\partial P}\right)_T = -\left(\frac{\partial V}{\partial T}\right)_P \tag{I-29}$$

These express isothermal variations of entropy in terms of derivatives that can be obtained from equation of state data.

The temperature variation of S can be found by measuring appropriate heat capacities, since the fundamental definition of a heat capacity C_z along path z is:

$$C_z = \frac{dQ_{rev}(z)}{dT} \tag{I-30}$$

Heat provided to a system from a reservoir that effectively does no work on the system, as in the usual experimental mode, will be the same as the heat it would have provided in a hypothetical reversible manner. Hence dQ_{rev} appears in Eq. (I-30). Then using Eq. (I-7) we find

$$C_z = T\left(\frac{\partial S}{\partial T}\right)_z \tag{I-31}$$

The heat capacity will depend on the path by which the heat is delivered. The two most common heat capacities are those along constant volume or constant pressure paths at constant mole numbers. Hence from Eq. (I-12)

$$C_V = \left(\frac{\partial U}{\partial T}\right)_V = T\left(\frac{\partial S}{\partial T}\right)_V \tag{I-32}$$

and from Eq. (I-20)

$$C_P = \left(\frac{\partial H}{\partial T}\right)_P = T\left(\frac{\partial S}{\partial T}\right)_P \tag{I-33}$$

Standard transformations show that

$$C_P - C_V = \frac{TV\alpha^2}{\beta_T} \tag{I-34}$$

where α is the coefficient of thermal expansion

$$\alpha \equiv \frac{1}{V}\left(\frac{\partial V}{\partial T}\right)_P \tag{I-35}$$

and β_T is the isothermal compressibility

$$\beta_T = -\frac{1}{V}\left(\frac{\partial V}{\partial P}\right)_T \tag{I-36}$$

To obtain a general expression for C in terms of C_V, write

$$dQ_{rev} = dU + P\,dV = \left(\frac{\partial U}{\partial T}\right)_V dT + \left[\left(\frac{\partial U}{\partial V}\right)_T + P\right]dV \tag{I-37}$$

and use Eq. (I-28) in Eq. (I-12) for constant mole number:

$$\left(\frac{\partial U}{\partial V}\right)_T = T\left(\frac{\partial S}{\partial V}\right)_T - P = T\left(\frac{\partial P}{\partial T}\right)_V - P \tag{I-38}$$

to find

$$C = \frac{dQ_{rev}}{dT} = C_V + T\left(\frac{\partial P}{\partial T}\right)_V \frac{dV}{dT} \tag{I-39}$$

Then for any particular path z, the factor dV/dT becomes $(\partial V/\partial T)_z$.

One-component single-phase stability conditions arising from Eq. (I-8) require that

$$C_V > 0 \tag{I-40}$$

and

$$\left(\frac{\partial P}{\partial V}\right)_T \leq 0 \tag{I-41}$$

such that β_T is positive and becomes infinite at the limit of phase stability. Then from Eq. (I-34), $C_P \geq C_V$.

Equilibrium between two phases α, β of one component at a given T and P is expressed by

$$\Delta G = \Delta H - T\Delta S = 0 \tag{I-42}$$

and its variation is controlled by the condition

$$d\mu_\alpha = d\mu_\beta$$
$$-S_{m_\alpha} dT + V_{m_\alpha} dP = -S_{m_\beta} dT + V_{m_\beta} dP$$

whence

$$\left(\frac{dP}{dT}\right)_\sigma = \frac{S_{m_\alpha} - S_{m_\beta}}{V_{m_\alpha} - V_{m_\beta}} = \frac{\Delta S}{\Delta V} = \frac{\Delta H}{T\Delta V} \tag{I-43}$$

Equation (I-43) is the Clapeyron equation for the slope of the equilibrium line (subscript σ) between two phases of one component in the P-T plane. To obtain the heat capacity of either equilibrium phase as it is heated with both pressure and volume change along the equilibrium line, adapt Eq. (I-39) using the chain rule

$$\left(\frac{\partial P}{\partial T}\right)_V \left(\frac{\partial T}{\partial V}\right)_P \left(\frac{\partial V}{\partial P}\right)_T = -1$$

to find

$$C_\sigma = C_V - T\left(\frac{\partial P}{\partial V}\right)_T\left(\frac{\partial V}{\partial T}\right)_P\left(\frac{\partial V}{\partial T}\right)_\sigma \tag{I-44}$$

For multicomponent systems the fundamental equation, the starting point for deriving necessary relationships between all sorts of properties—which is what the science of thermodynamics is fundamentally about—is the Gibbs-Duhem equation (I-15). Taking as independent variables T, P, and $(c - 1)$ mole fractions $X_2, \ldots, X_c$ (X_1 is determined by these) one may show that there exist $(c - 1)$ independent relations

$$\sum_{i=1}^{c} X_i\left(\frac{\partial \mu_i}{\partial X_j}\right)_{T,P} = 0 \qquad \text{for } j = 2, \ldots, c \tag{I-45}$$

at fixed T and P. Thus in a binary system there is one relation:

$$X_1\left(\frac{\partial \mu_1}{\partial X_2}\right)_{T,P} + X_2\left(\frac{\partial \mu_2}{\partial X_2}\right)_{T,P} = 0 \tag{I-46}$$

Writing Eq. (I-27) for one total mole of mixture we have

$$G_m = \sum_{i=1}^{c} X_i\mu_i \tag{I-47}$$

whence, using Eq. (I-45), we find

$$\left(\frac{\partial G_m}{\partial X_i}\right)_{T,P} = \mu_i - \mu_1 \qquad i = 2, \ldots, c \tag{I-48}$$

In particular for a binary mixture there is one equation:

$$\left(\frac{\partial G_m}{\partial X_2}\right)_{T,P} = \mu_2 - \mu_1 \tag{I-49}$$

which upon a second differentiation gives

$$\left(\frac{\partial^2 G_m}{\partial X_2^2}\right)_{T,P} = \left(\frac{\partial \mu_2}{\partial X_2}\right)_{T,P} - \left(\frac{\partial \mu_1}{\partial X_2}\right)_{T,P} \tag{I-50}$$

Using Eq. (I-46) in Eq. (I-50) we obtain

$$\left(\frac{\partial \mu_1}{\partial X_2}\right)_{T,P} = -X_2\left(\frac{\partial^2 G_m}{\partial X_2^2}\right)_{T,P} \tag{I-51}$$

and

$$\left(\frac{\partial \mu_2}{\partial X_2}\right)_{T,P} = (1 - X_2)\left(\frac{\partial^2 G_m}{\partial X_2^2}\right)_{T,P} \qquad (I\text{-}52)$$

For a single binary phase to be stable, in addition to the necessary conditions given by Eqs. (I-40) and (I-41) there is a condition of stability with respect to change of concentration (possible phase separation into two phases of different concentrations), which is

$$\left(\frac{\partial^2 G_m}{\partial X_2^2}\right)_{T,P} \geq 0 \qquad (I\text{-}53)$$

At the critical point for phase separation in binary liquid mixtures there are two conditions:

$$\left(\frac{\partial^2 G_m}{\partial X_2^2}\right)_{T,P} = 0 \qquad (I\text{-}54)$$

and

$$\left(\frac{\partial^3 G_m}{\partial X_2^3}\right)_{T,P} = 0 \qquad (I\text{-}55)$$

For the expression of the chemical potentials of components in terms of standard state properties and activity functions any of the listed references here (or indeed any textbook of physical chemistry) should be consulted.

References for Thermodynamics

H. B. Callen. *Thermodynamics*. Wiley, New York, 1960.

E. A. Guggenheim. *Thermodynamics*, 5th ed. North Holland, Amsterdam, 1967.

M. L. McGlashan. *Chemical Thermodynamics*. Academic Press, New York, 1979.

A. B. Pippard. *The Elements of Classical Thermodynamics*. Cambridge Univ. Press, Cambridge, 1966.

I. Prigogine and R. Defay. *Chemical Thermodynamics*. Longmans Green, London, 1954.

H. Reiss. *Methods of Thermodynamics*. Blaisdell, New York, 1965.

M. N. Zemansky and R. H. Dittman. *Heat and Thermodynamics*, 6th ed. McGraw-Hill, New York, 1981.

II

Mathematical Notes

It is assumed that the reader can make use of standard tables of series and of definite and indefinite integrals. A brief set of sources of these includes:

R. C. Weast, ed. *Handbook of Chemistry and Physics*. Chemical Rubber Pub. Co. (annual editions).

M. Abramowitz and I. A. Stegun. *Handbook of Mathematical Functions*. Reprint, Dover, New York, 1965.

I. S. Gradshteyn and I. M. Ryzhik. *Table of Integrals, Series and Products*. Academic Press, New York, 1965.

1. The Delta (δ) Function

The δ function was originally introduced by P. A. M. Dirac in his development of the general theory of quantum mechanics. As a function of x, $\delta(x)$ is zero everywhere except at $x = 0$, where it is infinitely large in such a way that any integral of it whose range of integration includes $x = 0$ will give

$$\int \delta(x)\, dx = 1 \qquad (\text{II-1})$$

Clearly also,

$$\int f(x)\delta(x)\, dx = f(0) \qquad (\text{II-2})$$

and by change of origin

$$\int f(x)\delta(x - a)\, dx = f(a) \qquad (\text{II-3})$$

Notice that the dimensions of $\delta(x)$ are those of x^{-1}.

An alternative way of defining the δ function is as the derivative* $s'(x)$ of a unit step function $s(x)$ given by

$$s(x) = 0 \qquad (x < 0) \qquad (\text{II-4})$$
$$ = 1 \qquad (x > 0)$$

This is proved by substituting $s'(x)$ for $\delta(x)$ on the left-hand side of Eq. (II-2) and integrating by parts to obtain the result on the right-hand side. We find for q_1 and q_2, any two positive numbers,

$$\int_{-q_1}^{+q_2} f(x)s'(x)\, dx = f(x)s(x) \Big]_{-q_1}^{+q_2} - \int_{-q_1}^{+q_2} f'(x)s(x)\, dx$$
$$= f(q_2) - \int_{0}^{q_2} f'(x)\, dx = f(0)$$

It is easy to verify the following, all of which have the meaning that both sides give the same result as factors in an integrand

$$\delta(-x) = \delta(x) \qquad (\text{II-5.1})$$

$$x\delta(x) = 0 \qquad (\text{II-5.2})$$

$$\delta(ax) = a^{-1}\delta(x) \qquad a > 0 \qquad (\text{II-5.3})$$

$$\int \delta(a - x)\delta(x - b)\, dx = \delta(a - b) \qquad (\text{II-5.4})$$

Equation (II-5.1) means that $\delta(x)$ is an even function.

---

* A prime denotes differentiation with respect to the argument.

The derivative of the δ function has the property

$$\int_{-q_1}^{+q_2} f(x)\delta'(x-a)\,dx = -f'(a) \equiv -\int_{-q_1}^{+q_2} \delta(x-a)f'(x)\,dx \qquad \text{(II-6)}$$

since $\delta(q_2 - a)$ and $\delta(-q_1 - a)$ both are zero in the partial integration linking the first and third terms. Similarly, using Eq. (II-5.2) we have the further equalities (as factors in an integrand):

$$\delta'(x) = -\delta'(-x) \qquad \text{(II-7.1)}$$

$$x\delta'(x) = -\delta(x) \qquad \text{(II-7.2)}$$

It can be shown by contour integration in the complex plane that an integral representation of the δ function is

$$\delta(a) = \frac{1}{2\pi}\int_{-\infty}^{+\infty} e^{iax}\,dx \qquad \text{for real } a \qquad \text{(II-8)}$$

We can qualitatively see that for $a = 0$ the integral is infinite and for nonzero a the integral is zero because

$$e^{iax} = \cos ax + i \sin ax$$

and the area under the curve for both cosine and sine terms must be as often positive as negative over any symmetrical range $-q$ to $+q$ in x. The three-dimensional δ function has for integral representation:

$$\delta(a) = \frac{1}{(2\pi)^3}\int d^3x e^{i\bar{x}\cdot\bar{a}} \qquad \text{(II-9)}$$

where $\bar{x}\cdot\bar{a}$ is a vector dot product.

Accepting Eq. (II-8) and noting

$$\int_{-g}^{+g} e^{iax}\,dx = \frac{1}{ia}\left(e^{iga} - e^{-iga}\right) = \frac{2}{a}\sin ga$$

we can represent the δ function as the following limit:

$$\delta(a) = \lim_{g\to\infty}\left(\frac{\sin ga}{\pi a}\right) \qquad \text{(II-10)}$$

2. The Stirling Approximation

The generalization of the factorial function is called the gamma function, $\Gamma(z)$, and by definition has the following integral representation:

$$\Gamma(z) = \int_0^\infty e^{-t} t^{z-1}\, dt \tag{II-11}$$

Performing a partial integration,

$$\Gamma(z) = \left. \frac{t^{z-1} e^{-t}}{-1} \right]_0^\infty + (z-1) \int_0^\infty t^{z-2} e^{-t}\, dt \tag{II-12}$$

and for $z > 1$ we have the recurrence relation

$$\Gamma(z) = (z-1)\Gamma(z-1) \qquad z > 1 \tag{II-13}$$

Directly from Eq. (II-11):

$$\Gamma(1) = 1 \tag{II-14}$$

and using Eq. (II-13):

$$\Gamma(2) = 1\Gamma(1) = 1$$
$$\Gamma(3) = 2\Gamma(2) = (2)(1)$$
$$\Gamma(4) = 3\Gamma(3) = (3)(2)(1)$$

Clearly, for z equal to an integer, N, we have

$$\Gamma(N) = (N-1)!$$

or

$$N! = \Gamma(N+1) = \int_0^\infty e^{-t} t^N\, dt \tag{II-15}$$

Substitute $t = qN$ to obtain

$$N! = N^{N+1} \int_0^\infty e^{-qN} q^N\, dq = N^{N+1} \int_0^\infty e^{Nf(q)}\, dq \tag{II-16}$$

in which

$$f(q) = \ln q - q \tag{II-17}$$

We can obtain the asymptotic value of the integral in Eq. (II-16) for very large N considering only the region of $f(q)$ near its maximum $f(q_0)$, because $e^{Nf(q)}$ will be a very strongly peaked function there. At $q = q_0, f'(q_0) = 0$ and the Taylor series for $f(q)$ will be

$$f(q) = f(q_0) + \frac{(q - q_0)^2}{2} f''(q_0) + \dots \tag{II-18.1}$$

$$= -1 - \frac{(q - 1)^2}{2} + \dots \tag{II-18.2}$$

where the second line follows from the form of Eq. (II-17). Introducing a new variable, $\zeta = q - 1$, in Eq. (II-16) and noting that $\zeta = -1$ at the lower limit,

$$N! \cong N^{N+1} \int_{-1}^{\infty} \exp\left[N\left(-1 - \frac{1}{2}\zeta^2 + \dots \right) \right] d\zeta$$

$$= N^{N+1} e^{-N} \int_{-1}^{\infty} \exp\left(-\frac{1}{2} N\zeta^2 \right) d\zeta \tag{II-19}$$

For large N the lower limit can be set equal to $-\infty$ since the integrand is so small whenever $\zeta \neq 0$. Hence, performing the standard integral,

$$N! \cong N^{N+1} e^{-N} \int_{-\infty}^{+\infty} \exp\left(-\frac{1}{2} N\zeta^2 \right) d\zeta = \sqrt{2\pi} N^{N+1/2} e^{-N} \tag{II-20}$$

Equation (II-20) is the Stirling approximation for the factorial of a large number. It is usually written in its logarithmic form:

$$\ln (N!) \simeq N \ln N - N + \tfrac{1}{2} \ln N + \tfrac{1}{2} \ln (2\pi) \tag{II-21}$$

For most purposes only the terms proportional to N need be retained.

3. The Multinomial Expansion

If we consider the expansion of

$$(a_1 + a_2 + \dots + a_k)^N = (a_1 + a_2 + \dots + a_k)(a_1 + a_2 + \dots + a_k) \dots$$
$$(a_1 + a_2 + \dots + a_k)$$

with N like factors on the right, every term in the resulting sum must consist of a product of powers of the a_i (taking one a_i from each of the N factors), which powers

n_i must add to N:

$$\prod_{i=1}^{k} a_i^{n_i} \qquad \text{for which} \qquad \sum_{i=1}^{k} n_i = N$$

The coefficients of each particular term must be the number of ways we can divide the N factors into sets of n_1 (from which we took the factor a_1), n_2 (from which we took the factor a_2), $n_3 \ldots n_k$, with no particular order of arrangement in any set; that is, the coefficient is $N!/(n_1! n_2! \ldots n_k!)$. Thus, summing over all possible choices of the numbers n_i we obtain

$$(a_1 + a_2 + \ldots + a_k)^N = \sum \frac{N!}{n_1! n_2! \ldots n_k!} (a_1)^{n_1} (a_2)^{n_2} \ldots (a_k)^{n_k} \qquad \text{(II-22)}$$

$$\left(\sum_{i=1}^{k} n_i = N \right)$$

This is the multinomial expansion formula and the sum is over all possible sets n_i that satisfy the condition in the parenthesis.

4. Euler's Theorem of Homogeneous Functions

A function $f(x, y, z, \ldots)$ is homogeneous of degree n if, when each independent variable is multiplied by an arbitrary parameter k, the function itself is multiplied by k^n. That is,

$$f(kx, ky, kz, \ldots) = k^n f(x, y, z, \ldots) \qquad \text{(II-23)}$$

Since k is not related to the variables $x, y, z, \ldots$ we may differentiate both sides of Eq. (II-23) with respect to k while keeping $x, y, z, \ldots$ constant. To facilitate this, write

$$u_1 = kx, \qquad u_2 = ky, \qquad u_3 = kz, \ldots \qquad \text{(II-24)}$$

and

$$f(u_1, u_2, u_3, \ldots) = k^n f(x, y, z, \ldots) \qquad \text{(II-25)}$$

The total differential of the left-hand side of Eq. (II-25) is

$$df = \left(\frac{\partial f}{\partial u_1} \right) du_1 + \left(\frac{\partial f}{\partial u_2} \right) du_2 + \left(\frac{\partial f}{\partial u_3} \right) du_3 + \ldots$$

and so

$$\left(\frac{\partial f}{\partial k}\right)_{x,y,z,\ldots} = \left(\frac{\partial f}{\partial u_1}\right)\left(\frac{\partial u_1}{\partial k}\right) + \left(\frac{\partial f}{\partial u_2}\right)\left(\frac{\partial u_2}{\partial k}\right) + \left(\frac{\partial f}{\partial u_3}\right)\left(\frac{\partial u_3}{\partial k}\right) + \ldots$$

which by Eq. (II-24) becomes

$$\left(\frac{\partial f}{\partial k}\right)_{x,y,z,\ldots} = x\left(\frac{\partial f}{\partial u_1}\right) + y\left(\frac{\partial f}{\partial u_2}\right) + z\left(\frac{\partial f}{\partial u_3}\right) + \ldots \tag{II-26}$$

The derivative of the right-hand side of Eq. (II-25) is

$$\left(\frac{\partial f}{\partial k}\right)_{x,y,z,\ldots} = nk^{n-1}f(x, y, z, \ldots) \tag{II-27}$$

Equating the right-hand side of Eq. (II-26) and Eq. (II-27) we have

$$x\left(\frac{\partial f}{\partial u_1}\right) + y\left(\frac{\partial f}{\partial u_2}\right) + z\left(\frac{\partial f}{\partial u_3}\right) + \ldots = nk^{n-1}f(x, y, z, \ldots) \tag{II-28}$$

Since k is arbitrary, Eq. (II-28) must hold for any value of k and in particular for $k = 1$, for which $u_1 = x$, $u_2 = y$, $\ldots$ so that

$$x\left(\frac{\partial f}{\partial x}\right) + y\left(\frac{\partial f}{\partial y}\right) + z\left(\frac{\partial f}{\partial z}\right) + \ldots = nf(x, y, z, \ldots) \tag{II-29}$$

Equation (II-29) is Euler's theorem of homogeneous functions.

5. Euler-Maclaurin and Other Summation Formulas

By contour integration in the complex plane it is possible to prove a useful formula due to Euler and Maclaurin that expresses the sum

$$\sum_{i=m}^{i=n} F(i) = F(m) + F(m + 1) + \ldots + F(n) \tag{II-30}$$

as the integral of $F(x)$ between the limits m and n plus a series of correction terms involving the function and its derivatives at the limits. We use the symbol

$$F_r(a) = \left(\frac{d^r F(x)}{dx^r}\right)_{x=a} \tag{II-31}$$

to mean the rth derivative of the function at the value $x = a$. The result is

$$\sum_{i=m}^{i=n} F(i) = \int_m^n F(x)\, dx + \frac{1}{2}\left[F(n) + F(m) \right]$$
$$+ \sum_{k \geq 1} (-1)^k \frac{B_k}{(2k)!}\left[F_{2k-1}(m) - F_{2k-1}(n) \right] \qquad \text{(II-32)}$$

where the B_k are the Bernoulli numbers,

$$B_1 = \frac{1}{6}, \qquad B_2 = \frac{1}{30}, \qquad B_3 = \frac{1}{42}, \qquad B_4 = \frac{1}{30} \ldots$$

The B_k also enter into the evaluation of the Riemann zeta functions $\zeta(s)$ of even integer argument ($s = 2k$ with $k = 1, 2, \ldots$) such that with the definition

$$\zeta(s) \equiv \sum_{n=1}^{\infty} n^{-s} \qquad \text{(II-33)}$$

the result is

$$\zeta(2k) = \sum_{n=1}^{\infty} \frac{1}{(n)^{2k}} = \frac{(2\pi)^{2k} B_k}{2(2k)!} \qquad k = 1, 2, \ldots \qquad \text{(II-34)}$$

The numerical evaluation of the first zeta functions of odd integer argument gives (for $s = 1$ the sum diverges)

$$\zeta(3) = 1.2020569$$
$$\zeta(5) = 1.0369278$$
$$\vdots$$

6. Newton's Method of Approximate Solution of Equations in One Variable

This method is presented in most calculus texts and will be merely stated. The equation whose root (or roots) we desire is written in the form $F(x) = 0$ and its derivative, $F'(x)$, evaluated analytically. Then a guess x_0 is made [for which $F(x_0)$ is small]. The improved root will be given by

$$x_1 = x_0 - \frac{F(x_0)}{F'(x_0)} \qquad \text{(II-35)}$$

Generally, the process must be repeated several times until convergence is achieved.

III

Answers to Problems and Comments on Their Solutions

Chapter 1

3b. No, since in the fixed wall case all improbable distributions will contribute additively to $\overline{(P_1 - P_2)^2}$.

4. $W'/W = \exp\left[(|Q|/kT_2 T_1)(T_1 - T_2)\right]$ For the case given $\ln(W'/W) = 3.26 \times 10^{18}$.

5. $W = \dfrac{10!}{4!3!2!1!} = 12,600$

8. From Problem 7, if $p = 1$, $\beta\varepsilon = \ln 2$, $q = 2$:

$$\mathcal{N}_i = \frac{9}{2^{i+1}} \qquad i = 0, 1, 2, \ldots$$

9a.
$$a(y + 2z) = \frac{2}{3}a \qquad \frac{y}{z} = \frac{2}{3}e^{a/kt} = \frac{2}{3}w$$

$$x + y + z = 1 \qquad \frac{y}{x} = \frac{2}{w}$$

9c. $T = 5.9 \times 10^2$ K.

9d. Average energy per particle at infinite temperature is $\frac{4}{3}a$ and remains finite because there is a limited number of energy levels available.

12. The larger the molecular mass the smaller is the spacing between translational energy levels and so the larger will be the number of such levels between the ground state and kT.

16. From Problem 15, the relative deviation, d, is

$$d = \frac{\sqrt{Np(1 - p)}}{Np} = \frac{(1 - p)^{1/2}}{(Np)^{1/2}} = (Np)^{-1/2} \qquad \text{for small } p$$

At STP $N = 6.02 \times 10^{23}$ in $V = 22.4 \times 10^3$ cm^3

If $\omega = 1$ mm^3, $\quad p = 4.46 \times 10^{-8}$ and $d = 6 \times 10^{-9}$

The macroscopic density scarcely varies from one cell to another if ω is any reasonable size. For d of unity, at STP ω will be a cube of side 3.3 nm with an average molecule count of $\bar{N}_1 = Np = 1$.

22. Use $z = e^{\mu/kT}$ and $U = \frac{3}{2}PV$ in

$$G = N\mu = U - TS + PV$$

to obtain $S/Nk = \frac{5}{2}(PV/NkT) - \ln z$. Then use power series given in the text.

23. z becomes unity for ideal fermions at $\lambda^3\rho \simeq 0.78$.

24. Phosphorus data given would give $C_{P_m} = 2.2R$ if gas were monatomic, which violates the minimum value of $2.5R$ for C_{P_m}. Actual vibrational contribution is $4.7R$.

26. The velocity distribution curve for the flowing gas will be a bell-shaped curve centered at $v_x = v_{x_0}$ instead of at $v_x = 0$.

27.
$$\bar{c}_1 = \left(\frac{2kT}{\pi m}\right)^{1/2} \qquad \overline{c_1^2} = \frac{kT}{m} \qquad \bar{c}_{1\rightarrow} = \frac{1}{2}\bar{c}_1 = \left(\frac{kT}{2\pi m}\right)^{1/2}$$

28. 4.13×10^{23}.

29. Ratio is $s^2 e^{-(s^2 - 1)}$.

32. On the assumptions of this problem the reactive molecules are those with speeds greater than $(4.67)c_p$.

34. Fraction in range required $= 9.25 \times 10^{-4}$.

Chapter 2

1. 1.64×10^4 K.

2. 4.13×10^{-2}.

3. X is a linear molecule with four atoms.

5. For F the fraction is 0.067, 0.22, and 0.31 at 300, 1,000, and 5,000 K, respectively.

6. For maximum $\mathcal{N}_J$, $\mathcal{J} = (T/2\theta_r)^{1/2} - \frac{1}{2}$.

10. $S_m = 219.7$ J·K^{-1}·mol^{-1}; $(C_P)_m = 43.76$ J·K^{-1}·mol^{-1}.

11. $\mathcal{J}'$ for rotational energy to equal kT is

$$\mathcal{J}' = \frac{(1 + 4T/\theta_r)^{1/2} - 1}{2}$$

at 150 K, $q_{rot} = 52.7$; $\mathcal{J}' = 6.78$; $\sum_{J=0}^{6} (2\mathcal{J} + 1) = 49$.

13. Although very few of the undissociated O_2 molecules have much vibrational excitation, those few that do have enough to break up into atoms will very rarely succeed in recombining because a rare three-body—or two-body at a wall—collision is needed for them to recombine. (A third body is needed to carry off the potential energy released on molecule formation lest the molecule immediately dissociate.)

15. The minimum D_0 varies little with atomic mass, being in the range 0.24–0.28 eV. For really high M_a the $\theta_v < 300$ K and this will cause the minimum D_0 to fall again toward 0.24 eV.

16. $K_P = (1.73)e^{424/T}$.

17. $D_0 = 0.755$ eV.

18. $K = 4$ for the oxygen isotopic reaction. For the hydrogen case the θ_r values will not cancel and the vibratory zero point energies differ significantly.

19. 6725 J·mol^{-1}.

21. $K = 0.057$ at 298 K. The K will increase with temperature increase since the number of energy states available to B will increase more than for A (because of the difference of their spacings).

22. $C_1 = 5042$ eV^{-1}·K; $C_2 = 6.483$.

23. $T/\theta_{ion} = 0.144$. Excited electronic states will be appreciably occupied only when $T \cong \theta_{ion}$.

24. $T = 7.8 \times 10^3$ K for 50 percent ionization. $T = 1.0 \times 10^4$ K for 99 percent ionization.

27. $224.3 \ J \cdot K^{-1} \cdot mol^{-1}$, 298.15 K; $247.2 \ J \cdot K^{-1} \cdot mol^{-1}$, 500 K.

28. 8.88×10^{10} atm^{-1} at 298.15 K; 6.66×10^{3} atm^{-1} at 500 K.

29.

g_1/g_0	x	max $(C_V)_m/R$	
1	2.40	0.439	
2	2.66	0.762	
9	3.50	2.06	(see Problem 2-41).

30. 13 kiloton: $T = 8.6 \times 10^{6}$ K, $\lambda_{max} = 3.4 \times 10^{-10}$ m. 2 megaton: $T = 30 \times 10^{6}$ K, $\lambda_{max} = 0.95 \times 10^{-10}$ m. Thermal energy spreads out over a spherical area of $4\pi r^2$ at distance r from the source.

31.

Star temp., K	f visible
2500	0.055
5800	0.45
10000	0.40
30000	0.044

See J. Pyle. *J. Chem. Ed.* **62**, 488 (1985).

32. Use $\bar{N} = \sum_i \bar{n}_i$ and replace the sum with an integral to find $\bar{N}$ proportional to T^3. Then PV, which is proportional to T^4, is also proportional to $\bar{N}T$.

35. For gas $P = 1.36$ atm, $C_V = 2.1 \ J \cdot K^{-1}$. For radiation $P = 0.25$ atm, $C_V = 3.0 \ J \cdot K^{-1}$.

37. Normal metastable D_2 at high temperature will be $\frac{2}{3}$ *ortho-* (with only even J levels occupied).

38. At any temperature, considering both forms to follow the ideal gas law, $\Delta H = \Delta U$. For $T < \theta_r$ the *ortho-* are practically all in the $J = 1$ level while the *para-* are practically all in the $J = 0$, level causing ΔU_m to become constant $(2R\theta_r)$.

39. The total entropy of one mole of the metastable mixture is $113.6 \ J \cdot K^{-1} \cdot mol^{-1}$.

41. Thermodynamically, the C_P is not merely the sum of the C_P values of the two species but has a contribution due to the change in the number of moles of each present as the equilibrium composition changes with temperature change. For $T < \theta_r$ this situation is essentially the same as the two-level model treated in Problem 2-29 with $g_1 = 9$ and $g_0 = 1$. The value 9 arises from the product of the three nuclear spin orientations times the rotational degeneracy of 3 for the *ortho-* species.

42. $K_1/K_2 = (0.531)e^{985/T}$, $\Delta H_{D_2O} = \Delta H_{H_2O} + 8.2$ kJ/mol independent of temperature. Experimentally, at 25°C $\Delta H_{D_2O} = \Delta H_{H_2O} + 3.6$ kJ/mol.

43. For fermions

$$\bar{n}_i = \frac{1}{e^{\mu/kT} e^{\varepsilon_i/kT} + 1} \qquad e^{\varepsilon_i/kT} = \frac{1 - \bar{n}_i}{\bar{n}_i e^{-\mu/kT}}$$

At equilibrium all fermions have the same μ value and so energy conservation leads to the following detailed balance condition: $\bar{n}_1 \bar{n}_2 (1 - \bar{n}_3)(1 - \bar{n}_4) = \bar{n}_3 \bar{n}_4 (1 - \bar{n}_1)(1 - \bar{n}_2)$. Since for fermions $\bar{n}_i$ can only range from 0 to 1, $\bar{n}_i$ is the probability that state i is occupied and $(1 - n_i)$ is the probability that state i is empty. Hence, in an energy interaction process $1 + 2 \rightarrow 3 + 4$ the above means the rate is proportional to the product of the probabilities of states 1 and 2 being occupied *and* to the product of the probabilities of states 3 and 4 being empty. If either state 3 or 4 is occupied there is no possibility of the process occurring because of the Pauli principle.

A similar analysis for bosons gives $\bar{n}_1 \bar{n}_2 (1 + \bar{n}_3)(1 + \bar{n}_4) = \bar{n}_3 \bar{n}_4 (1 + \bar{n}_1) \cdot (1 + \bar{n}_2)$. Now the $\bar{n}_i$ can be greater than unity and must be interpreted as being proportional to the probability of state i being occupied. For bosons the final state does not have to be empty in order for it to accept more particles. In fact $(1 + \bar{n}_i)$ must be interpreted as being proportional to the probability of state i accepting more particles.

44c. Use from Problem 2-32:

$$\frac{U}{\mathcal{N}} = \text{average photon energy} = \frac{3PV}{\mathcal{N}} = 3(0.900) \, kT = m_e c^2$$

$$T = 2.2 \times 10^9 \text{ K}$$

45a. For matter at the current epoch:

$$4 \times 10^{-27} \text{ kg/m}^3 \approx \frac{4 \times 10^{-27}}{1.67 \times 10^{-27}} = 2.4 \text{ m}^{-3}$$

number density of matter assumed as protons.

Energy density of matter $= (4 \times 10^{-27})c^2 = 3.6 \times 10^{-10} \text{ J} \cdot \text{m}^{-3}$; $c =$ speed of light.

45b. For radiation from Problem 2-32

$$\frac{\bar{\mathcal{N}}}{V} = 16\pi \left(\frac{kT}{hc}\right)^3 (1.202)$$

For $T = 2.7$ K,

$$\frac{\bar{\mathcal{N}}}{V} = 4.0 \times 10^8 \text{ m}^{-3}$$

with energy density

$$\left(\frac{U}{V}\right) = 3P = (2.7)\left(\frac{\overline{N}}{V}\right)kT = 4.0 \times 10^{-14}\,\text{J}\cdot\text{m}^{-3}$$

Hence at the present epoch energy resides mainly in matter by a factor of 9×10^3 but photons outnumber protons by a factor of 2×10^8.

45c. The common energy density of matter and radiation on the assumptions stated would have been 2.6×10^2 J·m^{-3} at a temperature of 2.4×10^4 K at an earlier time. At still earlier times the temperature would have been higher and the energy of the universe predominantly in radiation.

Chapter 3

3. Incorrect use of the Debye equation gives $D = -9.5$. Use of the improved equation gives $D = 6.9$ in reasonable accord with experiment.

4. 1.369.

5. 2.14

7. Best form is $T_b = 57.56(10^{30}\alpha/4\pi\varepsilon_0)^{0.762} + 0.03$.

8a.

$$a = \frac{2/3(\pi N_A^2 A)}{\sigma^3}$$

From LJ parameters, $A/\sigma^3 = 4\varepsilon^*\sigma^3$. From molecular data, $\sigma^3 = 8(\alpha/4\pi\varepsilon_0)$ and A may be calculated from listings in Table 3-4.

9. The temperature is high if $kT \gg \varepsilon^*$, where ε^* is the maximum depth of the potential well for the interaction of two molecules. For a hard sphere gas $B_3 = \frac{5}{8}B_2^2$.

10. For hard spheres, $B_2 = b_0$. For a square well, $B_2 = b_0[1 - \Delta(s^3 - 1)]$, From the van der Waals equation using a, b on a per molecule basis, $B_2 = b - a/kT$.

11. The relevant condition is that $(r^3/3)\{\exp[-\varepsilon(r)/kT] - 1\}$ must vanish at $r = \infty$, where $\varepsilon(r)/kT \to 0$, whence the condition becomes $(r^3/3)[-\varepsilon(r)/kT] \to 0$ as $r \to \infty$. This is satisfied if $\varepsilon(r)$ goes to zero faster than $1/r^3$, which is true for usual molecule-molecule interactions. It is not true for a Coulomb interaction going as $1/r^2$.

12. A form that fits is $f(n) = n^2 + n - 2$. Proceed using, $\sum_{n=0}^{\infty} y^n = 1/(1 - y)$:

$$\frac{\partial}{\partial y}\sum_{n=0}^{\infty} y^n = \sum_{n=0}^{\infty} ny^{n-1} = \frac{\partial}{\partial y}\left(\frac{1}{1-y}\right) = \frac{1}{(1-y)^2}, \text{ and so on.}$$

13.

$$U = U_{id} - NkT^2 \sum_{n=2} \left(\frac{1}{n-1}\right)\left(\frac{d}{dT}B_n\right)\rho^{n-1}$$

$$S = S_{id} - Nk \sum_{n=2} \left(\frac{1}{n-1}\right)\left(B_n + T\frac{d}{dT}B_n\right)\rho^{n-1}$$

$$C_V = (C_V)_{id} - 2NkT \sum_{n=2} \left(\frac{1}{n-1}\right)\left(\frac{d}{dT}B_n\right)\rho^{n-1}$$

$$- NkT^2 \sum_{n=2} \left(\frac{1}{n-1}\right)\left(\frac{d^2}{dT^2}B_n\right)\rho^{n-1}$$

14. Values (Δ) of the actual molar quantity minus the ideal molar quantity:

	1 atm	100 atm	1 atm[a]	100 atm[a]
$\Delta S_m/\text{J}\cdot\text{K}^{-1}\cdot\text{mol}^{-1}$	−0.013	−1.40	−0.015	−1.47
$\Delta U_m/\text{J}\cdot\text{mol}^{-1}$	−5.46	−570	—	—
$\Delta C_{V_m}/\text{J}\cdot\text{K}^{-1}\cdot\text{mol}^{-1}$	−0.01	−0.8	—	—

[a] [From Eq. (2-57)]

15. In a liquid, V_N is a sum of pair potentials that are functions of radial separation only (or at least an effective averaging equivalent to this can be accomplished). In a crystal, the pair potentials will have a strong polar angle dependence as well as dependence on radial separation.

17. 10.8 at $V/V_0 = 1.60$; 2.0 at $V/V_0 = 3.00$.

19. 4.96 at $V/V_0 = 1.60$; 2.05 at $V/V_0 = 3.00$.

Largest possible value is 36.04.

21. $$\frac{PV_m}{RT} = 1 + \frac{2.96192}{X} + \frac{5.48311}{X^2} + \frac{7.4550}{X^3} + \frac{8.466}{X^4} + \frac{8.799}{X^5} - \frac{7.89846}{T^*X}$$

22. $X_c = 5.665$; $T_c^* = 1.01$; $(PV_m/RT)_c = 0.359$.

24. At close packing (cp) there are 18 neighbors within 1.5σ of any central species, so

$$U_{cp} = \frac{3}{2}NkT - N\frac{(18)\varepsilon}{2}$$

27. Use $$\frac{\omega_i}{N_A} = \frac{1}{\rho_i} - \frac{1}{\rho_c} \qquad \text{such that} \qquad \rho_i = \frac{N_A}{\omega_i + V_{m_c}}$$

28. 4.8 decades; 7.9 decades.

29. Consider the implications of a negative α in the magnetic case.

30. ΔH_{vap_m} vanishes with exponent β as the critical point is approached. From results in Example 3-9 for the van der Waals fluid

$$\frac{\Delta H_{vap_m}}{RT} = 6\left(\frac{T_c - T}{T_c}\right)^{1/2}$$

32. $q = 2/5$ for the van der Waals fluid.

33.
$$\omega'_s = \left(-\frac{4}{3}t_s\right)^{1/2} \qquad \frac{P_s}{P_c} = 1 - 3(\omega'_s)^2 + 3(\omega'_s)^3$$

ω'_s is negative on the superheated liquid side of the spinodal and P_s can become negative, that is, a metastable liquid under tension. See *R. Reid. Am. Scientist* **64**, 146 (1976).

34.
$$\Delta\mu_m = \frac{P_c}{\rho_{m_c}}\left(\frac{\Delta\rho_m}{\rho_{m_c}}\right)\left(\frac{\Delta\rho_m}{\rho_{m_c}}\right)^2\left[\frac{3}{2} + \frac{6t}{(\Delta\rho_m/\rho_{m_c})^2} + \frac{3}{2}t - \frac{3}{8}(t+1)\left(\frac{\Delta\rho_m}{\rho_{m_c}}\right)\right]$$

Since $(\Delta\rho_m)^2$ is proportional to t, the first two terms are of comparable magnitude whereas the others are small in the critical region. If we neglect them and compare this result with Eq. (3.A-22) we can identify $\gamma/\beta = 2$, $1/\beta = 2$, and thus $\beta = 1/2$, $\gamma = 1$, and $\delta = (\gamma/\beta) + 1 = 3$, which are the expected classical exponents. Nevertheless, the van der Waals equation does not precisely follow Widom's equation (as real fluids seem to do) because the added terms show that we do not have a function of $t/(\Delta\rho_m/\rho_{m_c})^2$ only. They also show that $\Delta\mu_m$ for van der Waals is not strictly an odd function of $\Delta\rho_m$. Real fluids are much more closely antisymmetric about $\rho_m = \rho_{m_c}$ than is the van der Waals fluid.

35a. The exponent is $2\beta - 2 + \gamma_-$.
35b. The identity is $-\alpha_- = 2\beta - 2 + \gamma_-$.
35c. The inequality is $\alpha_- > 2 - 2\beta - \gamma_-$.

37. Do not neglect the temperature dependence of ΔH_{vap_m}. Integrate the Clapeyron equation to obtain as the P_σ of water

$$\ln(P_\sigma/1 \text{ atm}) = -\frac{6770}{T} - 5.05\ln T + 48.1$$

At the temperature T_f when all the liquid has just fully vaporized

$$P_\sigma = \frac{2RT_f}{30.6} = 0.00536\,T_f \qquad \text{(in atm)}$$

Solving numerically, $T_f = 395$ K.

Rewrite the gaseous term in C_{V_2} using $\rho_g = P/kT$ and the Clapeyron equation, since P will be P_σ until the last drop of liquid vaporizes. At any point prior to complete evaporation, n_g (moles of gas), is given by

$$n_g = \frac{P_\sigma V}{RT} = 373\frac{P_\sigma}{T} \qquad (P_\sigma \text{ in atm})$$

Results are

$t/°C$:	25	50	75	100	110	122	123
$10^{-3}C_V/\text{J}\cdot\text{K}^{-1}$	0.246	0.420	0.772	1.31	1.68	2.09	0.050

There is a discontinuity in C_V of $2.04 \times 10^3\ \text{J}\cdot\text{K}^{-1}$ at 122°C.

38. Any change in the model to limit randomness must give a negative contribution to $\Delta S^E_{m,mxg}$ and thus make it negative since it was zero in the fully random case.

41. Use conditions of critical demixing given by Eqs. (I-54, I-55). Consult a physical chemistry text for definitions and conventions regarding activities and activity coefficients.

42. The slope of $\ln(X''_b - X'_b)$ versus $\ln(1 - T/T_c)$ for $T/T_c > 0.90$ is about 0.48, and indeed with this classical theory we except the exponent β to be $\frac{1}{2}$.

43a. The Henry law constant k_a of species a is $k_a = f_a^* e^{zw'/RT}$.
43b. $T/T_c = 0.75$ and the coexisting phases have $X''_b = 0.888$; $X'_b = 0.112$.

44b. 0.755; 1.78.
44c. 3.95.

45. 4.0×10^{-10} m.

46. $C' = 0.364$; $\gamma_\pm = 0.664$ at minimum;
$I = 1.96$ for $\gamma_\pm = 1.000$.

47.
$$\pi^{el} = \frac{G^{el} - A^{el}}{V} = -\frac{kT\kappa^3}{24\pi}\left[\frac{3}{1 + \kappa\sigma} - 6s(\kappa\sigma)\right]$$

Consistent with the Debye-Hückel theory, which ascribes all nonideal solution effects to electrical charge interactions, there will be a major term π^{id} that looks like the ideal gas law $\pi^{id} = kT\sum_i \mathcal{N}_i$, where $\mathcal{N}_i$ is the number density of ions of type i. The total π is: $\pi = \pi^{id} + \pi^{el}$.

49. Fraction beyond $= (2/e)(e^{\kappa\sigma}/1 + \kappa\sigma)$.

53. There is no ionic association if $q < \sigma$. For 1-1 salts in water $q = 3.6 \times 10^{-10}$ m $< \sigma$; $\therefore$ no association. For 2-2 salts in water $q = 14.3 \times 10^{-10}$ m $> \sigma$.

$$(1 - \alpha) = 32\pi N_i q^3 (0.94) = (2.8 \times 10^{-25}) N_i$$

$$N_i / \text{m}^{-3} = M N_A (1000) \qquad M \text{ is molarity}$$

M	N_i/m^{-3}	$(1 - \alpha)$
10^{-4}	6.02×10^{22}	0.017
10^{-3}	6.02×10^{23}	0.17
10^{-2}	6.02×10^{24}	1.7

For $(1 - \alpha)$ larger than unity you must consider ionic triplets of net charge. For $(1 - \alpha) = 1$ (100 percent association) the $\gamma_\pm = 1$ since the species are without charge. This theory is rather ad hoc.

54. ΔA^{el} is always negative. For aqueous solutions in the full range 0 to 100°C $(1/T + (3/D)(\partial D/\partial T))$ is negative, so ΔS^{el} is positive. Certainly ΔU^{el} must be positive since there must be a net energy input to the solution to make up for the potential energy drop upon charge formation:

$$\Delta U^{el} = \Delta A^{el} + T\Delta S^{el} = \frac{3}{2} \Delta A^{el} \left(1 - \frac{T}{219}\right)$$

Note however that the process is a hypothetical thought experiment and we have not taken into account the entropy-reducing effect of solvation of the newly formed ions by the solvent.

55a. $$\Delta \bar{S}_2 = \nu R \ln \frac{(m_\pm)_b}{(m_\pm)_a} + \frac{3}{2}\left(\frac{1}{T} + \frac{3}{D}\frac{\partial D}{\partial T}\right)\left(\frac{1}{2}\right)\left(\frac{\Delta A^{el}(m_a)}{n_2} - \frac{\Delta A^{el}(m_b)}{n_2}\right)$$

55b. $$\Delta \bar{S}_2 = 38.29 - 2.05 = 36.2 \text{ J} \cdot \text{K}^{-1} \cdot \text{mol}^{-1}$$

The charge interaction term here is only about 5 percent of the main ideal solution result.

Chapter 4

1. $C_{V_m}/3$; $0.286 \text{ J} \cdot \text{K}^{-1} \cdot \text{mol}^{-1}$; 420 K.

2. $\theta_D = 1824$ K; $\theta_E = 1368$ K; $1.47 \text{ J} \cdot \text{K}^{-1} \cdot \text{mol}^{-1}$.

4. Using the full sum, $r_e = (1.090)\sigma$

$$\left(\frac{N_A \phi(0)}{2}\right)_{eq} = -71.59 \frac{\varepsilon^*}{k} \Big/ \text{J} \cdot \text{mol}^{-1}$$

5.
$$P = \frac{1}{2}\left(\frac{\partial \phi(0)}{\partial v}\right)_{N,T} - \frac{9}{8}k\left(\frac{\partial \theta_D}{\partial v}\right)_{N,T} - \frac{3kT}{\theta_D}\left(\frac{\partial \theta_D}{\partial v}\right)_{N,T}D(u)$$

Neglecting the PV/N term it is accurate to use

$$\mu_c \cong \frac{A_c}{N} = \frac{\phi(0)}{2} + \frac{9}{8}k\theta_D + 3kT\ln(1 - e^{-u}) - kTD(u)$$

6a. Use the chain rule

$$\left(\frac{\partial V}{\partial T}\right)_P\left(\frac{\partial T}{\partial P}\right)_V\left(\frac{\partial P}{\partial V}\right)_T = -1$$

and note that at V,N constant, $U_c(0)$ is constant since it is only a function of V/N. Then one can derive

$$\gamma_G = V\left(\frac{\partial P}{\partial U_c}\right)_{V,N}$$

and note that it is a measure of pressure variation with change of energy at constant volume. D. Wallace [*Thermodynamics of Crystals*. Wiley, New York, 1972. p. 5] suggests that a way to determine γ_G would be to add a known amount of energy with a laser pulse to a clamped solid and measure the pressure increment generated with transducers at the clamped surfaces.

6d. $\frac{1}{3}$. This poor result means the speed of sound is very dependent on V/N.

7. Use Eq. (I-34)

T/K	50	100	150	200	250	300	500
θ_D^*/K	309	313	311	296	275	292	300

The increase of C_{V_m} above $3R$ at the highest temperatures in this case is due to the electronic contribution.

8. Except at $T = 50$ K, γ_G is about constant at 1.91.

9. The calculated C_{P_m} is 83.3 J·K^{-1}·mol^{-1}.

10.
$$\ln\left(\frac{P_s}{P^0}\right) = \ln\left(\frac{kTg_{el}}{\lambda^3 P^0}\right) + \frac{\phi(0)}{2kT} + \frac{3}{2}\left(\frac{h\nu_E}{kT}\right) + 3\ln(1 - e^{-\theta_E/T})$$

$$P^0 = 1 \text{ atm.}$$

11. Calculated value is 456 torr.

14.
$$n_x^2 + n_y^2 + n_z^2 = \left(\frac{3}{\pi} N_A\right)^{2/3} = 6.916 \times 10^{15}$$

15. Below 0.33 K theoretically and below 0.37 K actually, since the electrons have an effective mass of $1.25m_e$.

17. Equations (1-162) and (1-163) each must be multiplied by 2 to take into account the spin $\frac{1}{2}$ property. Write

$$z = \frac{\lambda^3\rho}{2} + a_1\rho^2 + a_2\rho^3 + \dots$$

and determine a_1 and a_2. The results are

$$z = \frac{\lambda^3\rho}{2}\left[1 + \frac{\lambda^3\rho}{2^{5/2}} + (\lambda^3\rho)^2\left(\frac{1}{2^4} - \frac{1}{(4)3^{3/2}}\right) + \dots\right]$$

and

$$\frac{P}{\rho kT} = 1 + \frac{\lambda^3\rho}{2^{7/2}} + \frac{(\lambda^3\rho)^2}{4}\left(\frac{1}{8} - \frac{2}{3^{5/2}}\right) + \dots$$

Insert these into the result of Problem 1-22. Use the Sackur-Tetrode equation that gives $(S/Nk)_{c\ell} = \frac{5}{2} - \ln(\lambda^3\rho)$.

Because the coefficients of the expansion for C_V/Nk have such high powers of 2 in the denominators (and also in higher orders cancelations between terms of different signs), the high-temperature, low-density series will converge for $\rho\lambda^3 > 1$ or for $T < T_F$. It may be used for $T = \frac{1}{2}T_F$ and higher.

The low-temperature, high-density expansion in Eq. (4-69) can be seen to really be an expansion in powers of $\pi(T/T_F)$ and will not be accurate unless $T/T_F < 1/\pi = 0.32$. At $T/T_F = 0.50$ it gives $C_V/Nk = 0.64$, which is not a good value. It may be used for $T/T_F = 0.20$ and below.

18a. 5.47×10^4 K; 4.72 eV; 0.22 J·K^{-1}·mol^{-1}.
18b. 3.4×10^{13} K; 2.9×10^9 eV; 3.6×10^{-10} J·K^{-1}·mol^{-1}.
18c. 3.88 K; 3.34×10^{-4} eV; 21.5 J·K^{-1}·mol^{-1}.

In part **c** one can use the *high* temperature expansion result for the entropy since $T/T_F = 0.83 > \frac{1}{2}$.

20. $C_1 = 3Rs$, where s is the number of atoms in the unit cell of the lattice.

22. $\varepsilon_i = 1.2$ eV per defect; $\Delta S_{vib_m} = 14R$.

Chapter 5

3. Hits per atom per second are 1.7×10^8, 1.7×10^2, and 0.017, respectively, in the three cases.

4. 6.4×10^{-12} torr.

5. 8.7×10^{-8} seconds.

6.
$$P_1(t) = \frac{1}{2}[P_1(0) + P_2(0)] + \frac{1}{2}[P_1(0) - P_2(0)] \exp\left(-\frac{a\bar{c}t}{V}\right)$$

7. 389 seconds.

8.
$$t = \frac{1}{q}\ln\left(\frac{P_0}{P}\right) \qquad q = \frac{a}{V}\left(\frac{RT}{2\pi M}\right)^{1/2}$$

9.
$$\frac{P_{Xe}}{P_{He}} = 1.77; \qquad X_{Xe} = 0.639.$$

10. 1151 stages.

11a. One needs to average $(\rho c/4)\frac{1}{2}mc^2 = \rho mc^3/8$. The result for the average kinetic energy removed per unit area per unit time is $\sqrt{2\pi}\rho m(kT/\pi m)^{3/2}$.

11b. The number of molecules removing this kinetic energy is

$$\frac{\rho\bar{c}}{4} = \frac{\rho}{4}\left(\frac{8kT}{\pi m}\right)^{1/2}$$

Thus the kinetic energy per molecule in the escaping beam is $2kT$.

11c. Faster molecules preferentially effuse.

12. $\lambda_1 = 9.55 \times 10^{-8}$ m; $\qquad \lambda_2 = 4.55 \times 10^{-8}$ m.

$Z_{11} = 4.2 \times 10^{33}$ s^{-1}·m^{-3}; $\qquad Z_{22} = 32.0 \times 10^{33}$ s^{-1}·m^{-3}.

$Z_{12} = 29.2 \times 10^{33}$ s^{-1}·m^{-3}.

13. The average time between collisions is about 10^{-10} s at 1 atm and 10^{-8} s at 0.01 atm (when the collisions are not effective in removing energy from the excited molecules). Thus one can estimate the lifetime of the excited state to be $\sim 10^{-9}$ s.

15. $\sigma_0 = 317$ pm; $\qquad S = 118$ K.

16. For equality of pressure throughout the experimental system effusive effects must be absent. If they dominate, $P_1/\sqrt{T_1} = P_2/\sqrt{T_2}$ such that the measured P_2 at the high temperature will be far higher than the vapor pressure P_1. For the setup shown, at some temperature between 1.0 and 0.5 K and at all temperatures below this temperature the vapor pressure will not equal the high-temperature pressure reading.

17. 2.85.

18. 29 percent.

19.
$$P = \frac{(2\pi MRT_g)^{1/2}(pI)}{a(\tfrac{3}{2}R)(T_w - T_g)}$$
pI = power input in energy/s
a = wire surface area = πdl

20. 4.0×10^{-3} torr.

22b. 1.70×10^{-5} m²·s⁻¹ for O_2-N_2; 0.75×10^{-5} m²·s⁻¹ for O_2-CCl_4.

23. 1.30.

26. 0.802.

27.
$$\left(\frac{u_1}{c_m}\right) = \left(\frac{5\mathcal{M}^2}{9 + 3\mathcal{M}^2}\right)^{1/2}$$

28. If a fluid is truly incompressible, $(\partial \rho'/\partial P)_S = 0$ and $v_s = \infty$. Thus $\mathcal{M} = u/v_s$ will be zero for any nonzero u value.

29. 1.0×10^{-8} atm.

30. For $\mathcal{M} = 30$ (far beyond the throat), Reynolds number $\simeq (6.4 \times 10^5)r$. For $\mathcal{M} - 0.20$ (before the nozzle throat) Reynolds number $\simeq (2.2 \times 10^7)r$, where r = nozzle radius in meters at the position.

Chapter 6

1.

N:	3	5	15	25	51
P_{approx}	0.0071	0.0293	0.0895	0.0968	0.08744
P_{exact}	0	0.0312	0.0916	0.0974	0.08739

It is easy to show that the approximate probability function has a maximum at $N = 25$.

2a. 1.49×10^{-53}.
2b. 9.78×10^{-10}.

3. In the Gaussian approximation take the product of the individual probabilities to get to m in N steps and integrate over all m with result $1/\sqrt{\pi N}$. The exact result can be shown to be

$$\left(\frac{1}{2}\right)^{2N} \frac{(2N)!}{(N!)^2}$$

which reduces to $1/\sqrt{\pi N}$ with use of Eq. (6-9). Of course, for $N = 1$ the exact answer is $\frac{1}{2}$, since after the first particle moves once there is one chance in two that the second particle jumps to the same position.

4. $D = 4.7 \times 10^{-11}$ m^2/s.

5. When $(\overline{R_N^2})^{1/2}$ for the added molecules is equal to the longest length of motion open to them (in this example 80 cm), the molecules should be essentially uniformly mixed.

8. Two-dimensional systems cannot show diffusive behavior because of the logarithmic divergence of the diffusion constant.

10. 16.4×10^3 J·mol^{-1} for H$_2$O; 2.37×10^3 J·mol^{-1} for Hg.

12. 140 cm^3·g^{-1}; 5.6×10^5 g/mol.

13. In solvent (1) $a = 0.870$ (random coil). In solvent (2) $a = 1.75$ (long rod).

14. 1.5 kg·m^{-1}·s^{-1}.

15. 0.37 s (larger bubbles move faster).

16. M/kg·mol$^{-1} = 9.913 \times 10^{-49} (T/\eta D)^3 \rho'$, all units SI.

17a. 151 kg/mol.
17b. 44.7×10^{-10} m.
17c. No.

18. 24.4×10^{-13} s.

19. 1.4×10^{-12} m^2/s; 4.2 days.

20. 2.05×10^3 rpm.

22. Let $\phi = \text{erf}(x/\sqrt{4Dt})$. Then the solution is

$$c = \frac{c_0''}{2}(1 + \phi) + \frac{c_0'}{2}(1 - \phi)$$

Chapter 7

6. As the side ε of the black square is divided by 3 each black area is divided by 9 but their number is multiplied by 8. Hence $a(\varepsilon/3) = {}^8\!/_9 a(\varepsilon)$.

9b. The two points that move in one step to unity are $^2\!/_3$ and $^1\!/_3$. Their precursors are 0.81914, 0.18086, and 0.91944, 0.08056, respectively.
9c. If $\lambda = 4.00$ exactly, *all* the points on the line segment 0 to 1 iterate to bounded values and their $D = D_T = 1$. If $\lambda = \infty$ the inverse mapping shows that only the two points zero and unity iterate to bounded values and their $D = D_T = 0$.

11. To estimate α one iterates with a given Λ_n starting from $\bar{x} = {}^1\!/_2$ and calculates the x value for the point halfway through the cycle of 2^n. The absolute value of the difference of this x and $\frac{1}{2}$ is d_n. For example, in the period of 2^4, $x_8 = 0.4816738$ and $d_4 = 0.0183262$.

12. Use the method of Example 7-3.

14. $\bar{x} = 1/\lambda$ causes the function to be at a maximum as long as $\lambda > 0$. The only stable fixed point of cycle 1 is $x^* = 1$. The stable cycle of 2 starts at $\lambda = 2$ with $y_1 = y_2 = 1$ and ends at $\lambda = 2.5264$ with $y_1 = 1.7223$ and $y_2 = 2 - y_1$.

15. With this function the x value of the maximum depends on the λ value. We find, for example, $\Lambda_2 = 2.593518$. Then $\bar{x} = 1/\Lambda_2$, and acting twice on this value using $\lambda = \Lambda_2$ one obtains 0.1850722 such that $d_2 = |0.3855767 - 0.1850722| = 0.2005045$.

16. Notice that the sum of the x values of the cycle points always equals the number of points in the cycle.

18. For Eq. (7-45) the approximate $\lambda_c = 3.5616$. In the case of Eq. (7-59) equating the slopes gives a quadratic equation for y_1 in terms of λ_c, which then can be explicitly solved to obtain

$$y_1 = 1 + \sqrt{1 - (1/\lambda_c)}; \qquad \frac{dy_1}{d\lambda_c} = \frac{1}{2\lambda_c^2 \sqrt{1 - (1/\lambda_c)}}$$

Then guess a λ_c and solve numerically the additional requirement for y_1 as a cycle point of period 2:

$$y_1 \exp\left[\lambda_c(1 - y_1)\right] - 2 + y_1 = 0$$

The resulting approximate $\lambda_c = 2.7340$.

20. Prior triples still control the time evolution but now these triples will be represented by the 27 numbers 0 to 26 in base 3. The total number of rules is $3^{27} = 7.626 \times 10^{12}$. The number of legal rules is $3^{17} = 129, 140, 163$.

21a. A cycle of 4 appears after five steps.
21b. A cycle of 1 appears after three steps.
21c. A cycle of 6 appears after seven steps.

22. After the second step, configuration (0) occurs once and the other 31 configurations are all type (31).

23. The answers to this problem are illustrated in M. Gardner, Scientific American **223**, 120 (Oct. 1970).

24. In four moves the glider regains its original orientation and moves one cell diagonally down and to the right of its starting positions.

25. Make the approximation that at the eight-piece stage and beyond the pieces do not further sensibly shrink as they undergo removals of smaller and smaller gaps. (However do add up all the tiny gap sizes to infinite order.)

26. $x = \bar{x}$ is at about the middle of every stable periodic window.

27. The pairs of iterates that successively straddle the $x = \bar{x} = 0.5$ value are $f^{(4)} - f^{(2)}$; $f^{(4)} - f^{(8)}$; $f^{(16)} - f^{(8)}$; and $f^{(32)} - f^{(16)}$; $s_1 = 2.50304$; $s_2 = 5.80363$; $D = 0.5369$. A more accurate calculation gives $D = 0.5380$.

29. The step $(0/1)$ ranges from $\Omega = 0$ to $\Omega = 1/2\pi = 0.159155$. The step $(1/1)$ ranges from $\Omega = 1 - (1/2\pi)$ to 1. For $\Omega = 0.1600$, $P = 1$ and $Q = 60$. The width in Ω (extending very slightly to either side of 0.1600) for which $W = \frac{1}{60}$ is not easy to calculate.

Index

A CATALOG OF SELECTED
DOVER BOOKS
IN ALL FIELDS OF INTEREST

A CATALOG OF SELECTED DOVER
BOOKS IN ALL FIELDS OF INTEREST

CONCERNING THE SPIRITUAL IN ART, Wassily Kandinsky. Pioneering work by father of abstract art. Thoughts on color theory, nature of art. Analysis of earlier masters. 12 illustrations. 80pp. of text. 5⅜ x 8½.   23411-8 Pa. $3.95

ANIMALS: 1,419 Copyright-Free Illustrations of Mammals, Birds, Fish, Insects, etc., Jim Harter (ed.). Clear wood engravings present, in extremely lifelike poses, over 1,000 species of animals. One of the most extensive pictorial sourcebooks of its kind. Captions. Index. 284pp. 9 x 12.   23766-4 Pa. $12.95

CELTIC ART: The Methods of Construction, George Bain. Simple geometric techniques for making Celtic interlacements, spirals, Kells-type initials, animals, humans, etc. Over 500 illustrations. 160pp. 9 x 12. (USO)   22923-8 Pa. $9.95

AN ATLAS OF ANATOMY FOR ARTISTS, Fritz Schider. Most thorough reference work on art anatomy in the world. Hundreds of illustrations, including selections from works by Vesalius, Leonardo, Goya, Ingres, Michelangelo, others. 593 illustrations. 192pp. 7⅛ x 10¼.   20241-0 Pa. $9.95

CELTIC HAND STROKE-BY-STROKE (Irish Half-Uncial from "The Book of Kells"): An Arthur Baker Calligraphy Manual, Arthur Baker. Complete guide to creating each letter of the alphabet in distinctive Celtic manner. Covers hand position, strokes, pens, inks, paper, more. Illustrated. 48pp. 8¼ x 11.   24336-2 Pa. $3.95

EASY ORIGAMI, John Montroll. Charming collection of 32 projects (hat, cup, pelican, piano, swan, many more) specially designed for the novice origami hobbyist. Clearly illustrated easy-to-follow instructions insure that even beginning papercrafters will achieve successful results. 48pp. 8¼ x 11.   27298-2 Pa. $3.50

THE COMPLETE BOOK OF BIRDHOUSE CONSTRUCTION FOR WOODWORKERS, Scott D. Campbell. Detailed instructions, illustrations, tables. Also data on bird habitat and instinct patterns. Bibliography. 3 tables. 63 illustrations in 15 figures. 48pp. 5¼ x 8½.   24407-5 Pa. $2.50

BLOOMINGDALE'S ILLUSTRATED 1886 CATALOG: Fashions, Dry Goods and Housewares, Bloomingdale Brothers. Famed merchants' extremely rare catalog depicting about 1,700 products: clothing, housewares, firearms, dry goods, jewelry, more. Invaluable for dating, identifying vintage items. Also, copyright-free graphics for artists, designers. Co-published with Henry Ford Museum & Greenfield Village. 160pp. 8¼ x 11.   25780-0 Pa. $10.95

HISTORIC COSTUME IN PICTURES, Braun & Schneider. Over 1,450 costumed figures in clearly detailed engravings–from dawn of civilization to end of 19th century. Captions. Many folk costumes. 256pp. 8⅜ x 11¾.   23150-X Pa. $12.95

STICKLEY CRAFTSMAN FURNITURE CATALOGS, Gustav Stickley and L. & J. G. Stickley. Beautiful, functional furniture in two authentic catalogs from 1910. 594 illustrations, including 277 photos, show settles, rockers, armchairs, reclining chairs, bookcases, desks, tables. 183pp. 6½ x 9¼. 23838-5 Pa. $9.95

AMERICAN LOCOMOTIVES IN HISTORIC PHOTOGRAPHS: 1858 to 1949, Ron Ziel (ed.). A rare collection of 126 meticulously detailed official photographs, called "builder portraits," of American locomotives that majestically chronicle the rise of steam locomotive power in America. Introduction. Detailed captions. xi + 129pp. 9 x 12. 27393-8 Pa. $12.95

AMERICA'S LIGHTHOUSES: An Illustrated History, Francis Ross Holland, Jr. Delightfully written, profusely illustrated fact-filled survey of over 200 American lighthouses since 1716. History, anecdotes, technological advances, more. 240pp. 8 x 10¾. 25576-X Pa. $12.95

TOWARDS A NEW ARCHITECTURE, Le Corbusier. Pioneering manifesto by founder of "International School." Technical and aesthetic theories, views of industry, economics, relation of form to function, "mass-production split" and much more. Profusely illustrated. 320pp. 6⅛ x 9¼. (USO) 25023-7 Pa. $9.95

HOW THE OTHER HALF LIVES, Jacob Riis. Famous journalistic record, exposing poverty and degradation of New York slums around 1900, by major social reformer. 100 striking and influential photographs. 233pp. 10 x 7⅞. 22012-5 Pa. $10.95

FRUIT KEY AND TWIG KEY TO TREES AND SHRUBS, William M. Harlow. One of the handiest and most widely used identification aids. Fruit key covers 120 deciduous and evergreen species; twig key 160 deciduous species. Easily used. Over 300 photographs. 126pp. 5⅜ x 8½. 20511-8 Pa. $3.95

COMMON BIRD SONGS, Dr. Donald J. Borror. Songs of 60 most common U.S. birds: robins, sparrows, cardinals, bluejays, finches, more—arranged in order of increasing complexity. Up to 9 variations of songs of each species. Cassette and manual 99911-4 $8.95

ORCHIDS AS HOUSE PLANTS, Rebecca Tyson Northen. Grow cattleyas and many other kinds of orchids—in a window, in a case, or under artificial light. 63 illustrations. 148pp. 5⅜ x 8½. 23261-1 Pa. $4.95

MONSTER MAZES, Dave Phillips. Masterful mazes at four levels of difficulty. Avoid deadly perils and evil creatures to find magical treasures. Solutions for all 32 exciting illustrated puzzles. 48pp. 8¼ x 11. 26005-4 Pa. $2.95

MOZART'S DON GIOVANNI (DOVER OPERA LIBRETTO SERIES), Wolfgang Amadeus Mozart. Introduced and translated by Ellen H. Bleiler. Standard Italian libretto, with complete English translation. Convenient and thoroughly portable—an ideal companion for reading along with a recording or the performance itself. Introduction. List of characters. Plot summary. 121pp. 5¼ x 8½. 24944-1 Pa. $2.95

TECHNICAL MANUAL AND DICTIONARY OF CLASSICAL BALLET, Gail Grant. Defines, explains, comments on steps, movements, poses and concepts. 15-page pictorial section. Basic book for student, viewer. 127pp. 5⅜ x 8½. 21843-0 Pa. $4.95

BRASS INSTRUMENTS: Their History and Development, Anthony Baines. Authoritative, updated survey of the evolution of trumpets, trombones, bugles, cornets, French horns, tubas and other brass wind instruments. Over 140 illustrations and 48 music examples. Corrected and updated by author. New preface. Bibliography. 320pp. 5⅜ x 8½. 27574-4 Pa. $9.95

HOLLYWOOD GLAMOR PORTRAITS, John Kobal (ed.). 145 photos from 1926-49. Harlow, Gable, Bogart, Bacall; 94 stars in all. Full background on photographers, technical aspects. 160pp. 8⅜ x 11¼. 23352-9 Pa. $12.95

MAX AND MORITZ, Wilhelm Busch. Great humor classic in both German and English. Also 10 other works: "Cat and Mouse," "Plisch and Plumm," etc. 216pp. 5⅜ x 8½. 20181-3 Pa. $6.95

THE RAVEN AND OTHER FAVORITE POEMS, Edgar Allan Poe. Over 40 of the author's most memorable poems: "The Bells," "Ulalume," "Israfel," "To Helen," "The Conqueror Worm," "Eldorado," "Annabel Lee," many more. Alphabetic lists of titles and first lines. 64pp. 5 3⁄16 x 8¼. 26685-0 Pa. $1.00

PERSONAL MEMOIRS OF U. S. GRANT, Ulysses Simpson Grant. Intelligent, deeply moving firsthand account of Civil War campaigns, considered by many the finest military memoirs ever written. Includes letters, historic photographs, maps and more. 528pp. 6⅛ x 9¼. 28587-1 Pa. $11.95

AMULETS AND SUPERSTITIONS, E. A. Wallis Budge. Comprehensive discourse on origin, powers of amulets in many ancient cultures: Arab, Persian Babylonian, Assyrian, Egyptian, Gnostic, Hebrew, Phoenician, Syriac, etc. Covers cross, swastika, crucifix, seals, rings, stones, etc. 584pp. 5⅜ x 8½. 23573-4 Pa. $12.95

RUSSIAN STORIES/PYCCKNE PACCKA3bl: A Dual-Language Book, edited by Gleb Struve. Twelve tales by such masters as Chekhov, Tolstoy, Dostoevsky, Pushkin, others. Excellent word-for-word English translations on facing pages, plus teaching and study aids, Russian/English vocabulary, biographical/critical introductions, more. 416pp. 5⅜ x 8½. 26244-8 Pa. $8.95

PHILADELPHIA THEN AND NOW: 60 Sites Photographed in the Past and Present, Kenneth Finkel and Susan Oyama. Rare photographs of City Hall, Logan Square, Independence Hall, Betsy Ross House, other landmarks juxtaposed with contemporary views. Captures changing face of historic city. Introduction. Captions. 128pp. 8¼ x 11. 25790-8 Pa. $9.95

AIA ARCHITECTURAL GUIDE TO NASSAU AND SUFFOLK COUNTIES, LONG ISLAND, The American Institute of Architects, Long Island Chapter, and the Society for the Preservation of Long Island Antiquities. Comprehensive, well-researched and generously illustrated volume brings to life over three centuries of Long Island's great architectural heritage. More than 240 photographs with authoritative, extensively detailed captions. 176pp. 8¼ x 11. 26946-9 Pa. $14.95

NORTH AMERICAN INDIAN LIFE: Customs and Traditions of 23 Tribes, Elsie Clews Parsons (ed.). 27 fictionalized essays by noted anthropologists examine religion, customs, government, additional facets of life among the Winnebago, Crow, Zuni, Eskimo, other tribes. 480pp. 6⅛ x 9¼. 27377-6 Pa. $10.95

FRANK LLOYD WRIGHT'S HOLLYHOCK HOUSE, Donald Hoffmann. Lavishly illustrated, carefully documented study of one of Wright's most controversial residential designs. Over 120 photographs, floor plans, elevations, etc. Detailed perceptive text by noted Wright scholar. Index. 128pp. 9¼ x 10¾. 27133-1 Pa. $11.95

THE MALE AND FEMALE FIGURE IN MOTION: 60 Classic Photographic Sequences, Eadweard Muybridge. 60 true-action photographs of men and women walking, running, climbing, bending, turning, etc., reproduced from rare 19th-century masterpiece. vi + 121pp. 9 x 12. 24745-7 Pa. $10.95

1001 QUESTIONS ANSWERED ABOUT THE SEASHORE, N. J. Berrill and Jacquelyn Berrill. Queries answered about dolphins, sea snails, sponges, starfish, fishes, shore birds, many others. Covers appearance, breeding, growth, feeding, much more. 305pp. 5¼ x 8¼. 23366-9 Pa. $8.95

GUIDE TO OWL WATCHING IN NORTH AMERICA, Donald S. Heintzelman. Superb guide offers complete data and descriptions of 19 species: barn owl, screech owl, snowy owl, many more. Expert coverage of owl-watching equipment, conservation, migrations and invasions, etc. Guide to observing sites. 84 illustrations. xiii + 193pp. 5⅜ x 8½. 27344-X Pa. $8.95

MEDICINAL AND OTHER USES OF NORTH AMERICAN PLANTS: A Historical Survey with Special Reference to the Eastern Indian Tribes, Charlotte Erichsen-Brown. Chronological historical citations document 500 years of usage of plants, trees, shrubs native to eastern Canada, northeastern U.S. Also complete identifying information. 343 illustrations. 544pp. 6½ x 9¼. 25951-X Pa. $12.95

STORYBOOK MAZES, Dave Phillips. 23 stories and mazes on two-page spreads: Wizard of Oz, Treasure Island, Robin Hood, etc. Solutions. 64pp. 8¼ x 11. 23628-5 Pa. $2.95

NEGRO FOLK MUSIC, U.S.A., Harold Courlander. Noted folklorist's scholarly yet readable analysis of rich and varied musical tradition. Includes authentic versions of over 40 folk songs. Valuable bibliography and discography. xi + 324pp. 5⅜ x 8½. 27350-4 Pa. $9.95

MOVIE-STAR PORTRAITS OF THE FORTIES, John Kobal (ed.). 163 glamor, studio photos of 106 stars of the 1940s: Rita Hayworth, Ava Gardner, Marlon Brando, Clark Gable, many more. 176pp. 8⅜ x 11¼. 23546-7 Pa. $12.95

BENCHLEY LOST AND FOUND, Robert Benchley. Finest humor from early 30s, about pet peeves, child psychologists, post office and others. Mostly unavailable elsewhere. 73 illustrations by Peter Arno and others. 183pp. 5⅜ x 8½. 22410-4 Pa. $6.95

YEKL and THE IMPORTED BRIDEGROOM AND OTHER STORIES OF YIDDISH NEW YORK, Abraham Cahan. Film Hester Street based on Yekl (1896). Novel, other stories among first about Jewish immigrants on N.Y.'s East Side. 240pp. 5⅜ x 8½. 22427-9 Pa. $6.95

SELECTED POEMS, Walt Whitman. Generous sampling from *Leaves of Grass.* Twenty-four poems include "I Hear America Singing," "Song of the Open Road," "I Sing the Body Electric," "When Lilacs Last in the Dooryard Bloom'd," "O Captain! My Captain!"–all reprinted from an authoritative edition. Lists of titles and first lines. 128pp. 5$\frac{3}{16}$ x 8¼. 26878-0 Pa. $1.00

THE BEST TALES OF HOFFMANN, E. T. A. Hoffmann. 10 of Hoffmann's most important stories: "Nutcracker and the King of Mice," "The Golden Flowerpot," etc. 458pp. 5⅜ x 8½. 21793-0 Pa. $9.95

FROM FETISH TO GOD IN ANCIENT EGYPT, E. A. Wallis Budge. Rich detailed survey of Egyptian conception of "God" and gods, magic, cult of animals, Osiris, more. Also, superb English translations of hymns and legends. 240 illustrations. 545pp. 5⅜ x 8½. 25803-3 Pa. $13.95

FRENCH STORIES/CONTES FRANÇAIS: A Dual-Language Book, Wallace Fowlie. Ten stories by French masters, Voltaire to Camus: "Micromegas" by Voltaire; "The Atheist's Mass" by Balzac; "Minuet" by de Maupassant; "The Guest" by Camus, six more. Excellent English translations on facing pages. Also French-English vocabulary list, exercises, more. 352pp. 5⅜ x 8½. 26443-2 Pa. $8.95

CHICAGO AT THE TURN OF THE CENTURY IN PHOTOGRAPHS: 122 Historic Views from the Collections of the Chicago Historical Society, Larry A. Viskochil. Rare large-format prints offer detailed views of City Hall, State Street, the Loop, Hull House, Union Station, many other landmarks, circa 1904-1913. Introduction. Captions. Maps. 144pp. 9⅜ x 12¼. 24656-6 Pa. $12.95

OLD BROOKLYN IN EARLY PHOTOGRAPHS, 1865-1929, William Lee Younger. Luna Park, Gravesend race track, construction of Grand Army Plaza, moving of Hotel Brighton, etc. 157 previously unpublished photographs. 165pp. 8⅞ x 11¾. 23587-4 Pa. $13.95

THE MYTHS OF THE NORTH AMERICAN INDIANS, Lewis Spence. Rich anthology of the myths and legends of the Algonquins, Iroquois, Pawnees and Sioux, prefaced by an extensive historical and ethnological commentary. 36 illustrations. 480pp. 5⅜ x 8½. 25967-6 Pa. $8.95

AN ENCYCLOPEDIA OF BATTLES: Accounts of Over 1,560 Battles from 1479 B.C. to the Present, David Eggenberger. Essential details of every major battle in recorded history from the first battle of Megiddo in 1479 B.C. to Grenada in 1984. List of Battle Maps. New Appendix covering the years 1967-1984. Index. 99 illustrations. 544pp. 6½ x 9¼. 24913-1 Pa. $14.95

SAILING ALONE AROUND THE WORLD, Captain Joshua Slocum. First man to sail around the world, alone, in small boat. One of great feats of seamanship told in delightful manner. 67 illustrations. 294pp. 5⅜ x 8½. 20326-3 Pa. $5.95

ANARCHISM AND OTHER ESSAYS, Emma Goldman. Powerful, penetrating, prophetic essays on direct action, role of minorities, prison reform, puritan hypocrisy, violence, etc. 271pp. 5⅜ x 8½. 22484-8 Pa. $6.95

MYTHS OF THE HINDUS AND BUDDHISTS, Ananda K. Coomaraswamy and Sister Nivedita. Great stories of the epics; deeds of Krishna, Shiva, taken from puranas, Vedas, folk tales; etc. 32 illustrations. 400pp. 5⅜ x 8½. 21759-0 Pa. $10.95

BEYOND PSYCHOLOGY, Otto Rank. Fear of death, desire of immortality, nature of sexuality, social organization, creativity, according to Rankian system. 291pp. 5⅜ x 8½. 20485-5 Pa. $8.95

A THEOLOGICO-POLITICAL TREATISE, Benedict Spinoza. Also contains unfinished Political Treatise. Great classic on religious liberty, theory of government on common consent. R. Elwes translation. Total of 421pp. 5⅜ x 8½. 20249-6 Pa. $9.95

MY BONDAGE AND MY FREEDOM, Frederick Douglass. Born a slave, Douglass became outspoken force in antislavery movement. The best of Douglass' autobiographies. Graphic description of slave life. 464pp. 5⅜ x 8½. 22457-0 Pa. $8.95

FOLLOWING THE EQUATOR: A Journey Around the World, Mark Twain. Fascinating humorous account of 1897 voyage to Hawaii, Australia, India, New Zealand, etc. Ironic, bemused reports on peoples, customs, climate, flora and fauna, politics, much more. 197 illustrations. 720pp. 5⅜ x 8½. 26113-1 Pa. $15.95

THE PEOPLE CALLED SHAKERS, Edward D. Andrews. Definitive study of Shakers: origins, beliefs, practices, dances, social organization, furniture and crafts, etc. 33 illustrations. 351pp. 5⅜ x 8½. 21081-2 Pa. $8.95

THE MYTHS OF GREECE AND ROME, H. A. Guerber. A classic of mythology, generously illustrated, long prized for its simple, graphic, accurate retelling of the principal myths of Greece and Rome, and for its commentary on their origins and significance. With 64 illustrations by Michelangelo, Raphael, Titian, Rubens, Canova, Bernini and others. 480pp. 5⅜ x 8½. 27584-1 Pa. $9.95

PSYCHOLOGY OF MUSIC, Carl E. Seashore. Classic work discusses music as a medium from psychological viewpoint. Clear treatment of physical acoustics, auditory apparatus, sound perception, development of musical skills, nature of musical feeling, host of other topics. 88 figures. 408pp. 5⅜ x 8½. 21851-1 Pa. $10.95

THE PHILOSOPHY OF HISTORY, Georg W. Hegel. Great classic of Western thought develops concept that history is not chance but rational process, the evolution of freedom. 457pp. 5⅜ x 8½. 20112-0 Pa. $9.95

THE BOOK OF TEA, Kakuzo Okakura. Minor classic of the Orient: entertaining, charming explanation, interpretation of traditional Japanese culture in terms of tea ceremony. 94pp. 5⅜ x 8½. 20070-1 Pa. $3.95

LIFE IN ANCIENT EGYPT, Adolf Erman. Fullest, most thorough, detailed older account with much not in more recent books, domestic life, religion, magic, medicine, commerce, much more. Many illustrations reproduce tomb paintings, carvings, hieroglyphs, etc. 597pp. 5⅜ x 8½. 22632-8 Pa. $11.95

SUNDIALS, Their Theory and Construction, Albert Waugh. Far and away the best, most thorough coverage of ideas, mathematics concerned, types, construction, adjusting anywhere. Simple, nontechnical treatment allows even children to build several of these dials. Over 100 illustrations. 230pp. 5⅜ x 8½. 22947-5 Pa. $7.95

DYNAMICS OF FLUIDS IN POROUS MEDIA, Jacob Bear. For advanced students of ground water hydrology, soil mechanics and physics, drainage and irrigation engineering, and more. 335 illustrations. Exercises, with answers. 784pp. 6⅛ x 9¼. 65675-6 Pa. $19.95

SONGS OF EXPERIENCE: Facsimile Reproduction with 26 Plates in Full Color, William Blake. 26 full-color plates from a rare 1826 edition. Includes "The Tyger," "London," "Holy Thursday," and other poems. Printed text of poems. 48pp. 5¼ x 7. 24636-1 Pa. $4.95

OLD-TIME VIGNETTES IN FULL COLOR, Carol Belanger Grafton (ed.). Over 390 charming, often sentimental illustrations, selected from archives of Victorian graphics—pretty women posing, children playing, food, flowers, kittens and puppies, smiling cherubs, birds and butterflies, much more. All copyright-free. 48pp. 9¼ x 12¼. 27269-9 Pa. $7.95

PERSPECTIVE FOR ARTISTS, Rex Vicat Cole. Depth, perspective of sky and sea, shadows, much more, not usually covered. 391 diagrams, 81 reproductions of drawings and paintings. 279pp. 5⅜ x 8½. 22487-2 Pa. $7.95

DRAWING THE LIVING FIGURE, Joseph Sheppard. Innovative approach to artistic anatomy focuses on specifics of surface anatomy, rather than muscles and bones. Over 170 drawings of live models in front, back and side views, and in widely varying poses. Accompanying diagrams. 177 illustrations. Introduction. Index. 144pp. 8⅜ x11¼. 26723-7 Pa. $8.95

GOTHIC AND OLD ENGLISH ALPHABETS: 100 Complete Fonts, Dan X. Solo. Add power, elegance to posters, signs, other graphics with 100 stunning copyright-free alphabets: Blackstone, Dolbey, Germania, 97 more–including many lower-case, numerals, punctuation marks. 104pp. 8⅛ x 11. 24695-7 Pa. $8.95

HOW TO DO BEADWORK, Mary White. Fundamental book on craft from simple projects to five-bead chains and woven works. 106 illustrations. 142pp. 5⅜ x 8. 20697-1 Pa. $4.95

THE BOOK OF WOOD CARVING, Charles Marshall Sayers. Finest book for beginners discusses fundamentals and offers 34 designs. "Absolutely first rate . . . well thought out and well executed."–E. J. Tangerman. 118pp. 7¾ x 10⅝. 23654-4 Pa. $6.95

ILLUSTRATED CATALOG OF CIVIL WAR MILITARY GOODS: Union Army Weapons, Insignia, Uniform Accessories, and Other Equipment, Schuyler, Hartley, and Graham. Rare, profusely illustrated 1846 catalog includes Union Army uniform and dress regulations, arms and ammunition, coats, insignia, flags, swords, rifles, etc. 226 illustrations. 160pp. 9 x 12. 24939-5 Pa. $10.95

WOMEN'S FASHIONS OF THE EARLY 1900s: An Unabridged Republication of "New York Fashions, 1909," National Cloak & Suit Co. Rare catalog of mail-order fashions documents women's and children's clothing styles shortly after the turn of the century. Captions offer full descriptions, prices. Invaluable resource for fashion, costume historians. Approximately 725 illustrations. 128pp. 8⅜ x 11¼. 27276-1 Pa. $11.95

THE 1912 AND 1915 GUSTAV STICKLEY FURNITURE CATALOGS, Gustav Stickley. With over 200 detailed illustrations and descriptions, these two catalogs are essential reading and reference materials and identification guides for Stickley furniture. Captions cite materials, dimensions and prices. 112pp. 6½ x 9¼. 26676-1 Pa. $9.95

EARLY AMERICAN LOCOMOTIVES, John H. White, Jr. Finest locomotive engravings from early 19th century: historical (1804–74), main-line (after 1870), special, foreign, etc. 147 plates. 142pp. 11⅜ x 8¼. 22772-3 Pa. $10.95

THE TALL SHIPS OF TODAY IN PHOTOGRAPHS, Frank O. Braynard. Lavishly illustrated tribute to nearly 100 majestic contemporary sailing vessels: Amerigo Vespucci, Clearwater, Constitution, Eagle, Mayflower, Sea Cloud, Victory, many more. Authoritative captions provide statistics, background on each ship. 190 black-and-white photographs and illustrations. Introduction. 128pp. 8⅞ x 11¾. 27163-3 Pa. $13.95

EARLY NINETEENTH-CENTURY CRAFTS AND TRADES, Peter Stockham (ed.). Extremely rare 1807 volume describes to youngsters the crafts and trades of the day: brickmaker, weaver, dressmaker, bookbinder, ropemaker, saddler, many more. Quaint prose, charming illustrations for each craft. 20 black-and-white line illustrations. 192pp. 4⅝ x 6. 27293-1 Pa. $4.95

VICTORIAN FASHIONS AND COSTUMES FROM HARPER'S BAZAR, 1867–1898, Stella Blum (ed.). Day costumes, evening wear, sports clothes, shoes, hats, other accessories in over 1,000 detailed engravings. 320pp. 9⅜ x 12¼. 22990-4 Pa. $14.95

GUSTAV STICKLEY, THE CRAFTSMAN, Mary Ann Smith. Superb study surveys broad scope of Stickley's achievement, especially in architecture. Design philosophy, rise and fall of the Craftsman empire, descriptions and floor plans for many Craftsman houses, more. 86 black-and-white halftones. 31 line illustrations. Introduction 208pp. 6½ x 9¼. 27210-9 Pa. $9.95

THE LONG ISLAND RAIL ROAD IN EARLY PHOTOGRAPHS, Ron Ziel. Over 220 rare photos, informative text document origin (1844) and development of rail service on Long Island. Vintage views of early trains, locomotives, stations, passengers, crews, much more. Captions. 8⅞ x 11¾. 26301-0 Pa. $13.95

THE BOOK OF OLD SHIPS: From Egyptian Galleys to Clipper Ships, Henry B. Culver. Superb, authoritative history of sailing vessels, with 80 magnificent line illustrations. Galley, bark, caravel, longship, whaler, many more. Detailed, informative text on each vessel by noted naval historian. Introduction. 256pp. 5⅜ x 8½. 27332-6 Pa. $7.95

TEN BOOKS ON ARCHITECTURE, Vitruvius. The most important book ever written on architecture. Early Roman aesthetics, technology, classical orders, site selection, all other aspects. Morgan translation. 331pp. 5⅜ x 8½. 20645-9 Pa. $8.95

THE HUMAN FIGURE IN MOTION, Eadweard Muybridge. More than 4,500 stopped-action photos, in action series, showing undraped men, women, children jumping, lying down, throwing, sitting, wrestling, carrying, etc. 390pp. 7⅞ x 10⅝. 20204-6 Clothbd. $25.95

TREES OF THE EASTERN AND CENTRAL UNITED STATES AND CANADA, William M. Harlow. Best one-volume guide to 140 trees. Full descriptions, woodlore, range, etc. Over 600 illustrations. Handy size. 288pp. 4½ x 6⅜. 20395-6 Pa. $6.95

SONGS OF WESTERN BIRDS, Dr. Donald J. Borror. Complete song and call repertoire of 60 western species, including flycatchers, juncoes, cactus wrens, many more–includes fully illustrated booklet. Cassette and manual 99913-0 $8.95

GROWING AND USING HERBS AND SPICES, Milo Miloradovich. Versatile handbook provides all the information needed for cultivation and use of all the herbs and spices available in North America. 4 illustrations. Index. Glossary. 236pp. 5⅜ x 8½. 25058-X Pa. $6.95

BIG BOOK OF MAZES AND LABYRINTHS, Walter Shepherd. 50 mazes and labyrinths in all–classical, solid, ripple, and more–in one great volume. Perfect inexpensive puzzler for clever youngsters. Full solutions. 112pp. 8⅛ x 11. 22951-3 Pa. $4.95

CATALOG OF DOVER BOOKS

PIANO TUNING, J. Cree Fischer. Clearest, best book for beginner, amateur. Simple repairs, raising dropped notes, tuning by easy method of flattened fifths. No previous skills needed. 4 illustrations. 201pp. 5⅜ x 8½. 23267-0 Pa. $6.95

A SOURCE BOOK IN THEATRICAL HISTORY, A. M. Nagler. Contemporary observers on acting, directing, make-up, costuming, stage props, machinery, scene design, from Ancient Greece to Chekhov. 611pp. 5⅜ x 8½. 20515-0 Pa. $12.95

THE COMPLETE NONSENSE OF EDWARD LEAR, Edward Lear. All nonsense limericks, zany alphabets, Owl and Pussycat, songs, nonsense botany, etc., illustrated by Lear. Total of 320pp. 5⅜ x 8½. (USO) 20167-8 Pa. $6.95

VICTORIAN PARLOUR POETRY: An Annotated Anthology, Michael R. Turner. 117 gems by Longfellow, Tennyson, Browning, many lesser-known poets. "The Village Blacksmith," "Curfew Must Not Ring Tonight," "Only a Baby Small," dozens more, often difficult to find elsewhere. Index of poets, titles, first lines. xxiii + 325pp. 5⅝ x 8¼. 27044-0 Pa. $8.95

DUBLINERS, James Joyce. Fifteen stories offer vivid, tightly focused observations of the lives of Dublin's poorer classes. At least one, "The Dead," is considered a masterpiece. Reprinted complete and unabridged from standard edition. 160pp. 5³⁄₁₆ x 8¼. 26870-5 Pa. $1.00

THE HAUNTED MONASTERY and THE CHINESE MAZE MURDERS, Robert van Gulik. Two full novels by van Gulik, set in 7th-century China, continue adventures of Judge Dee and his companions. An evil Taoist monastery, seemingly supernatural events; overgrown topiary maze hides strange crimes. 27 illustrations. 328pp. 5⅜ x 8½. 23502-5 Pa. $8.95

THE BOOK OF THE SACRED MAGIC OF ABRAMELIN THE MAGE, translated by S. MacGregor Mathers. Medieval manuscript of ceremonial magic. Basic document in Aleister Crowley, Golden Dawn groups. 268pp. 5⅜ x 8½. 23211-5 Pa. $8.95

NEW RUSSIAN-ENGLISH AND ENGLISH-RUSSIAN DICTIONARY, M. A. O'Brien. This is a remarkably handy Russian dictionary, containing a surprising amount of information, including over 70,000 entries. 366pp. 4½ x 6⅛. 20208-9 Pa. $9.95

HISTORIC HOMES OF THE AMERICAN PRESIDENTS, Second, Revised Edition, Irvin Haas. A traveler's guide to American Presidential homes, most open to the public, depicting and describing homes occupied by every American President from George Washington to George Bush. With visiting hours, admission charges, travel routes. 175 photographs. Index. 160pp. 8¼ x 11. 26751-2 Pa. $11.95

NEW YORK IN THE FORTIES, Andreas Feininger. 162 brilliant photographs by the well-known photographer, formerly with *Life* magazine. Commuters, shoppers, Times Square at night, much else from city at its peak. Captions by John von Hartz. 181pp. 9¼ x 10¾. 23585-8 Pa. $12.95

INDIAN SIGN LANGUAGE, William Tomkins. Over 525 signs developed by Sioux and other tribes. Written instructions and diagrams. Also 290 pictographs. 111pp. 6⅛ x 9¼. 22029-X Pa. $3.95

ANATOMY: A Complete Guide for Artists, Joseph Sheppard. A master of figure drawing shows artists how to render human anatomy convincingly. Over 460 illustrations. 224pp. 8⅜ x 11¼. 27279-6 Pa. $10.95

MEDIEVAL CALLIGRAPHY: Its History and Technique, Marc Drogin. Spirited history, comprehensive instruction manual covers 13 styles (ca. 4th century thru 15th). Excellent photographs; directions for duplicating medieval techniques with modern tools. 224pp. 8⅜ x 11¼. 26142-5 Pa. $12.95

DRIED FLOWERS: How to Prepare Them, Sarah Whitlock and Martha Rankin. Complete instructions on how to use silica gel, meal and borax, perlite aggregate, sand and borax, glycerine and water to create attractive permanent flower arrangements. 12 illustrations. 32pp. 5⅜ x 8½. 21802-3 Pa. $1.00

EASY-TO-MAKE BIRD FEEDERS FOR WOODWORKERS, Scott D. Campbell. Detailed, simple-to-use guide for designing, constructing, caring for and using feeders. Text, illustrations for 12 classic and contemporary designs. 96pp. 5⅜ x 8½. 25847-5 Pa. $2.95

SCOTTISH WONDER TALES FROM MYTH AND LEGEND, Donald A. Mackenzie. 16 lively tales tell of giants rumbling down mountainsides, of a magic wand that turns stone pillars into warriors, of gods and goddesses, evil hags, powerful forces and more. 240pp. 5⅜ x 8½. 29677-6 Pa. $6.95

THE HISTORY OF UNDERCLOTHES, C. Willett Cunnington and Phyllis Cunnington. Fascinating, well-documented survey covering six centuries of English undergarments, enhanced with over 100 illustrations: 12th-century laced-up bodice, footed long drawers (1795), 19th-century bustles, 19th-century corsets for men, Victorian "bust improvers," much more. 272pp. 5⅝ x 8¼. 27124-2 Pa. $9.95

ARTS AND CRAFTS FURNITURE: The Complete Brooks Catalog of 1912, Brooks Manufacturing Co. Photos and detailed descriptions of more than 150 now very collectible furniture designs from the Arts and Crafts movement depict davenports, settees, buffets, desks, tables, chairs, bedsteads, dressers and more, all built of solid, quarter-sawed oak. Invaluable for students and enthusiasts of antiques, Americana and the decorative arts. 80pp. 6½ x 9¼. 27471-3 Pa. $8.95

HOW WE INVENTED THE AIRPLANE: An Illustrated History, Orville Wright. Fascinating firsthand account covers early experiments, construction of planes and motors, first flights, much more. Introduction and commentary by Fred C. Kelly. 76 photographs. 96pp. 8¼ x 11. 25662-6 Pa. $8.95

THE ARTS OF THE SAILOR: Knotting, Splicing and Ropework, Hervey Garrett Smith. Indispensable shipboard reference covers tools, basic knots and useful hitches; handsewing and canvas work, more. Over 100 illustrations. Delightful reading for sea lovers. 256pp. 5⅜ x 8½. 26440-8 Pa. $7.95

FRANK LLOYD WRIGHT'S FALLINGWATER: The House and Its History, Second, Revised Edition, Donald Hoffmann. A total revision–both in text and illustrations–of the standard document on Fallingwater, the boldest, most personal architectural statement of Wright's mature years, updated with valuable new material from the recently opened Frank Lloyd Wright Archives. "Fascinating"–*The New York Times*. 116 illustrations. 128pp. 9¼ x 10¾. 27430-6 Pa. $11.95

PHOTOGRAPHIC SKETCHBOOK OF THE CIVIL WAR, Alexander Gardner. 100 photos taken on field during the Civil War. Famous shots of Manassas Harper's Ferry, Lincoln, Richmond, slave pens, etc. 244pp. 10⅝ x 8¼.   22731-6 Pa. $9.95

FIVE ACRES AND INDEPENDENCE, Maurice G. Kains. Great back-to-the-land classic explains basics of self-sufficient farming. The one book to get. 95 illustrations. 397pp. 5⅜ x 8½.   20974-1 Pa. $7.95

SONGS OF EASTERN BIRDS, Dr. Donald J. Borror. Songs and calls of 60 species most common to eastern U.S.: warblers, woodpeckers, flycatchers, thrushes, larks, many more in high-quality recording.   Cassette and manual 99912-2 $9.95

A MODERN HERBAL, Margaret Grieve. Much the fullest, most exact, most useful compilation of herbal material. Gigantic alphabetical encyclopedia, from aconite to zedoary, gives botanical information, medical properties, folklore, economic uses, much else. Indispensable to serious reader. 161 illustrations. 888pp. 6½ x 9¼. 2-vol. set. (USO)   Vol. I: 22798-7 Pa. $9.95
Vol. II: 22799-5 Pa. $9.95

HIDDEN TREASURE MAZE BOOK, Dave Phillips. Solve 34 challenging mazes accompanied by heroic tales of adventure. Evil dragons, people-eating plants, blood-thirsty giants, many more dangerous adversaries lurk at every twist and turn. 34 mazes, stories, solutions. 48pp. 8¼ x 11.   24566-7 Pa. $2.95

LETTERS OF W. A. MOZART, Wolfgang A. Mozart. Remarkable letters show bawdy wit, humor, imagination, musical insights, contemporary musical world; includes some letters from Leopold Mozart. 276pp. 5⅜ x 8½.   22859-2 Pa. $7.95

BASIC PRINCIPLES OF CLASSICAL BALLET, Agrippina Vaganova. Great Russian theoretician, teacher explains methods for teaching classical ballet. 118 illus-trations. 175pp. 5⅜ x 8½.   22036-2 Pa. $5.95

THE JUMPING FROG, Mark Twain. Revenge edition. The original story of The Celebrated Jumping Frog of Calaveras County, a hapless French translation, and Twain's hilarious "retranslation" from the French. 12 illustrations. 66pp. 5⅜ x 8½.   22686-7 Pa. $3.95

BEST REMEMBERED POEMS, Martin Gardner (ed.). The 126 poems in this superb collection of 19th- and 20th-century British and American verse range from Shelley's "To a Skylark" to the impassioned "Renascence" of Edna St. Vincent Millay and to Edward Lear's whimsical "The Owl and the Pussycat." 224pp. 5⅜ x 8½.   27165-X Pa. $4.95

COMPLETE SONNETS, William Shakespeare. Over 150 exquisite poems deal with love, friendship, the tyranny of time, beauty's evanescence, death and other themes in language of remarkable power, precision and beauty. Glossary of archaic terms. 80pp. 5³⁄₁₆ x 8¼.   26686-9 Pa. $1.00

BODIES IN A BOOKSHOP, R. T. Campbell. Challenging mystery of blackmail and murder with ingenious plot and superbly drawn characters. In the best tradition of British suspense fiction. 192pp. 5⅜ x 8½.   24720-1 Pa. $6.95

THE INFLUENCE OF SEA POWER UPON HISTORY, 1660–1783, A. T. Mahan. Influential classic of naval history and tactics still used as text in war colleges. First paperback edition. 4 maps. 24 battle plans. 640pp. 5⅜ x 8½. 25509-3 Pa. $12.95

THE STORY OF THE TITANIC AS TOLD BY ITS SURVIVORS, Jack Winocour (ed.). What it was really like. Panic, despair, shocking inefficiency, and a little heroism. More thrilling than any fictional account. 26 illustrations. 320pp. 5⅜ x 8½. 20610-6 Pa. $8.95

FAIRY AND FOLK TALES OF THE IRISH PEASANTRY, William Butler Yeats (ed.). Treasury of 64 tales from the twilight world of Celtic myth and legend: "The Soul Cages," "The Kildare Pooka," "King O'Toole and his Goose," many more. Introduction and Notes by W. B. Yeats. 352pp. 5⅜ x 8½. 26941-8 Pa. $8.95

BUDDHIST MAHAYANA TEXTS, E. B. Cowell and Others (eds.). Superb, accurate translations of basic documents in Mahayana Buddhism, highly important in history of religions. The Buddha-karita of Asvaghosha, Larger Sukhavativyuha, more. 448pp. 5⅜ x 8½. 25552-2 Pa. $12.95

ONE TWO THREE . . . INFINITY: Facts and Speculations of Science, George Gamow. Great physicist's fascinating, readable overview of contemporary science: number theory, relativity, fourth dimension, entropy, genes, atomic structure, much more. 128 illustrations. Index. 352pp. 5⅜ x 8½. 25664-2 Pa. $8.95

ENGINEERING IN HISTORY, Richard Shelton Kirby, et al. Broad, nontechnical survey of history's major technological advances: birth of Greek science, industrial revolution, electricity and applied science, 20th-century automation, much more. 181 illustrations. ". . . excellent . . ."–*Isis.* Bibliography. vii + 530pp. 5⅜ x 8¼. 26412-2 Pa. $14.95

DALÍ ON MODERN ART: The Cuckolds of Antiquated Modern Art, Salvador Dalí. Influential painter skewers modern art and its practitioners. Outrageous evaluations of Picasso, Cézanne, Turner, more. 15 renderings of paintings discussed. 44 calligraphic decorations by Dalí. 96pp. 5⅜ x 8½. (USO) 29220-7 Pa. $4.95

ANTIQUE PLAYING CARDS: A Pictorial History, Henry René D'Allemagne. Over 900 elaborate, decorative images from rare playing cards (14th–20th centuries): Bacchus, death, dancing dogs, hunting scenes, royal coats of arms, players cheating, much more. 96pp. 9¼ x 12¼. 29265-7 Pa. $11.95

MAKING FURNITURE MASTERPIECES: 30 Projects with Measured Drawings, Franklin H. Gottshall. Step-by-step instructions, illustrations for constructing handsome, useful pieces, among them a Sheraton desk, Chippendale chair, Spanish desk, Queen Anne table and a William and Mary dressing mirror. 224pp. 8⅛ x 11¼. 29338-6 Pa. $13.95

THE FOSSIL BOOK: A Record of Prehistoric Life, Patricia V. Rich et al. Profusely illustrated definitive guide covers everything from single-celled organisms and dinosaurs to birds and mammals and the interplay between climate and man. Over 1,500 illustrations. 760pp. 7½ x 10⅛. 29371-8 Pa. $29.95

Prices subject to change without notice.

Available at your book dealer or write for free catalog to Dept. GI, Dover Publications, Inc., 31 East 2nd St., Mineola, N.Y. 11501. Dover publishes more than 500 books each year on science, elementary and advanced mathematics, biology, music, art, literary history, social sciences and other areas.